Nanotechnology in Plant Sciences

Volume 7

Series Editors

Jameel M. Al-Khayri, Department of Agricultural Biotechnology, King Faisal University, Al-Ahsa, Saudi Arabia

Shri Mohan Jain, Department of Agricultural Sciences, University of Helsinki, Helsinki, Finland

This book series provides a platform for researchers to update advances in nanotechnology research and applications relevant to various fields of plant sciences. This series is an excellent reference source of cutting-edge knowledge for agriculturists, breeders, biotechnologists, and molecular biologists interested in plant nanotechnology. It is essential for professionals, researchers, policymakers, and commercial entrepreneurs concerned with understanding the mechanisms and innovative applications of nanotechnology to achieve enhanced crop production improvement, agricultural sustainability, environment protection, and food and nutritional security.

Ramesh Namdeo Pudake • Utkarsh Jain •
Jameel M. Al-Khayri
Editors

Phyto Nanotoxicology

Mechanisms and Analytical Methodologies of Plant Nanotoxicity

Springer

Editors
Ramesh Namdeo Pudake
Amity Institute of Nanotechnology
Amity University Uttar Pradesh
Noida, Uttar Pradesh, India

Jameel M. Al-Khayri
Department of Agricultural Biotechnology
King Faisal University
Al-Ahsa, Saudi Arabia

Utkarsh Jain
School of Health Science and Technology (SoHST)
University of Petroleum and Energy Studies (UPES)
Bidholi Dehradun, Uttarakhand, India

ISSN 3004-9334 ISSN 3004-9342 (electronic)
Nanotechnology in Plant Sciences
ISBN 978-3-032-24517-5 ISBN 978-3-032-24518-2 (eBook)
https://doi.org/10.1007/978-3-032-24518-2

This Springer imprint is published by the registered company Springer Nature Switzerland AG
The registered company address is: Gewerbestrasse 11, 6330 Cham, Switzerland

Preface

Engineered nanomaterials have become relevant in large-scale industrial applications and are used in many common devices and products. Their intentional and non-intentional release into the environment may have negative effects on the health of humans and plants. There is increasing concern about nanotoxicity in plants because they are not only a fundamental part of the ecology but also the basis of trophic systems. Nanotoxicology is the branch of toxicology responsible for describing the effects of nanomaterials on living organisms and the environmental consequences resulting from their use. Thorough studies are important because nanomaterials have stronger toxicological effects than those reported for the same materials in bulk or in solution. The main toxicological factor observed in various organisms is oxidative stress. Many techniques have been used to evaluate the toxicity of nanomaterials in plants; however, there is no consensus on the optimal method for assessing toxicity in plants. A concern is that bioaccumulation of nanoparticles in plants can enter the food chain of animals and humans, constituting a potential health problem. Therefore, extensive safety research projects and regulations are urgently needed. To understand some toxic effects that nanomaterials may have on plants, this two-volume book set describes the basic concepts of plant nanotoxicology, including properties of nanomaterials and interaction with the anatomy and physiology of plants. In addition, the methodologies to evaluate plant toxicity to nanoparticles as well as the resistance and responses of plant cells to nanomaterials are discussed.

The first volume focuses on the mechanisms and analytical methodologies that are important in plant nanotoxicity. This volume reviews the present status of nanomaterials that might be directly and indirectly encountered by plants. This book discusses the characterization of nanomaterials in pristine form and in complex biomolecules. The properties of engineered nanomaterials (ENM) that are important in relation to plants and the techniques to study the physiological and molecular impacts of ENM on plant bioprocesses are summarized in this volume. Studies on new omics technologies like transcriptomics, proteomics, and metabolomics for a deeper understanding of plant nanotoxicology are also discussed in this volume.

The current book consists of 16 chapters, supplemented with 14 tables and 56 figures, all precisely planned to enhance readers' understanding and engagement throughout the book. These peer-reviewed chapters are written by 57 recognized researchers from reputed institutes of five countries. The chapters open with an introduction that provides essential background, followed by an in-depth exploration of the topic supported by high-quality color figures and tables. To guide future research, each chapter concludes with thoughtful recommendations for further study. The book serves as a valuable reference source for graduate students, researchers, academicians, and scientists of plant science and nanotechnology. It is a reference book for undergraduate and graduate students of life science disciplines as a bridge between the textbooks and recent scientific literature, providing in-depth knowledge in nanotechnology.

Noida, Uttar Pradesh, India — Ramesh Namdeo Pudake
Bidholi Dehradun, Uttarakhand, India — Utkarsh Jain
Al-Ahsa, Saudi Arabia — Jameel M. Al-Khayri

Acknowledgment

The editors extend their appreciation for the support of the Deanship of Scientific Research, Vice Presidency for Graduate Studies and Scientific Research, King Faisal University, Saudi Arabia (Grant No. KFU262381).

Contents

Editors and Contributors

About the Editors

Ramesh Namdeo Pudake is an Assistant Professor in Amity University Uttar Pradesh—one of the top-ranked private universities in India. He has a PhD in Crop Genetics and Breeding from China Agricultural University, Beijing, PRC. After his PhD, he engaged in research in a range of organisms but with a focus on genomics and nanobiotechnology. He has also worked in the Department of Agronomy at Iowa State University Ames, IA, USA, on host-pathogen interaction and gene mapping. Currently, he is focusing on research on different applications of genomics, and nanomaterials in agriculture. He has published 50 research publications, 10 books, and 32 book chapters. He also has one scholarship award from the Chinese government for pursuing his doctoral studies. He is a guest editor and expert reviewer for several journals of repute.

Utkarsh Jain achieved his PhD in Medical Microbiology from Ludwig Maximilian University of Munich (LMU), Germany, in 2014. After his PhD, he joined the Indian Institute of Science (IISc), Bangalore, India, as a postdoctoral research fellow. Later, he contributed as a Group Leader—Cell Culture, Navya Biologicals, and as an Application Specialist at Carl Zeiss, Germany, in the Microscopy products division. He worked at Amity University Uttar Pradesh, Noida, India, as Assistant Professor during 2016–2022. Presently he serves as

Senior Associate Professor, UPES, Dehradun, India. He has published more than 100 research papers and has secured 9 research projects from notable funding agencies in India and the UK including BIRAC, DST, SERB, UKEREI, and DBT. He was listed among the Top 2% of World Scientists by Stanford University. He has presented his work at more than 50 international conferences in different countries, filed 20 patents, published 9 book chapters, and edited 2 books. In addition, he was awarded Outstanding Plus Researcher and Academician, the Young Scientist Award, the Best Extramural Research Award, and Best Poster and Oral Presentation awards in Paris and India. He has supervised 4 PhD students and 25 research students for their various Dissertation projects. Presently, 6 PhD students are working under his supervision. He is a reviewer of various international and national high-quality research journals. His field of research is the development of various biosensing-based detection platforms for the detection of pathogenic diseases.

Jameel M. Al-Khayri is affiliated with the Department of Agricultural Biotechnology, King Faisal University, Saudi Arabia. He received a BS in Biology in 1984 from the University of Toledo, an MS in Agronomy in 1988, and a PhD in Plant Science in 1991 from the University of Arkansas. He is the National Correspondent of the International Association of Plant Tissue Culture and Biotechnology. He has authored 130 research articles and 70 chapters, and has edited several journal special issues. In addition, he has edited 30 reference books on plant biotechnology, genetic resources, breeding, genomics, and nanotechnology. In addition to teaching, advising, and research, he held administrative responsibilities as Assistant Director of the Date Palm Research Center, Head of the Department of Plant Biotechnology, and Vice Dean for Development and Quality Assurance. He served as a Member of Majlis Ash Shura (Saudi Legislative Council) for the 2009–2012 term. He is interested in agricultural innovations and sustainability.

Contributors

Aditi Amity Institute of Biotechnology, Amity University Jharkhand, Ranchi, Jharkhand, India

Sana Awan Department of Earth and Environment, Florida International University, Miami, FL, USA

Fatima Batool Department of Environmental Sciences, Government College University, Faisalabad, Pakistan

Anjali Bharti School of Health Sciences & Technology (SoHST), UPES, Dehradun, Uttarakhand, India

Sougta Bhattacharjee ICAR-Indian Agricultural Research Institute, Gauria karma, Hazaribagh, Jharkhand, India

Goutam Biswas Department of Chemistry, Cooch Behar Panchanan Barma University, Cooch Behar, West Bengal, India

Jutishna Bora Amity Institute of Biotechnology, Amity University Jharkhand, Ranchi, Jharkhand, India

Silvia Carlo Department of Chemistry, Life Sciences and Environmental Sustainability, University of Parma, Parma, Italy

Payal Chakraborty Amity Institute of Organic Agriculture, Amity University, Noida, Uttar Pradesh, India

Nidhi Chauhan School of Health Sciences & Technology (SoHST), UPES, Dehradun, Uttarakhand, India

Chanchal Das Department of Chemistry, Saheed Kshudiram College, Alipurduar, India

Archna Dashmana Himalayan School of Biosciences, Swami Rama University, Dehradun, Uttarakhand, India

Debojyoti Datta Department of Microbiology, St. Xavier's College (Autonomous), Kolkata, West Bengal, India

Caleb Fernandes Department of Biotechnology, School of Applied Sciences, REVA University, Bengaluru, Karnataka, India

Cannon Antony Fernandes Department of Life Sciences, School of Sciences, Christ University, Bengaluru, India

Ankita Ghosh School of Health Sciences & Technology (SoHST), UPES, Dehradun, Uttarakhand, India

Vinita Hooda Department of Botany, Maharshi Dayanand University, Rohtak, Haryana, India

Sabir Hussain Department of Environmental Sciences, Government College University, Faisalabad, Pakistan

Aneeza Ishfaq Department of Environmental Sciences, Government College University, Faisalabad, Pakistan

Utkarsh Jain School of Health Sciences & Technology (SoHST), UPES, Dehradun, Uttarakhand, India

Hem C. Joshi Department of Botany, D.S.B. Campus, Kumaun University, Nainital, Uttarakhand, India

Richa Joshi Department of Botany, D.S.B. Campus, Kumaun University, Nainital, Uttarakhand, India

Harshan Mahesh Kumar Department of Life Sciences, School of Sciences, Christ University, Bengaluru, India

Asha Kumari ICAR-Indian Agricultural Research Institute, Gauria karma, Hazaribagh, Jharkhand, India

Jaya Kumari Amity Institute of Biotechnology, Amity University Jharkhand, Ranchi, Jharkhand, India

Vasantha Veerappa Lakshmaiah Department of Life Sciences, School of Sciences, Christ University, Bengaluru, India

Smita Lata Amity Institute of Biotechnology, Amity University Jharkhand, Ranchi, Jharkhand, India

Mathiazhakan Lavanya Department of Botany, Periyar University, Salem, Tamil Nadu, India

Tanushree Maharana Department of Biotechnology, School of Applied Sciences, REVA University, Bengaluru, Karnataka, India

Faisal Mahmood Department of Environmental Sciences, Government College University, Faisalabad, Pakistan

Sumira Malik Amity Institute of Biotechnology, Amity University Jharkhand, Ranchi, Jharkhand, India

University Center for Research & Development (UCRD), Chandigarh University, Mohali, Punjab, India

Sampreet Manna Department of Biotechnology, St. Xavier's College (Autonomous), Kolkata, West Bengal, India

Marta Marmiroli Department of Chemistry, Life Sciences and Environmental Sustainability, University of Parma, Parma, Italy

Nelson Marmiroli Consorzio Interuniversitario Nazionale per le Scienze Ambientali (CINSA), University of Parma, Parma, Italy

Shalini Mehta Ranchi Women's College, Ranchi University, Ranchi, Jharkhand, India

Prabhakar Mishra Department of Biotechnology, School of Applied Sciences, REVA University, Bengaluru, Karnataka, India

Ridhhi Mishra Amity Institute of Biotechnology, Amity University Jharkhand, Ranchi, Jharkhand, India

Yuvashree Muralidaran Department of Biotechnology, School of Applied Sciences, REVA University, Bengaluru, Karnataka, India

Sadia Mustafa Department of Environmental Sciences, Government College University, Faisalabad, Pakistan

Nelson Mwebesa Department of Chemistry, Makerere University, Kampala, Uganda

Praveen Nagella Department of Life Sciences, School of Sciences, Christ University, Bengaluru, India

Ivan Oyege Department of Earth and Environment, Florida International University, Miami, FL, USA

Department of Chemistry, Busitema University, Tororo, Uganda

Luca Pagano Consorzio Interuniversitario Nazionale per le Scienze Ambientali (CINSA), University of Parma, Parma, Italy

Saheb Pal ICAR-Indian Agricultural Research Institute, Gauria karma, Hazaribagh, Jharkhand, India

Prabha Pant Department of Botany, D.S.B. Campus, Kumaun University, Nainital, Uttarakhand, India

Babita Patni High Altitude Plant Physiology Research Centre, Hemvati Nandan Bahuguna Garhwal University, Srinagar, Garhwal, Uttarakhand, India

Tehziba Raza Department of Environmental Sciences, Government College University, Faisalabad, Pakistan

Aryadeep Roychoudhury Discipline of Life Sciences, School of Sciences, Indira Gandhi National Open University, New Delhi, India

Kuppusamy Selvam Department of Botany, Periyar University, Salem, Tamil Nadu, India

Subhadeep Sen Department of Chemistry, Cooch Behar Panchanan Barma University, Cooch Behar, West Bengal, India

Muhammad Shahid Department of Bioinformatics and Biotechnology, Government College University, Faisalabad, Pakistan

Tanvir Shahzad Department of Environmental Sciences, Government College University, Faisalabad, Pakistan

Sandeep Sharma Department of Botany and Plant Sciences, University of California, Riverside, CA, USA

Muthugounder Subramanian Shivakumar Department of Biotechnology, Periyar University, Salem, Tamil Nadu, India

Sweta Sinha Amity Institute of Biotechnology, Amity University Jharkhand, Ranchi, Jharkhand, India

Dennis Ssekimpi Department of Civil and Environmental Engineering, Florida State University, Tallahassee, FL, USA

Chapter 1
Status of Nanomaterial Uses in Agriculture and Environment

Aryadeep Roychoudhury, Debojyoti Datta, and Sampreet Manna

Abstract Nanotechnology is being applied in agricultural and environmental contexts to solve global problems such as food security and controlling food waste and pollution. Due to their exceptional physicochemical properties, nanomaterials can be used in agriculture as nano fertilizers, nanopesticides, and nanosensors to help with nutrient use efficiency, minimise chemical runoff, and mitigate environmental pollution. Veterinary and human health management uses real-time monitoring technologies that directly improve pest control, allow monitoring of soil and crop health in real-time, and enable the site-specific release of nutrients and agrochemicals, thereby reducing wastage, optimizing resource utilization, and promoting precision agriculture. Nanotechnology also helps in environmental remediation by providing advanced water purification systems, pollutant degradation, and restoration of soil health. However, there are some problems associated with nanoparticle toxicity, ecological impact, and cost-effectiveness. The chapter underscores the central role that nanotechnology plays in the innovation of agriculture and the restoration of the environment. The unique properties of nanomaterials can be exploited by scientists and practitioners to design solutions to enhance food security, mitigate climate change impacts, and address environmental degradation. The full potential of nanotechnology for a sustainable future requires continued interdisciplinary research.

Keywords Nanotechnology · Nanomaterials · Agriculture · Environment · Food Security · Precision Agriculture · Pollution Control · Real-time Monitoring

---

A. Roychoudhury (✉)
Discipline of Life Sciences, School of Sciences, Indira Gandhi National Open University, New Delhi, India

D. Datta
Department of Microbiology, St. Xavier's College (Autonomous), Kolkata, West Bengal, India

S. Manna
Department of Biotechnology, St. Xavier's College (Autonomous), Kolkata, West Bengal, India

R. N. Pudake et al. (eds.), *Phyto Nanotoxicology*, Nanotechnology in Plant Sciences 7, https://doi.org/10.1007/978-3-032-24518-2_1

1.1 Introduction

Nanomaterials or nanoparticles are materials in the size range of 1–100 nanometres (Ganji and Kachapi 2015) and have novel characteristics because of their small dimensions and being different from coarse particles or massive materials. In their structure, nanoparticles can be solid or semi-solid; their forms are diverse, viz., spherically shaped as nanospheres, reef-like nano reefs, tubular-like nanotubes, prolate-like nanorods and fibers, and cubelike-nanoboxes (Kumari et al. 2023a, 2023b). The word 'nano' is derived from the Greek and Roman words *nannos* or *nanus,* which were used to refer to very small living structures. When techniques of microscopy evolved to a further level, more profound instruments were developed, including the transmission electron microscope, added by Ernst Ruska in 1931, and the scanning tunnelling microscope designed by Gerd Binnig and Heinrich Rohrer during the 1980s (Robinson 1986); such achievements provided ground that enabled researchers to investigate, categorize and apply nanoparticles. Nanoparticles can be prepared by many processes, including hydrothermal synthesis, attrition, pyrolysis, and radiation technologies, depending on the specific application. Another method for making nanoparticles from metals of low melting point includes the use of inert gas condensation (Al-Harbi and Abd-Elrahman 2024). There are semi-solid particles as well or small liposomes which can be used for medical purposes since they can easily penetrate tissues. The functionality of nanoparticles can be tuned, e.g., by coating nanoparticles with red blood–cell membranes that help nanoparticles in avoiding recognition by the immune system when being developed for drug delivery. Implementing particles at a nanoscale is one of the most progressive advancements of recent years in science and technology, and has found its applications in medicine, electronics, and the environment (Ghidan & Antary 2019). Nanoparticles are important in several areas with a predilection towards biology and medicine (Jiménez-Jiménez et al. 2020). For instance, the amphiphilic Janus particles, which have both aqueous and non-aqueous surfaces, are used for stabilising emulsions and as drug carriers. Other nanoparticles that are used are iron oxide nanoparticles or polymeric nanoparticles that have different biomedical uses (Mehlhorn 2016).

Nanotechnology refers to a special kind of technology that involves the deliberate mastering of those materials that exist in the atomic and molecular columns to be used to develop other new materials, structures and devices with improved or altogether different functions. It has been discovered that when existing materials are sized to nanoscales, behaviours like electrical and optical characteristics, mechanical properties, and chemical activity may also vary significantly. While nanomedical application involves the use of nanomaterials for drug delivery, electronics employs it for miniaturisation and energy harnessing for better storage and transfer of energy (Fu et al. 2024). Nanotechnology is a multidisciplinary science that combines chemistry, physics, material science, and engineering and holds promise for transforming key industries like electronics and semiconductors, biotechnology, manufacturing, and energy (Bhushan 2017).

1.1.1 Unique Characteristics of Nanomaterials

Based on the guidelines provided by the National Nanotechnology Initiative (NNI), nanotechnology can be described in terms of criteria for particles. First, it deals with substances or apparatus that have one of their dimensions with a size of between 1 and 100 nanometers (Ganji and Kachapi 2015). This size range is important because most nanomaterials have characteristics that differ significantly from those of the same material in a macroscopic form. Furthermore, nanotechnology must take advantage of the phenomena that can occur only at a nanoscale; so, the impacts must not surpass this level. The final factor is the possibility of quantitatively controlling these nanostructures for specific applications, which gives the possibility to predict their performance characteristics. Moreover, the engineering of nanoparticles should also enable them to be incorporated into larger structures, but at the same time retain the characteristics of nanostructures to function as intended (Roco 2005). Nanomaterials are characterized by substantial transformations in their physical, chemical, and biological features if their dimensions range from 1 to 100 nm. Some of these changes result from the highly developed surface area-to-volume ratio and surface activity that characterize nanoscale materials. Physically, nanoparticles exemplify different optical properties; for instance, the color of gold nanoparticles depends on the size of the nanoparticles. In the mechanical aspect, nanoparticles are also endowed with higher levels of toughness, and hence they can be applied in coating and structural applications. The high surface-to-volume ratio in nanoparticles makes the substances highly reactive to other substances as compared to the bulk substances (Khan et al. 2022). These are enhanced reactivity due to sharper edges and vertices that promote use in catalysis and other reactions, e.g., to the improved chemical characteristics of the nanoparticles and their ability to act as an environmental pollutant adsorbent or as an active catalyst for reduction of emissions in exhaust systems of vehicles. From the biological standpoint, the nanoparticles have the opportunity to interact with biological systems due to their size (Narayanan and El-Sayed 2005). That is why these are particularly useful for designing new applications in drug delivery, imaging, and diagnostics due to their ability to enter cell membranes and tissues. Nevertheless, these properties also make these materials suspect of their toxicity issues. High reactivity, which makes nanoparticles suitable for numerous uses, means that they can react with living tissues and nature in inconceivable ways, which is dangerous to human health and ecology. These properties, augmented by the dexterity offered by the size of nanomaterials, have made it possible for nanomaterials to be used in many different fields in medicine, electronics, energy production, and management of the environment, whilst at the same time challenging safety implications in environmental bodies (Khan et al. 2017).

1.1.2 Demands of Nano-Biotechnology

Due to the importance of nano-biotechnology in medicine as a diagnostic, therapeutic modality, and drug delivery platform, nanobiotechnology serves as a revolutionary platform in medicine. Because of their very small size and high surface area, nanoparticles are good for the targeted manipulation of biological molecules and cells; as a result, they find immense application in medicine. In diagnostics, it is also used for the screening of diseases, for instance, cancers (Malik et al. 2023). When bound to specific biomarkers, nanoparticles deliver incredibly high sensitivity, helping to detect diseases in their infancy. Quantum dots and nanoscale cantilevers improve the capacity to see molecular alterations even at subcellular levels, increasing opportunities for early diagnosis tenfold. Therapeutically, nanotechnology brings improvement to the drug delivery system through the policy of targeted delivery. Another application is in drug delivery by attaching nanoparticles to drugs; medicine can be targeted at specific tissues or cells, thus minimizing the impact on the whole system and making treatment more effective (Swierczewska et al. 2011). This is especially beneficial in cancer treatment whereby through the use of nanoparticles, therapeutic agencies may be taken directly to the cancer cells without affecting the whole cells in the vicinity, for instance, dendrimers, a nanoshell, and liposomes help the delivery of the drug with a lot of precision and are biodegradable so that the toxic effects can be pre-programmed to release at a later date. Nanotechnology brings opportunities in regenerative medicine for managing diseases and chronic disorders such as diabetes and AIDS. Nanoparticle systems are developed to provide insulin in diabetic patients or control the immune response in AIDS patients (Navalakhe and Nandedkar 2007; Graham et al. 2025).

1.1.3 Nanotechnology in the Agricultural Field

Nanobiotechnology also has revolutionary uses in the agricultural field, improving crop productivity and pest control, and providing solutions to environmental issues. One area of importance is nano-fertilisers, which prolong the nutrient effectiveness, reduce the number of applications, and help to avoid the polluting effects of overuse of fertiliser (Jha et al. 2023). These fertilizers are within nanomaterials and release nutrients gradually, which enables plants to absorb the nutrients with no wastage in the environment. In pesticide delivery, it makes formulations with increased surface area to interact with the surroundings, enhancing the solubility and ease of movement of the pesticide. The self-degradable nano-encapsulated pesticide can be used in an on-demand release following the environmental conditions, thereby offering prolonged pest control with no detrimental influence of pesticidal residuals (Nongbet et al. 2022). Precision farming is also a part of nanotechnology, where nanosensors are used to monitor the status of the soil and plants, their health, and the climate in real time. Such sensors provide the farmers with an informed decision-making platform so that there is an efficient use of water, fertilizers, and pesticides. Nanotechnology supports the identification and management of plant diseases. In this case, nanosensors are used to recognize disease-causing pathogens at pre-contamination levels, thereby avoiding widespread epidemics. Nanoparticles play a role in transferring genetic components for creating crops that have desirable characteristics, including abiotic stress tolerance and the ability to better nutrient acquisition (Manjunatha et al. 2019). Smart textile or nano-textiles that can sense and respond to changes in the environment in the textile industry is another innovation in nanobiotechnology. The bactericidal properties of silver nanoparticles or UV and infrared protective properties of metal oxide nanoparticles give the fabrics higher wear resistance, self-cleaning abilities, as well as flame retardance (Syduzzaman et al. 2023). Various techniques, such as electrospinning and nanocoating, enable the application of nanomaterials on textiles, making the textiles suitable for multiple uses (Shah et al. 2022). Modern farming systems have drawbacks in that they use a mainstream business approach that pollutes soil and water and disrupts species diversity. On the other hand, nanotechnology is cleaner, more efficient, and has fewer side effects than most traditional technologies. It opens opportunities to design and use nanoparticles, as well as nanocarriers for the precise application of chemicals like pesticides, herbicides, and fertilizers (Xuan et al. 2023). This approach is in contrast with other approaches, where a lot of chemicals are used and released into the environment due to improper application. Nanotechnology enables the release of minute amounts of chemicals on a point-of-need basis; for instance, nano-encapsulated or nano-carriers can transport the pesticide to the specific areas of the plant that have been affected by the pests, avoiding any chance of spreading the chemicals to other organisms (Fajardo et al. 2022). The other benefit is that nano-structured catalysts and nano-filters play a significant role in tackling pollution and strengthening the protection of the environment. Nano-filters can be implemented in water treatment techniques to filter dangerous elements originating from industries and urban areas before they find entry in the agricultural systems. Likewise, the use of nanoparticles is in dealing with unpolluted soil to improve the quality of the agricultural fields. Nanotechnology also supports what is known as controlled agriculture, where the use of nano-sensors is used to track the states of plant health, soil, and nutrients. This makes it easier for the farmers to decide when and how much water, fertilizers, or pesticides are to be used to minimize wastage in farming (Gehrke et al. 2015). It helps to solve the problems of greenhouse gas emissions and deterioration of soil. Nanotechnology ensures that the crops receive the nutrients for an extended period, as opposed to when the nutrients are released at once, and hence does not pollute the water sources. The application of nano-coatings to crops helps avoid frequent replacements, resulting in efficient utilization of resources (Mousavi and Rezaei 2011).

1.1.4 Synthesis of Nanomaterials

Nanomaterials can be synthesized by both top-down and bottom-up approaches (Fig. 1.1). The first method of synthesis, or the top-down method, includes the use of physical or chemical methods to reduce a larger bulk material into a nanoparticle form, including milling or lithography. One disadvantage of this method is that it may bring surface imperfections on the nanoparticles, which may result in changes to their physical and chemical properties. These surface irregularities can affect how nanoparticles respond to surrounding conditions or biological systems, and may dampen their utility. The bottom-up approach, on the other hand, involves the buildup of nanoparticles from the bottom through a chemical method where the

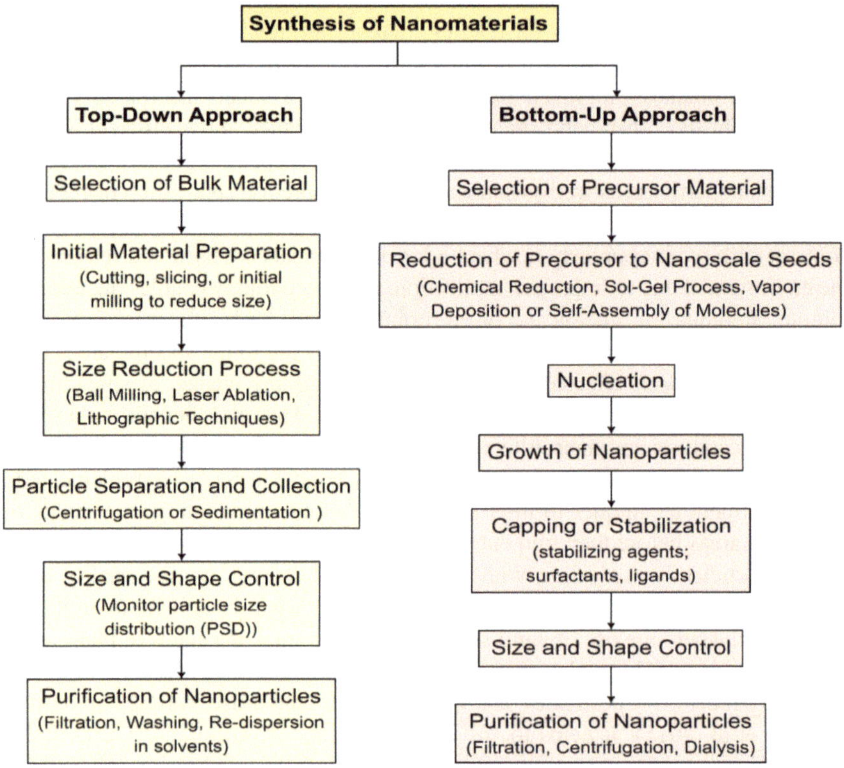

Fig. 1.1 Synthesis of Nanomaterials. The image shows two primary methods for synthesising nanomaterials: the top-down and bottom-up approaches, depicted as flowcharts. In the top-down approach, bulk materials are gradually broken down into nanoscale particles through physical methods such as mechanical milling or lithography followed by size reduction processes and the particles are then purified. In the bottom-up approach, the process begins with atoms, molecules or ions, which assemble into nanostructures through chemical or physical methods. This involves nucleation, growth and stabilisation steps, with controlled parameters to define the final nanoparticle characteristics

constituent is made from atomic or molecular structures. This method can be better controlled than the other methods of synthesis and has a lower probability of defects. It is especially useful in agriculture because it is beneficial for synthesizing nanoparticles with biological agents such as plant extracts without polluting the environment. It is called the green synthesis technique as no toxic reagent enters the process and is highly suitable for agricultural practices. This approach means that nanoparticles created in this process can be designed to specific applications, for instance, in increasing nutrient absorption in plants, pesticide delivery systems, or to promote the health of the soil. It is for this reason that the bottom–up approach has considerable potential in agriculture for the development of highly functional nanoparticles that are environmentally benign (Sood and Chopra 2017).

1.1.5 Types of Nanomaterials

Organic nanomaterials/nanoparticles are mainly based on carbon structures and are well regarded to be biocompatible, non-toxic and biodegradable. Examples include micelles and liposomes, nanogels and polymeric nanoparticles and so on (Fig. 1.2). Currently, organic nanoparticles are preferred to inorganic nanoparticles in biomedical applications such as drug delivery because they degrade naturally in the environment and their toxicity levels are lower. One of the developments where organic nanoparticles are administered is to transport substances such as drugs to specific areas in the body for better effect and with minimal side effects. Due to the fact that they have hollowness or encapsulation in their structures, their drug release is controlled, making them proper for therapeutic purposes. The inorganic nanoparticles are based on metals or metal oxides and unlike organic nanoparticles are more stable and hydrophilic in nature. These are used in biomedical industries because of their diverse characteristics like magnetism in some of them, specific optical characteristics and higher reactivity. Iron oxide-based magnetic nanoparticles (mNPs) are one of the widely known and applied MRI contrast agents for tumor detection and drug delivery systems. Some inorganic nanoparticles like zinc oxide and iron oxide have shown a lot of antibacterial potential. Both types of nanoparticles are highly significant in biomedicine, catalysis, as sensors, and for energy storage, owing to the differences in their physical, chemical, and biological properties (Alshammari et al. 2023).

1.1.6 Potential Uses and Benefits of Nanomaterials

Nanotechnology facilitates nano fertilizers of improved nutrient utilization and contains nutrients in concentrated form to enable plants to absorb the nutrients optimally and minimize wastage. Moreover, the successful application of nanotechnology helps to develop nanoparticles that could remove filterable and nonfilterable

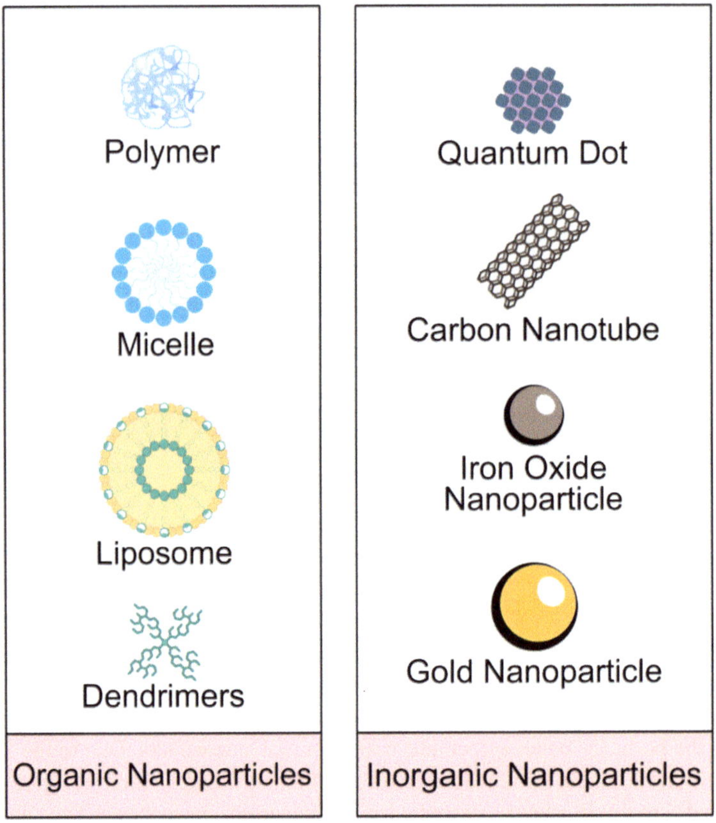

Fig. 1.2 Types of Nanoparticles. The image represents various organic and inorganic nanoparticles. These are widely used in medicine, diagnostics, and materials science. In agriculture, polymer-based nanoparticles can be used to deliver fertilisers or pesticides in a controlled, target-dependent manner, reducing other chemical wastes and environmental hazards. Liposomes and micelles can encapsulate nutrients or agrochemicals, so that efficient delivery to plants can occur. Carbon nanotubes and quantum dots can enhance environmental monitoring and photosynthesis, while iron oxide and gold nanoparticles are useful in soil remediation, biosensing, and real-time monitoring for pollution detection

pollutants and treat water for further usage in irrigation systems (Khan and Rizvi 2014). Figure 1.3 depicts a basic flowchart of the key aspects of nanoparticles.

1.1.7 Limitations of Conventional Agricultural Practices

Organic farming as well as conventional farming has a long way to go in trying to solve many issues related to the inefficiencies of modern agriculture. Organic farming is considered to be friendly to the environment, while it may yield less

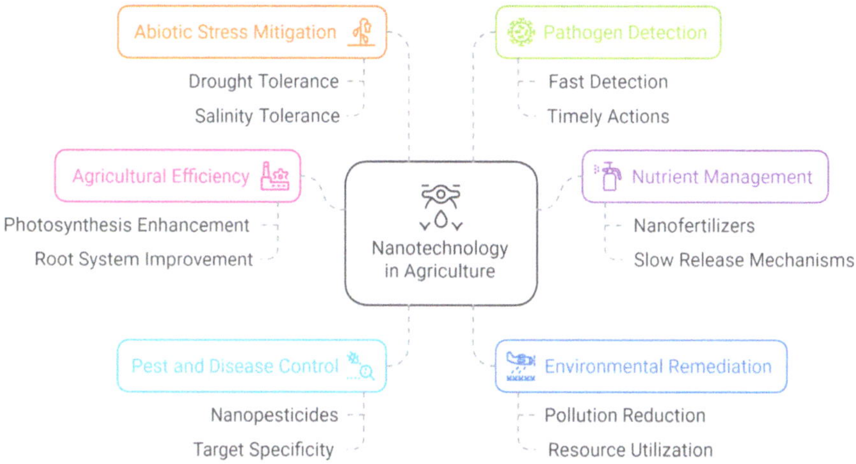

Fig. 1.3 The multifaceted applications of nanotechnology in agriculture. The figure presents a central hub representing key areas such as abiotic stress mitigation, agricultural efficiency, pathogen detection, nutrient management, pest and disease control, and environmental remediation. Nanotechnology can be utilized to enhance plant resilience to drought and salinity, improve photosynthesis and root systems, enable rapid disease detection, optimize nutrient delivery, develop targeted pesticides, and reduce pollution

compared with conventional farming. High-scale organic farming often tends to produce lower yield than using these methods in smaller extent. Organic farming lacks the kind of precision needed in the process for optimum use of resources like water and nutrients.

One of the major inefficiencies in traditional agriculture is low fertiliser nutrient use efficiency (NUE). It is said that 50% or more of the nutrients applied through fertilisers are lost due to volatilization, leaching or runoff and eventually contribute to environmental pollution and resource wastage (Nongbet et al. 2022). Excessive application of nitrogen and phosphorus through fertilisers coming from the agricultural sector may cause water bodies to become contaminated, leading to eutrophication and damaging aquatic ecosystems.

Nanotechnology provides solutions in the form of nanofertilizers, which are highly efficient compared to traditional fertilisers. These fertilisers could be engineered so that they respond to moisture or pH, and thus incorporate less fertiliser and minimise losses. Nanofertilizers are developed to provide fertilisers in a slow and controlled release. This will, in turn, enhance NUE by reducing nutrient loss while at the same time ensuring that the nutrients required are adequately supplied to plants. This kind of controlled release reduces the leaching into groundwater and also prevents many problems such as eutrophication. Nanofertilizers also offer a better supply of deficient micronutrients like zinc (Zn) and iron (Fe), which are important determinants for crop health as well as nutritional quality. Studies show that ZnO nanoparticles raise crop productivity by as much as 91% for barley, while the yield raised with conventional fertilizer $ZnSO_4$ is just 31% (Liu and Lal 2015).

Besides, nano fertilizers can interact more significantly and effectively with both the organic and inorganic portions of soil, thereby affecting their behavior and bioavailability, and thus are effective in various types of soil. Nanomaterials such as iron oxide nanoparticles might immobilize heavy metals in the soil. The rejuvenation of soil health can thus be achieved. Carbon-based nanomaterials can also degrade organic pollutants, providing another dimension to a cleaner and more sustainable agricultural system (Xu et al. 2021).

1.2 Benefits of Nanomaterial Use for Agriculture and the Environment

1.2.1 Agricultural Improvement

Nanomaterials are now powerful agents that help improve the productivity of agricultural crops, accelerate the necessary biological processes in plants, such as photosynthesis or the formation of root systems. Concerning the application of nanotechnology in agriculture, nanomaterials have several benefits in enhancing crop productivity. These materials enhance the efficiency of nutrient delivery, thereby enhancing the chances of nutrient uptake by the plant and enhancing germination. Nanomaterials have a broad surface capable of interacting with plant cells, thus improving the uptake of nutrients (Panpatte et al. 2016). Another important advantage of nanomaterials in agriculture is that they can increase the efficiency of photosynthesis. The uptake of TiO_2 nanoparticles has been reported to enhance the photosynthesis process and light absorption (Zheng et al. 2005) so that the plants enjoy better biomass productivity and yield improvement. In the germination process and root formation, nanomaterials also have a major influence. Multiwalled carbon nanotubes (MWCNTs) and other nanoparticles such as ZnO have been found to facilitate faster germination processes and enhanced germination rates (Lahiani et al. 2013; Joshi et al. 2017). These materials invade through the seed coat to effect water absorption and other enzymatic activities required for germination. They increase the capability of the root system to uptake nutrients and improve soil status for better growth of the root system.

1.2.2 Enhanced Nutrient Management

Nanotechnology equally enables the controlled and precise delivery of agents such as fertilizers and pesticides in a plant (Petosa et al. 2016). It lowers the cost and emission of pollutants to the environment since only the necessary quantity of chemicals reaches the appropriate plant tissues or soil sections. This method brings improvement in the ways and manner by which agricultural inputs are used, thereby improving the overall performance of plants with limited inputs, for example, nano fertilizers can be designed to have slow-releasing nutrients for a continuing release of plant growth and development process, without the need for constant application of the nutrient (Shang et al. 2019). Nano fertilizers are significant forms of nanomaterials that have increased nutrient efficiency by diminishing the rates of nutrient losses. Organic fertilizers pose some common issues like low nutrient uptake by the plants, and hence high amounts are wasted through leaching, volatilization, as well as fixation. However, there are difficulties in the application of concentrated mineral fertilizers. Nano-fertilizers can avoid these problems by increasing the accuracy and efficiency of the introduction of the necessary nutrients. Since nano-fertilizers stand out by their higher surface area, more contact area between the fertilizer and the plant is established. Nano-fertilizers are finer than most commercial fertilizers and are smaller than 100 nm; the size enables the fertilizer to penetrate deeper into the root and leaves of the plant, improving overall nutrient absorption (Qureshi et al. 2018). A second practical advantage of nano-fertilizers is the slow release of nutrients in the soil (Sharma et al. 2022). Nano-fertilizers can be designed to deliver nutrients to plants in a sustained manner to meet the elemental requirements of the plant. Such a slow-release mechanism minimizes nutrient losses and ensures nutrient availability, enhancing nutrient uptake and improving crop yield and utilization of available resources (Gade et al. 2023). Most particularly, they assist in minimizing environmental pollution that is commonly caused by the use of commercial fertilizers. Due to low nutrient losses, nano-fertilizers can help eliminate cases of soil and water pollution, thereby enhancing agricultural practices (Meena et al. 2017).

1.2.3 Improved Pest and Disease Management

New pest and disease control is promoted by nanomaterials through increasing efficiency of pesticides, as well as by decreasing the potential adverse effects on the environment. Nanopesticides make it possible to have extended effectiveness of active ingredients and keep the concentration of chemicals as low as possible. These nanoformulations enhance dispersion of active ingredients and enhance compatibility with pests, thereby improving upon target specificity and reducing side effects. Nanomaterials also shield active compounds from degradation and hence the need to apply the control agents is less (Ragaei et al. 2014).

1.2.4 Environmental Remediation and Drug Delivery

Nanopesticides in general prove a remarkable effectiveness in pest and disease control. They deliver the active ingredients in a slow and regulated manner, increasing the durability of the products without the need for frequent use. Due to the higher formulations of the preparation system, nanopesticides reduce chemical consumption, reducing the likelihood of soil and water pollution (Yadav et al. 2022). Nanoencapsulation techniques help in shielding the active components from degradative factors which consequently increases the shelf life of the pesticides, enhancing their performance. Pesticide efficiency is achieved by enhancing its localization so that other undesired organisms or environment are not affected. This mode of delivery makes sure that non-pest organisms are least affected and that contact between humans and animals with dangerous chemicals is limited. An integrated nanotechnology approach to pest control is more efficient or sustainable and encourages safer uses of agricultural inputs to make agriculture smarter and environment-friendly (Ragaei and Sabry 2014).

1.2.5 *Mitigation of Abiotic Stress in Plants*

Nanomaterials substantially contribute to the alleviation of abiotic stress in plants due to improved stress tolerance and increased growth (Fig. 1.4). These materials

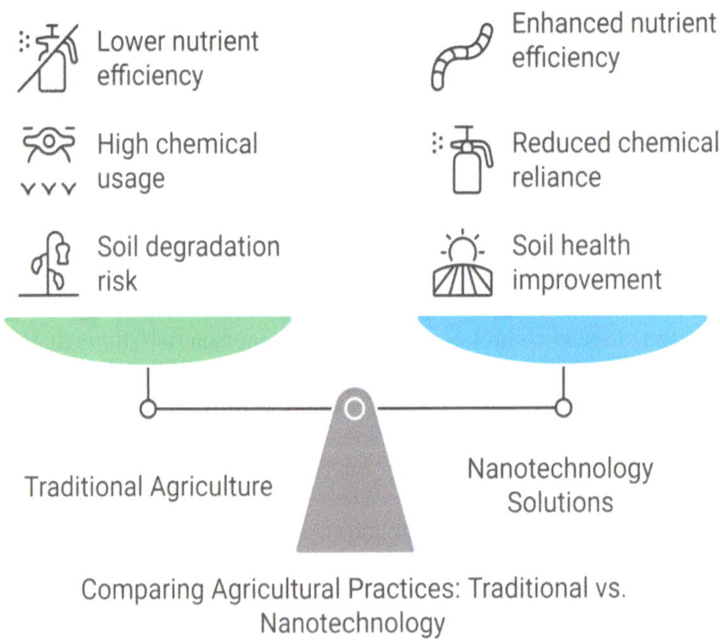

Fig. 1.4 A visual representation of the potential advantages of nanotechnology over traditional practices in agriculture. The figure illustrates a comparison between traditional agricultural practices and those incorporating nanotechnological solutions. Traditional agriculture is associated with lower nutrient efficiency, high chemical usage and a risk of soil degradation. In contrast, nanotechnological solutions are depicted as offering enhanced nutrient efficiency, reduced reliance on chemicals and improved soil health

because of their particle size and surface properties have high compatibility that enhances plant productivity against unfavorable environmental factors such as drought and salinity, for example, silicon nanoparticles (SNPs) increase drought tolerance because of increase in the capacity for water absorption via an increase in the content of leaf water and turgor pressure (Wang et al. 2021). They also activate antioxidant enzymes which help plants combat oxidative stress. It has been established that nanoparticles like Nano-SiO_2 and ZnO are of paramount importance in increasing the plant ability to resist salt stress (Fu et al. 2023). They increase chlorophyll content, seed emergence, rooting pattern, and shoot emergence under salinity stress. They control ionic balance, particularly by reduction of Na^+ influx, which is crucial for cell viability. TiO_2 nanoparticles have the potential to increase stress tolerance through stabilization of the photosystems and photosynthetic pigments as well as improved response to both drought and cold stress (Saxena et al. 2016).

1.2.6 Pathogen Detection

Various properties of nanomaterials contribute to the improvement of diagnostic procedures, making the detection of plant pathogens more effective (Fig. 1.4). Nanomaterials have a size range of 1–100 nm and thus have a high surface area to volume ratio enabling a better interaction with the target pathogens. This results in better sensing mechanisms than the usually employed conventional methods. In biosensors, nanostructures arise as efficient agents in shortening the time required to identify the pathogen (Castillo-Henríquez et al. 2020), for example, the systems containing carbon nanomaterials and metal nanoparticles enable the fast detection of pathogens, thus promoting the timely actions. These nanomaterials can physically transform and alter their characteristics when a stimulus is applied and this feature is beneficial for sensing (Khazaei et al. 2023). That versatility creates a way of coming up with different kinds of sensors, depending on the pathogen in question. Some of the nano biosensors can be developed to be very selective to specific pathogens, thereby minimizing the chances of false alarms. This specificity is attained by the application of recognition ligands for example the antibodies or the DNA oligos. Conventional practices may use costly instrumentation and professional operators as compared to using nanotechnology which can make the process cheaper and easier (Takallu et al. 2024). Portability makes it easier to apply portable biosensors that enable in-situ analysis of the levels of the nanomaterials in plants and the surrounding soil. Colorimetric biosensors which use colloidal gold nanoparticles show the presence of pathogens visually through a change of color; this is critical in precision agriculture and disease outbreak identification (Thakur et al. 2022). This method is easy to perform and does not entail the use of complicated apparatus (Rajpal et al. 2024). Nanomaterials can be used to identify a multitude of pathogens such as viruses, bacteria, and fungi, which makes the tools vital in agricultural diagnostics (Shivashakarappa et al. 2022).

1.2.7 Sustainable Practices (Food Quality, Safety, and Biofortification)

Nanotechnology is relevant in improving sustainable agriculture production for food quality, safety, and biofortification. Several researchers have indicated that nanomaterials with special focus on nanofertilizers increase the rate of nutrient uptake and utilization in plants. One benefit of the targeted delivery system is that it increases the nutrient absorption rate and nutritional values. Micronutrient-fortified nano fertilizer could increase the concentration of the nutrients in the crops, notably in the regions where resource constraints are rampant (El-Saadony et al. 2021). It is possible to enhance the level of nutritional quality of food crops through the use of nanotechnology in agronomic biofortification. Through enhancement of micronutrient density of edible crops like Zn and Fe, nanomaterials contribute to counter the problem of micronutrient malnutrition in human diets. This is important in that in most third-world countries, people suffer from deficiencies of essential micronutrients (Bhardwaj et al. 2022). Chemical fertilizers cause toxicity in soil whenever they are applied in large quantities. The controlled release mechanism ensures that the nutrients required by the plants will be available when needed, thereby promoting healthier soil and crop systems (Bisane et al. 2023). The use of nanotechnology in agriculture promotes sustainable practices by enhancing the utilization of nutrients. This not only improves the quality of food being consumed but also has a positive effect on the environment through a decrease in the overuse of important chemicals such as fertilizers. The approach of integrating nanoparticles into biofortification shows a potential way to fight against nutritional deficiencies. Therefore, the utilization of nanomaterials in improving the nutrient density of crops offers significant potential for food security and human health (Chandra et al. 2023).

1.3 Applications in Agricultural Nanobiotechnology

1.3.1 Nanofertilizers

Currently, nanotechnology has been widely accepted as a novel technology in agriculture, especially through nanofertilizers. These new fertilizers improve NUE and reduce nutrient leaching, which are some of the issues of interest in conventional farming systems. The first benefit of using nanofertilizers involves increased NUE and this is considered to be the greatest benefit of the nanofertilizers (Gade et al. 2023). This is made possible in such a way that the nutrients are delivered in a controlled fashion, hence there is no leaching and volatilization. It allows specific feeding of plants at the right intervals for improved growth and yield of crops. Such slow-release mechanisms can be designed to be triggered by physical factors such as pH, temperature, and humidity, making it easy to have a specific rate of release of nutrients that corresponds to the different stages of plant growth (Yadav et al.

2023a, b, c). The use of metallic nanoparticles as inputs in the agricultural sector has also been considered. Metallic nanoparticles have been reported to improve plant growth and mitigate selected nutrient deficiencies, for example, ZnO nanoparticles have been used to overcome symptoms of zinc deficiency in plants, particularly rice (Mi et al. 2023). Utilization of metallic nanoparticles has been associated with enhanced growth parameters including plant height, biomass, and yield in different crops as revealed by cucumber and black-eyed beans. Compared to other fertilizers, the slow release of nano fertilizers greatly reduces the possibility of nutrient runoff and pollution, which are significant problems with conventional fertilizers. This contributes to the conservation and management of the soil for sustainable agricultural practices so that farming can support the increasing population (Easwaran et al. 2024). Consequently, the development of nanofertilizers is imperative to proper nutrient management. Methods of preparation of nanofertilizers, including the formation of nanocomposites and thin coatings have resulted in improved and efficient delivery of nutrients as compared to routine fertilizers. Nanofertilizers are used for improving crop yield and quality which makes them most suitable for food security in the future (Yadav et al. 2023a, b, c). With respect to the nutrient dimension of the soil, the use of nanofertilizers has several advantages. It acts in soil maintaining the right pH and levels of organic matter crucial for the plants (Saurabh et al. 2024). Nanofertilizers are most suitable for the improvement of deficiencies of essential micronutrients like zinc (Zn), iron (Fe), manganese (Mn), and copper (Cu). Their particles are nanoscale in size which when applied in the soil makes them easily available to plants through improvement of nutrient bioavailability. Such micronutrient deficiencies are addressed appropriately by nanofertilizers, thereby enhancing crop resistance to diseases and other forms of environmental stress. The effectiveness of nanofertilizers offers a solution to enhance efficiency in the use of nutrients and minimizing losses, ensuring micronutrient supply to the plants (Haydar et al. 2024).

1.3.2 Nanopesticides

Nanopesticides are less likely to enter massively into the ecosystem to cause higher bioaccumulation and are more effective than regular pesticides. They increase the dissolution of the ingredients which gives effective control of pests, and at the same time reduces runoff (Chaud et al. 2021). The nanopesticides can be segmented into encapsulation, adsorption and nanoparticle-based delivery systems. Delivery methods include the incorporation of active substances into nanocarriers like poly-ε-caprolactone (PCL), or nanostructured lipid carriers (NLC). This approach not only protects the active ingredients against degradation, but also controls the manner of release, thereby enhancing the shelf life. Nanopesticides possess greater effectiveness against pests such as nematodes than other conventional chemical pesticides (Eghbalinejad et al. 2024).

1.3.3 Nanofungicide and Nanophytovirology

The nano-fungicides like nano selenium (nano-Se) and nano silica (nano-SiO_2) have a desirable impact on the reduction of leaf spot disease in common beans infected by *Alternaria alternata*. These nano-fungicides also significantly affect disease intensity and growth characteristics such as fresh leaf weight and chlorophyll content. They also improve enzymatic and antioxidant actions and offer defense against the pathogen. In broader agricultural use, there is evidence that both Nano-Se and nano-SiO_2 have great potential as a substitute fungicide (Taha et al. 2023). Nanophytovirology is the use of nanotechnology in the diagnosis, identification, and control of plant viral diseases which are key challenges to crop production. Nanoparticles with great reactivity due to their size are used as means to transport genetic structures or to inactivate viruses, providing new ways of controlling plant and even human viruses. Some limitations of the method include the toxicity of nanoparticles and low titer virus detection (Marwal and Gaur 2019).

1.3.4 Phytoremediation

Phytoremediation, the use of plants to clean up contaminated soil, is improved by nanotechnology. It has been observed that incorporating nanomaterials through this process offers the potential to enhance the effectiveness of pollutant degradation and provides a new approach to tackling pollution, for instance, bimetallic nanoparticles (Pd/Fe) as well as nano zero-valent iron (nZVI)-based nanoparticles are outstanding in degrading organic pollutants, particularly chlorpyrifos and atrazine (Aseri et al. 2024). Such nanoparticles help degrade other molecules more quickly, converting them into forms that have minimal negative effects and are less likely to threaten the environment. Nanoparticles assist phytoremediation by enhancing nutrient availability and increasing the rate of enzymes and detoxification processes in plants. They enable processes that are crucial in the uptake and transformation followed by the elimination of pollutants from the rhizosphere (Nedjimi 2021).

1.4 Current Challenges in Agriculture and Its Nanotechnological Solutions

The traditional agricultural practice encounters several challenges due to the oversized population and climate change. The world population keeps on growing and there is an ever-increasing requirement for food production that can feed the world without compromising our natural environment. Nanotechnology has promised solutions that could mitigate such dire consequences for many of these truly pressing issues (Fig. 1.5).

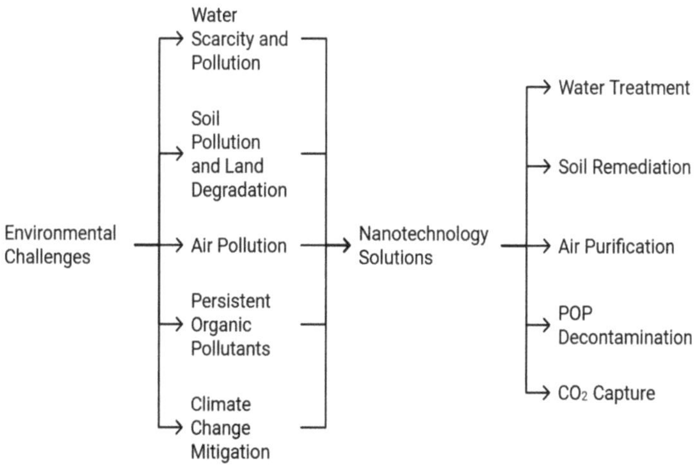

Fig. 1.5 Visual representation of the potential of nanotechnology to provide innovative and effective solutions for a wide range of environmental issues. Environmental challenges including water scarcity and pollution, soil pollution and land degradation, air pollution, persistent organic pollutants, and climate change can be managed effectively through water treatment, soil remediation, air purification, POPS decontamination, and CO_2 capture

1.4.1 Population Growth and Demand for Increased Food Production

The world population is increasing rapidly at an exponential rate; however, the availability of arable land is limited and hence the land-to-population size is decreasing rapidly. Crop produced per living being is drastically decreased due to limited production, supply, and availability. The world population will reach 9.7 billion by 2050, and India contributes 17.78% of the total population. The increment of population is creating immense pressure on agricultural systems to produce more food as well as to the environment because of the lack of sustainability. This problem was not addressed using conventional farming methods (extensive farming with local knowledge) for several constraints like soil degradation (soil erosion resulting in topsoil loss, acidification of soil, and infertility of soil), land availability (extensive agricultural land development results in deforestation), and also because lower yield occurs due to environmental stress. Climate change has intensified these stressors due to which traditional farming methods may struggle even more to cope with the changing conditions. Increasing food production without expanding the area for cultivation is crucial to avoid further deforestation and loss of biodiversity (AbdelRahman 2023). Nanotechnology facilitates precision agriculture, increasing crop production, improving NUE, and reducing the amount lost in harvest. Nano-sized fertilizers and pesticides may be formulated to release the active ingredient more effectively. Fewer ingredients mean less environmental pollution.

Plant-derived nanomaterials (PDNMs) might revolutionize how current traditional practices have been linked to solving the various seemingly impossible issues associated with these practices. This is because the nanomaterials offer novel solutions for stimulating crop production, resistance to disease, and soil health, thus decoupling the use of synthetic chemicals and reducing the utilization of land. The efficiency of the uptake of nutrients through PDNMs would be improved greatly while minimizing the use of fertilizer. This is because plant material-based zinc and iron nanoparticles are used to mitigate nutrient deficiencies in crops (Okeke et al. 2023). Ultimately, it enhances growth without harming soil health. Moreover, conventional pest and disease management practices depend highly on chemical pesticides, which are toxic to the soil and environment. Such toxic pesticides are sharply opposed by PDNMs, among which silver (Ag) nanoparticles have a high antimicrobial activity and can be applied to controlling pathogenic microbes with minimal negative effects on the environment (Abdel-Mohsen et al. 2013; Angelini et al. 2022). This not only minimizes reliance on other lethal pesticides but also promotes more sustainable agricultural practices. Water scarcity continues to be the main challenge to traditional agriculture. PDNMs enhance water retention in soil, and thus, water use is maximized. Their function as a carrier for stress-resistant compounds continues to position plants to cope with greater tolerance toward drought and salinity, brought about by environmental stressors (Abdel-Mohsen et al. 2013).

Bactericidal mechanisms of PDNMs can be extracellular or intracellular. Extracellularly, these nanoparticles accumulate onto the bacterial surface where they disrupt the signaling pathways and cell transport. Intracellularly, these nanoparticles may permeate into microbial cytoplasm and release their contents, thus disturbing the functionalities of the bacteria, for example, green synthesized positively charged silver nanoparticles (AgNPs) are quite more effective against the negative charge of bacterial cells (Banasiuk et al. 2017). Similarly, iron nanoparticles extracted from *Hibiscus rosa sinensis* exhibited antibiotic action against various strains of bacteria via free radicals that enhanced the disrupting action of microbial membrane integrity (Buarki et al. 2022). Nickel (Ni^{2+}) nanoparticles extracted from *Stevia rebaudiana* have also been synthesized with impressive antimicrobial and antioxidant activity towards a significant number of bacterial strains (Srihasam et al. 2020). Thus, with the introduction of PDNMs in agriculture, we may be enabling more robust and sustainable agricultural systems that will more readily meet the prospects needed to feed a growing population.

1.4.2 Climate Change Effects on Crop Yield

Climate change appears destructive for agriculture through the effects brought by changed rainfall patterns, increased frequency of extreme weather events, and modified growing seasons. Some of the major crops expected to lose yield because of temperature stress and drought conditions caused by climate-related factors include cereals like wheat, rice, and maize. One of the significant effects of climate change is enhanced drought stress through increased temperature and changes in rainfall patterns. Drought stress at the critical reproductive development stages may cause a significant reduction in maize yield. A study established that a water loss of 40% can result in a decrease of 39.3% in yield (Kim and Lee 2023). Maize is highly sensitive to drought stress at the pollination time, which can result in a low water supply causing wilt, smaller ears, and fewer kernels per ear. This reduces potential yield by as much as 8% per day. Other negative impacts include those on reproduction; high temperatures during reproductive growth can have negative impacts on the viability of pollen. Research has indicated that temperature higher than 32 °C have resulted in serious damage both to the pollen, thus disrupting the critical synchrony necessary for proper fertilisation to take place (Kim and Lee 2023). Changes in precipitation patterns exacerbate conditions for moisture stress, delaying critical growth stages and reducing overall kernel weight.

Nanotechnology provides several solutions for the attenuation of climate-related impacts on crops. Nanoparticles can form nano-hydrogels and super absorbent polymers (SAPs) helping in the absorption of large quantities of water supplied slowly into the soil for greater water use efficiency and imposing lesser drought impact on crops, for example, silver-coated hydrogels have been shown to exhibit excellent water retention, absorbing as much as 190 times their weight of water, thereby significantly protracting irrigation cycles (Vundavalli et al. 2015). This capacity to conserve water makes them highly beneficial for rainfed agriculture, where water availability is inconsistent. Biochar is an example of carbon-based nanoparticles that are produced from biomass through pyrolysis; they help sequester carbon in soil by improving soil structure, nutrient retention, and water management while fostering beneficial microbial activity (Yadav et al. 2023a, b, c). It acts as a long-term carbon sink, sequestering carbon and reducing greenhouse gas emissions like methane and nitrous oxide.

1.4.3 Precision Agriculture and Real-Time Monitoring

Traditional farming methods follow the uniform applications of water, pesticides, fertilisers, as well as ethnic methods of farming across a whole field, but this has proven to be inefficient and wasteful due to high demand of crops and population increment. Increment in production efficiency involves utilization of technology such as global positioning system (GPS), sensors, Unmanned Aerial Vehicles (UAVs), satellite imaging, remote sensing and IoT (Internet of Things), in the collection of crop information in a more accurate way and on a reasonably high scale (Zain et al. 2023). In precision farming, nanotechnology could be used in conjunction with real time monitoring (by advanced biosensors) of delivering nutrients and agrochemicals using nanoencapsulated fertilisers, pesticides and herbicides more efficiently to plants (Zain et al. 2023).

Iqbal et al. (2022) formulated a protocol for nanotechnology-based elicitors to enhance the multiplication of shoots and callus induction in mung bean varieties. The results considerably showed that nanoparticles may be used as elicitors in further enhancing the production of secondary metabolites in plant cells. The use of nanoparticles as elicitors to enhance secondary metabolite production in mung bean cells relates to precision medicine by enabling the targeted production of bioactive plant compounds that can be used in developing plant-derived pharmaceuticals. This approach allows for controlled, efficient, and scalable production of specific compounds with medicinal properties, aligning with the principles of personalized and precision drug development.

Another key finding is that growth responses in mustard significantly improved after treatment with zinc oxide nanoparticles (ZnONPs) synthesised from the leaves of *Carya illinoinensis*. The study showed the capability of ZnONPs in enhancing plant growth, nutritional content and antioxidant property, thus forming a basis for the use of ZnONPs in biofortification to not only improve crop quality but also crop resilience to environmental threats (Geremew et al. 2023).

1.4.4 Post-Harvest Management and Food Preservation

Post harvest food losses hold a massive place globally. In fact, 30% of food lost after harvest goes into spoilage and degradation (Fung 2009). Thus, the utilisation of nanotechnology in extending shelf life for agricultural products and improving food preservation constitute one of the huge applications in agriculture and packaging-preservation industry (Neme et al. 2021). Nanocoatings which may include nanoparticles and nano emulsions, applied on packaging materials, will reduce spoilage and contamination due to antimicrobial properties during storage and transport (Mohammad and Ahmad 2024). This has been designed to improve the performance and durability of the coatings. This is performed in a range of industries that include electronics, food packaging, insecticides, etc. Nanotechnology has led to the development of nano insecticides and nano fertilisers. These are applicable in much lesser quantities compared to the traditional applications as they encapsulate pesticides in nano capsules or emulsify them into nano emulsions. Examples include nano capsules which have a core pesticide covered by an outer polymeric shell, managing the release of the active ingredient; this can be activated only through specific environmental conditions such as moisture or pH. Nano-encapsulations ensure longer activities of the pesticide, to achieve better pest control with fewer chemicals (Mohammad and Ahmad 2024). Polysaccharides and polyesters like chitosan, alginates, starch and polyethylene glycol are commonly used in nano encapsulations to enhance the effectiveness of pesticide and adhesion to plant surfaces (Sirohi et al. 2021).

Another application of nanotechnology, viz., nanosensors can also be used to monitor pathogens or spoilage indicators in real-time, enabling rapid implementation of remedial actions and reducing food losses. When these markers are detected, nanosensors produce a detectable signal that permits monitoring of food products (Aphibanthammakit and Kasemwong 2021). Nanosensors can also be designed for embedding in packaging or direct application on crops to detect plant pathogens, thereby resulting in a more rapid and focused response to emerging threats.

The various applications of nanotechnology as a tool to mitigate challenges in agriculture are listed in Table 1.1.

Table 1.1 The applications of nanotechnology in agriculture as a solution to various challenges

Agricultural challenge	Nanotechnology Solution	Benefits	References
Population Growth and Demand for Increased Food Production	**Nanofertilizers and Nano-pesticides**—Formulated for controlled release, enhanced nutrient efficiency and reduced environmental pollution	Increased productivity, reduced environmental pollution, higher crop yield per unit of arable land.	Abdel-Mohsen et al. (2013), Okeke et al. (2023)
	Plant-Derived Nanomaterials (PDNMs)—Green-synthesized nanomaterials like Ag and Fe nanoparticles enhance nutrient uptake and pest resistance.	Improves crop resistance to stress, reduces use of synthetic pesticides, enhances nutrient uptake efficiency.	Banasiuk et al. (2017), Srihasam et al. (2020)
Climate Change Effects on Crop Yield	**Nano-Hydrogels and Super Absorbent Polymers (SAPs)**—Nanoparticles like silver-coated hydrogels enhance water retention in soil.	Helps in drought tolerance, conserves water and maintains soil moisture during dry periods.	Vundavalli et al. (2015)
	Biochar (Carbon-based Nanomaterial)—Improves soil structure, sequesters carbon and enhances nutrient and water management.	Carbon sequestration, reduction in greenhouse gases, enhanced soil quality.	Yadav et al. (2023a, b, c)
Limitations of Conventional Agricultural Practices	**Nanofertilizers with Controlled Release**—Delivers nutrients on-demand, minimizing volatilization, leaching and runoff.	Higher nutrient use efficiency (NUE), reduced groundwater contamination and prevention of eutrophication.	Nongbet et al. (2022), Liu and Lal (2015)
	Metal and Carbon-Based Nanoparticles—Iron oxide and carbon-based nanoparticles to immobilize heavy metals in soil, maintaining soil health.	Enhanced soil quality, prevention of heavy metal uptake by plants restoration of soil health.	Xu et al. (2021)

(continued)

Table 1.1 (continued)

Agricultural challenge	Nanotechnology Solution	Benefits	References
Precision Agriculture and Real-Time Monitoring	**Nanoencapsulated Fertilizers and Pesticides**—Allows precise nutrient and pesticide delivery based on plant needs.	Reduced chemical input, minimized environmental impact and improved crop health.	Zain et al. (2023)
	Nanosensors—Enables real-time monitoring of soil, crop health and environmental conditions.	Enhances precision in nutrient and water management, early detection of pests and diseases and real-time decision-making.	Zain et al. (2023), Aphibanthammakit and Kasemwong (2021)
Post-Harvest Management and Food Preservation	**Nanocoatings on Packaging**—Nanoparticles used in packaging materials prevent spoilage and microbial contamination.	Longer shelf life, reduced spoilage, enhanced food safety during transport and storage.	Mohammad and Ahmad (2024)
	Nano-encapsulated Pesticides—Nano capsules containing pesticides are activated only under specific environmental conditions, such as moisture or pH changes.	Improved pest control with lower quantities of pesticides, reduced chemical exposure in the environment and extended pesticide activity.	Sirohi et al. (2021)
	Nanosensors for Pathogen Detection—Monitors food spoilage indicators and detects pathogens in real-time.	Allows rapid responses to spoilage or contamination, reducing post-harvest losses and ensuring food quality throughout the supply chain.	Aphibanthammakit and Kasemwong (2021)

1.5 Aspects of Environmental Nanotechnology (Problems and Remedy)

The scarcity of potable water and water pollution constitute continuing global challenges, which gain momentum with the contamination of water bodies with heavy metals, organic pollutants, pesticides, and emerging contaminants such as pharmaceuticals and endocrine disruptors. These act as significant risks to humans and aquatic ecosystems. Nanotechnology offers innovative solutions such as nano adsorbents like carbon nanotubes (CNTs), graphene oxide (GO), and zeolites, which exhibit exceptional adsorption capabilities for removing harmful substances from water. A water treatment PVDF membrane containing graphene oxide (GO) and multi-walled carbon nanotubes (MWCNTs) were used to improve the filtration performance (Chae et al. 2021). In addition, nZVI particles offer high effectiveness in reducing and immobilizing toxic contaminants from water sources, such as arsenic and chromium (Zeng et al. 2023). Nanofiltration membranes, made with nanoscale precision, are one of the efficient and sustainable methods of filtering salts and microorganisms along with impurities that make clean, safe, and healthy drinking water available to everyone around the globe. These sophisticated nanotechnological approaches therefore mark a transformational approach toward improving water security as well as overcoming water pollution.

Soil pollution and land degradation can be caused by heavy metals, pesticides, hydrocarbons, and industrial waste, posing significant threats to agricultural productivity and ecosystem health. Conventional remediation techniques are often expensive and environmentally invasive. Nano-phytoremediation, a cutting-edge nanotechnological approach, offers an eco-friendly solution by leveraging the synergistic effects of plants and nanoparticles. Nanoparticles such as iron oxide (Fe_2O_3) and carbon-based nanomaterials enhance the bioavailability of pollutants, making them more accessible for uptake and degradation by plants. This method not only accelerates the removal of contaminants from the soil but also restores soil health and fertility, providing a sustainable and cost-effective strategy to combat soil pollution and rehabilitate degraded land (Dhanapal et al. 2024).

Air pollution includes the emission of greenhouse gases such as CO_2 and methane and air pollutants like volatile organic compounds and particulate matter, all of which contribute to global warming and respiratory illnesses (Jiang et al. 2016). Materials that can decompose harmful air pollutants, (for example VOCs and NOx) include materials like TiO_2, ZnO, and CeO_2 (Laxmi et al. 2023). Catalysts of nanogold and nanosilver are being used for the efficient conversion of greenhouse gases into valuable chemicals or less harmful compounds (Laxmi et al. 2023). Advanced nanosorbents offer high surface area and selective adsorption abilities to capture CO_2 and other harmful gases.

Materials such as polychlorinated biphenyls (PCBs), dioxins, and organochlorine pesticides are known to be bioaccumulative, non-biodegradable, and environmentally persistent; these are called persistent organic pollutants (POPs). As a result, they contribute to environmental pollution affecting global public health severely and causing environmental damage in the long run. These pollutants are spread in land and water, interfere with the balance of natural processes in the biosphere, and move up along the food web. The use of nanoscale materials enables the application of new approaches to the problem of the addition of POPs. The degradation and detoxification of such contaminants are greatly aided by titanium dioxide (TiO_2) and nanoparticle compounds such as Fe/Pd (Boulkhessaim et al. 2022). Graphene and carbon nanotube membranes have also been reported to remove and sequester POPs, thereby minimizing their release and bioavailability. These strategies at the nanoscale level present suitable and long-term strategies for the decontamination of the environment, polluted with POPs, ensuring the safety of humans and natural ecosystems.

Mitigating climate change effects is thus one of the most important global challenges. It has its antecedent in the rising call to reduce greenhouse gases. Thus, MOFs and graphene-based nanostructures enable much better CO_2 capture

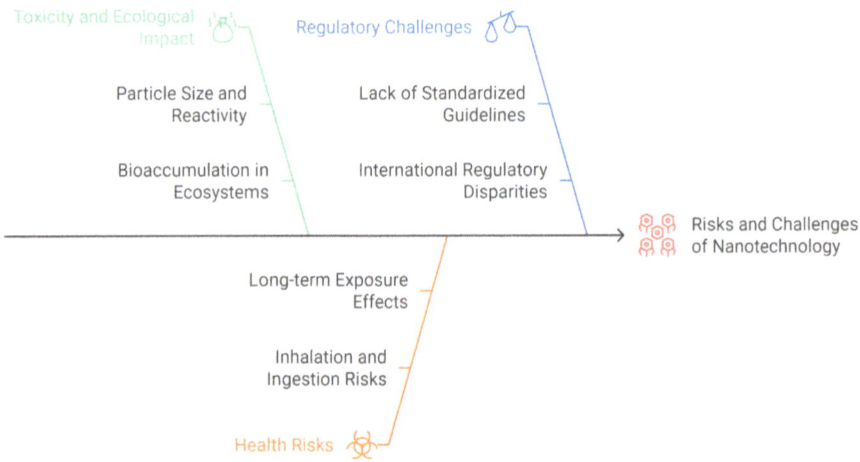

Fig. 1.6 Risks and challenges associated with nanotechnology. The figure highlights concerns related to toxicity and ecological impact, such as the potential for particle reactivity, bioaccumulation, concerns related to health risks, such as long-term exposure effects and inhalation/ingestion risks. Additionally, it addresses regulatory challenges like the lack of standardized guidelines and international regulatory disparities. These factors underscore the need for careful consideration and robust risk assessment to ensure safe and responsible development or application of nanotechnology

and sequestration. MOFs offer a high porosity rate and the ability to possess a structure that is capable of growth (Zhang et al. 2022). Due to the remarkable structure of its surface, graphene is good at absorbing CO_2. It can also help turn captured CO_2 into useful products as well, for instance (Kraegeloh et al. 2018). These tiny materials have a lot of potential just as the researchers are discovering them. They might be able to reduce the quantity of CO_2 in their environment, mitigate the impact of climate change, and contribute to the attainment of sustainable development goals (Fig. 1.6).

1.6 Risks and Challenges of Using Nanotechnology

Nanotechnology offers transformative potential in agriculture, such as targeted pesticide delivery and energy efficiency. However, adoption of it also invites critical challenges, especially related to health and the environment as well as certain ethical issues. The genotoxicity and carcinogenicity of nanomaterials are complicated by particle size, surface chemistry, and coating materials. It is already known that nanoparticles can interact with chromosomal or mitochondrial DNA and can cause genotoxic effects (Muzammil et al. 2023). Thus, the induced breakdown of biological components and functions of plants can arise from oxidative stress (triggered by nanoparticles) that causes chromosome aberrations, micronucleus formation, or altered gene expression. Knowledge of such mechanisms is necessary for assessing the risks of nanomaterials and the safe application of nanomaterials. Nanotoxicology concerns itself with the effects of engineered nanomaterials (ENMs) on human beings and the environment. Some of the key problems encompassing ENM toxicity include conventional ENM toxicity pathways and the approach to designing the ENM that would exert minimal toxic effects, especially through oxidative stress (Corsi et al. 2022). Engineered nanoparticles released unintentionally into the environment can contaminate groundwater and also lead to detrimental impacts on biodiversity. In addition, some studies indicate that nanoparticles can penetrate through plant cellular systems, become translocated to various plant parts, and influence multiple trophic levels. Consequently, nanoparticles can bioaccumulate in the food chain. This bioaccumulation can interfere with major physiological processes in plants, viz., transpiration, respiration, and photosynthesis, leading to a general decline in plant health and productivity (Iavicoli et al. 2017). Additionally, the oxidative stress induced by nanoparticle exposure emphasizes the need to explore the molecular aspects of plant response to such stressors (Muzammil et al. 2023). Although nanotechnology shows great promise, there are significant knowledge gaps about its long-term impacts on agriculture. Figure 1.7 presents a general overview of the risks of nanoparticle use. There are often insufficient comprehensive assessments of the combined effects of nanomaterials on ecosystems and public health in current research. Consequently, it underscores the need for interdisciplinary collaboration among scientists in addressing these challenges, to guarantee environmental safety for nanotechnology applications in agriculture, and to achieve a holistic understanding of the impacts of nanotechnology in agriculture. Table 1.2 shows different nano-based products that focus on overcoming these challenges and these aim to achieve better yield, reduce environmental impact from chemical run-off, conserve resources, and provide more precise management tools.

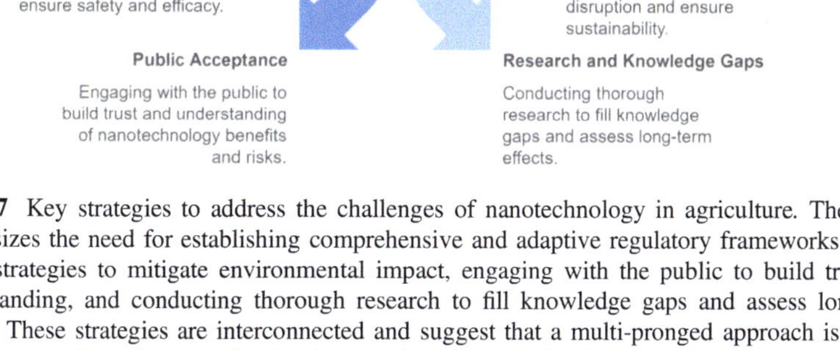

Fig. 1.7 Key strategies to address the challenges of nanotechnology in agriculture. The figure emphasizes the need for establishing comprehensive and adaptive regulatory frameworks, developing strategies to mitigate environmental impact, engaging with the public to build trust and understanding, and conducting thorough research to fill knowledge gaps and assess long-term effects. These strategies are interconnected and suggest that a multi-pronged approach is crucial for the successful and responsible implementation of nanotechnology in agriculture

Table 1.2 Nano-based products in agriculture and environment

Category	Product name	Description	Company/ developer	Source
Nano-Fertilizers	Nano Urea & Nano DAP	Liquid nano-fertilizers to enhance nutrient uptake and reduce quantity used	IFFCO, India	Sharma et al. (2024)
	Nano Green	Organic nanoparticle-based plant tonic for growth improvement	Nano Green Sciences, USA	Kaushik et al. (2023)
	Nano Max NPK	Nutrient-rich fertilizer in nano form for better yield	JU Agri Sciences, India	Basavegowda and Baek (2021)
Nano-Pesticides	Karate Zeon	Nano-encapsulated insecticide for controlled release	Syngenta	Gomes et al. (2023)
	Nanosept Aqua Super	Silver nanoparticle-based broad-spectrum plant protection	Nanosept, Hungary	Rajput et al. (2021)
Nanosensors	Graphene Sensors	Real-time soil nutrient monitoring (e.g., nitrate, phosphate)	Various (R&D & startups)	Parameswari et al. (2024)
	Zinc Oxide Sensors	Detect nitrogen deficiency in crops	Various	Hanif et al. (2024)
Nano-Coated Seeds	Silver nanoparticle Seed Coating	Antimicrobial seed coating to enhance germination	Multiple academic/commercial groups	Khan et al. (2023)
Nano Hydrogels	Cellulose-Based Hydrogels	Superabsorbent, biodegradable gels for soil water retention	Multiple academic/commercial groups	Das et al. (2025)
Water Purification	Tata Swach	Silver nanoparticle-based water purifier using rice husk ash	Tata Chemicals, India	https://www.tatachemicals.com
Irrigation Water Treatment	Nano Membranes	Treats low-quality water by removing salt and pathogens	Used in precision irrigation setups	Joseph et al. (2023)

1.7 Conclusion

Sophisticated technologies like omics, for instance, can have beneficial impacts on biological systems to allow safer use of ENM in medicine and agriculture. The potential of sustainable development in the use of ENMs depends on the evaluation of the ENM impact on the environment and humans during the life cycle of the product without discouraging innovation (Majumdar and Keller 2020). The application of nanotechnology in agriculture presents several innovations in crop production as well as crop protection. Thus, the expansion of complex short formulations has overwhelmed new agricultural products, based on nanoparticles and has prepared only rough scientific evaluations and standardized checklists for their application, imposing great difficulties in assessing the safety of such products (Shukla et al. 2024). Realizing this shortcoming, several countries have introduced regulatory steps to assess nano-based products in agriculture, including guidelines and legislation for safety assessment. For example, India's "Guidelines for Evaluation of Nano-based Agri-input and Food Products" is intended to evaluate nano-containing agri-input and food preparation, These shall provide better benefits and ensure safety (Government of India 2020). Globally, bodies such as the Organisation for Economic Co-operation and Development (OECD) have started defining test methods, exclusively only for nanomaterials. OECD Test Guideline No. 318 is the first TTG established for nanomaterials which indicates the advancement of standard safety testing methods (Monikh et al. 2018). Nevertheless, the major problem of inconsistency in testing procedures is the absence of internationally recognized procedures in standardized testing, for example, the European Union has pointed out a lack of sound standard test methods for nano-sizes regarding current definitions; harmonization of testing protocols is needed (Parisi et al. 2014). Nanotechnology has substantial potential in agriculture for improving yield and soil treatment, but its positive effects on the stable characteristics and detrimental effects on non-target species are partially understood. They can shift microbial formations that affect soil health, be toxic to aquatic life, and can biomagnify, thereby disrupting food chains (Iavicoli et al. 2017). The likely risks are those of unintentional toxicity to soil-friendly microorganisms and disruption of nutrient cycles. Furthermore, the chronic impacts of nanomaterials deposited in the soil environment in the long run have not been adequately explored. Intensive toxicological tests on ENPs in real settings are essential to comprehend the influence of ENPs on microbial richness as well as the balance in the soil ecosystem (Tran et al. 2023). Different countries have developed recommendations for conducting risk estimates, environmental effects, and human health consequences. In the United States, pesticides and agricultural products use nanotechnology that has been controlled by agencies such as EPA and USDA with a focus on toxicity and labeling. Europe to some extent has strict legislation under EFSA and ECHA; they emphasize food safety. It is, however, worth noting that India has working guidelines under the Environment Protection Act and Food Safety and Standards Act to approve most products, based on nanomaterials for the market based on safety benchmarked results (Kumari et al. 2023a, b). "Safe-by-design" is the social practice of comprehending the existence of material characteristics, biological outcomes, and exposure models by means of common instruments such as hazard evaluation and risk assessment. It also encourages partnership through 'trusted spaces' that provide for safe means by which innovators and regulators share information. This has the advantage of producing responsible innovation and goes further to harmonize safety, functionality and sustainability of the environment (Van Gelder et al. 2021). Harnessing the full potential of nanotechnology will therefore be instrumental in achieving a sustainable future where agricultural practices are resource-efficient and capable of meeting the demands of a growing population.

References

Abdel-Mohsen AM, Abdel-Rahman RM, Fouda MMG, Vojtova L, Uhrova L, Hassan AF, Al-Deyab SS, El-Shamy IE, Jancar J (2013) Preparation, characterization and cytotoxicity of schizophyllan/silver nanoparticle composite. Carbohydr Polym 102:238–245. https://doi.org/10.1016/j.carbpol.2013.11.040

AbdelRahman MAE (2023) An overview of land degradation, desertification and sustainable land management using GIS and remote sensing applications. Rendiconti Lincei. Scienze Fisiche E Naturali 34(3):767–808. https://doi.org/10.1007/s12210-023-01155-3

Al-Harbi N, Abd-Elrahman NK (2024) Physical methods for preparation of nanomaterials, their characterization and applications: a review. J Umm Al-Qura Univ Appl Sci. https://doi.org/10.1007/s43994-024-00165-7

Alshammari BH, Lashin MMA, Mahmood MA, Al-Mubaddel FS, Ilyas N, Rahman N, Sohail M, Khan A, Abdullaev SS, Khan R (2023) Organic and inorganic nanomaterials: fabrication, properties and applications. RSC Adv 13(20):13735–13785. https://doi.org/10.1039/d3ra01421e

Angelini J, Klassen R, Široká J, Novák O, Záruba K, Siegel J, Novotná Z, Valentová O (2022) Silver nanoparticles alter microtubule arrangement, dynamics and stress phytohormone levels. Plants 11(3):313. https://doi.org/10.3390/plants11030313

Aphibanthammakit C, Kasemwong K (2021) Nanoencapsulation in agricultural applications. In: Elsevier eBooks, pp 359–382. https://doi.org/10.1016/b978-0-12-821506-7.00015-6

Aseri V, Chopade RL, Godara V, Lohar S, Kumari P, Mavry B, Nagar V, Sharma V, Awasthi G, Awasthi KK, Sankhla MS (2024) Advancing phytoremediation of contaminated soil through nanoparticle-based approaches: current progress and future perspectives. Lett Appl NanoBioSci 13(2):57. https://doi.org/10.33263/LIANBS132.057

Banasiuk R, Krychowiak M, Swigon D, Tomaszewicz W, Michalak A, Chylewska A, Ziabka M, Lapinski M, Koscielska B, Narajczyk M, Krolicka A (2017) Carnivorous plants used for green synthesis of silver nanoparticles with broad-spectrum antimicrobial activity. Arab J Chem 13(1): 1415–1428. https://doi.org/10.1016/j.arabjc.2017.11.013

Basavegowda N, Baek K (2021) Current and future perspectives on the use of nanofertilizers for sustainable agriculture: the case of phosphorus nanofertilizer. 3 Biotech 11(7). https://doi.org/10.1007/s13205-021-02907-4

Bhardwaj AK, Chejara S, Malik K, Kumar R, Kumar A, Yadav RK (2022) Agronomic biofortification of food crops: an emerging opportunity for global food and nutritional security. Front Plant Sci 13. https://doi.org/10.3389/fpls.2022.1055278

Bhushan B (2017) Introduction to nanotechnology. In: Springer handbooks, pp 1–19. https://doi.org/10.1007/978-3-662-54357-3_1

Bisane M, Chakravarty I, Mehetre S, Mukherjee P, Mandavgane SA (2023) Biomass ash as a source of nutrient; a case study on cotton (Gossypium). In: Elsevier eBooks, pp 301–308. https://doi.org/10.1016/b978-0-323-91743-8.00005-8

Boulkhessaim S, Gacem A, Khan SH, Amari A, Yadav VK, Harharah HN, Elkhaleefa AM, Yadav KK, Rather S, Ahn H, Jeon B (2022) Emerging trends in the remediation of persistent organic pollutants using nanomaterials and related processes: a review. Nanomaterials 12(13):2148. https://doi.org/10.3390/nano12132148

Bruna T, Maldonado-Bravo F, Jara P, Caro N (2021) Silver nanoparticles and their antibacterial applications. Int J Mol Sci 22(13):7202. https://doi.org/10.3390/ijms22137202

Buarki F, AbuHassan H, Hannan FA, Henari FZ (2022) Green synthesis of iron oxide nanoparticles using *Hibiscus rosa sinensis* flowers and their antibacterial activity. J Nanotechnol 2022:1–6. https://doi.org/10.1155/2022/5474645

Castillo-Henríquez L, Brenes-Acuña M, Castro-Rojas A, Cordero-Salmerón R, Lopretti-Correa M, Vega-Baudrit JR (2020) Biosensors for the detection of bacterial and viral clinical pathogens. Sensors 20(23):6926. https://doi.org/10.3390/s20236926

Chae J, Lim T, Cheng H, Hu J, Kim S, Jung W (2021) Graphene oxide and carbon nanotubes-based polyvinylidene fluoride membrane for highly increased water treatment. Nanomaterials 11(10): 2498. https://doi.org/10.3390/nano11102498

Chandra MS, Jayashankar Kadam PV, Chand SW, Geethanjali D (2023) Role of nano-particles in agronomic biofortification. In: Sustainable agriculture and food security. Deepika Book Agency

Chaud M, Souto EB, Zielinska A, Severino P, Batain F, Oliveira-Junior J, Alves T (2021) Nanopesticides in agriculture: benefits and challenge in agricultural productivity, toxicological risks to human health and environment. Toxics 9(6):131. https://doi.org/10.3390/toxics9060131

Corsi I, Desimone MF, Cazenave J (2022) Editorial: Nanotoxicology: challenges and solutions to safeguard human health and the environment. Front Bioeng Biotechnol 10. https://doi.org/10.3389/fbioe.2022.1110246

Das D, Chingakham N, Sarma M, Basu S, Bhaladhare S (2025) Cellulose-based biodegradable superabsorbent hydrogel: a sustainable approach for water conservation and plant growth in agriculture. Int J Biol Macromol 305:141176. https://doi.org/10.1016/j.ijbiomac.2025.141176

Dhanapal AR, Thiruvengadam M, Vairavanathan J, Venkidasamy B, Easwaran M, Ghorbanpour M (2024) Nanotechnology approaches for the remediation of agricultural polluted soils. ACS Omega 9(12):13522–13533. https://doi.org/10.1021/acsomega.3c09776

Easwaran C, Moorthy G, Christopher SR, Mohan P, Marimuthu R, Koothan V, Nallusamy S (2024) Nano hybrid fertilizers: a review on the state of the art in sustainable agriculture. Sci Total Environ 172533. https://doi.org/10.1016/j.scitotenv.2024.172533

Eghbalinejad M (2024) Nanopesticides in the environment–understanding their fate, bioavailability, effects, and risks in soil and water [PhD dissertation]. In: RECETOX. Faculty of Science, Masaryk University. (Original work published 2024). https://doi.org/10.13140/RG.2.2.33888.29447

Eghbalinejad M, Hofman J, Kotouček J, Grillo R, Bílková ZH, Reiff N, Höss S (2024) Nano-enabled pesticides: a comprehensive toxicity assessment of tebuconazole nanoformulations with nematodes at single species and community level. Environ Sci Eur 36(1). https://doi.org/10.1186/s12302-024-00879-9

El-Saadony MT, Almoshadak AS, Shafi ME, Albaqami NM, Saad AM, El-Tahan AM, Desoky EM, Elnahal AS, Almakas A, El-Mageed TAA, Taha AE, Elrys AS, Helmy AM (2021) Vital roles of sustainable nano-fertilizers in improving plant quality and quantity-an updated review. Saudi J Biol Sci 28(12):7349–7359. https://doi.org/10.1016/j.sjbs.2021.08.032

Fajardo C, Martinez-Rodriguez G, Blasco J, Mancera JM, Thomas B, De Donato M (2022) Nanotechnology in aquaculture: applications, perspectives and regulatory challenges. Aquacult Fish 7(2):185–200. https://doi.org/10.1016/j.aaf.2021.12.006

Fu C, Khan MN, Yan J, Hong X, Zhao F, Chen L, Ma H, Li Y, Li J, Wu H (2023) Mechanisms of nanomaterials for improving plant salt tolerance. Crop Environ 2(2):92–99. https://doi.org/10.1016/j.crope.2023.03.002

Fu Y, Liu T, Wang H, Wang Z, Hou L, Jiang J, Xu T (2024) Applications of nanomaterial technology in biosensing. J Sci Adv Mater Dev 9(2):100694. https://doi.org/10.1016/j.jsamd.2024.100694

Fung D (2009) Food spoilage, preservation and quality control. In: Elsevier eBooks, pp 54–79. https://doi.org/10.1016/b978-012373944-5.00122-x

Gade A, Ingle P, Nimbalkar U, Rai M, Raut R, Vedpathak M, Jagtap P, Abd-Elsalam KA (2023) Nanofertilizers: the next generation of agrochemicals for long-term impact on sustainability in farming systems. Agrochemicals 2(2):257–278. https://doi.org/10.3390/agrochemicals2020017

Ganji DD, Kachapi SHH (2015) Introduction to nanotechnology, nanomechanics, micromechanics, and nanofluid. In: Elsevier eBooks, pp 1–11. https://doi.org/10.1016/b978-0-323-35237-6.00001-7

Gehrke I, Geiser A, Somborn-Schulz A (2015) Innovations in nanotechnology for water treatment. Nanotechnol Sci Appl 1. https://doi.org/10.2147/nsa.s43773

Geremew A, Carson L, Woldesenbet S, Wang H, Reeves S, Brooks N Jr, Saganti P, Weerasooriya A, Peace E (2023) Effect of zinc oxide nanoparticles synthesized from *Carya illinoinensis* leaf extract on growth and antioxidant properties of mustard (*Brassica juncea*). Front Plant Sci 14:1108186. https://doi.org/10.3389/fpls.2023.1108186

Ghidan AY, Antary TMA (2019) Applications of nanotechnology in agriculture. In: IntechOpen eBooks. https://doi.org/10.5772/intechopen.88390

Gomes SI, Chidiamassamba SB, Scott-Fordsmand JJ, Amorim MJ (2023) Environmental hazards of nanopesticides to non-target soil species–commercial nanoformulation versus its active substance (Karate Zeon® and lambda-cyhalothrin). Sci Total Environ 891:164664. https://doi.org/10.1016/j.scitotenv.2023.164664

Government of India, Department of Biotechnology, Ministry of Science & Technology, Ministry of Agriculture and Farmers' Welfare, Ninawe, S., & Krishna, V (2020) In: Adholeya A, Dinda A, Subramanian KS, Anandharamakrishnan C (eds) Guidelines for evaluation of nano-based agri-input and food products in India. Department of Biotechnology, New Delhi. https://dbtindia.gov.in/sites/default/files/Guidlines%20Document_0.pdf

Graham W, Torbett-Dougherty M, Islam A, Soleimani S, Bruce-Tagoe TA, Johnson JA (2025) Magnetic nanoparticles and drug delivery systems for anti-cancer applications: a review. Nanomaterials 15(4):285. https://doi.org/10.3390/nano15040285

Hanif S, Javed R, Cheema M, Kiani MZ, Farooq S, Zia M (2024) Harnessing the potential of zinc oxide nanoparticles and their derivatives as nanofertilizers: trends and perspectives. Plant Nano Biol 10:100110. https://doi.org/10.1016/j.plana.2024.100110

Haydar MS, Ghosh D, Roy S (2024) Slow and controlled release nanofertilizers as an efficient tool for sustainable agriculture: recent understanding and concerns. Plant Nano Biol 7:100058. https://doi.org/10.1016/j.plana.2024.100058

Iavicoli I, Leso V, Beezhold DH, Shvedova AA (2017) Nanotechnology in agriculture: opportunities, toxicological implications, and occupational risks. Toxicol Appl Pharmacol 329:96–111. https://doi.org/10.1016/j.taap.2017.05.025

India population (2024) - Worldometer (n.d.). https://www.worldometers.info/world-population/india-population/

Iqbal Z, Javad S, Naz S, Shah AA, Shah AN, Paray BA, Gulnaz A, Abdelsalam NR (2022) Elicitation of the in vitro cultures of selected varieties of *Vigna radiata* L. with zinc oxide and copper oxide nanoparticles for enhanced phytochemicals production. Front Plant Sci:13. https://doi.org/10.3389/fpls.2022.908532

Jha A, Pathania D, Sonu N, Damathia B, Raizada P, Rustagi S, Singh P, Rani GM, Chaudhary V (2023) Panorama of biogenic nano-fertilizers: a road to sustainable agriculture. Environ Res 235:116456. https://doi.org/10.1016/j.envres.2023.116456

Jiménez-Jiménez C, Manzano M, Vallet-Regí M (2020) Nanoparticles coated with cell membranes for biomedical applications. Biology 9(11):406. https://doi.org/10.3390/biology9110406

Joseph TM, Al-Hazmi HE, Śniatała B, Esmaeili A, Habibzadeh S (2023) Nanoparticles and nanofiltration for wastewater treatment: from polluted to fresh water. Environ Res 238:117114. https://doi.org/10.1016/j.envres.2023.117114

Joshi A, Kaur S, Dharamvir K, Nayyar H, Verma G (2017) Multi-walled carbon nanotubes applied through seed-priming influence early germination, root hair, growth and yield of bread wheat (*Triticum aestivum* L.). J Sci Food Agric 98(8):3148–3160. https://doi.org/10.1002/jsfa.8818

Kaushik A, Singh RK, Tyagi PK (2023) Green synthesized nanoparticle-based drug delivery: recent trends and future prospects. Precis Nanomed 6–4:1109–1132. https://doi.org/10.33218/001c.89165

Khan MR, Rizvi TF (2014) Nanotechnology: scope and application in plant disease management. Plant Pathol J 13(3):214–231. https://doi.org/10.3923/ppj.2014.214.231

Khan I, Saeed K, Khan I (2017) Nanoparticles: properties, applications and toxicities. Arab J Chem 12(7):908–931. https://doi.org/10.1016/j.arabjc.2017.05.011

Khan Y, Sadia H, Shah SZA, Khan MN, Shah AA, Ullah N, Ullah MF, Bibi H, Bafakeeh OT, Khedher NB, Eldin SM, Fadhl BM, Khan MI (2022) Classification, synthetic, and characterization approaches to nanoparticles, and their applications in various fields of nanotechnology: a review. Catalysts 12(11):1386. https://doi.org/10.3390/catal12111386

Khan S, Zahoor M, Khan RS, Ikram M, Islam NU (2023) The impact of silver nanoparticles on the growth of plants: the agriculture applications. Heliyon 9(6):e16928. https://doi.org/10.1016/j.heliyon.2023.e16928

Khazaei M, Hosseini MS, Haghighi AM, Misaghi M (2023) Nanosensors and their applications in early diagnosis of cancer. Sens Bio-Sens Res 41:100569. https://doi.org/10.1016/j.sbsr.2023.100569

Kim KH, Lee BM (2023) Effects of climate change and drought tolerance on maize growth. Plants (Basel, Switz) 12(20):3548. https://doi.org/10.3390/plants12203548

Kraegeloh A, Suarez-Merino B, Sluijters T, Micheletti C (2018) Implementation of safe-by-design for nanomaterial development and safe innovation: why we need a comprehensive approach. Nanomaterials 8(4):239. https://doi.org/10.3390/nano8040239

Kumari R, Suman K, Karmakar S, Mishra V, Lakra SG, Saurav GK, Mahto BK (2023a) Regulation and safety measures for nanotechnology-based agri-products. Front Genome Edit 5:1200987. https://doi.org/10.3389/fgeed.2023.1200987

Kumari S, Raturi S, Kulshrestha S, Chauhan K, Dhingra S, András K, Thu K, Khargotra R, Singh T (2023b) A comprehensive review on various techniques used for synthesizing nanoparticles. J Mater Res Technol 27:1739–1763. https://doi.org/10.1016/j.jmrt.2023.09.291

Lahiani MH, Dervishi E, Chen J, Nima Z, Gaume A, Biris AS, Khodakovskaya MV (2013) Impact of carbon nanotube exposure to seeds of valuable crops. ACS Appl Mater Interfaces 5(16): 7965–7973. https://doi.org/10.1021/am402052x

Laxmi V, Singhvi N, Ahmad N, Sinha S, Negi T, Gupta V, Mubashshir M, Ahmad A, Sharma S (2023) Emerging field of nanotechnology in environment. Indian J Microbiol 63(3):244–252. https://doi.org/10.1007/s12088-023-01092-7

Liu R, Lal R (2015) Potentials of engineered nanoparticles as fertilizers for increasing agronomic productions. The Science of the Total Environment 514:131–139. https://doi.org/10.1016/j.scitotenv.2015.01.104

Majumdar S, Keller AA (2020) Omics to address the opportunities and challenges of nanotechnology in agriculture. Crit Rev Environ Sci Technol. https://doi.org/10.1080/10643389.2020.1785264

Malik S, Muhammad K, Waheed Y (2023) Emerging applications of nanotechnology in healthcare and medicine. Molecules 28(18):6624. https://doi.org/10.3390/molecules28186624

Manjunatha R, Naik D, Usharani K (2019) Nanotechnology application in agriculture: a review. J Pharm Phytochem 8(3):1073–1083

Marwal A, Gaur R (2019) Nanophytovirology: an emerging field for disease management. In: IntechOpen eBooks. https://doi.org/10.5772/intechopen.86653

Meena DS, Gautam C, Patidar OP, Meena HM, Prakasha G, Vishwajith. (2017) Nano-fertilizers is a new way to increase nutrients use efficiency in crop production. Int J Agric Sci 9(7):3831–3833. https://ureaknowhow.com/wp-content/uploads/2020/09/2017-Singh-Nano-Fertilizers-is-a-new-way-to-increase-nutrients-use-efficiency-in-crop-production.pdf

Mehlhorn H (2016) Nanoparticles—definitions. In: Parasitology research monographs, pp 1–14. https://doi.org/10.1007/978-3-319-25292-6_1

Mi K, Yuan X, Wang Q, Dun C, Wang R, Yang S, Yang Y, Zhang H, Zhang H (2023) Zinc oxide nanoparticles enhanced rice yield, quality, and zinc content of edible grain fraction synergistically. Front Plant Sci 14. https://doi.org/10.3389/fpls.2023.1196201

Mohammad ZH, Ahmad F (2024) Nanocoating and its application as antimicrobials in the food industry: a review. Int J Biol Macromol 254(Pt 2):127906. https://doi.org/10.1016/j.ijbiomac.2023.127906

Monikh FA, Praetorius A, Schmid A, Kozin P, Meisterjahn B, Makarova E, Hofmann T, Von Der Kammer F (2018) Scientific rationale for the development of an OECD test guideline on engineered nanomaterial stability. NanoImpact 11:42–50. https://doi.org/10.1016/j.impact.2018.01.003

Mousavi SR, Rezaei M (2011) Nanotechnology in agriculture and food production. J Appl Environ Biol Sci 1–10:414–419. https://www.researchgate.net/publication/216346568

Muzammil S, Ashraf A, Siddique MH, Aslam B, Rasul I, Abbas R, Afzal M, Faisal M, Hayat S (2023) A review on toxicity of nanomaterials in agriculture: current scenario and future prospects. Sci Prog 106(4):368504231221672. https://doi.org/10.1177/00368504231221672

Narayanan R, El-Sayed MA (2005) Catalysis with transition metal nanoparticles in colloidal solution: nanoparticle shape dependence and stability. J Phys Chem B 109(26):12663–12676. https://doi.org/10.1021/jp051066p

Nedjimi B (2021) Phytoremediation: a sustainable environmental technology for heavy metals decontamination. SN Appl Sci 3(3). https://doi.org/10.1007/s42452-021-04301-4

Neme K, Nafady A, Uddin S, Tola YB (2021) Application of nanotechnology in agriculture, postharvest loss reduction and food processing: food security implication and challenges. Heliyon 7(12):e08539. https://doi.org/10.1016/j.heliyon.2021.e08539

Nongbet A, Mishra AK, Mohanta YK, Mahanta S, Ray MK, Khan M, Baek KH, Chakrabartty I (2022) Nanofertilizers: a smart and sustainable attribute to modern agriculture. Plants (Basel, Switz) 11(19):2587. https://doi.org/10.3390/plants11192587

Okeke ES, Nweze EJ, Anaduaka EG, Okoye CO, Anosike CA, Joshua PE, Ezeorba TPC (2023) Plant-derived nanomaterials (PDNM): a review on pharmacological potentials against pathogenic microbes, antimicrobial resistance (AMR) and some metabolic diseases. 3 Biotech 13(9): 291. https://doi.org/10.1007/s13205-023-03713-w

Panpatte DG, Jhala YK, Shelat HN, Vyas RV (2016) Nanoparticles: the next generation technology for sustainable agriculture. In: Panpatte DG, Jhala YK (eds) Nanotechnology for agriculture: crop production & protection. Springer, pp 289–300. https://doi.org/10.1007/978-81-322-2644-4_18

Parameswari P, Belagalla N, Singh BV, Abhishek G, Rajesh G, Katiyar D, Hazarika B, Paul S (2024) Nanotechnology-based sensors for real-time monitoring and assessment of soil health and quality: a review. Asian J Soil Sci Plant Nutr 10(2):157–173. https://doi.org/10.9734/ajsspn/2024/v10i2272

Parisi C, Vigani M, Rodríguez-Cerezo E, European Commission (2014) Proceedings of a workshop on "Nanotechnology for the agricultural sector: from research to the field.". In: JRC SCIENTIFIC AND POLICY REPORTS (Report EUR 26625 EN). European Commission. https://doi.org/10.2791/80497. (Original work published 2014)

Petosa AR, Rajput F, Selvam O, Öhl C, Tufenkji N (2016) Assessing the transport potential of polymeric nanocapsules developed for crop protection. Water Res 111:10–17. https://doi.org/10.1016/j.watres.2016.12.030

Qureshi A, Singh D, Dwivedi S (2018) Nano-fertilizers: a novel way for enhancing nutrient use efficiency and crop productivity. Int J Curr Microbiol App Sci 7(2):3325–3335. https://doi.org/10.20546/ijcmas.2018.702.398

Ragaei M, Sabry AH (2014) Nanotechnology for insect pest control. Int J Sci Environ 3(2):528–545. https://www.researchgate.net/profile/Kazafy_Sabry/publication/261365123_Nanotechnology_FOR_INSECT_PEST_CONTROL/links/0a85e534131e922f2e000000.pdf. https://www.ijset.net

Rajpal VR, Mehta S, Minkina T, Rajput VD, Husen A (2024) Exploring metal and metal-oxide nanoparticles for nanosensing and biotic stress management in plant systems. Curr Res Biotechnol 7:100219–100219. https://doi.org/10.1016/j.crbiot.2024.100219

Rajput VD, Singh A, Minkina T, Rawat S, Mandzhieva S, Sushkova S, Shuvaeva V, Nazarenko O, Rajput P, Komariah N, Verma KK, Singh AK, Rao M, Upadhyay SK (2021) Nano-enabled products: challenges and opportunities for sustainable agriculture. Plants 10(12):2727. https://doi.org/10.3390/plants10122727

Robinson AL (1986) Electron microscope inventors share Nobel physics prize. Science 234(4778): 821–822. https://doi.org/10.1126/science.234.4778.821

Roco MC (2005) Environmentally responsible development of nanotechnology. Environ Sci Technol 39(5):106A–112A. https://doi.org/10.1021/es053199u

Saxena R, Tomar RS, Kumar M (2016) Exploring Nanobiotechnology to mitigate abiotic stress in crop plants. J Pharm Sci Res 8(9):974

Shah MA, Pirzada BM, Price G, Shibiru AL, Qurashi A (2022) Applications of nanotechnology in smart textile industry: a critical review. J Adv Res 38:55–75. https://doi.org/10.1016/j.jare.2022.01.008

Shang Y, Hasan MK, Ahammed GJ, Li M, Yin H, Zhou J, Ahammed GJ, Zhou J (2019) Applications of nanotechnology in plant growth and crop protection: a review. Molecules 24:2558. https://doi.org/10.3390/molecules24142558

Sharma V, Javed B, Byrne H, Curtin J, Tian F (2022) Zeolites as carriers of nano-fertilizers: from structures and principles to prospects and challenges. Appl Nano 3(3):163–186. https://doi.org/10.3390/applnano3030013

Sharma S, Kumar A, Singh K, Suman S, Pathak SO (2024) Implementation of nano-urea and nano-DAP for increasing crop growth, yield and soil fertility in cauliflower and tomato. Res Square (Research Square). https://doi.org/10.21203/rs.3.rs-2960032/v2

Shivashakarappa K, Reddy V, Tupakula VK, Farnian A, Vuppula A, Gunnaiah R, Department of Food, Environment and Nutritional Sciences, University of Milan, Italy, Department of Agriculture Biotechnology, University of Agricultural Sciences Dharwad, India, & Department of Horticulture, University of Horticultural Sciences Bagalkot, India (2022) Nanotechnology for the detection of plant pathogens [Agriculture]. Plant Nano Biol 100018. https://doi.org/10.1016/j.plana.2022.100018

Shukla K, Mishra V, Singh J, Varshney V, Verma R, Srivastava S (2024) Nanotechnology in sustainable agriculture: a double-edged sword. J Sci Food Agric 104(10):5675–5688. https://doi.org/10.1002/jsfa.13342

Sirohi R, Tarafdar A, Kumar Gaur V, Singh S, Sindhu R, Rajasekharan R, Madhavan A, Binod P, Kumar S, Pandey A (2021) Technologies for disinfection of food grains: advances and way forward. Food Res Int (Ottawa, Ont.) 145:110396. https://doi.org/10.1016/j.foodres.2021.110396

Sood R, Chopra DS (2017) Metal–plant frameworks in nanotechnology: an overview. Phytomedicine 50:148–156. https://doi.org/10.1016/j.phymed.2017.08.025

Srihasam S, Thyagarajan K, Korivi M, Lebaka VR, Mallem SPR (2020) Phytogenic generation of NiO nanoparticles using stevia leaf extract and evaluation of their in-vitro antioxidant and antimicrobial properties. Biomolecules 10(1):89. https://doi.org/10.3390/biom10010089

Swierczewska M, Liu G, Lee S, Chen X (2011) High-sensitivity nanosensors for biomarker detection. Chem Soc Rev 41(7):2641–2655. https://doi.org/10.1039/c1cs15238f

Syduzzaman M, Hassan A, Anik HR, Akter M, Islam MR (2023) Nanotechnology for high-performance textiles: a promising frontier for innovation. ChemNanoMat 9(9). https://doi.org/10.1002/cnma.202300205

Taha NA, Hamden S, Bayoumi YA, Elsakhawy T, El-Ramady H, Solberg SØ (2023) Nanofungicides with selenium and silicon can boost the growth and yield of common bean (*Phaseolus vulgaris* L.) and control *Alternaria* leaf spot disease. Microorganisms 11(3):728. https://doi.org/10.3390/microorganisms11030728

Takallu S, Aiyelabegan HT, Zomorodi AR, Alexandrovna KV, Aflakian F, Asvar Z, Moradi F, Behbahani MR, Mirzaei E, Sarhadi F, Vakili-Ghartavol R (2024) Nanotechnology improves the detection of bacteria: recent advances and future perspectives. Heliyon 10(11):e32020. https://doi.org/10.1016/j.heliyon.2024.e32020

Thakur M, Wang B, Verma ML (2022) Development and applications of nanobiosensors for sustainable agricultural and food industries: recent developments, challenges and perspectives. Environ Technol Innovation 26:102371. https://doi.org/10.1016/j.eti.2022.102371

Tran T, Nguyen M, Lin C, Hoang T, Nguyen T, Lone AM, Khedulkar AP, Gaballah MS, Singh J, Chung WJ, Nguyen DD (2023) Review on fate, transport, toxicity and health risk of nanoparticles in natural ecosystems: emerging challenges in the modern age and solutions toward a sustainable environment. Sci Total Environ 912:169331. https://doi.org/10.1016/j.scitotenv.2023.169331

Van Gelder P, Klaassen P, Taebi B, Walhout B, Van Ommen R, Van De Poel I, Robaey Z, Asveld L, Balkenende R, Hollmann F, Van Kampen EJ, Khakzad N, Krebbers R, De

Lange J, Pieters W, Terwel K, Visser E, Van Der Werff T, Jung D (2021) Safe-by-design in engineering: an overview and comparative analysis of engineering disciplines. Int J Environ Res Public Health 18(12):6329. https://doi.org/10.3390/ijerph18126329

Vundavalli R, Vundavalli S, Nakka M, Rao DS (2015) Biodegradable nano-hydrogels in agricultural farming–alternative source for water resources. Procedia Mater Sci 10:548–554. https://doi.org/10.1016/j.mspro.2015.06.005

Wang M, Wang R, Mur LAJ, Ruan J, Shen Q, Guo S (2021) Functions of silicon in plant drought stress responses. Hortic Res 8(1). https://doi.org/10.1038/s41438-021-00681-1

Xu W, Yang T, Liu S, Du L, Chen Q, Li X, Dong J, Zhang Z, Lu S, Gong Y, Zhou L, Liu Y, Tan X (2021) Insights into the Synthesis, types and application of iron Nanoparticles: The overlooked significance of environmental effects. Environment International 158:106980. https://doi.org/10.1016/j.envint.2021.106980

Xuan L, Ju Z, Skonieczna M, Zhou P, Huang R (2023) Nanoparticles-induced potential toxicity on human health: applications, toxicity mechanisms, and evaluation models. MedComm 4(4). https://doi.org/10.1002/mco2.327

Yadav J, Jasrotia P, Kashyap PL, Bhardwaj AK, Kumar S, Singh M, Singh GP (2022) Nanopesticides: current status and scope for their application in agriculture. Plant Protect Sci 58. https://doi.org/10.17221/102/2020-PPS

Yadav A, Yadav K, Abd-Elsalam KA (2023a) Nanofertilizers: types, delivery and advantages in agricultural sustainability. Agrochemicals 2(2):296–336. https://doi.org/10.3390/agrochemicals2020019

Yadav SPS, Bhandari S, Bhatta D, Poudel A, Bhattarai S, Yadav P, Ghimire N, Paudel P, Paudel P, Shrestha J, Oli B (2023b) Biochar application: a sustainable approach to improve soil health. J Agric Food Res 11:100498. https://doi.org/10.1016/j.jafr.2023.100498

Yadav A, Yadav K, Abd-Elsalam KA (2023c) Exploring the potential of nanofertilizers for a sustainable agriculture. Plant Nano Biol 5:100044. https://doi.org/10.1016/j.plana.2023.100044

Zain M, Ma H, Rahman SU, Nuruzzaman M, Chaudhary S, Azeem I, Mehmood F, Duan A, Sun C (2023) Nanotechnology in precision agriculture: advancing towards sustainable crop production. Plant Physiol Biochem 206:108244. https://doi.org/10.1016/j.plaphy.2023.108244

Zeng G, He Y, Wang F, Luo H, Liang D, Wang J, Huang J, Yu C, Jin L, Sun D (2023) Toxicity of nanoscale zero-valent iron to soil microorganisms and related defense mechanisms: a review. Toxics 11(6):514. https://doi.org/10.3390/toxics11060514

Zhang Q, Yang H, Zhou T, Chen X, Li W, Pang H (2022) Metal-organic frameworks and their composites for environmental applications. Adv Sci (Weinheim, Baden-Wurttemberg, Germany) 9(32):e2204141. https://doi.org/10.1002/advs.202204141

Zheng L, Hong F, Lu S, Liu C (2005) Effect of nano-TiO_2 on strength of naturally aged seeds and growth of spinach. Biol Trace Elem Res 104(1):083–092. https://doi.org/10.1385/bter:104:1:083

Chapter 2
Overview of the Fundamentals of the Synthesis of Nanoparticles and Characterization Techniques

Subhadeep Sen, Chanchal Das, and Goutam Biswas

Abstract Researchers are more interested in nanomaterials because of their special or novel physical and chemical properties unlike the larger materials. These exceptional qualities have enabled numerous cutting-edge applications in electronics, the food industry, chemical catalysis, medicine and pharmaceuticals, and agriculture. Nanoparticles are commonly synthesized using physical, chemical and biological methods. Physical and chemical methods are sometimes costly, energy-intensive, and potentially hazardous. Biological methods use plants or microorganisms, offering an environmentally sustainable alternate to other popular methods. Nanotechnologists must comprehend and be proficient in the intricate methods required to accurately characterize these processes to employ this multidisciplinary approach to nanoparticle manufacturing. Various types of nanoparticles, their synthesis, and characterization processes are discussed in this chapter.

Keywords Nanoparticles · Nanotechnology · Green synthesis · Characterization · Phytogenic nanoparticles · Surface-coated nanoparticles.

2.1 Introduction

In the current century, nanotechnology has developed as a significant scientific achievement. The interdisciplinary field of nanotechnology encompasses the fabrication, functionalization, and application of engineered materials with dimensions smaller than 100 nm in all three dimensions. The term "nano" is derived from the Greek word "nanos," that means "a dwarf". It is being commonly used in the technical literature to describe small things and processes in contemporary science

S. Sen · G. Biswas (✉)
Department of Chemistry, Cooch Behar Panchanan Barma University, Cooch Behar, West Bengal, India
e-mail: goutam@cbpbu.ac.in

C. Das
Department of Chemistry, Saheed Kshudiram College, Alipurduar, India

R. N. Pudake et al. (eds.), *Phyto Nanotoxicology*, Nanotechnology in Plant Sciences 7, https://doi.org/10.1007/978-3-032-24518-2_2

(Mekuye and Abera 2023). Nanoscience, nanotechnology, nanorobots, nanomagnets, and nanoelectronics exemplify these concepts. People have used nano-level processes and encountered nano-sized items before the advent of nanotechnology.

Nanoparticles have noteworthy applications in several fields, including environmental science, agriculture, food science, biotechnology, biomedicine, and medicine, such as wastewater treatment, environmental monitoring, and antibacterial agents. The biotechnological applications of nanoparticles have expanded owing to their characteristics, such as compatibility with biological systems, anti-inflammatory activity, efficient drug delivery, tumour targeting, and bioabsorption. Green technology using nanomaterials helps reduce global warming and climate change risks. Technologies that utilize nanoparticles are more efficient than those that utilize bulk materials. In 2019, nanoparticles helped identify and treat COVID-19 and later aided in addressing monkeypox worldwide ("Let's talk about lipid nanoparticles," 2021; "Nanomedicine and the COVID-19 vaccines," 2020; Skwarek et al. 2023; Talebian et al. 2020). Nanomaterials react differently from bulk materials because the quantum confinement effect enhances its properties at the nanoscale, and smaller materials have high surface area-to-volume ratios, that affect reaction speeds. The development of nanoparticle products requires extensive planning and faces challenges such as high production costs and specific reaction conditions. Researchers have focused on developing post-synthetic modification, functionalization, and characterization techniques for these materials to improve their performance. Nanomaterials can be created using solid-, liquid-, and gas-phase techniques, with liquid-phase methods being straightforward and cost-effective. Scientists continue to work on improving the synthesis through a better understanding of nucleation and crystal growth (Joudeh and Linke 2022; Mekuye and Abera 2023).

The advent of nanotechnology in the ancient age

Black hair dyeing was popular in ancient Egypt and was thought to have been derived from plant materials such as henna. Ancient hair samples from Egyptian grave sites were colored using a paste of lime and lead oxide mixed with water. During the dyeing process galenite (lead sulfide, PbS) NPs were created. By reacting the dye with sulfur from hair keratin, ancient Egyptians created tiny PbS nanoparticles for consistent coloring (Joudeh and Linke 2022). The Lycurgus Cup from the fourth century demonstrates the early application of nanotechnology. This Roman cup has unique optical properties, with its colour changing depending on the light source location. It appears green under natural light but turns red when illuminated from within. Recent examinations have proven that it has Au and Ag nanoparticles of around 50–100 nm, which give the cup color through plasmon excitation of electrons (Barber and Freestone 1990). The use of nanotechnology is also found in ancient Mesopotamia, Indian, and the Maya civilizations (Joudeh and Linke 2022).

2.2 Classification of Nanoparticles

Nanomaterials can be grouped depending on the materials employed in the synthesis process, starting materials, and most importantly, their structures or dimensions.

2.2.1 Classification Based on Dimensions

Nanomaterials are grouped based on their dimensions: zero, one, two, and three dimensions.

- **Zero-Dimensional (0D) nanomaterials** are either non-dimensional outside the nanometric range (<100 nm) or have all three axis dimensions (x, y, and z) lies in a nanoscale range. 0D nanomaterials comprises quantum dots (QDs), nanospheres, nanoparticles, carbon nanodots and fullerenes. They can have many shapes and forms, such as metallic or ceramic, amorphous or crystalline, and single or polycrystalline forms (Mekuye and Abera 2023; Singh et al. 2020).
- **One-dimensional (1D) nanomaterials: Two** among the three dimensions (x and y) are in the nanoscale range for these materials, while the third dimension (>10 nm) is outside the nanometric range. Illustrations of needle-shaped nanomaterials include nanotubes, nanofibers, nanorods, thin films, and nanowires. These materials can be standalone, amorphous or crystalline, single or polycrystalline, chemically pure or impure, or implanted in a new medium, such as metals, ceramics, or polymers. Metallic, ceramic, and polymeric one-dimensional nanoparticles are all possible (Mekuye and Abera 2023; Singh et al. 2020).
- **Two-dimensional (2D) nanomaterials** are flat and have dimensions larger than nanometers. Examples of these materials are coatings, thin-film multilayers, tubes, nanosheets or nanowalls, fibers, free particles, ultrafine-grained overlayers, wires, and platelets. They can be crystalline or amorphous, with different chemical compositions, integrated into polymeric, metallic, or surrounding matrix materials, or coated on a substrate (Mekuye and Abera 2023; Singh et al. 2020).
- **Three-dimensional (3D) nanomaterials** are those materials, which are not restricted to the nanoscale range in all dimensions. Since the bulk material is composed of distinct blocks that are on the nanometer scale (1–100 nm), but otherwise in the nanometer range or larger than 100 nm, all 3D material measurements are over 100 nm. The multiple nano layers included in these materials create an interface by the intimate contact of 0D, 1D, and 2D structural elements, dispersed of nanoparticles, and bundles of nanowires and nanotubes. Colloids, thin films that have atomic-scale porosity, and free nanoparticles in different forms are a few types of 3D nanomaterials (Mekuye and Abera 2023; Singh et al. 2020).

2.2.2 *Classification Based on the Synthesis of Nanoparticles*

Various synthesis techniques have been developed to fulfil the objectives of nanotechnology. Engineered nanomaterials have been synthesized for multifunctional systems and devices. Novel manufacturing and synthesis technologies have enabled advanced utilization in electronics, sensors, and nanomedicine. The cost of the production process is critical for commercialization and is an important obstacle in designing new nanomaterials. The two primary types of currently accessible and widely used synthesis techniques are top-down and bottom-up techniques and are illustrated in Fig. 2.1.

(a) **Top-down approach:**

Top-down synthesis involves the disintegration of the initial materials into micromaterials and then nanomaterials through a destructive process. This method reduces bulk materials to the nanometric scale primarily through sputtering, physical vapor deposition, ball milling, laser ablation and thermal decomposition.

Sputtering is a surface event in which material atoms are ejected layer-by-layer and deposited onto a substrate. This process is driven by the plasma energy acting on the cathode surface. The materials used for deposition are blocks or discs, called targets. Gas used in sputtering is typically an inert gas, such as argon, krypton, or xenon, for sputtering heavy element compounds, whereas neon is used for lighter elements. This process occurs on cold substrates at low pressures. Owing to their low-temperature operation, it is ideal to deposit thin films with various constituents. The target material quality, surface structure, reflection angle of the incident particles, particle energy, and other factors influence the interaction between the incident particles and the surface atoms during sputtering. Thin films, a product of contemporary engineering, are used in integrated optics, optoelectronics, waveguides, microelectronics, solar panels, protective coatings, oxidation resistance, medicinal applications, and antireflective coatings for jewelry, cars, mirrors, and photovoltaic

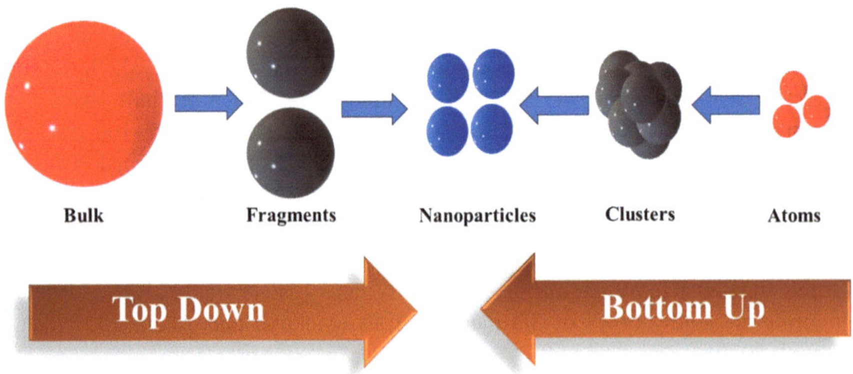

Fig. 2.1 Schematic showing two approaches for nanoparticle synthesis

systems. Thin films have also been deposited on silicon wafers and glass using this technique for semiconductor and optical applications (Altammar 2023; Jain et al. 2023).

Physical vapor deposition (PVD) is a popular method for creating surface coatings and thin (sub-micron to nano) or thick (>5 mm) films on appropriate surfaces. Using a cathodic arc as the source, the process was conducted in a vacuum chamber at a high vacuum of 10^{-6} Torr. This method deposits the coating components evenly over the entire surface, rather than only in specific areas. This process has several benefits, including (i) better coating properties than the substrate material, (ii) environmentally friendly compared to another processes, (iii) production of all kinds of inorganic materials, including metals and alloys, and some organic materials on a substrate, and (iv) excellent coating adhesion. Thin films used in optical, optoelectronic, magnetic, and microelectronic devices are the main applications of this technique. The primary drawback of PVD is that it requires sophisticated machinery, which makes it expensive and results in poor results. (Bahadori et al. 2020; Jain et al. 2023).

Laser ablation is commonly used method for creating different nanomaterials and nanofilms is laser ablation, which involves removing tiny portions of solid material (occasionally liquid) from the surface by treating it with a high energy laser beam. Ruby lasers were the first to observe this phenomenon in the 1960s (Jain et al. 2023). The incident laser penetrates the sample surface according to the laser's wavelength and the target material's refractive index. NPs of different sizes and shapes can be formed by adjusting the laser beam's wavelength, peak power, irradiation period, repetition rate, focusing condition, and other parameters. When utilizing this process to create nanoparticles, stabilizing chemicals are not needed. Various carbon-based (fullerenes, different carbon nanotubes), metal and semiconductor-based nanomaterials are all fabricated using this technique (Jain et al. 2023; Qiu et al. 2021).

Mechanical milling is most straightforward top-down approach for making different phase mixes is mechanical milling, which is extensively used to fabricate nanomaterials, nanoalloys, nanograins, nanoquasi-crystalline materials, and nanocomposites. This process is distinct because it incorporates solid-state interactions among the reactants at ambient temperature. Spray coatings that are resistant to wear, complex metal nanoparticles, carbide- and oxide-strengthened aluminum alloys, and metal and non-metal nanocomposites have been produced using this technique by milling under an inert atmosphere. The primary goal of milling is to reduce the size of particles and mix them with other phases. A very fine powder with a size range of 2–20 nm can be produced using this method. Powder features, including particle shape, surface characteristics, and aggregate disintegration, can be altered using this method. These methods have the advantages of being easy to use and requiring little expensive equipment. However, this method has the drawbacks of irregularly shaped nanomaterials and crystal flaws (Altammar 2023; Jain et al. 2023).

Lithography is a flexible and effective tool for creating patterns of nanosizes. Through a multi-step procedure, this technology can be applied to create precise and uniform 2D metal arrays of NPs on a substrate with controlled size, order, and spacing. Triangular nanocrystals of Cu, Al, and Ag were formed using this technique. This method is used to build laser discs, microfluidics, organic and bio-organic circuits, and information storage devices, such as molecular electronics. This method can be used to prepare various devices, including sensors and transducers (Naikoo et al. 2022).

(b) **Bottom-up approach**

Small particles coalesce to form nanoscale crystals through bottom-up methods, which are constructive approaches based on the formation of nanocrystals from particles at the atomic level. Various synthesis processes, such as co-precipitation, spinning, sol-gel, and chemical vapour deposition (CVD), are primarily used to synthesize nanomaterials.

Sol-gel synthesis is a wet process that employs chemical reagents as the precursors for synthesizing nanoparticles. It is widely used because of its simplicity and less requirement of energy. This technique uses metal oxides and chlorides extensively. Heating was used to distribute these metal oxides and chlorides into a liquid medium, which was subsequently shaken, stirred, or sonicated. The result consisted of both solid and liquid phases. The synthesized nanoparticles can be harvested by centrifugation, filtration, or sedimentation (Bokov et al. 2021; Mackenzie and Bescher 2007).

Co-precipitation method
is a chemical process in which several chemicals simultaneously precipitate from the solutions. This entails the simultaneous creation of solid particles and is often used to create materials and nanoparticles with specific compositions. Particle size and characteristics can be customized by regulating the process using precipitating agents, temperature, and pH adjustments (Das et al. 2020; El Hachmi et al. 2021, 2023; Ghosh et al. 2021; Sen et al. 2024).

Hydrothermal method
Nanomaterials can be synthesized using a hydrothermal approach, which includes dissolving the precursor materials in water under high temperature and pressure environment, prior to crystallization. This solution-based method facilitates accurate control over the morphology and composition of the nanomaterials. Within an autoclave and sealed container, aqueous solutions of the reactants are subjected to temperatures exceeding the boiling point of water (100 °C) and pressures exceeding 1 atm. The hydrothermal method is a valuable tool across numerous scientific and industrial fields because it is a robust and versatile technique for producing a wide range of materials with controlled properties (Bertelsen et al. 2024; Duraisamy et al. 2021; Gan et al. 2020).

Spinning
Nanoparticles can be synthesized by spinning them in a spinning disc reactor (SDR). An SDR consists of a rotary disc inside a reactor with regulated physical factors, such as temperature and humidity. Inert gases like Nitrogen, Argon and others, are used to remove oxygen from the reaction chamber, and create an inert environment within the chamber. Oxygen can initiate various chemical reactions during nanoparticle production and reduce the synthesis efficiency; therefore, it must be eliminated from the reaction chamber. Precipitation occurs when tiny molecules combine because of the spinning. The precipitated NPs were collected, repeatedly washed, and oven dried. Numerous SDR parameters, including temperature, rotation speed, precursor-to-liquid ratio, and disc surface area, affect the properties of nanomaterials (Singh et al. 2020).

Chemical Vapor Deposition
This process relies on the collection of molecules of material in gaseous onto a solid substrate in a reaction chamber at an ideal temperature. In the reaction chamber gaseous molecules are deposited on a heated substrate. The formation of extremely pure, homogeneous, and hard NPs is one of the many advantages of this process. Temperature significantly affects nanoparticle synthesis reactions. However, the hazardous gaseous byproducts are produced during the process; and also it need specialized equipment for the synthesis process (Bhaviripudi et al. 2007; Mekuye and Abera 2023; Singh et al. 2020).

Pyrolysis is the most crucial technique for fabrication of NPs and is generally used in industrial level synthesis. Pyrolysis is based on the high-temperature and high-pressure burning of precursor materials. This technique involves the use of liquid or gaseous precursor ingredients introduced into the furnace through a tiny hole. Compared with other methods, this method offers several advantages, including simplicity, profitability, and a continuous production of nanoparticles with high yield (Jain et al. 2023; Mekuye and Abera 2023; Singh et al. 2020).

Biosynthesis
Because of the abundance of carbohydrates, proteins, enzymes, organic acids, vitamins, flavonoids, tannins, phenols, and terpenoids, which play a role in the bio-reduction of metallic ions, various plant parts and their extracts, are widely used for the production of environmentally friendly nanoparticles (Table 2.1). Numerous investigations have been conducted on the utilization of plants and their biomass for the synthesis of nanoparticles (Fig. 2.2). Green nanoparticles can be created from plants using three different techniques: intracellular, extracellular, and phytochemical-mediated.

Intracellular: Because synthesis takes place inside the plant cell and nanoparticles are obtained by dismantling the structure, the intracellular approach is somewhat comparable to the intracellular method using microorganisms. It is necessary to control the growth factors of plant species to prevent interference with synthesis.

Table 2.1 Green synthesis of NPs using various plants

Plant	Source of plant extract	Nanoparticles	References
Jasmin sambac	Leaf	Copper oxide	Nouren et al. (2024)
Cynara scolymus L	Flower petals	Silver	Saygi et al. (2024)
Nigella sativa	Seed	Copper	Kumar et al. (2023)
Actinidia ch'inensis	Fruit Peel	Silver	Bharathi et al. (2024)
Euterpe oleracea Mart	Seed	Zinc oxide	Vieira et al. (2024)
Abrus precatorius	Bark	Magnesium oxide	Ali et al. (2023)
Capsicum annum	Fruit	Gold	Patil et al. (2023)
Anogeissus latifolia	Gum ghatti	Palladium	Kora and Rastogi (2018)
Cochlospermum gossypium	Tree gum	Selenium	Kora (2018)
Cinnamomum tamala	Leaf	Iron oxide	Das et al. (2020)
Oryza sativa	Husk	Silica	Usgodaarachchi et al. (2021)

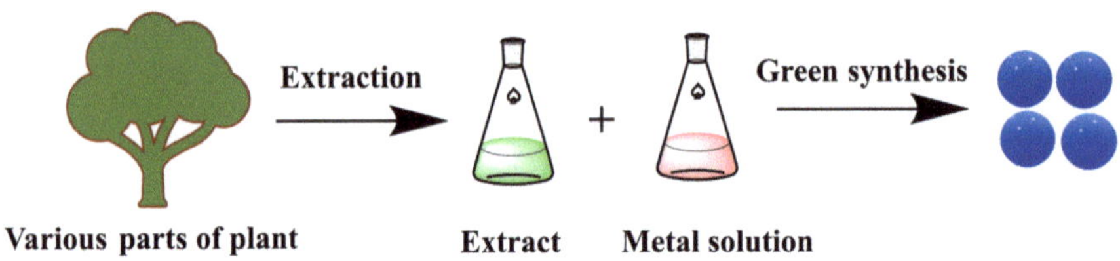

Fig. 2.2 General representation of nanoparticle biosynthesis

Extracellular: This method is the most widely used because it is rapid and convenient to perform. The process begins with plant extraction, which is typically supplemented with a water-based metal salt precursor. In a single process, different extract ingredients combine to form and stabilize the nanoparticles.

Phytochemically mediated: This method is similar to the extracellular strategy; however, it stabilizes nanoparticles using different phytochemical compounds and other substances. Although more steps and components are involved, the synthesis can be controlled. Plant extracts rich in phytocompounds can be utilized as bioreactors, stabilizing agents, and bioactivators to efficiently capture and trap extracellular NPs. Different plant species and different plant parts of same species contain varying amounts of active reducing compounds, and that alter the reduction capabilities and significantly affect NP synthesis. Additionally, the calcination temperature, duration, pH, and amount of plant extract (active phytocompound) all affect NP production (Das et al. 2020, 2022; Ghosh et al. 2023).

Table 2.2 Green synthesis of organic/polymeric nanoparticles using plant extracts

Nanoparticles	Source	References
Cellulose	Cotton	Pandi et al. (2021)
	Cashew Apple	De Araújo et al. (2024)
Starch	Horse chestnut	Ahmad et al. (2020)
	Lotus seeds	Lin et al. (2020)
Lignin	Fig tree bark	Shafiee et al. (2024)

In addition to inorganic, many organic polymeric NPs have been fabricated from various parts of trees, including cellulose, starch, and lignin. The synthesis of these nanoparticles typically follows a top-down approach (Table 2.2).

2.2.3 *Surface Modification of Nanoparticles*

Fine tuning of nanoparticle's surface properties is crucial for optimising their performance across various applications, ranging from advanced drug delivery systems to high-efficiency catalysts. Although conventional surface modification strategies, whether implemented in situ or post-synthesis, offer valuable avenues for tailoring nanoparticle characteristics, they often fail to achieve dynamic responsivity and synergistic interfacial interactions in nanocomposites. These approaches enable the fine-tuning of characteristics, such as biocompatibility, stability, dispersibility, and targeting capabilities, which are crucial for biomedical, catalytic, and industrial applications.

- **In-situ surface modifications**

In situ surface modification of nanoparticles involves modifying their surface during synthesis, offering a one-step surface functionalization method.

Methods of In-Situ Surface Modification
In situ modification involves the merging of materials, such as alkoxysilanes, ions, and nanoparticles, directly into the nanomaterials during synthesis (Trayford and Van Rijt 2024). Surface functionalization uses covalent and non-covalent bonds, like hydrogen bonds, electrostatic forces, and van der Waals interactions, to combine various molecules of organic and inorganic origin at the nanoscale level (Ahmad et al. 2022; Ngouangna et al. 2022). Various linker molecules are often used to connect ligands to nanoparticle surfaces. Although non-covalent interactions are simpler, it might be affected by factors like pH and ionic strength (Sanità et al. 2020).

Materials for Surface Modification
Organic molecules namely phosphates, citrates, thiols, amines, and polymers such as polyethylene glycol (PEG), chitosan, polyvinyl acetate (PVA), dextran, and poly (lactic-co-glycolic acid) (PLGA), can be utilized for surface modification (Kaymaz et al. 2023). Inorganic molecules, such as silica, metals, and metal oxides, are also used. Organic compounds are biocompatible and biodegradable, whereas inorganic

compounds exhibit improved stability and dispersity. Metallic elements such as gold are used to dope iron oxide NPs and make them more stable and biocompatible (Lau et al. 2023).

Advantages of Surface Functionalization

Surface functionalization stabilizes nanoparticles against agglomeration and oxidation, making them compatible with subsequent phases and facilitating their organization. When organic and inorganic materials are added to the surface of nanoparticles, the resulting increase in physicochemical properties can overcome these obstacles. Altering the surface of nanoparticles is a successful strategy for adjusting their physicochemical and biological characteristics and attaining the intended therapeutic efficacy in vivo (Ly et al. 2024). By inhibiting aggregation and protein adsorption, functionalizing the surfaces of nanoparticles with polymers like PEG greatly increases stability, preserves a uniform size distribution, and extends retention time in biological settings.

Post surface modifications

Post-synthesis surface adjustment is a crucial process for altering the properties of NPs to enhance their performance in diverse applications. This capability to modify the outer surface of NPs post-synthesis provides a versatile means of optimizing their functionality without altering their core structure.

Methods for Post-Synthesis Surface Modification

Post-synthetic modification of NP surfaces can be achieved using various techniques, which can be majorly classified into covalent and non-covalent conjugation. Covalent modifications involve the formation of strong chemical bonds between the NP surface and the modifying agent, ensuring high stability and robustness. Common covalent modification strategies include the following.

(a) **Salinization:**

This process is often used for silica NPs, employing aminosilanes to introduce amino groups onto the NP surface, thereby facilitating subsequent bioconjugation reactions (An et al. 2014).

Cross-coupling Reactions: These reactions are used to covalently bind organic molecules to the NP surface, utilizing homo- or hetero-bifunctional crosslinkers to add functional groups, such as R-NH_2 or R-COOH.

(b) **Ligand Exchange:**

This strategy is particularly useful for metal oxides, where the original surface ligands are replaced with functional groups, such as diols, amines, carboxylic acids, or thiols. In contrast, non-covalent modifications rely on weaker interactions, including van der Waals forces, electrostatic forces, hydrogen bonding, and hydrophobic interactions (Manoj et al. 2024). These approaches are beneficial because of their simplicity and minimal impact on the structures of the modified molecules. However, they are more prone to reaction conditions, such as pH and ionic strength, that can affect the stability of the modification.

Materials for Post-Synthesis Surface Modification
The choice of materials for post-synthetic surface modification depends on the desired functionality and application of NPs. The commonly used materials include the following:

(a) **Polymers:** PEG is frequently used to enhance biocompatibility and stability by creating a hydrophilic layer on the NP surface, reducing protein adsorption and immune response (Suk et al. 2016). Dextran is a complex branched polysaccharide, is used to modify iron oxide NPs and increase their biocompatibility for use in biomedical applications. Chitosan, an oligosaccharide, reduces toxicity when coated with NPs (Shukla et al. 2015).
(b) **Proteins and peptides:** Antibodies and their antigen-binding fragments can be utilized for targeted drug delivery due of their high specificity and binding affinity to specific receptors (Kumari et al. 2023). Peptides with high affinity for specific receptors can also be used to modify NPs and enhance drug uptake into cancer cells.
(c) **Small Molecules:** Zwitterionic ligands, such as carboxybetaine and sulfobetaine, are used to modulate surface charge, optimize solubility, minimize protein corona formation, and enhance stability in biological fluids (Ouyang et al. 2024). Cell-penetrating peptides (CPPs) with specific amino acid sequences have been used to enhance cellular NP uptake.

Applications and Advantages of Post-Synthesis Surface Modification
Post-synthetic surface modification has been used in various fields and offers numerous advantages.

(a) **Biomedical Applications:** Modified NPs can be used in cancer diagnosis and therapy, drug delivery, imaging, and biomolecule detection. Surface modification enhances biocompatibility, reduces toxicity, and enables targeted delivery of therapeutic agents. For example, modifying NPs with human serum albumin (HSA) enhances their cellular uptake by breast cancer cells.
(b) **Improved Stability and Dispersibility:** Surface modification can prevent NP aggregation and improve dispersion in various solvents. This is particularly important for applications that require stable and uniform NP suspension.
(c) **Modulation of Immune Response:** Surface modification can promote immune escape, increasing the half-life of nanomaterials in the circulatory system by reducing their clearance by macrophages. Conversely, during vaccine development, surface modifications can trigger specific immune responses.

Overall, post-synthetic surface modification is a powerful and versatile strategy for optimizing NP properties, rendering them appropriate for several advanced applications. By carefully selecting modification methods and materials, researchers can tailor NPs to meet the specific conditions of their intended uses.

- **Phytogen-Based Surface Modification of Nanoparticles**

The clinical use of NPs is severely constrained by their toxicity and absorption efficiency. As toxicity and absorption are intimately linked to the surface composition of NPs, its surface modification is a potent technique for resolving both problems. Phytogen-based surface modification, which utilizes plant-extract compounds, is a favorable approach for enhancing the biocompatibility, stability, and functionality of nanoparticles. This approach leverages the diverse array of bioactive molecules found in plants and can be a green and sustainable alternate to conventional chemical modification methods.

Advantages of Phytogen-Based Surface Modification

Phytogens or plant extract-based compounds are generally considered biocompatible and non-toxic, reducing the potential for adverse effects associated with synthetic modifiers. The use of phytogens can make stable and good dispersion of NPs in biological system, preventing aggregation and improving circulation time. Furthermore, phytogens can impart specific functionalities to NPs, such as targeting ligands for enhanced cellular uptake and antioxidant properties for therapeutic applications. This green synthesis approach, which utilizes plant extracts, is environmentally friendly and cost-effective, reducing the need for strong reactive chemicals and high energy consuming processes. Plant-derived compounds offer a diverse range of chemical functionalities that can be exploited for surface modification. Polyphenols possess antioxidant and metal-chelating properties, making them suitable for stabilizing metallic nanoparticles and preventing oxidation. Polysaccharides such as pectins and chitosan can improve the biocompatibility and stability of NPs and provide a platform for further functionalization. Terpenoids and alkaloids exhibit various biological activities, allowing the creation of multifunctional NPs with therapeutic and diagnostic capabilities. The versatility of phytogens enables the design of NPs with required properties for various biomedical applications.

Techniques for Phytogen-Based Surface Modification

The choice of phytogen-based surface modification method depends on the type of NPs, desired functionality, and chemical properties of the phytogen. In the case of the post-synthetic method, adsorption is a simple and widely used technique in which phytogens are physically adsorbed onto the NP surface via weak interactions namely van der Waals forces, hydrogen bonding, or electrostatic interactions. It is particularly suitable for metallic and silica NPs, where phytogens form coatings on the surface. Covalent conjugation involves the formation of more stable chemical bonding between the phytogen and nanoparticle surface, providing a more robust and durable modification. This can be achieved through the use of crosslinkers or by activating functional groups on both phytogens and nanoparticles.

Another approach involves in situ synthesis, in which the nanoparticles are formed in the presence of phytogen, that acts as a reducing agent (wherever required) and, also as a capping agent. This method is commonly applied for the green synthesis of metallic nanoparticles, such as silver and gold, where plant extracts reduce metal salts to form stable nanoparticles. Layer-by-layer assembly is a method in which alternating layers of positively and negatively charged phytogen are deposited onto a nanoparticle surface, creating a multilayer coating with a controlled thickness and composition. In this process, multiple phytogens are incorporated with different functionalities to create complex multifunctional nanoparticles.

2.3 Characterizations

Nanoparticles have gained significant attention these days because of their distinctive size-dependent characteristics and extensive use in various sectors, including environmental science, electronics, and medicine. Understanding and controlling these properties necessitates the precise characterization of NPs, which involves determining their physical, chemical, and structural characteristics. Several advanced techniques can be used for this, providing distinct insights into different aspects of NP characteristics. This section provides a summary of the key nanoparticle characterization methods and their significance in advancing nanoscience.

(a) **Size and Morphology Characterization**

The size and morphology of engineered nanomaterials are critical parameters that significantly influence their properties and uses in various fields. Several techniques are available to analyse these aspects, with Transmission Electron Microscopy (TEM) and Scanning Electron Microscopy (SEM) being the most prominent. TEM involves the transmission of a high-energy electron beam through a sample, providing high-resolution images of the internal structure, particle morphology, crystal structure, lattice spacing, and defects of the sample. TEM can also be used for elemental analysis using energy-dispersive X-ray spectroscopy (EDS or EDX). Both TEM and SEM are invaluable for directly measuring the nanoparticle size, size distribution, grain size, and their morphology. Dynamic light Scattering (DLS) is another vital technique used in determining NPs' size distribution suspended in liquids. DLS measures the intensity fluctuations of scattered light, providing information on particle size, polydispersity, and aggregation state. Atomic Force Microscopy (AFM) probes the surface properties of NPs using a sharp tip attached to a cantilever, providing topographical information at nanometer resolution and measuring the particle size, height, roughness, and mechanical properties. The size and morphological attributes of NPs can be comprehensively understood by these techniques. In studies by Hemlata and team (2020), biosynthesized silver nanoparticles (AgNPs) were used for biological applications. Characterization by DLS revealed an average hydrodynamic size of 90 nm and a polydispersity index (PDI) of 0.3 (Hemlata et al. 2020). They concluded that because the hydrodynamic size was below 150 nm and the PDI was 0.3, the synthesized nanoparticles were suitable for cellular uptake. Morphological analysis via SEM demonstrated that the nanoparticles exhibited polymorphic shapes, including irregular granulated forms, ellipsoids, and significant aggregation, with sizes ranging from 35 to 50 nm. Rajeshkumar and team used *Persea americana* seed extract to synthesize copper nanoparticles (CuNPs), and examined their morphology using TEM and AFM (Rajeshkumar and Rinitha 2018). They observed that the NPs were well separated, spherical in shape, and ranged from 40 to 90 nm in size, with AFM analysis further corroborating the spherical morphology of the synthesized CuNPs.

(b) **Crystal Structure, Skeletal and Composition Analysis**

The characterization of the crystal structure and elemental composition of nanomaterials is essential for understanding their intrinsic characteristics and behaviour. X-ray Diffraction (XRD) is an important tool for studying the crystalline structure of NPs (Abreu et al. 2015). By exposing the sample to an X-ray beam, the generated diffraction pattern provides valuable data on the atomic layer spacing and crystal alignment. This information is instrumental in identifying the nanoparticles' composition, assessing their phase purity, and evaluating their crystallinity. XRD is widely used because it provides detailed structural information. For a sensitive assessment of the elemental composition of samples, Inductively Coupled Plasma Mass Spectrometry (ICP-MS) is a very good technique. It can analyse metals and selected nonmetals with atomic masses from 7 to 250 (Kuznetsova et al. 2020; Rua-Ibarz et al. 2020). During ICP-MS, liquid samples are aerosolized and sent through a high-temperature heat source to ionize the elements, which are then separated and detected using a mass spectrometer. For metallic elements, ICP-MS testing enables sensitive detection over an 8-order concentration range. Similarly, X-ray photoelectron or photoemission spectroscopy (XPS) can be utilized to measure the percentage of elements present on a surface-coated nanoparticle (Radu et al. 2017). Fourier-transform infrared spectroscopy (FTIR) is another important technique that can discover the chemical bonds and functional groups in NPs by measuring the transmission and absorption of infrared light, providing information on their surface chemistry, composition, and molecular interactions (Sandler et al. 2019). Together, these tools can help in a comprehensive analysis of the crystal structure and chemical composition of the NPs. For example, AgNPs were synthesized using mango seed extract and assessed their crystallinity using powder XRD (Donga and Chanda 2021). The resulting diffractogram displays diffraction peaks at 2θ values of 37.95°, 43.08°, 64.48°, and 77.40°, corresponding to the (111), (200), (220), and (311) planes, respectively. They compared their findings with the Joint Committee on Powder Diffraction Standards (JCPDS) file no. 040783 and concluded that the produced AgNPs exhibited a crystalline structure with a face-centred-cubic lattice. The Selected Area Electron Diffraction (SAED) pattern of the AgNPs proved the presence of (111), (200), (220), and (311) planes, indicating a highly crystalline nature and supporting the XRD findings. Additionally, they obtained FTIR spectra of AgNPs and compared these results with spectra of pure plant extract. The distinct stretching frequencies observed in the extract corresponded well with those of the synthesized AgNPs, indicating the occurrence of polyphenols, flavonoids, and proteins on NPs. Furthermore, EDX analysis was performed to validate the nanoparticle composition.

(c) **Surface Charge and Stability**

The charge on surface of NPs plays a vital role in its stability and interactions with colloidal solutions. Zeta potential (ζ) measurement is the most commonly used technique for analysing the surface charge of NPs and quantifying the charge stability of colloidal NPs. It measures the potential difference between the main fluid and the layer of fluid which is connected to oppositely charged ions of the nanoparticle surface. Particle stability can be inferred from the magnitude of the zeta potential; higher potentials signify stronger electrostatic repulsion, and hence, greater particle stability (Das et al. 2022). Zeta potentials of NPs greater than ±30 mV were regarded as highly stable, whereas those with zeta potentials between −10 and +10 mV were regarded as neutral. It is crucial to remember that the surface charge of the NPs is dependent on the pH of the solution and can be zero at a particular pH, known as the isoelectric point. Techniques such as capillary isoelectric focusing are emerging as new methods; however, zeta potential remains the standard for measuring surface charge. Determining the particle size and surface charge is crucial for the proper characterization and application of NPs. Das and team prepared *Cinnamomum tamala* leaf extract-coated magnetite nanoparticles for the removal of U(VI) from water (Das et al. 2023a). They conducted a zeta potential study to evaluate the surface charge across a pH range of 1–12 and discovered that the material exhibited zero charge at pH 3.5, known as the point of zero charge (PZC). The highest negative charge was recorded at pH 8 ($\zeta = -24$ mV), whereas the highest positive charge was observed at pH 2 ($\zeta = +15$ mV). Given that the targeted contaminant was uranyl cation (UO_2^{2+}), a pH of 8 was selected for the removal process.

(d) **Optical Properties**

These properties are highly sensitive to their size, shape, composition, and surrounding environment, making their characterization essential. Extinction is the total amount of light absorbed and dispersed by a material, which is measured using UV-visible (UV-Vis) spectroscopy (Aly et al. 2024; El Hachmi et al. 2025). In its most basic form, a sample is kept between a photodetector and a light source, and the intensity of UV/visible light is determined both before and after it passes through the sample. To ensure accuracy, the data were shown as extinction as a function of wavelength, with a background correction applied to each spectrum. UV/Vis spectroscopy is a useful technique for the identification, characterization, and research of nanomaterials because plasmonic nanoparticles have optical characteristics that are sensitive to their size, shape, concentration, agglomeration state, and refractive index, which are close to the nanoparticle surface. A single- or dual-beam diode array spectrometer was used to collect spectra from the ultraviolet to the visible region for UV-Vis analysis (Mandru et al. 2023). Dark-field microscopy is another light microscopy method that images samples using oblique illumination. Because the scattered light from individual NPs is faint compared with the incident beam, this approach is commonly employed to analyze plasmonic NPs. Because light microscopy is diffraction-limited, NPs appear as brilliant point sources. High-quality objectives and a white light source were used for dark-field microscopy. These optical characterization methods provide detailed insights into the behavior and potential applications of nanomaterials in various fields of study. According to Vinay et al. (2019) *Rauvolfia tetraphylla* extract-synthesized AgNPs exhibit optical properties such as fluorescence emission (Vinay et al. 2019). UV-Vis and PL spectra were obtained to validate the results. The UV-Vis spectra indicated a λ_{max} of 463 nm, confirming the presence of AgNPs. Additionally, the analysis of the spectra allowed for the determination of the band gap, which was found to be 2.74 eV based on the Tauc plot. The excitation spectrum of AgNPs showed a strong excitation peak at 390 nm, with an emission wavelength of 484 nm. They also showed that the emission spectra of AgNPs were excitation dependent. It was proposed that the photoluminescence observed in noble metals could result from electrons being excited from filled d-bands to energy levels above the Fermi level. These excited carriers then undergo scattering with phonons, both electrons and holes, causing energy dissipation. Eventually, an electron from the occupied sp band recombines with a hole, producing a photoluminescence.

(e) **Surface Area Analysis**

The surface area is a critical factor that affects the reactivity, adsorption, and catalytic activity of NPs. The commonly used method Brunauer-Emmett-Teller (BET) determine the specific surface area of NPs (Bardestani et al. 2019). BET analysis involves the physical adsorption of a gas, generally nitrogen at cryogenic level onto the surface of a material (Bardestani et al. 2019; Schlemminger et al. 2015). The amount of gas adsorbed was measured as a function of the relative pressure, and the surface area was calculated based on the BET theory. It extends Langmuir's adsorption theory to multilayer adsorption and accounts for the interactions between adsorbate molecules. The process involves several steps: sample preparation (degassing to remove adsorbed contaminants), adsorption of nitrogen gas at the liquid nitrogen temperature, and data analysis using the BET equation. The BET equation relates the amount of gas adsorbed to the pressure and monolayer capacity, from which the surface area can be calculated. This technique is applicable to a wide range of materials, including meso- and microporous NPs, and bulk materials. The accuracy of BET analysis depends on factors such as the purity of the gas, precision of the pressure measurements, and uniformity of the sample. Sacko et al. synthesized magnetite nanoparticles coated with tannic acid, a polyphenolic acid derived from plants, analyzed their surface area using BET analysis, and compared their findings with those of uncoated magnetite nanoparticles (Sacko et al. 2024). They determined that the surface area of the tannic acid-coated magnetite nanoparticles was 99 m^2/g, in contrast to 10.1 m^2/g for the bare magnetite nanoparticles. The results implied that the surface area of the bare magnetite nanoparticles increased by approximately tenfold after surface modification. Additionally, they measured the pore volume and pore size, finding that the tannic acid-coated magnetite nanoparticles had a pore volume and size of 0.08 cm^3/g and 9.4 nm, respectively, which were higher than those of the bare magnetite nanoparticles (pore volume and size = 0.24 cm^3/g and 4.4 nm, respectively).

(f) **Thermal Stability Analysis**

Thermal stability is an important characteristic in determining the suitability of NPs for high-temperature applications. Differential scanning calorimetry (DSC), Differential thermal analysis (DTA), and Thermogravimetric analysis (TGA) are some of the thermal analysis methods that can be used to evaluate the thermal behavior of

NPs (Cao et al. 2006; Wen and Huang 2019). In a controlled environment, TGA is used to calculate the change in the mass of a sample with temperature and time. The thermal decomposition temperature, content, and stability of the NPs were determined using this technique. DSC quantifies the heat flow related with the thermal transitions of a material as a function of temperature. DSC can be used to detect phase transitions, melting points, crystallization temperatures, and glass transition temperatures, providing valuable information regarding the thermal behavior of NPs. DTA estimates the difference between the temperature of sample and reference material as a function of the temperature. DTA was used to identify thermal events, such as phase transitions, melting, and decomposition. These techniques provide complementary information regarding the thermal properties of NPs. For example, in TGA the amount of organic coating on NPs can be determined, whereas DSC can be implicated to study the melting behavior of the NP core material. Understanding the thermal stability of NPs is critical for several fields like catalysis, drug delivery, and the development of high-temperature coatings. The selection of thermal analysis tools is done on the basis of specific information required and the nature of the material being studied. Das and team produced magnetite nanoparticles using an extract from *Terminalia arjuna* bark and conducted TGA over a temperature range of 25–900 °C to assess the thermal stability and percentage of coating (Das et al. 2022). Weight loss occurred in three separate stages. The first phase: weight loss of 12.6% occurs when the temperature was in between 25 and 180 °C, primarily involving the release of adsorbed water (H_2O) and hydroxide ions (OH^-). The second phase (occurring between 180 and 500 °C) accounts for a weight loss of 26.6% and is attributed to the degradation and release of surface coatings, especially various polyphenolic compounds. The final phase spans–500–900 °C and results in an additional weight loss of 4.1%, further contributing to the degradation and loss of the surface coatings. This structured breakdown highlights the sequential nature of the weight-loss process and the specific chemical species involved at each stage. From these findings, the team concluded that the nanoparticles were coated with 30.7% of the material (excluding H_2O and OH^-). Overall, the TGA results confirmed that the nanoparticles were thermally stable.

(g) **Magnetic Nature Analysis**

The magnetic attributes of nano materials are central to their applications in magnetic resonance imaging (MRI), magnetic hyperthermia, and targeted drug delivery (TDL). Various techniques are available to study the magnetic properties of NPs, including superconducting quantum interference device (SQUID) magnetometry, vibrating sample magnetometry (VSM), and ferromagnetic resonance (FMR) (Blachowicz et al. 2022; Marinica 2017; Škrátek et al. 2020; Wernsdorfer et al. 1995). A very sensitive technique -SQUID magnetometry measures the magnetic moment of a sample using superconducting quantum interference devices. SQUID magnetometers can measure extremely small magnetic moments, making them ideal for characterizing the magnetic properties of nanomaterials. VSM measures the magnetic moment of a sample by mechanically vibrating it in a uniform magnetic field and detecting the induced voltage in a pickup coil. VSM is a versatile technique that can be used to measure a wide range of magnetic properties, including the magnetization, coercivity, and remanence of the samples. FMR is a spectroscopic technique which measures the absorption of microwave radiation by a sample in a magnetic field. It is sensitive to the magnetic anisotropy and relaxation processes of magnetic materials. DC magnetometry techniques are commonly used to describe the magnetic behavior of materials. These techniques measure the response of a material to a static magnetic field and provide information on its magnetic ordering, saturation magnetization, and coercivity. The characterization of magnetic NPs often involves measuring their size distribution, shape, and surface charge. These parameters influence the magnetic properties of NPs and their behavior in various applications. Das et al. created magnetite nanoparticles using extracts from *Nigella sativa* seeds and conducted alternate gradient magnetometry (AGM) to assess their magnetic properties (Das et al. 2023b). The findings indicated key characteristics, including the absence of a hysteresis loop, zero coercivity, remnant magnetization of 20 emu/g, and saturation magnetization of 57.87 emu/g. Given the zero coercivity and lack of a hysteresis loop, they concluded that the nanoparticles exhibited superparamagnetic properties at the room temperature. Furthermore, owing to the significant magnetization value of 57.87 emu/g, they suggested potential applications in biomedical fields and water purification processes.

A summary of the various characterization techniques for different NPs is presented in a tabular format in Table 2.3. Characterization is essential for identifying nanoparticles and their surface modifiers and is mandatory for choosing suitable applications. In conclusion, it can be said that nanoparticles with different coatings require a few similar and alternate characterization techniques, as shown in Fig. 2.3.

2.4 Applications of Phytogen-Modified Nanoparticles

Nanoparticles modified using phytogenic compounds exhibit versatile functionality and find applications across diverse domains, particularly in biomedicine. In drug delivery, phytogen-modified NPs can improve drug loading, enhance targeted delivery to specific cells or tissues, and control drug release kinetics (Gulia et al. 2022). For example, NPs modified with curcumin, a polyphenol with anti-inflammatory and anticancer properties, can be used to deliver chemotherapeutic drugs directly to tumor cells, thereby reducing side effects and improving therapeutic efficacy. In diagnostics, phytogen-modified NPs can be used as contrast agents for medical imaging, enhancing the sensitivity and specificity of the detection methods. Gold nanoparticles modified with plant-derived antibodies can be used to target specific cancer markers, allowing for the early and accurate diagnosis of cancer.

Phytogen-modified NPs can also be used in the field of tissue engineering and regenerative medicine to foster cell adhesion, proliferation, and differentiation. Polysaccharide-modified nanoparticles, such as those modified with chitosan, can serve as scaffolds for bone regeneration, offering a biocompatible and biodegradable matrix for cell growth (Tian et al. 2022). In agriculture, phytogen-modified NPs can

Table 2.3 Different characterization techniques used for various nanoparticles

Techniques	Information
DLS	Size distribution. It measures scattered light when a laser passes through a colloid.
Microscopy (TEM, SEM, AFM, STEM)	SEM and TEM are useful tools to check the purity of the material and exfoliation of bundles. Size, morphology, crystal structure, spreading of NPs in, biological or other matrices/supports, quick examination–elemental composition. AFM also evaluates 3D shape and surface .
EDX or EDS or its mapping analysis	Elemental components and colorful mapping of elements that are present in NPs.
XPS	Applied to analyse the surface of a material, elemental composition and oxidation states.
CHNS analyser	Determine the percentage of carbon, hydrogen, nitrogen and sulfur.
Atomic Emission Spectroscopy (AES)	Elemental composition.
ICP-MS	It identifies and quantifies the elemental composition of samples. Size, size distribution, and concentration can be determined from this technique.
Powdered X-ray diffractometry (pXRD)	Crystal structure, elemental composition, oxidation states, electronic structure, crystalline grain size, particle size, defect, growth kinetics.
Selected area electron diffraction (SAED)	Can provide crystal structure, diffraction pattern and orientation.
Laser Diffraction (Static Laser Light Scattering, SLS)	Detects and measures particles of sizes ranging from nm to mm, particle size distribution. To know particle size, this tool estimates angular variation of light scatter.
Grazing Incidence Small-Angle X-Ray Scattering (GISAXS)	Information about the structure, morphology, correlations, and mutual orientation.
X-ray Absorption Near-Edge Structure Spectroscopy (XANES)	Local carbon environment, including graphitic carbon, diamond, and carbon nitride. This characteristic of XANES spectroscopy results from the absorption transition's angular dependency.
Zeta Potential	Surface charge, stability of colloidal suspensions, isoelectric point. Higher value of potential indicates increased stability and electrostatic repulsion.
UV-Vis Spectroscopy	Detection of unique optical properties, shape, sensitivity to size, concentration, and agglomeration. It measures analyte concentration based on the intensity of light absorbed by the sample. Measures the band gap for the semiconducting NPs
Photoluminescence (PL) spectroscopy	Optical properties. Biomaterial conformational changes and their conjugation with NPs, encapsulation efficiency and its release.
Cathodoluminescence (CL)	An emission spectroscopy that measures the light emission spectrum under free electron, to estimate optical properties.

(continued)

Table 2.3 (continued)

Techniques	Information
Time Resolve fluorescence (TRF) spectroscopy	Estimates the gradual decay of fluorescence after excitation.
Raman Spectroscopy	Provide information about defect levels and graphitization.
Infrared Spectroscopy (FTIR, NIR)	Types of functional groups (e.g., polymeric, organic, and inorganic materials) that are existing on the surface.
DSC/TGA	Compositional purity and thermal stability.
BET surface area analysis	Can estimate surface area, porosity, density, and pore volume and size
Four-Point Probe Resistivity Measurement	Provides quantitative data of conductive properties.
Magnetic moment measurements (VSM, AGM, SQUID)	Provide magnetic nature. Superconducting quantum inference device (SQUID) is the most sensitive and vibrating sample magnetometry (VSM) is the least sensitive technique, while alternating gradient magnetometry (AGM) lies in between them.

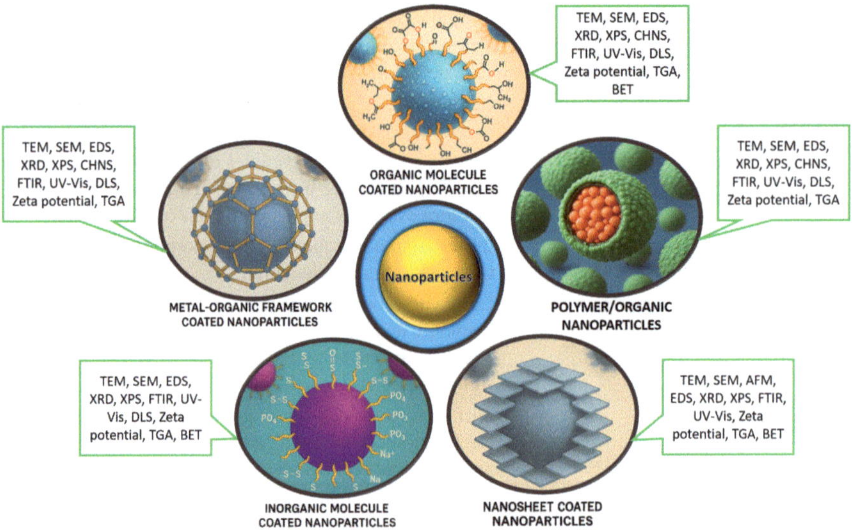

Fig. 2.3 Various coatings on nanoparticle surfaces with their essential characterization techniques

be used as nanofertilizers to deliver nutrients directly to plants and improve crop yields (Yadav 2025). Silver nanoparticles modified with plant extract-modified AgNPs can be used as antimicrobial agents to protect crops from pathogens and reduce the need for synthetic pesticides (Tariq et al. 2022).

2.5 Conclusion

Despite the numerous advantages of phytogen-based surface modifications, some limitations need to be addressed. The stability and reproducibility of phytogen-modified NPs can be changed due to the variations in the composition of plant extracts, environmental conditions, and extraction methods used. Characterization of phytogen-modified NPs can be challenging because plant extracts are biochemically complex and variable between batches. The potential toxicity of certain phytogens, particularly at high concentrations, should be carefully evaluated. The long duration impacts of phytogen-modified NPs on human health and the environment remain largely unknown and require further investigation.

Further research should prioritize the development of uniform and validated protocols for extraction processes to ensure consistency, reproducibility, and scalability across applications, as well as the purification and characterization of phytogens. More advanced techniques, such as chromatography and spectroscopy, should be used to identify and quantify the active compounds in plant extracts. Computational models can be develop to predict the relations between phytogens and nanoparticle surfaces can aid in the rational design of modified NPs. Furthermore, *in vivo* trials are needed to evaluate the efficacy, toxicity, and long-term effects of phytogen-modified NPs in relevant animal models. By addressing these limitations and pursuing future directions, phytogen-based surface modification can become a widely adopted and sustainable approach for nanoparticle functionalization, unlocking its full potential for various applications.

References

Abreu CM, Souza DN, Paschoal CMM (2015) A brief overview in the study of archaeological materials through the technique of x-ray diffraction (XRD). In: Proceedings of the international symposium on crystallography. Presented at the international symposium on crystallography. Editora Edgard Blücher, Fortaleza, Brasil, pp 7–7. https://doi.org/10.5151/phypro-sic100-007

Ahmad M, Gani A, Hassan I, Huang Q, Shabbir H (2020) Production and characterization of starch nanoparticles by mild alkali hydrolysis and ultra-sonication process. Sci Rep 10:3533. https://doi.org/10.1038/s41598-020-60380-0

Ahmad F, Salem-Bekhit MM, Khan F, Alshehri S, Khan A, Ghoneim MM, Wu H-F, Taha EI, Elbagory I (2022) Unique properties of surface-functionalized nanoparticles for bio-application: functionalization mechanisms and importance in application. Nanomaterials 12:1333. https://doi.org/10.3390/nano12081333

Ali S, Sudha KG, Thirumalaivasan N, Ahamed M, Pandiaraj S, Rajeswari VD, Vinayagam Y, Thiruvengadam M, Govindasamy R (2023) Green synthesis of magnesium oxide nanoparticles by using Abrus precatorius Bark extract and their photocatalytic, antioxidant, antibacterial, and cytotoxicity activities. Bioengineering 10:302. https://doi.org/10.3390/bioengineering10030302

Altammar KA (2023) A review on nanoparticles: characteristics, synthesis, applications, and challenges. Front Microbiol 14:1155622. https://doi.org/10.3389/fmicb.2023.1155622

Aly ST, Saed A, Mahmoud A, Badr M, Garas SS, Yahya S, Hamad KH (2024) Preparation of magnetite nanoparticles and their application in the removal of methylene blue dye from wastewater. Sci Rep 14. https://doi.org/10.1038/s41598-024-69790-w

An SSA, Kim K, Kim HM, Lee W, Lee C, Kim T, Lee J-K, Jeong J, Paek S, Shin J-H (2014) Surface treatment of silica nanoparticles for stable and charge-controlled colloidal silica. Int J Nanomedicine 29. https://doi.org/10.2147/ijn.s57922

Bahadori E, Ramis G, Rossetti I (2020) Matching nanotechnologies with reactor scale-up and industrial exploitation. In: Nanomaterials for the detection and removal of wastewater pollutants. Elsevier, pp 407–442. https://doi.org/10.1016/B978-0-12-818489-9.00015-3

Barber DJ, Freestone IC (1990) An investigation of the origin of the colour of the Lycurgus cup by analytical transmission electron microscopy. Archaeometry 32:33–45. https://doi.org/10.1111/j.1475-4754.1990.tb01079.x

Bardestani R, Patience GS, Kaliaguine S (2019) Experimental methods in chemical engineering: specific surface area and pore size distribution measurements—BET, BJH, and DFT. Can J Chem Eng 97:2781–2791. https://doi.org/10.1002/cjce.23632

Bertelsen AD, Kløve M, Broge NLN, Bondesgaard M, Stubkjær RB, Dippel A-C, Li Q, Tilley R, Vogel Jørgensen MR, Iversen BB (2024) Formation mechanism and hydrothermal synthesis of highly active $Ir_{1-x}Ru_xO_2$ nanoparticles for the oxygen evolution reaction. J Am Chem Soc 146:23729–23740. https://doi.org/10.1021/jacs.4c04607

Bharathi D, Lee J, Karthiga P, Mythili R, Devanesan S, AlSalhi MS (2024) Kiwi fruit peel biowaste mediated green synthesis of silver nanoparticles for enhanced dye degradation and antibacterial activity. Waste Biomass Valoriz 15:1859–1868. https://doi.org/10.1007/s12649-023-02328-9

Bhaviripudi S, Mile E, Steiner SA, Zare AT, Dresselhaus MS, Belcher AM, Kong J (2007) CVD synthesis of single-walled carbon nanotubes from gold nanoparticle catalysts. J Am Chem Soc 129:1516–1517. https://doi.org/10.1021/ja0673332

Blachowicz T, Grzybowski J, Ehrmann A (2022) Influence of agglomerations on magnetic properties of polymer matrices filled with magnetic nanoparticles. Mater Today Proc 67:792–796. https://doi.org/10.1016/j.matpr.2022.07.362

Bokov D, Turki Jalil A, Chupradit S, Suksatan W, Javed Ansari M, Shewael IH, Valiev GH, Kianfar E (2021) Nanomaterial by sol-gel method: synthesis and application. Adv Mater Sci Eng 2021:5102014. https://doi.org/10.1155/2021/5102014

Cao LF, Xu GY, Xie D, Guo MX, Luo L, Li Z, Wang MP (2006) Thermal stability of Fe, Co, Ni metal nanoparticles. Phys Status Solidi B 243:2745–2755. https://doi.org/10.1002/pssb.200642014

Das C, Sen S, Singh T, Ghosh T, Paul SS, Kim TW, Jeon S, Maiti DK, Im J, Biswas G (2020) Green synthesis, characterization and application of natural product coated magnetite nanoparticles for wastewater treatment. Nanomaterials 10:1615. https://doi.org/10.3390/nano10081615

Das C, Singh S, Bhakta S, Mishra P, Biswas G (2022) Bio-modified magnetic nanoparticles with Terminalia arjuna bark extract for the removal of methylene blue and lead (II) from simulated wastewater. Chemosphere 291:132673. https://doi.org/10.1016/j.chemosphere.2021.132673

Das C, Ghosh NN, Pulhani V, Biswas G, Singhal P (2023a) Bio-functionalized magnetic nanoparticles for cost-effective adsorption of U(VI): experimental and theoretical investigation. RSC Adv 13:15015–15023. https://doi.org/10.1039/D3RA00799E

Das C, Nath Ghosh N, Bhardwaj R, Narula K, Mishra P, Biswas G (2023b) Enhanced photocatalytic degradation of a hydrocortisone by biomodified and biocompatible magnetite nanoparticles and its mechanistic assessment. J Ind Eng Chem 128:369–382. https://doi.org/10.1016/j.jiec.2023.08.001

De Araújo LGS, Rodrigues THS, Rates ERD, Alencar LMR, Rosa MDF, Ponte Rocha MV (2024) Production of cellulose nanoparticles from cashew apple bagasse by sequential enzymatic hydrolysis with an ultrasonic process and its application in biofilm packaging. ACS Omega 9:50671–50684. https://doi.org/10.1021/acsomega.4c08702

Donga S, Chanda S (2021) Facile green synthesis of silver nanoparticles using *Mangifera indica* seed aqueous extract and its antimicrobial, antioxidant and cytotoxic potential (3-in-1 system). Artif Cells Nanomed Biotechnol 49:292–302. https://doi.org/10.1080/21691401.2021.1899193

Duraisamy S, Ganguly A, Sharma PK, Benson J, Davis J, Papakonstantinou P (2021) One-step hydrothermal synthesis of phase-engineered MoS_2 /MoO_3 electrocatalysts for hydrogen evolution reaction. ACS Appl Nano Mater 4:2642–2656. https://doi.org/10.1021/acsanm.0c03274

El Hachmi A, Manoun B, Sajieddine M, Tamraoui Y, El Ouahbi S (2021) Synthesis, structural and optical properties of perovskites-type: Sr3Fe2+Mo1−O9−3/2 (x = 0.30, 0.45, 0.60, 0.75 and 1.00). Polyhedron 200:115133. https://doi.org/10.1016/j.poly.2021.115133

El Hachmi A, Sen S, Mondal R, Paul M, Saha A, Im J, Biswas G (2023) Structural, morphological, magnetic and optical properties of Jeanbandyite prepared by the co-precipitation method. Mater Today Commun 34:105358. https://doi.org/10.1016/j.mtcomm.2023.105358

El Hachmi A, Biswas G, Sen S, Manoun B, Draoui K, Sadoune Z (2025) Linear and non-linear optical properties of $FeSnO(OH)_5$ oxyhydroxide perovskite. J Mater Chem C 13:6630–6640. https://doi.org/10.1039/D4TC04040F

Gan YX, Jayatissa AH, Yu Z, Chen X, Li M (2020) Hydrothermal synthesis of nanomaterials. J Nanomater 2020:1–3. https://doi.org/10.1155/2020/8917013

Ghosh N, Sen S, Biswas G, Singh LR, Haldar PK (2021) Synthesis and characterization of CuO nanoparticles and CuO/PVA-PVP nanocomposites: adsorption activity for Malachite Green (MG) dye and antibacterial property in wastewater treatment (preprint). In Review. https://doi.org/10.21203/rs.3.rs-1089880/v1

Ghosh N, Sen S, Biswas G, Saxena A, Haldar PK (2023) Adsorption and desorption study of reusable magnetic iron oxide nanoparticles modified with Justicia adhatoda leaf extract for the removal of textile dye and antibiotic. Water Air Soil Pollut 234:202. https://doi.org/10.1007/s11270-023-06217-8

Gulia K, James A, Pandey S, Dev K, Kumar D, Sourirajan A (2022) Bio-inspired smart nanoparticles in enhanced cancer theranostics and targeted drug delivery. J Funct Biomater 13:207. https://doi.org/10.3390/jfb13040207

Hemlata MPR, Singh AP, Tejavath KK (2020) Biosynthesis of silver nanoparticles using *Cucumis prophetarum* aqueous leaf extract and their antibacterial and antiproliferative activity against cancer cell lines. ACS Omega 5:5520–5528. https://doi.org/10.1021/acsomega.0c00155

Jain S, Kumar N, Tahir M, Garg S (2023) Nanoparticle synthesis, characterization and applications. In: The impact of nanoparticles on agriculture and soil. Elsevier, pp 13–40. https://doi.org/10.1016/B978-0-323-91703-2.00007-5

Joudeh N, Linke D (2022) Nanoparticle classification, physicochemical properties, characterization, and applications: a comprehensive review for biologists. J Nanobiotechnol 20:262. https://doi.org/10.1186/s12951-022-01477-8

Kaymaz SV, Nobar HM, Sarıgül H, Soylukan C, Akyüz L, Yüce M (2023) Nanomaterial surface modification toolkit: principles, components, recipes, and applications. Adv Colloid Interf Sci 322:103035. https://doi.org/10.1016/j.cis.2023.103035

Kora AJ (2018) Tree gum stabilised selenium nanoparticles: characterisation and antioxidant activity. IET Nanobiotechnol 12:658–662. https://doi.org/10.1049/iet-nbt.2017.0310

Kora AJ, Rastogi L (2018) Green synthesis of palladium nanoparticles using gum ghatti (Anogeissus latifolia) and its application as an antioxidant and catalyst. Arab J Chem 11: 1097–1106. https://doi.org/10.1016/j.arabjc.2015.06.024

Kumar M, Kaushik D, Kumar A, Gupta P, Proestos C, Oz E, Orhan E, Kaur J, Khan MR, Elobeid T, Bordiga M, Oz F (2023) Green synthesis of copper nanoparticles from *Nigella sativa* seed extract and evaluation of their antibacterial and antiobesity activity. Int J Food Sci Technol 58: 2883–2892. https://doi.org/10.1111/ijfs.16359

Kumari M, Acharya A, Krishnamurthy PT (2023) Antibody-conjugated nanoparticles for target-specific drug delivery of chemotherapeutics. Beilstein J Nanotechnol 14:912–926. https://doi.org/10.3762/bjnano.14.75

Kuznetsova OV, Mokhodoeva OB, Maksimova VV, Dzhenloda RK, Jarosz M, Shkinev VM, Timerbaev AR (2020) High-resolution ICP-MS approach for characterization of magnetic nanoparticles for biomedical applications. J Pharm Biomed Anal 189:113479. https://doi.org/10.1016/j.jpba.2020.113479
Lau ECHT, Åhlén M, Cheung O, Ganin AY, Smith DGE, Yiu HHP (2023) Gold-iron oxide (Au/Fe3O4) magnetic nanoparticles as the nanoplatform for binding of bioactive molecules through self-assembly. Front Mol Biosci 10. https://doi.org/10.3389/fmolb.2023.1143190
Let's talk about lipid nanoparticles (2021) Nat Rev Mater 6:99–99. https://doi.org/10.1038/s41578-021-00281-4
Lin X, Sun S, Wang B, Zheng B, Guo Z (2020) Structural and physicochemical properties of lotus seed starch nanoparticles prepared using ultrasonic-assisted enzymatic hydrolysis. Ultrason Sonochem 68:105199. https://doi.org/10.1016/j.ultsonch.2020.105199
Ly P-D, Ly K-N, Phan H-L, Nguyen HHT, Duong V-A, Nguyen HV (2024) Recent advances in surface decoration of nanoparticles in drug delivery. Front Nanotechnol 6. https://doi.org/10.3389/fnano.2024.1456939
Mackenzie JD, Bescher EP (2007) Chemical routes in the synthesis of nanomaterials using the Sol–Gel process. Acc Chem Res 40:810–818. https://doi.org/10.1021/ar7000149
Mandru A, Mane J, Mandapati R (2023) A review on UV-visible spectroscopy. https://doi.org/10.5281/ZENODO.10232708
Manoj GM, Shalini M, Thenmozhi K, Ponnusamy VK, Hari S (2024) Recent advancements in the surface modification and functionalization of magnetic nanomaterials. Appl Surf Sci Adv 21:100608. https://doi.org/10.1016/j.apsadv.2024.100608
Marinica OM (2017) Study of static magnetic properties of transformer oil based magnetic fluids for various technical applications using demagnetizing field correction. J Nanomater 2017:1–9. https://doi.org/10.1155/2017/5407679
Mekuye B, Abera B (2023) Nanomaterials: An overview of synthesis, classification, characterization, and applications. Nano Sel 4:486–501. https://doi.org/10.1002/nano.202300038
Naikoo GA, Arshad F, Almas M, Hassan IU, Pedram MZ, Aljabali AAA, Mishra V, Serrano-Aroca Á, Birkett M, Charbe NB, Goyal R, Negi P, El-Tanani M, Tambuwala MM (2022) 2D materials, synthesis, characterization and toxicity: a critical review. Chem Biol Interact 365:110081. https://doi.org/10.1016/j.cbi.2022.110081
Nanomedicine and the COVID-19 vaccines (2020) Nat Nanotechnol 15:963–963. https://doi.org/10.1038/s41565-020-00820-0
Ngouangna EN, Zaidi Jaafar M, Norddin M, Agi A, Oseh JO, Mamah S (2022) Surface modification of nanoparticles to improve oil recovery mechanisms: a critical review of the methods, influencing parameters, advances and prospects. J Mol Liq 360:119502. https://doi.org/10.1016/j.molliq.2022.119502
Nouren S, Bibi I, Kausar A, Sultan M, Nawaz Bhatti H, Safa Y, Sadaf S, Alwadai N, Iqbal M (2024) Green synthesis of CuO nanoparticles using Jasmin sambac extract: conditions optimization and photocatalytic degradation of Methylene Blue dye. J King Saud Univ - Sci 36:103089. https://doi.org/10.1016/j.jksus.2024.103089
Ouyang X, Liu Y, Zheng K, Pang Z, Peng S (2024) Recent advances in zwitterionic nanoscale drug delivery systems to overcome biological barriers. Asian J Pharm Sci 19:100883. https://doi.org/10.1016/j.ajps.2023.100883
Pandi N, Sonawane SH, Anand Kishore K (2021) Synthesis of cellulose nanocrystals (CNCs) from cotton using ultrasound-assisted acid hydrolysis. Ultrason Sonochem 70:105353. https://doi.org/10.1016/j.ultsonch.2020.105353
Patil TP, Vibhute AA, Patil SL, Dongale TD, Tiwari AP (2023) Green synthesis of gold nanoparticles via Capsicum annum fruit extract: characterization, antiangiogenic, antioxidant and anti-inflammatory activities. Appl Surf Sci Adv 13:100372. https://doi.org/10.1016/j.apsadv.2023.100372
Qiu J, Li Y, Jia Y (2021) Synthesis methods. In: Persistent phosphors. Elsevier, pp 31–67. https://doi.org/10.1016/B978-0-12-818637-4.00002-1

Radu T, Iacovita C, Benea D, Turcu R (2017) X-ray photoelectron spectroscopic characterization of iron oxide nanoparticles. Appl Surf Sci 405:337–343. https://doi.org/10.1016/j.apsusc.2017.02.002

Rajeshkumar S, Rinitha G (2018) Nanostructural characterization of antimicrobial and antioxidant copper nanoparticles synthesized using novel Persea americana seeds. OpenNano 3:18–27. https://doi.org/10.1016/j.onano.2018.03.001

Rua-Ibarz A, Bolea-Fernandez E, Pozo G, Dominguez-Benetton X, Vanhaecke F, Tirez K (2020) Characterization of iron oxide nanoparticles by means of single-particle ICP-mass spectrometry (SP-ICP-MS)—chemical *versus* physical resolution to overcome spectral overlap. J Anal At Spectrom 35:2023–2032. https://doi.org/10.1039/d0ja00183j

Sacko A, Nure JF, Nyoni H, Mamba B, Nkambule T, Msagati TAM (2024) The application of tannic acid-coated magnetite nanoparticles for recovery of microplastics from the water system. Water Conserv Sci Eng 9:41. https://doi.org/10.1007/s41101-024-00275-7

Sandler SE, Fellows B, Mefford OT (2019) Best practices for characterization of magnetic nanoparticles for biomedical applications. Anal Chem 91:14159–14169. https://doi.org/10.1021/acs.analchem.9b03518

Sanità G, Carrese B, Lamberti A (2020) Nanoparticle surface functionalization: how to improve biocompatibility and cellular internalization. Front Mol Biosci 7. https://doi.org/10.3389/fmolb.2020.587012

Saygi KO, Bayram HM, Bayram E (2024) Green synthesis of silver nanoparticles using artichoke flower petals and application in endodontic dentistry. Biomass Convers Biorefinery 14:5531–5539. https://doi.org/10.1007/s13399-022-02857-8

Schlemminger C, Næss E, Bünger U (2015) Adsorption hydrogen storage at cryogenic temperature—material properties and hydrogen ortho-para conversion matters. Int J Hydrog Energy 40:6606–6625. https://doi.org/10.1016/j.ijhydene.2015.03.104

Sen S, Basak S, Pandit S, Sarkar P, Kim TW, Ghosh N, Kong HJ, Ryu JS, Singh B, Roy R, Mondal R, Saha A, Jeon S, Bhattacharya S, Im J, Biswas G (2024) Cellular internalization, cytotoxicity and DNA binding property of 2,3-diaminophenazine, synthesized using Jeanbandyite, a heterogeneous catalyst. Sci Rep 14:29684. https://doi.org/10.1038/s41598-024-81330-0

Shafiee M, Sabouri Z, Jalili A, Darroudi M (2024) Eco-friendly synthesis of pure lignin nanoparticle and silver-doped lignin nanoparticle using fig tree bark for photocatalytic degradation of Rhodamine B dye and assessment of their anticancer and antibacterial performance. Inorg Chem Commun 170:113399. https://doi.org/10.1016/j.inoche.2024.113399

Shukla S, Jadaun A, Arora V, Sinha RK, Biyani N, Jain VK (2015) In vitro toxicity assessment of chitosan oligosaccharide coated iron oxide nanoparticles. Toxicol Rep 2:27–39. https://doi.org/10.1016/j.toxrep.2014.11.002

Singh V, Yadav P, Mishra V (2020) Recent advances on classification, properties, synthesis, and characterization of nanomaterials. In: Srivastava N, Srivastava M, Mishra PK, Gupta VK (eds) Green synthesis of nanomaterials for bioenergy applications. Wiley, pp 83–97. https://doi.org/10.1002/9781119576785.ch3

Škrátek M, Dvurečenskij A, Kluknavský M, Barta A, Bališ P, Mičurová A, Cigáň A, Eckstein-Andicsová A, Maňka J, Bernátová I (2020) Sensitive SQUID bio-magnetometry for determination and differentiation of biogenic iron and iron oxide nanoparticles in the biological samples. Nanomaterials 10:1993. https://doi.org/10.3390/nano10101993

Skwarek A, Gąsecka A, Jaguszewski MJ, Szarpak Ł, Dzieciątkowski T, Filipiak KJ (2023) Nanoparticles: a breakthrough in COVID-19 prevention, diagnosis and treatment. Arch Med Sci 19:1410–1420. https://doi.org/10.5114/aoms/142103

Suk JS, Xu Q, Kim N, Hanes J, Ensign LM (2016) PEGylation as a strategy for improving nanoparticle-based drug and gene delivery. Adv Drug Deliv Rev 99:28–51. https://doi.org/10.1016/j.addr.2015.09.012

Talebian S, Wallace GG, Schroeder A, Stellacci F, Conde J (2020) Nanotechnology-based disinfectants and sensors for SARS-CoV-2. Nat Nanotechnol 15:618–621. https://doi.org/10.1038/s41565-020-0751-0

Tariq M, Mohammad KN, Ahmed B, Siddiqui MA, Lee J (2022) Biological synthesis of silver nanoparticles and prospects in plant disease management. Molecules 27:4754. https://doi.org/10.3390/molecules27154754

Tian Y, Wu D, Wu D, Cui Y, Ren G, Wang Y, Wang J, Peng C (2022) Chitosan-based biomaterial scaffolds for the repair of infected bone defects. Front Bioeng Biotechnol 10. https://doi.org/10.3389/fbioe.2022.899760

Trayford C, Van Rijt S (2024) *In situ* modified mesoporous silica nanoparticles: synthesis, properties and theranostic applications. Biomater Sci 12:5450–5467. https://doi.org/10.1039/d4bm00094c

Usgodaarachchi L, Thambiliyagodage C, Wijesekera R, Bakker MG (2021) Synthesis of mesoporous silica nanoparticles derived from rice husk and surface-controlled amine functionalization for efficient adsorption of methylene blue from aqueous solution. Curr Res Green Sustain Chem 4:100116. https://doi.org/10.1016/j.crgsc.2021.100116

Vieira IRS, Da Silva AA, Da Silva BD, Torres Neto L, Tessaro L, Lima AKO, Garcia MP, Ribeiro JADA, Rodrigues CM, De Sousa AMF, Carvalho NMF, Rajkovic A, Conte-Junior CA (2024) Antioxidant, antimicrobial, and anticancer potential of green synthesized ZnO nanoparticles from Açaí (Euterpe oleracea Mart.) berry seed residue extract. Waste Biomass Valoriz 15:4717–4734. https://doi.org/10.1007/s12649-024-02485-5

Vinay SP, Udayabhanu NG, Chandrappa CP, Chandrasekhar N (2019) Enhanced photocatalysis, photoluminescence, and anti-bacterial activities of nanosize Ag: green synthesized via Rauvolfia tetraphylla (devil pepper). SN Appl Sci 1:477. https://doi.org/10.1007/s42452-019-0437-0

Wen Y-H, Huang R (2019) Effect of chemical ordering on thermal stability of Pt–Co nanoparticles. J Phys Chem C 123:12007–12014. https://doi.org/10.1021/acs.jpcc.9b02540

Wernsdorfer W, Hasselbach K, Mailly D, Barbara B, Benoit A, Thomas L, Suran G (1995) DC-SQUID magnetization measurements of single magnetic particles. J Magn Magn Mater 145:33–39. https://doi.org/10.1016/0304-8853(94)01621-6

Yadav M (2025) Nanoparticle-facilitated targeted nutrient delivery in plants: breakthroughs and mechanistic insights. Plant Nano Biol 12:100156. https://doi.org/10.1016/j.plana.2025.100156

Chapter 3
Physical and Chemical Properties of NPs Influencing Their Plant Toxicity

Aneeza Ishfaq, Sadia Mustafa, Sabir Hussain, Tanvir Shahzad, Muhammad Shahid, Fatima Batool, Tehziba Raza, and Faisal Mahmood

Abstract Nanoparticles have emerged as innovative in several fields, including agriculture, because of their unique physical and chemical properties. Nanoparticles measuring less than 100 nm enter plant cell membranes easily yet produce reactive oxygen species that cause cellular damage through oxidative stress. The unique physicochemical characteristics of nanoparticles affect the complex process of NP-plant interactions. This chapter discusses plant nanoparticle toxicity's main factors: size, shape, surface charge, coating, aggregation state, hydrophobicity, porosity, and ligand binding. The effects of these variables on oxidative stress, development, and metabolism are described in more detail below. This chapter clarifies the complex interplay between NP properties and plant responses, providing a framework for the development of nanotechnologies that balance agricultural benefits with environmental safety.

Keywords Nanoparticles · Agriculture · NPs plant interaction · Transport · Oxidative Stress

3.1 Introduction

Nanoparticles (NPs) are fine powders with a small size of 1–100 nm. The physicochemical characteristics of nanoparticles include a large surface area-to-volume ratio, adjustable reactivity, and solubility, as well as specificity in interacting with biological systems at the molecular level. Nanoparticles have received much

A. Ishfaq · S. Mustafa · S. Hussain · T. Shahzad · F. Batool · T. Raza · F. Mahmood (✉)
Department of Environmental Sciences, Government College University, Faisalabad, Pakistan
e-mail: aneeza.201703601@gcuf.edu.pk; sadiamustafa.201602598@gcuf.edu.pk; sabir.hussain@gcuf.edu.pk; tanvirshahzad@gcuf.edu.pk; Fatima.201602599@gcuf.edu.pk; faisalmahmood@gcuf.edu.pk

M. Shahid
Department of Bioinformatics and Biotechnology, Government College University, Faisalabad, Pakistan
e-mail: mshahid@gcuf.edu.pk

R. N. Pudake et al. (eds.), *Phyto Nanotoxicology*, Nanotechnology in Plant Sciences 7, https://doi.org/10.1007/978-3-032-24518-2_3

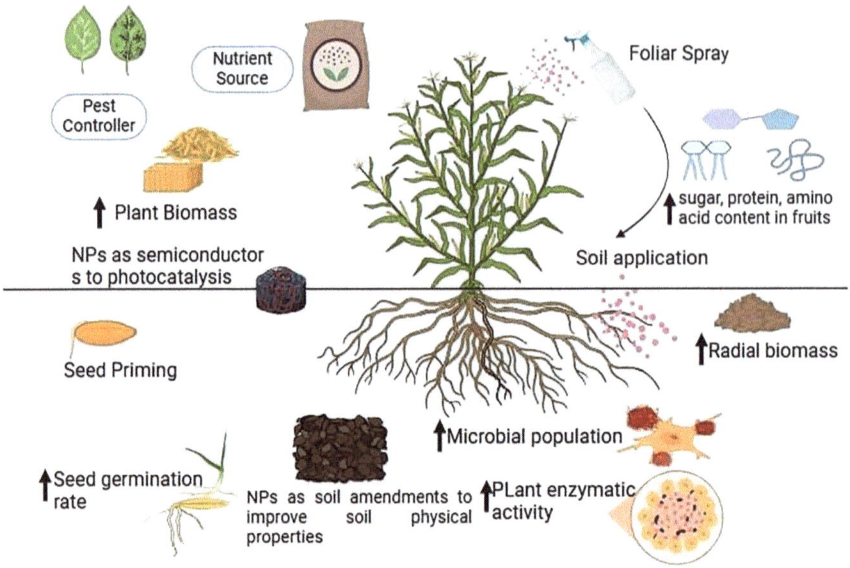

Fig. 3.1 Schematic showing the various benefits of nanoparticles in agriculture

attention in various fields including agriculture. Nanoparticles are categorized into metallic (Silver, Gold), polymeric, ceramic (Titanium dioxide, zinc oxide), and carbon-based (Ahmed et al. 2024). These nanoparticles have distinct characteristics that fulfill various demands of agriculture such as improving plant growth and enhancing pest resistance. As for agriculture, nanoparticles are used to increase the effectiveness of nanotechnology in traditional agriculture. NPs increase the delivery of specific active ingredients in pesticides and herbicides while reducing their lethal impacts on other forms of life (Hafez and Khalil 2024; Shafqat et al. 2023; Wahab et al. 2024). For example, AgNPs have antibacterial activity that helps control pathogens that attack plants. TiO_2 NPs enhance the photosynthesis rate in plants to mitigate the damaging ultraviolet light (Javan et al. 2024). Ishfaq et al. (2023) used biogenic MnO NPs in wheat plants to mitigate the wastewater. In agriculture, NPs are primarily employed as nano-pesticides and nano-fertilizers. Because of leaching and volatilization, chemical fertilizers are not very effective. Currently, farmers typically respond by applying excessive amounts of fertilizer that boost crop yields but have negative environmental effects (Joudeh and Linke 2022). However, nano fertilizers are substances that have higher efficiency than conventional chemical fertilizers but are administered in lower quantities (Fig. 3.1). Because they can deliver nutrients precisely according to plant needs, and they are more efficient than other methods. Some NPs are a possible substitute for chemical pesticides and may be less expensive than biopesticides because of their demonstrated antibacterial, insecticidal, and nematocidal properties (Shafqat et al. 2024). Additionally, NPs promote sustainable agriculture through increasing the balanced usage of water and fertilizers, which is crucial in areas of water and soil deficiencies. For instance, nanotechnology has emerged to reduce fertilizer use by 20–30%, thus reducing runoff and its impacts on the water systems (Singh et al. 2024).

Moreover, NPs are employed within soil to make better water-holding capacity and moisture content. The technological development of nanoparticles has greatly increased agricultural productivity and sustainability (Haidri et al. 2024b). One of the most important benefits is increasing nutrient transport and ensuring that plants receive nutrients with minimal losses. This targeted delivery not only increases the efficiency of nutrient usage but also increases yield of crops. Furthermore, the NPs reduce disease management by precise application of pesticides or biocontrol agents, paving the way for minimum chemical input. The bacterium that causes tomato spot disease, *Xanthomonas perforans*, is effectively inhibited by TiO_2 NPs due to their photocatalytic activity (Shafqat et al. 2023). African leafworm of cotton, or *Spodoptera littoralis*, is susceptible to the insecticidal effects of CuO NPs (Cromwell et al. 2014). Ag NPs have nematocidal action against root-knot nematodes and *Meloidogyne* spp (Shafqat et al. 2023). Overall, these achievements introduce NPs as an important factor in increasing agricultural yields while ensuring environmental sustainability.

This chapter aims to differentiate the toxicity of nanoparticles (NPs) for plants based on their physical and chemical characteristics. Past studies have identified nanoparticles' essential characteristics, including size, morphology, surface charge, and chemical composition because these features determine their plant system conduct. The chapter focuses on these properties to determine the possible threat of harming plants and ecosystems.

3.2 Importance of Studying NP-Induced Plant Toxicity

The potential impact of NP exposure on plants is tremendous, especially regarding bioaccumulation and food chain transfer. When plants uptake NPs from soil, water, or through the air, the particles can penetrate and localize in many parts of plant (Hussain et al. 2021). Such accumulation is not only toxic to plant metabolism, it interferes with various plant functions, including photosynthesis, respiration, and nutrient uptake, but also poses a threat to species that consume such plants (Mustafa et al. 2023, 2025). For instance, the nanoparticles, the silver nanoparticles (AgNPs), and the titanium dioxide nanoparticles (TiO_2 NPs) are bio-saturated in edible plant tissues to the food chain and toxic to animals and humans (Sheikh et al. 2024). Within the food chain, nanoparticles can have cytotoxic properties, causing oxidative stress, inflammation, and disruption of various cellular processes in higher organisms. Also, the dissolution of NPs into soil and water bodies is detrimental to the beneficial microorganisms in the soils, directly affecting soil productivity and ecosystem stability. The combination of risks exists because NPs stay persistent in the environment while also moving between ecosystems in ways that affect biodiversity. Understanding these risks is essential to minimize their impact on the environment and human health and achieve environmentally friendly application of nanotechnology in agriculture (Osman et al. 2024).

3.3 Physical and Chemical Properties of Nanoparticles That Impact Plant Toxicity

Physio-chemical properties like as size, shape, surface charge, surface coating, chemical compositions, and stability affect NPs' toxicity throughout the plant (Table 3.1). Recently, Oh et al. (2016) examined toxicity of CdSe quantum dots, specifically, utilizing the data meta-analysis of 307 publications that described 1741 cell viability-related data samples. It has been demonstrated that the toxicity of QD nanoparticles is strongly related to their diameter, toxicity assay type, exposure duration, and surface characteristics (such as shell, ligand, and surface modifications). The physicochemical properties are listed in Fig. 3.2.

Table 3.1 The physical and chemical properties of nanoparticles and their mechanism influence plant

Property	Description	Mechanism of Influence	Examples	References
Size	Particles have a size from 1–1000 nm affecting the surface area/volume ratio	Smaller sizes easily penetrate plant cells than intracellular toxicity easily occurs	Higher ROS generation in plants that were treated with <50 nm TiO_2.	Djanaguiraman et al. (2024)
Shape	Geometric shapes such as spheres, rods, and cubes change the surface reactivity	Geometries with high aspect ratios such as rods lead to higher mechanical damage to cell structures	Nanorod form of ZnO causes greater oxidative stress than spherical forms	Yang et al. (2017)
Surface Charge (Zeta Potential)	Defines the degree of interaction of the ion with cell membranes	Positive charges interfere with negatively charged plant cell wall and cell membrane	AgNPs with positive charges have a greater toxic effect as compared to neutral or negative charges on root cells	Rastogi et al. (2017)
Surface Coating/ Functionalization	Stabilizers or functional groups on NP surfaces modify interaction with plants	Functional groups can increase or decrease uptake and modify toxicity pathways	PEGylated NPs exhibit reduced toxicity compared to bare particles	Chen (2018)

(continued)

Table 3.1 (continued)

Property	Description	Mechanism of Influence	Examples	References
Aggregation State	Ability to aggregate in-water conditions/propensity to move in groups	Aggregation decreases bioavailability although it creates concentration zones of localized toxicity	Lowering the germination rates is a result of the clustering of Fe_3O_4 NPs	Huang et al. (2017)
Hydrophobicity	Frequency of contact with water; affects uptakes and translocation in plants	Some hydrophobic NPs may preferentially accumulate in lipid-rich areas thus producing local inflammation	Water-repellent carbon nanotubes limit water absorption in roots	Joudeh and Linke (2022)
Translocation	Transportation of nanoparticles from roots through the xylem or directly into the phloem of the shoot	NPs stored in plant tissues causing toxicity at localized or systemic level	ZnO NPs move to the leaves' surface affect the photosynthetic ability of chlorophyll and cause chlorosis	Chen (2018)
Porosity	Molecules react more if nanoparticles are porous or contain voids to hold and transport other molecules	High porosity increases the surface availability for the absorption of nutrients or toxins leading to an imbalance of the normal nutrients	Porous silica NPs disrupt the uptake of nutrients and slow down the growth of plants	Yang et al. (2017)
Hydrophilicity	The affinity of NPs for water, is often determined by surface functional groups	Hydrophilic NPs are readily soluble in water, which increases both plant uptake and transport within the system	Water soluble TiO_2 NPs pass the root cell membrane and influence water and nutrient conduction and absorption	Tripathi et al. (2017)

3.3.1 *Size and Morphology*

These size-dependent effects indicate that the size of the NPs is a critical factor to consider when designing NPs for employment in agriculture (Ishfaq et al. 2023). The effect of size on nanomaterial biodistribution, release kinetics, and cellular absorption is one of the significant problems in nanotechnology. Because of their small

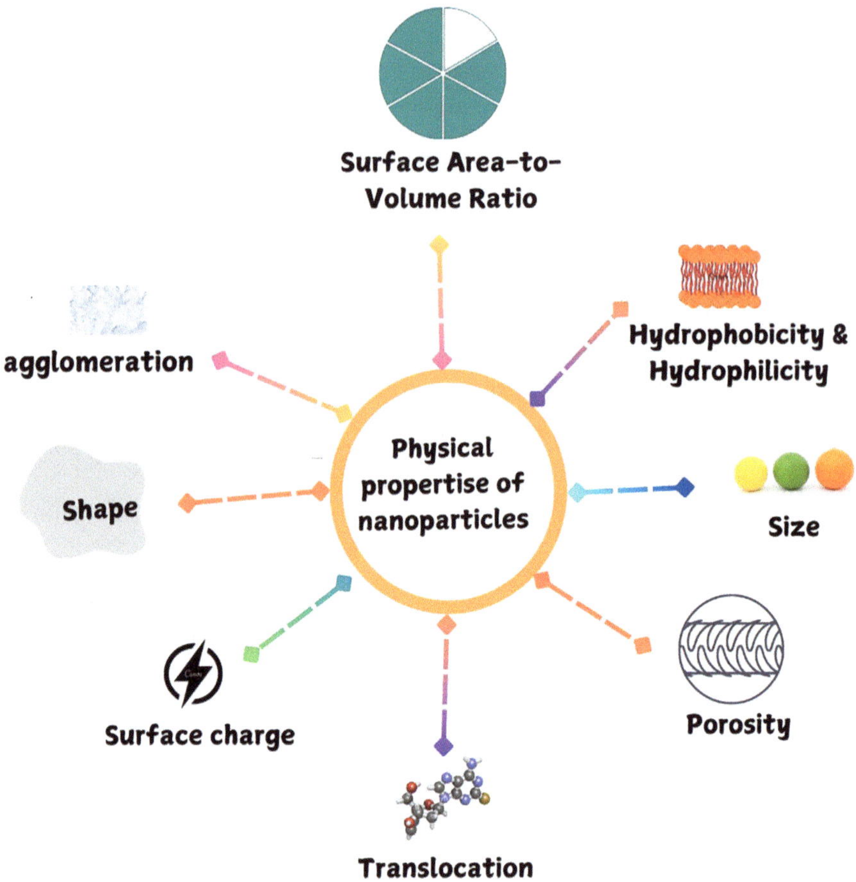

Fig. 3.2 The physicochemical properties of nanoparticles in plants

size, macrophages frequently do not identify nanoparticles as foreign substances, allowing them to enter macrophage cells through membrane pores. Conversely, reticuloendothelial systems quickly absorb nanoparticles (Raza et al. 2023). Particles' surface area increases as their size decreases, and a larger surface area promotes particle diffusion into cells (Zhang et al. 2020). Considerate NP size influences plant interactions for lowering negative impacts and enhancing their positive consequences in precision agriculture. Nanoparticles also exhibit new behaviors when in contact with other nanoparticles to form larger structures and their fate and interactions with soil microbes may also change. Carlson et al. (2008) examined the size-dependent impacts of silver particles on rat alveolar macrophages, whereas Park et al. (2011) examined its impact on the human leukocyte cell line U937. Both studies showed that the tiny silver nanoparticles used generated more reactive oxygen species (ROS) and a more pronounced inflammatory reaction than the larger particles.

Shape and morphology of NPs influence their contact with plant cells and affect their uptake, distribution, and potential toxicity. NPs with different shapes such as spherical, rod-like, cubic, or irregular forms possess different features of their interaction with plant tissues and cells. Previous research on spherical gold nanoparticles shows that they have a higher uptake efficiency than other shapes. Spherical shapes are transported more easily through the apoplast and symplast routes (Aseel et al. 2024). However, rod-like nanoparticles have elongated shapes, which can help to align with fibrous components of plant cell walls but significantly increase the risk of mechanically damaging plant tissues (Saod et al. 2022). Compared to spherical nanoparticles, elongated nanoparticles exhibit greater adherence to cells (Ishfaq et al. 2025). Because of their curved shape, spherical particles can only engage with target cell receptors through a small number of their binding sites. However, because the elongated NPs have a larger surface area, they could relate with cell surface more multivalently (Haidri et al. 2024a). For instance, it observed that TiO_2 NPs are even more disruptive to cell walls compared to spherical modification (Sheikh et al. 2024). In cerium oxide nanoparticles (CeO_2 NPs), cubic or sharp-edged nanoparticles are capable of giving localized mechanical stress to cell membranes and organelles (Yadav and Yadav 2024). The capacity of nanoparticles to produce ROS and attach to biomolecules, which alters toxic effects, is similarly influenced by their size and shape. For example, thin shell-like structures with small angles and high uncertainty, such as edge-angle and high aspect-ratio structures have more reactive surface atoms, which influence plant cells to oxidative stress (Naseem and Durrani 2021).

3.3.2 Surface Area to Volume Ratio

The surface area-to-volume ratio of nanoparticles determines reactivity and plant toxicity due to several reasons. The availability of nanoparticles depends on a large surface area-to-volume ratio that allows more available surface sites for interaction with biological molecules, such as proteins, lipids, and enzymes. The reactivity of nanoparticles increases with their surface area and having absorbed into plant cells. As the surface area of NPs increases, production of ROS interacts with plant tissue causing cellular damage such as membrane rupture, DNA damage, and reduced photosynthesis (Shafqat et al. 2023). Kohan-Baghkheirati and Geisler-Lee (2015) showed gold nanoparticles (AuNPs) and silver nanoparticles (AgNPs) have a small size and a larger surface area, which easily interact with the cell membranes and other organelles in plants. The studies have indicated that AgNPs have a larger surface area, leading to higher levels of toxicity than large silver particles. This is so because a higher surface area gives more reactive sites in which the ROS can be produced with negative effects such as weakening of the cell walls, the mitochondria, and chloroplasts in plants. According to Sun et al. (2025), the absorption of AgNPs in *Arabidopsis thaliana* plants caused accelerated damage to cell membranes and ultimately the death, especially by the smaller AgNPs, where there was a large surface area to volume ratio of the particles. In soil large surface area and porosity of nanoparticles increase their ability and allow them to separate in the soil structure and interact with soil microorganisms or plant roots. Higher interaction can also be detrimental in several ways, for example, soil fertility status may be affected or the microbial ecology of soil imbalances, impairing the toxic effects on plants. The surface area and porosity of the nanoparticles remain the most important parameters that interact with soil and plant systems (Ahmed et al. 2021). Since the NPs contain large ratio of surface area and volume, they have higher toxic reactions, the porous structure affects the nutrient ability and microbiological characteristics and is used accordingly in the agricultural field (Samal and Subudhi 2024).

3.3.3 Surface Coating

Such morphological modifications as coating or functionalization strongly affect both the nanoparticle behavior in the plant systems and their toxicity. Surface functionalization is a process by which a nanoparticle is coated with a particular chemical group, polymer, or biomolecule to change its properties like hydrophobicity, reactivity, and biocompatibility of NPs (Adhikari et al. 2024). The surface coating of NPs decreases or increases toxicity, depend on the particular kind of coating (Ishfaq et al. 2025). For example using biocompatible barriers like PEG, polysaccharides, or proteins diminishes the negative impact of NP hence moved of NP in plant tissues which may cause oxidative stress can be constrained by their interaction with the plant cell walls (Tripathi et al. 2017). These coatings develop a hindrance that minimizes interaction with the reactive core and minimizes the production of ROS. For coating and protecting nanoparticles, highly described, commercially available soil organic matter is utilized (Sarkar et al. 2025). Nature-friendly surface coatings like fulvic acid (FA), dissolved organic carbon, soluble humic acid and natural organic matter can improve NP accumulation resistance by improving steric interruption, electrostatic revulsion, and hydrophobicity. By applying such coatings in advance, the NPs' characteristics may be fine-tuned, NP stability is maintained during transportation, and nanosized application is made simpler at the moment of introduction. Any NPs added to soils are predicted to spontaneously form a certain amount of organic covering (Jeon et al. 2024). HA may be one of the popular natural organic coatings because of its adaptable chemical composition. Simple techniques for HA surface coating of NPs are available, and HA coverings can facilitate NP suspension and plant nutrient uptake in addition to providing steric stability. It has also been demonstrated that HA offers supplementary advantages through the removal of pollutants and heavy metals in wastewater treatment (Oriekhova and Stoll 2016). A similar method was employed by Alhaithloul et al. (2023) to regulate the dissolution, stability, and destiny of CuO NPs for agricultural distribution using EPS generated from phytoplankton. However, hydrophobic coatings decrease the diffusion rate and accumulation. (Ekim et al. 2024) produced Au NPs coated with albumin that held up well to drying and redispersion under physiologically relevant circumstances. The coating of NPs acts to slow down the release of Zn^{2+} ions, reduce phytotoxicity, and enhance root formation (Huang et al. 2017). No adverse effects from high iron or the polymer were noted following noticeable daphnid consumption. On the other hand, other coatings that were evaluated, including citrate, or ascorbate, had a greater impact on ROS formation, which increased the number of toxicity sources in the crustaceans (Dejene and Geletaw 2024). For agricultural applications, inorganic coatings are also taken into consideration. To lessen agglomeration, some researchers are adorning or decorating their principal nanoparticles with smaller NPs that have a higher surface potential or employing NPs with a more robust outer shell that are core-shell. Silica and gold are both excellent options for core-shell NPs due to their high solution resilience and simplicity of surface functionalisation. One-pot sol-gel synthesis, which may be used to create extremely homogeneous and single-disperse NPs or surface shells for a range of applications, is facilitated by SiO_2 (Sarkar et al. 2025).

3.3.4 Agglomeration State

The agglomeration state is the degree of co-clustering of nanoparticles in the environment. NPs can also exist as single, dispersed particles or as agglomerates (clusters of particles). Consequently, NPs self-assemble into large structures, the effective surface area is reduced, and their reactivity and uptake by plants may be mitigated. However, the size of the agglomerates, the type of material, and their stability in the environment can still affect the interaction of plants and the toxicity of these NPs (Pattnaik et al. 2024). For instance, large agglomerates cannot penetrate the cell wall and plant tissues. However, small agglomerates might still get through the cell wall or root epidermis, and source cell damage and toxicity. It is understood that agglomeration also affects the distribution of NPs in the plant system (Djanaguiraman et al. 2024). Charge screening can create ionic bridging or enhance agglomeration among normally stable NPs. As a result, less accessible surface area may be available for interactions between plants and NP or for solubilization. The zeta potential or effective surface charge of NPs also effect agglomeration (Aseel et al. 2024). They are more likely to aggregate at pH levels around the point of zero charge (PZC), whereas NPs exhibit a net-neutral surface charge. At very optimistic or very adverse zeta potential that is distant from the PZC, greater NP stability is seen. Therefore, varnishes meant to stabilise NPs by repelling electrostatic charges need to be adjusted to account for a range of soil pH values and electrical conductivities (Cartwright et al. 2020).

When NPs are highly agglomerated, their mobility in the soil matrix is also limited and their bioavailability is low; however, if the NPs find their way to the roots of plants, they can cause substantial toxicity. Larger agglomerates can be retained within the root zone or deposited in certain plant tissues, for example, the root cortex while smaller agglomerates are not very agglomerated and move easily to plant parts above the ground such as the leaves or stems. This dispersion pattern is important for establishing whether nanoparticles have differential toxic effects on parts of the plant since some plant types are more sensitive to nanoparticle toxicity. Liu et al. (2020) showed that aggregated Fe_3O_4 NPs lower the germination rate in rice seeds by altering iron availability. Rastogi et al. (2017) showed that SiO_2 NPs agglomerates generate oxidative stress in lettuce leaves because of uneven distribution.

3.3.5 Hydrophobicity and Hydrophilicity

Hydrophobicity and hydrophilicity are the two physical characteristics of the nanoparticles that determine their ability to interrelate with the plants and their toxicity. These properties determined the nanoparticles' affinity to water, soil, and plant cell walls, solubility, and dispersion in plant tissues as well as transportation in the plant systems. Many hydrophobic nanoparticles have poor affinity for water, meaning that they stick together in water do not dissolve well in body fluids, and do not circulate well (Kang et al. 2024). Nevertheless, when hydrophobic nanoparticles encounter the plant surface, they easily enter into waxy cuticles because leaves contain hydrophobic cuticular layers. Such results could lead to larger nutrient uptake by foliar routes that may result in localized toxicity. Furthermore, these hydrophobic NPs can also be adsorbed to hydrophobic contaminants in the soil and decrease their bioavailability for roots (Dev et al. 2018; Juárez-Maldonado et al. 2019). The hydrophobic property of NPs has a major impact on their interactions with cells. According to a variety of studies, NPs' surface hydrophobicity regulates their opsonization, biodistribution, and cell uptake (Li et al. 2020). Cell intake was enhanced by core-shell NPs with a hydrophobic PLGA surface, and their composition improved NP biodistribution in the tissues of the eyes. The conductance of the NPs with varying copolymer ratios about surface hydrophobicity fluctuated. Serum proteins improved the hydrophobicity of NPs, which in turn increased their cytotoxicity and cellular absorption (Bindra et al. 2023).

While hydrophilic nanoparticles get dispersed in water and show better mobility in the tissues of the plant when compared to hydrophobic nanoparticles. These nanoparticles may be translocated into the plant more easily through the roots because of their solubility within the apoplastic and symplastic pathways. Moreover, extreme hydrophilic leads to an undesired distribution of the physiological function of the plant and causes toxicity. Hydrophilic nanoparticles are known to have low affinity with the cell walls of plants; they therefore have a poor ability to deliver nutrients or pesticides to the intended areas. The hydrophobic to hydrophilic tendency of the particles plays a crucial role when designing nanoparticles for use in agriculture (Chen et al. 2025). Functionalization strategies may modulate such properties to improve the interaction of nanoparticles with specific plant tissues and ensure efficient delivery while minimizing harmful effects. While nanoparticles designed for foliar application require hydrophobic functionality while nanoparticles tailored for root system delivery need verification of hydrophilic functionality to facilitate absorption and translocation (Cartwright et al. 2020). Thus, knowing these properties, the researchers can guide the nanoparticle systems as safer for plant-related systems. Hydrophobic and hydrophilic NPs use different mechanisms to interact with the cells. According to molecular dynamics (MD), semi-hydrophilic nanoparticles can be immobilized on the membrane bilayer's surface, whereas hydrophobic NPs may penetrate the lipophilic core of the cell (Chen et al. 2025). Rapid penetration of the cell membrane allows hydrophobic NPs to get into cells. Nevertheless, after being absorbed on the surface of the bilayer, hydrophilic NPs passed across the membrane and entered the cell.

3.3.6 *Porosity*

Porous character is one of the most significant parameters of nanoparticles affecting their plant system interactions and their toxicity level. Nanoparticles with a highly porous structure that involve mesoporous silica nanoparticles (MSNs) or zeolites, demonstrate a larger surface area to volume ratio and thus better capability for adsorption and release of substances such as nutrients, pesticides, or contaminants (Wang et al. 2024). The porosity of nanoparticles plays a crucial role in soil and plant uptake. Silica nanoparticles (SiO_2 NPs) contain pores that can take in more ionic nutrients or organic molecules in the surrounding space and potentially affect the nutrient availability to plants (Gu et al. 2024). Entry of nanoparticles into plant roots modifies the plant's nutrient processes allowing heavy metals to combine with additional toxic elements. Plants use porous nanoparticles for active substance delivery through controlled release systems which reduce pollution and improve fertilization and pesticide effectiveness (Miguel Sábio et al. 2024). Nanoparticles can capture essential nutrients and biomolecules, but this process may produce toxic effects. The water-absorption capacity of hydrophilic porous materials in soil improves drought tolerance for plants. Plant uptake becomes restricted because porous materials retain both water and nutrients inside their structures which diminishes their availability for plant growth and development (Halfadji et al. 2024).

3.3.7 *Ligand Binding*

Scientists researched the cellular internalization of nanoparticles using various surface ligands and core materials which included platinum (Pt), palladium (Pd), and gold (Au). The cell intake process and surface hydrophobicity of nanoparticles depend on their ligands together with core materials (Djanaguiraman et al. 2024). The hydrophobic characteristic of gold nanoparticles (Au NPs) helps them uptake more efficiently compared to platinum nanoparticles (Pt NPs). Bai et al. (2019) explained that Pt NPs with their compact 10–15 times smaller size entered cells sparingly using non-specific phagocytosis and pinocytosis pathways. Nanoparticles that use ligands fulfill various functions in biomedical applications by improving both cellular intake and circulation stability while providing targeted drug delivery and reducing their harmful effects on cells (Wong et al. 2024).

While little research has looked at the effectiveness of ligand dispersion on cellular uptake, increasing the effectiveness of medication delivery is the primary objective of surface alteration of nanoparticles using ligands. The statistical dynamical model of enlargement shows that when the receptor density is sufficient, the uniform arrangement of ligands on the NP interface leads to maximum cellular intake (Chen et al. 2024). Whether the consistent arrangement of ligands on the viral capsule guarantees that the virus will be taken up by the host cell has not yet been demonstrated. NPs also show quick membrane wrapping and absorption when their ligand distribution is homogeneous. Since ligand-free patches reduces the uptake process, cellular absorption will be impeded if they cover more than 20% of the surface (Raza et al. 2023).

3.3.8 Translocation and Distribution

Once nanoparticles (NPs) enter a plant's root system, their movement within the plant is determined by their physical properties. The effective transport of small NPs occurs through the xylem to reach different tissues such as leaves and stems. The relocation of nanoparticles throughout plant tissues may generate toxic effects in different plant sections. The movement of NPs relies heavily on their stability as well as the environmental conditions of the soil near the roots (Mushtaq et al. 2020). Uptake of nanoparticles relies on the plant cell wall which restricts entry through chemical-based testing and solubility-based selection. The literature shows that the cell wall is the main factor that blocks NPs from entering plant tissues (Mustafa et al. 2023). Nanoparticles enter plant cells from the surface; they generally need to be within the size range of 40–50 nm. Additionally, NP properties influence their ability to bypass the cell wall and membrane or adhere to root surfaces. Positively charged NPs have a stronger tendency to bind to the cell wall (Irfan et al. 2023). Moreover, NP morphology and surface coatings significantly affect their behavior in the rhizosphere and interactions with plants. Once absorbed by the roots, NPs can be transported to newly developed seeds and aerial tissues. The translocation process is determined by both NP characteristics and specific plant traits (Gordillo-Delgado et al. 2020). Research has identified silicon dioxide (SiO_2) and titanium dioxide (TiO_2) as two extremely durable NPs that hold stable forms within plant tissues. Different from silicon dioxide and titanium dioxide NPs, nickel oxide (NiO), lanthanum oxide (La_2O_3), copper (II) oxide (CuO), and cerium dioxide (CeO_2) experience structural changes when incorporated into plant systems (Falco et al. 2020). The roots and shoots of *Zea mays* accumulate ZnO NPs especially in Zn-phosphate form because hydroponic exposure enhances dissolution in the rhizosphere as well as ionic zinc transport (Samadi et al. 2020). Similar Zn accumulation has been observed in soil-grown wheat crops. Additionally, ZnO and CeO_2 NPs have been translocated from the soil into Glycine max, with CeO_2 NPs remaining in nanoparticulate form (Rizwan et al. 2025). In Z. *mays*, CuO NPs move from roots to shoots via the xylem and may also be redistributed from shoots to roots through the phloem. Research indicates that TiO_2 NPs contribute to translocation in *Triticum aestivum*, but root accumulation is limited when NP size exceeds 140 nm. NPs around 36 nm can accumulate in the root parenchyma but may be unable to reach the stele, preventing their movement into the shoot (Zhang and Su 2024).

Plants undergo various physiological changes that affect NP toxicity and bioavailability (Aslam et al. 2023). Research shows that extremely small NPs (3–5 nm) can directly penetrate root epidermal cells or enter through capillary forces and osmotic pressure. Some NPs can create new pores in the epidermal cell wall, facilitating their entry into plant tissues. Its entry was made easier by certain NPs that created new holes in the epidermal cell wall (Djanaguiraman et al. 2024). The xylem can move upward unidirectionally once NPs have passed through the cell walls and are apoplastically transported through extracellular spaces to the central vascular cylinder (Ameen et al. 2021). However, NPs need to simultaneously cross the Casparian strip barrier in order to reach the core vascular cylinder. By attaching to the transporter proteins in the epidermal cell membrane through endocytosis, pore creation, and movement, this is achieved (Jeon et al. 2024). As NPs are internalized in the cytoplasm, they move between cells via the plasmodesmata. The NPs that are fail to internalise collect on the Casparian strip, whilst those that have made it to the xylem are moved to the branches and then returned to the roots via the phloem (Budhani et al. 2019). The cytoplasm of cortical cells, the nuclei, and the epidermal cell wall can all contain NPs that plants absorb (Fig. 3.3). NPs can be directly absorbed by seeds through both mobility in the cotyledon and penetration into the coat by parenchymatic intercellular gaps (Djanaguiraman et al. 2024). Through the stomata or cuticles, the NPs that the leaves apply can enter the leaves. The cuticle serves as the main leaf barrier, preventing NPs from entering the plant through stomata until they are 10 nm in size. Apoplastic and symplastic pathways allow NPs to enter the vascular system of plants (Rajput et al. 2020). The cytoplasm of the neighboring cell facilitates the transfer of NPs (between 10 and 50 nm) (symplastic route). As a result, the apoplastic route causes larger nanoparticles to move between the cells. Through the phloem sieve tubes, internalized NPs are carried with the sugar flow. Because roots, fruits, grains, and young leaves are strong sap sinks, NPs can travel both ways and accumulate in these organs to differing degrees as a result of vascular transport by phloem (Saxena et al. 2020). The term "nonselective direction of least resistance" refers to the apoplastic passage. It is generally accepted that the apoplastic route for translocation is preferred by numerous water nutrients and unnecessary metal complexes (Zhang and Su 2024).

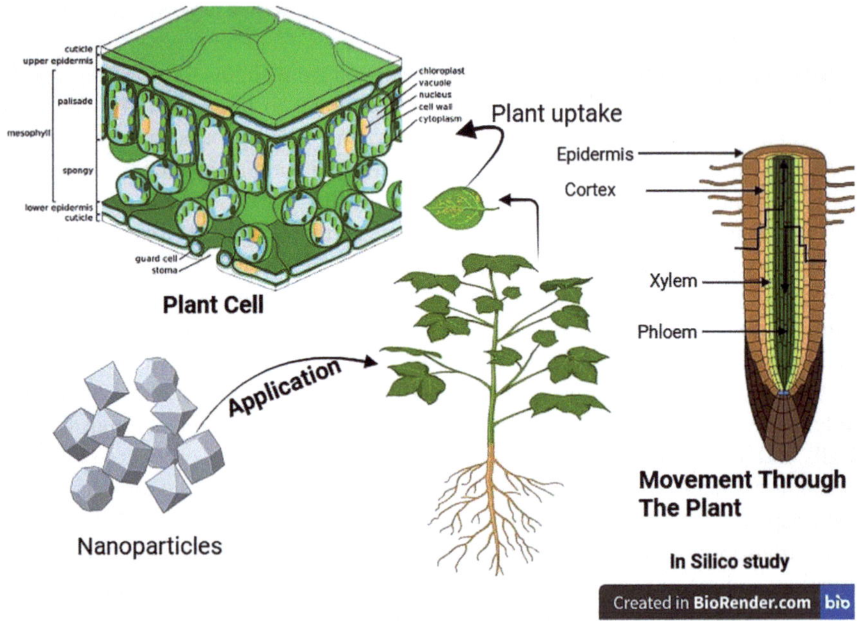

Fig. 3.3 Mechanisms of translocation of nanomaterials in plants

3.3.9 Surface Charge

Surface charge is another aspect that is equally responsible for the plant uptake and translocation of NPs. The surface charge contributes significantly to the physical, chemical, and biological properties of nanoparticles, particularly in terms of stability, dispersion, and the ability to interact with the environment. The source of this charge comes from functional groups exposed to the surface and the adsorption of ions from the environment. The surface charge is commonly determined from the zeta potential, which characterizes the slipping plane of the electrical double layer around the nanoparticles. Nanoparticles with zeta potentials above and below ± 30 mV are practically considered stable in suspension because electrostatic repulsion gives no chance to combine (Singh et al. 2019). The cellular uptake from cationic NPs shows enhanced behavior because of their strong attraction to negatively charged cell membranes; however, this prompts toxic effects (Kovalchuk et al. 2019). The cellular uptake of NPs depends on the way the surface properties of NPs interact with the membranes of cells. The amino groups reacted more intensely with cellular structures as the reason for enhanced activity. The stress response intensified after functionalization because enhanced uptake pathways and signaling mechanisms affected aerial tissues as well as root tissue. Liu et al. (2020) studied that the plants exposed to AC-SiO_2 NPs generated higher ROS content compared to untreated particles. Results showed increased reactivity of the amino groups with cellular structures. Functionalization led to a higher uptake of stress signaling with impacts on root and aerial tissue (Ishfaq et al. 2025). According to Samadi et al. (2020), the mucilaginous liquid from *Arabidopsis thaliana* roots stopped positively-charged gold nanoparticles (AuNPs) (~12 nm) from penetrating root tissue through hyperspectral imaging microscopy (HSI) and X-ray computed nanotomography (nano-CT) analysis. Variations in plant physiology and metabolism result in species-specific nanoparticle uptake capacities. Experimental evidence has revealed that dicotyledonous plants show better cerium (Ce) movement compared to monocotyledons according to Sanzari et al. (2019). According to Rastogi et al. (2019), the plant species *Azolla caroliniana* Willd successfully absorbed nanoparticles of both 4 nm and 18 nm sizes.

3.4 Biophysical Interactions and Stress Responses

Smaller and highly reactive nanoparticles could generate reactive oxygen species (ROS) upon contact with cellular components that enhanced oxidative stress, membrane lipid peroxidation, and DNA damage (Samal and Subudhi 2024). Nanoparticles are absorbed in plant cells, resulting in a variety of stress responses on cells and metabolism, and activation of signaling pathways. Due to their properties, nanoparticles can penetrate plant cells in several ways and cause various stress responses. Once plants are induced to stress, significant production of ROS happens, which results in oxidative damage, and cell death (Chu et al. 2020). ROS activity and its related effects can be directly measured to estimate this oxidative stress. Additionally, many cell organelles, including peroxisomes and mitochondria, produce ROS as a metabolic byproduct (Samal and Subudhi 2024). There have been repeated reports of oxidative bursts in plants exposed to harmful concentrations of NPs. According to Holghoomi and Colagar (2024), exposure to various NPs like Ag, ZnO, and Al_2O_3 could induced reactive nitrogen species and H_2O_2 in duckweed corn and tobacco bright yellow cells. In tobacco BY2 cells, Al_2O_3 NPs also induced the production of superoxide anion, one of the highly reactive forms of ROS (Samadi et al. 2020). Inhibition of germination, elimination of shoot elongation, and even plant death are just a few of the detrimental effects that NPs can have on plants (Lv et al. 2019). Seed germination is important for productivity and numerous studies have demonstrated the detrimental effects of NPs induced the seed germination inhibition (Kumar et al. 2019). Cucumber, lettuce, and radish seeds were incubated with metal oxide nanoparticles (NiO, CuO, Fe_2O_3, TiO_2, and Co_3O_4) to test their phytotoxic effects. All three plants demostrate a noticeable reduction in seed germination on exposure to NiO and CuO NPs (Khan et al. 2019). Nanoparticles (NPs) produce harmful effects on plants because they stick onto seed surfaces which release free metallic ions during this process. The application of NPs leads to seed germination reduction as well as detrimental effects on multiple plant growth characteristics including biomass production shoot and root elongation and leaf number (Samal and Subudhi 2024).

Scientific evaluation of plant toxicity depends heavily on the measurement of plant hormone levels and activities. Scientific studies which examined Bt-transgenic and non-transgenic cotton leaves as well as a control group revealed cerium oxide nanoparticles (CeO_2 NPs) did not produce significant changes in indole-3-acetic acid, abscisic acid, and gibberellic acid levels (Guo et al. 2024). Trans-zeatin-riboside (t-ZR) concentrations found in leaves of traditional cotton plants subjected to 500 mg/L CeO_2 NPs dropped by 25% in comparison to plants that were not treated. A research study found that γ-Fe_2O_3 nanoparticles decreased levels of IAA substances and ABA compounds in transgenic rice roots and non-transgenic rice roots (Dhahri et al. 2025).

3.5 Essential Factors That Determine the Effective Adsorption of NPs

The successful introduction of nanoparticles into physiological environments requires both high stability and clarity (Ma et al. 2024). The speed at which NPs accumulate within plant roots depends on elements found in the environment alongside the characteristics of these NPs (Aqeel et al. 2023). Studies showed that the mixture of silver sulfide (Ag_2S) nanoparticles, potassium chloride, and ammonium thiosulfate treatment led to higher Ag NP accumulation in *Lactuca sativa* roots and shoots (Fetyan et al. 2024). According to Mushtaq et al. (2020), soil organic matter has shown its ability to lower cerium dioxide (CeO_2) nanoparticles uptake in *Zea mays* roots.

3.6 Conclusion

Nanoparticles play an essential role in agriculture by enhancing nutrient transport as well as improving pest control abilities. The accumulation of nanoparticles inside plants leads to disruptions in functions such as photosynthesis, respiration and nutrient absorption, thus creating potential damage to food networks and ecosystems. The plant mobility and NP toxicity depend on four critical characteristics, including particle size, surface areas, encapsulation materials, and hydrophobicity. Despite their beneficial effects on agricultural production, the environmental persistence of NPs brings potential ecological issues. Efficient, sustainable applications need specific optimization techniques that focus on reducing toxicity risks. Researchers should focus on formulations of nanoparticles based on natural polymers and plant-based materials that are both bio-based and biodegradable with minimum negative environment effect. The use of friendly alternatives approaches rather conventional nanoparticle materials may cope both environmental challenges and production of toxic substances. Long-term field research must be evaluated to determining that how NPs may impact soil microorganisms and plant health together ensuring food safety through different agricultural systems. Nanotechnology requires corresponding adjustments to regulatory frameworks that should develop safe thresholds and standardized testing protocols to be used in modern agriculture. The development of sustainable nanoparticle applications required combined efforts between nanotechnology experts, professionals from environmental science, plant biology and policymakers for achieving objectives in both crop yields and ecosystem health protection. Future investigations must explore eco-friendly nanoparticle formulations because they may help in creating harmony between agricultural productivity and environmental safety.

References

Adhikari K, Timilsina A, Chen H (2024) The effects of size and surface-coating of CuO-nanoparticles on extractable Cu and enzyme activities in soil. Soil Environ Health 2(1): 100065

Ahmed B, Rizvi A, Ali K, Lee J, Zaidi A, Khan MS, Musarrat J (2021) Nanoparticles in the soil–plant system: a review. Environ Chem Lett 19:1545–1609

Ahmed SF, Kumar PS, Ahmed B, Mehnaz T, Shafiullah GM, Duong XQ, Mofijur M, Badruddin IA, Kamangar S (2024) Carbon-based nanomaterials: characteristics, dimensions, advances and challenges in enhancing photocatalytic hydrogen production. Int J Hydrog Energy 52:424–442

Alhaithloul HAS, Ali B, Alghanem SMS, Zulfiqar F, Al-Robai SA, Ercisli S, Yong JWH, Moosa A, Irfan E, Ali Q (2023) Effect of green-synthesized copper oxide nanoparticles on growth, physiology, nutrient uptake, and cadmium accumulation in Triticum aestivum (L.). Ecotoxicol Environ Saf 268:115701

Ameen F, Alsamhary K, Alabdullatif JA, ALNadhari, S. (2021) A review on metal-based nanoparticles and their toxicity to beneficial soil bacteria and fungi. Ecotoxicol Environ Saf 213:112027

Aqeel M, Khalid N, Nazir A, Irshad MK, Hakami O, Basahi MA, Alamri S, Hashem M, Noman A (2023) Foliar application of silver nanoparticles mitigated nutritional and biochemical perturbations in chilli pepper fertigated with domestic wastewater. Plant Physiol Biochem 194 (November 2022):470–479. https://doi.org/10.1016/j.plaphy.2022.12.005

Aseel DG, Ibrahim OM, Abdelkhalek A (2024) Biosynthesized silver nanoparticles mediated by Ammi visnaga extract enhanced systemic resistance and triggered multiple defense-related genes, including SbWRKY transcription factors, against tobacco mosaic virus infection. BMC Plant Biol 24(1):1–16. https://doi.org/10.1186/s12870-024-05449-y

Aslam MM, Okal EJ, Waseem M (2023) Cadmium toxicity impacts plant growth and plant remediation strategies. Plant Growth Regul 99(3):397–412

Bai X, Wang S, Yan X, Zhou H, Zhan J, Liu S, Sharma VK, Jiang G, Zhu H, Yan B (2019) Regulation of cell uptake and cytotoxicity by nanoparticle core under the controlled shape, size, and surface chemistries. ACS Nano 14(1):289–302

Bindra P, Sharma S, Sahu BK, Bagdwal H, Shanmugam V, Singh M (2023) Targeted nutrient application to tomato plant with MOF/Zeolite composite wrapped with stimuli-responsive biopolymer. Mater Today Commun 34(October 2022):105264. https://doi.org/10.1016/j.mtcomm.2022.105264

Budhani S, Egboluche NP, Arslan Z, Yu H, Deng H (2019) Phytotoxic effect of silver nanoparticles on seed germination and growth of terrestrial plants. J Environ Sci Health C 37(4):330–355

Carlson C, Hussain SM, Schrand AMK, Braydich-Stolle L, Hess KL, Jones RL, Schlager JJ (2008) Unique cellular interaction of silver nanoparticles: size-dependent generation of reactive oxygen species. The journal of physical chemistry B 112(43):13608–13619

Cartwright A, Jackson K, Morgan C, Anderson A, Britt DW (2020) A review of metal and metal-oxide nanoparticle coating technologies to inhibit agglomeration and increase bioactivity for agricultural applications. Agronomy 10(7):1018
Chen H (2018) Metal based nanoparticles in agricultural system: behavior, transport, and interaction with plants. Chem Spec Bioavailab 30(1):123–134
Chen Y-Q, Xue M-D, Li J-L, Huo D, Ding H-M, Ma Y (2024) Uncovering the importance of ligand mobility on cellular uptake of nanoparticles: insights from experimental, computational, and theoretical investigations. ACS Nano 18(8):6463–6476
Chen D, Cui L, Rong H, Liu D, Fang Y, Sithole TJ, Wang Y, Lv Y, Lv Z, Xiang T (2025) Effect of hydrophobic nanoparticle size on corrosion resistance of superhydrophobic mortar. Constr Build Mater 467:140282
Chu XY, Yang SZ, Zhu MQ, Zhang DY, Shi XC, Xia B, Yuan Y, Liu M, Wu JW (2020) Isorhapontigenin improves diabetes in mice via regulating the activity and stability of PPARγ in adipocytes. J Agric Food Chem 68(13):3976–3985
Cromwell WA, Yang J, Starr JL, Jo Y-K (2014) Nematicidal effects of silver nanoparticles on root-knot nematode in bermudagrass. J Nematol 46(3):261
Dejene BK, Geletaw TM (2024) A review of plant-mediated synthesis of zinc oxide nanoparticles for self-cleaning textiles. Res J Text Appar 28(4):879–892
Dev A, Srivastava AK, Karmakar S (2018) Nanomaterial toxicity for plants. Environ Chem Lett 16: 85–100
Dhahri R, Benamara M, Bouzidi S, Ben Moussa S, Alzahrani AYA, Nassar KI, Zahmouli N, Elkenany EB, Al-Syadi AM (2025) Effect of Gd doping on the microstructure and electrical characteristics of Maghemite (γ-Fe_2O_3) ceramics. J Sol-Gel Sci Technol 113(1):225–242
Djanaguiraman M, Anbazhagan V, Dhankher OP, Prasad PVV (2024) Uptake, translocation, toxicity, and impact of nanoparticles on plant physiological processes. Plants 13(22):3137
Ekim R, Arikan B, Alp-Turgut FN, Koyukan B, Ozfidan-Konakci C, Yildiztugay E (2024) Polyvinylpyrrolidone-coated copper nanoparticles dose-dependently conferred tolerance to wheat under salinity and/or drought stress by improving photochemical activity and antioxidant system. Environ Res 241:117681
Falco WF, Scherer MD, Oliveira SL, Wender H, Colbeck I, Lawson T, Caires ARL (2020) Phytotoxicity of silver nanoparticles on Vicia faba: evaluation of particle size effects on photosynthetic performance and leaf gas exchange. Sci Total Environ 701:134816
Fetyan NAH, Essa TA, Salem TM, Taha AA, Elgobashy SF, Tharwat NA, Elsakhawy T (2024) Promising eco-friendly nanoparticles for managing bottom rot disease in lettuce (Lactuca sativa var. Longifolia). Microbiol Res 15(1):196–212
Gordillo-Delgado F, Zuluaga-Acosta J, Restrepo-Guerrero G (2020) Effect of the suspension of Ag-incorporated TiO2 nanoparticles (Ag-TiO2 NPs) on certain growth, physiology and phytotoxicity parameters in spinach seedlings. PLoS One 15(12):e0244511
Gu H, Shui X, Zhang Y, Zeng T, Yang J, Wu Z, Zhang X, Yang N (2024) Porous carbon scaffolded Fe-based alloy nanoparticles for electrochemical quantification of acetaminophen and rutin. Carbon 221:118954
Guo Z-H, Liu H, Zheng S-Q, Wang X-P, Xie Z-H, Zhang Z-M, Liu J-R, Li J-M, Lin S-Y, Gu C (2024) Effects of exogenous GA, IAA, ABA and ethylene on pear (Pyrus pyrifolia) fruit during different development stages. Plant Growth Regul 104(1):549–561
Haidri I, Ishfaq A, Shahid M, Hussain S, Shahzad T, Shafqat U, Mustafa S, Mahmood F (2024a) Enhancement of antioxidants' enzymatic activity in the wheat crop by Shewanela sp. mediated zinc oxide nanoparticles against heavy metals contaminated wastewater. J Soil Sci Plant Nutr 24:1–22
Haidri I, Ishfaq A, Shahid M, Shahzad T, Hussain S, Mahmood F (2024b) Green nanoparticles for textile wastewater treatment: the current insights. In: Microbes based approaches for the management of hazardous contaminants. John Wiley & Sons, Inc., Hoboken, NJ, USA pp 277–292. https://doi.org/10.1002/9781119851158.ch18

Hafez M, Khalil HF (2024) Nanoparticles in sustainable agriculture: recent advances, challenges, and future prospects. Communications in Soil Science and Plant Analysis 55(14):2181–2196

Halfadji A, Naous M, Rajendrachari S, Ceylan Y, Ceylan KB, Shekar PVR (2024) Effective investigation of electro-catalytic, photocatalytic, and antimicrobial properties of porous CuO nanoparticles green synthesized using leaves of Cupressocyparis leylandii. J Mol Struct 1301: 137318

Holghoomi R, Colagar AH (2024) Applications of biocompatible nanoparticles in plant biotechnology for enhanced secondary metabolite biosynthesis. Inorg Chem Commun 167:112753

Huang Y-W, Cambre M, Lee H-J (2017) The toxicity of nanoparticles depends on multiple molecular and physicochemical mechanisms. Int J Mol Sci 18(12):2702

Hussain A, Rizwan M, Ali S, Rehman MZ, Ur Qayyum MF, Nawaz R, Ahmad A, Asrar M, Ahmad SR, Alsahli AA, Alyemeni MN (2021) Combined use of different nanoparticles effectively decreased cadmium (Cd) concentration in grains of wheat grown in a field contaminated with Cd. Ecotoxicol Environ Saf 215:112139. https://doi.org/10.1016/j.ecoenv.2021.112139

Irfan M, Aslam H, Maqsood A, Tazeen SK, Mahmood F, Shahid M (2023) Changes in plant microbiome in response to abiotic stress. In: Plant microbiome for plant productivity and sustainable agriculture. Springer, pp 99–119. Singapore: Springer Nature Singapore

Ishfaq A, Shahid M, Nawaz M, Ibrar D, Hussain S, Shahzad T, Mahmood F, Rais A, Gul S, Gaafar A-RZ (2023) Remediation of wastewater by biosynthesized manganese oxide nanoparticles and its effects on development of wheat seedlings. Front Plant Sci 14:1263813

Ishfaq A, Haidri I, Shafqat U, Khan I, Iqbal M, Mahmood F, Hassan MU (2025) Impact of biogenic zinc oxide nanoparticles on physiological and biochemical attributes of pea (Pisum sativum L.) under drought stress. Physiol Mol Biol Plants 0123456789. https://doi.org/10.1007/s12298-024-01537-3

Javan M, Ameri A, Selahvarzi Y, Sayyad-Amin P (2024) TiO2 NPs as a promising strategy for crop conservation resulting from deficit irrigation in Fragaria× ananassa Cv. Camarosa. Commun Soil Sci Plant Anal 56:1–17

Jeon S, Zhang Y, Castillo C, Nava V, Ristroph K, Therrien B, Meza L, Lowry GV, Giraldo JP (2024) Targeted delivery of sucrose-coated nanocarriers with chemical cargoes to the plant vasculature enhances long-distance translocation. Small 20(7):2304588

Joudeh N, Linke D (2022) Nanoparticle classification, physicochemical properties, characterization, and applications: a comprehensive review for biologists. J Nanobiotechnol 20(1):262

Juárez-Maldonado A, Ortega-Ortíz H, Morales-Díaz AB, González-Morales S, Morelos-Moreno Á, Cabrera-De la Fuente M, Sandoval-Rangel A, Cadenas-Pliego G, Benavides-Mendoza A (2019) Nanoparticles and nanomaterials as plant biostimulants. Int J Mol Sci 20(1):162

Kang M, Bai X, Liu Y, Weng Y, Wang H, Ye Z (2024) Driving role of zinc oxide nanoparticles with different sizes and hydrophobicity in metabolic response and eco-corona formation in sprouts (Vigna radiata). Environ Sci Technol 58(22):9875–9886

Khan Z, Shahwar D, Ansari MKY, Chandel R (2019) Toxicity assessment of anatase (TiO2) nanoparticles: a pilot study on stress response alterations and DNA damage studies in Lens culinaris Medik. Heliyon 5(7):e02069

Kohan-Baghkheirati E, Geisler-Lee J (2015) Gene expression, protein function and pathways of Arabidopsis thaliana responding to silver nanoparticles in comparison to silver ions, cold, salt, drought, and heat. Nanomaterials 5(2):436–467

Kovalchuk NM, Johnson D, Sobolev V, Hilal N, Starov V (2019) Interactions between nanoparticles in nanosuspension. Adv Colloid Interf Sci 272:102020

Kumar G, Mathimani T, Rene ER, Pugazhendhi A (2019) Application of nanotechnology in dark fermentation for enhanced biohydrogen production using inorganic nanoparticles. International Journal of Hydrogen Energy 44(26):13106–13113

Li K, Li H, Xiao T, Zhang G, Liang A, Zhang P, Lin L, Chen Z, Cao X, Long J (2020) Zero-valent manganese nanoparticles coupled with different strong oxidants for thallium removal from wastewater. Front Environ Sci Eng 14(2). https://doi.org/10.1007/s11783-019-1213-5

Liu Y, Pan B, Li H, Lang D, Zhao Q, Zhang D, Wu M, Steinberg CEW, Xing B (2020) Can the properties of engineered nanoparticles be indicative of their functions and effects in plants? Ecotoxicol Environ Saf 205:111128

Lv J, Christie P, Zhang S (2019) Uptake, translocation, and transformation of metal-based nanoparticles in plants: recent advances and methodological challenges. Environ Sci Nano 6(1):41–59

Ma X, Tian Y, Yang R, Wang H, Allahou LW, Chang J, Williams G, Knowles JC, Poma A (2024) Nanotechnology in healthcare, and its safety and environmental risks. J Nanobiotechnol 22(1): 715

Miguel Sábio R, Corrêa Carvalho G, Li J, Chorilli M, Santos HA (2024) Advanced porous materials for antimicrobial treatment. Nano Select 5(7–8):2300114

Mushtaq W, Shakeel A, Fazili MA, Chakrabartty I, Sevindik M (2020) Pros and cons of nanotechnology. In: Nanobiotechnology in agriculture: an approach towards sustainability. Springer, Cham, pp 207–222. https://doi.org/10.1007/978-3-030-39978-8_13

Mustafa S, Mahmood F, Shafqat U, Hussain S, Shahid M, Batool F, Elnour RO, Hashem M, Asseri TAY, Shahzad T (2023) The biosynthesis of nickel oxide nanoparticles: an eco-friendly approach for Azo Dye decolorization and industrial wastewater treatment. Sustainability 15(20):14965

Mustafa S, Hussain S, Shahzad T, Mahmood F (2025) Elucidating the phytotoxicity threshold of PGPR, ZnO/MnO-NCs and corncob biochar on physical and biochemical attributes of wheat under Cd stress. Physiol Mol Biol Plants 31:1–20

Naseem T, Durrani T (2021) The role of some important metal oxide nanoparticles for wastewater and antibacterial applications: a review. Environ Chem Ecotoxicol 3:59–75

Oh E, Liu R, Nel A, Gemill KB, Bilal M, Cohen Y, Medintz IL (2016) Meta-analysis of cellular toxicity for cadmium-containing quantum dots. Nat Nanotechnol 11(5):479–486

Oriekhova O, Stoll S (2016) Stability of uncoated and fulvic acids coated manufactured CeO2 nanoparticles in various conditions: from ultrapure to natural Lake Geneva waters. Sci Total Environ 562:327–334

Osman AI, Zhang Y, Farghali M, Rashwan AK, Eltaweil AS, Abd El-Monaem EM, Mohamed IMA, Badr MM, Ihara I, Rooney DW (2024) Synthesis of green nanoparticles for energy, biomedical, environmental, agricultural, and food applications: a review. Environ Chem Lett 22(2):841–887

Park JH, Choppala GK, Bolan NS, Chung JW, Chuasavathi T (2011) Biochar reduces the bioavailability and phytotoxicity of heavy metals. Plant and soil 348(1):439–451

Pattnaik BK, Sahu C, Moulick D (2024) Nanoparticles in soil–water–plant inference. In: The nanotechnology driven agriculture. CRC Press, Boca Raton, pp 85–105

Rajput V, Minkina T, Mazarji M, Shende S, Sushkova S, Mandzhieva S, Jatav H (2020) Accumulation of nanoparticles in the soil-plant systems and their effects on human health. Annals of Agricultural Sciences 65(2):137–143

Rastogi A, Zivcak M, Sytar O, Kalaji HM, He X, Mbarki S, Brestic M (2017) Impact of metal and metal oxide nanoparticles on plant: a critical review. Front Chem 5:78. https://doi.org/10.3389/fchem.2017.00078

Rastogi A, Zivcak M, Tripathi DK, Yadav S, Kalaji HM (2019) Phytotoxic effect of silver nanoparticles in Triticum aestivum: improper regulation of photosystem I activity as the reason for oxidative damage in the chloroplast. Photosynthetica 57(1):209

Raza MAS, Zulfiqar B, Iqbal R, Muzamil MN, Aslam MU, Muhammad F, Amin J, Aslam HMU, Ibrahim MA, Uzair M, Habib-ur-Rahman M (2023) Morpho-physiological and biochemical response of wheat to various treatments of silicon nano-particles under drought stress conditions. Sci Rep 13(1):1–13. https://doi.org/10.1038/s41598-023-29784-6

Rizwan HM, Shafqat U, Ishfaq A, Batool F, Mahmood F (2025) Identifying the phytotoxicity of biosynthesized metal oxide nanoparticles and their impact on antioxidative enzymatic activity in maize under drought stress. Plants 14:1–22

Samadi S, Saharkhiz MJ, Azizi M, Samiei L, Ghorbanpour M (2020) Multi-walled carbon nanotubes stimulate growth, redox reactions and biosynthesis of antioxidant metabolites in Thymus daenensis celak. in vitro. Chemosphere 249:126069

Samal RR, Subudhi U (2024) Biochemical and biophysical interaction of rare earth elements with biomacromolecules: a comprehensive review. Chemosphere 357:142090
Sanzari I, Leone A, Ambrosone A (2019) Nanotechnology in plant science: to make a long story short. Front Bioeng Biotechnol 7:120
Saod WM, Hamid LL, Alaallah NJ, Ramizy A (2022) Biosynthesis and antibacterial activity of manganese oxide nanoparticles prepared by green tea extract. Biotechnol Rep 34(April): e00729. https://doi.org/10.1016/j.btre.2022.e00729
Sarkar MM, Saha P, Karmakar B, Toppo P, Paul P, Dua TK, Mathur P, Roy S (2025) Sugar-coating on the surface of silica nanoparticles attenuates the dose-and size-dependent toxicity of the nanoparticles for plant-based applications. Plant Physiol Biochem 223:109778
Saxena P, Sangela V, Ranjan S, Dutta V, Dasgupta N, Phulwaria M, Rathore DS, Harish. (2020) Aquatic nanotoxicology: impact of carbon nanomaterials on algal flora. Energy Ecol Environ 5: 240–252
Shafqat U, Maqsood A, Ishfaq A, Mustafa S, Rasheed Y, Mahmood F, Hassan MU, Al-Khayri JM, Aldaej MI, Sattar MN, Rezk AAS, Almaghasla MI, Shehata WF, Shalaby TA (2023) Green nanotechnology for plant bacterial diseases management in cereal crops: a review on metal-based nanoparticles. Notulae Botanicae Horti Agrobotanici Cluj-Napoca 51(3):1–21. https://doi.org/10.15835/nbha51313333
Shafqat U, Yasin MU, Shahid M, Hussain S, Shahzad T, Mahmood F, Ishfaq A, Nawaz M, Shah AN, Ali HM (2024) Evaluating the impact of biogenic nanoparticles and pesticide application in controlling Cotton Leaf Curl Virus Disease (CLCuD) in Cotton (Gossypium hirsutum L.). Chem Biol Technol Agric 11:149
Sheikh M, Makkad S, Shende S, Deshmukh M (2024) Antimicrobial efficacy of metal-doped titanium dioxide nanoparticles: a comprehensive review. Int J Pharm Investig 14(4):1042
Singh AV, Laux P, Luch A, Sudrik C, Wiehr S, Wild A-M, Santomauro G, Bill J, Sitti M (2019) Review of emerging concepts in nanotoxicology: opportunities and challenges for safer nanomaterial design. Toxicol Mech Methods 29(5):378–387
Singh M, Goswami SP, Sachan P, Sahu DK, Beese S, Pandey SK (2024) Nanotech for fertilizers and nutrients-improving nutrient use efficiency with Nano-enabled fertilizers. J Exp Agric Int 46(5):220–247
Sun C, Zhao C, Wang G, Han R (2025) Cerium oxide nanoparticles ameliorate Arabidopsis thaliana root damage under UV-B stress by modulating the cell cycle and auxin pathways. Protoplasma 262:1–19
Tripathi DK, Singh S, Singh S, Pandey R, Singh VP, Sharma NC, Prasad SM, Dubey NK, Chauhan DK (2017) An overview on manufactured nanoparticles in plants: uptake, translocation, accumulation and phytotoxicity. Plant Physiol Biochem 110:2–12
Wahab A, Bibi H, Batool F, Muhammad M, Ullah S, Zaman W, Abdi G (2024) Plant growth-promoting rhizobacteria biochemical pathways and their environmental impact: a review of sustainable farming practices. Plant Growth Regulation 104(2):637–662
Wang A, Ma Y, Zhao D (2024) Pore engineering of porous materials: effects and applications. ACS Nano 18(34):22829–22854
Wong K, Nie Z, Wong M, Wang Y, Liu J (2024) Metal–drug coordination nanoparticles and hydrogels for enhanced delivery. Adv Mater 36(26):2404053
Yadav A, Yadav K (2024) Exploring the effect of engineered nanomaterials on soil microbial diversity and functions: a review. J Environ Nanotechnol 13(1):48–62
Yang J, Cao W, Rui Y (2017) Interactions between nanoparticles and plants: phytotoxicity and defense mechanisms. J Plant Interact 12(1):158–169
Zhang H-Y, Su W-H (2024) Classification, uptake, translocation, and detection methods of nanoparticles in crop plants: a review. Environ Sci Nano 11:1847
Zhang H, Xu F, Xue J, Chen S, Wang J, Yang Y (2020) Enhanced removal of heavy metal ions from aqueous solution using manganese dioxide-loaded biochar: behavior and mechanism. Sci Rep 10(1):1–13. https://doi.org/10.1038/s41598-020-63000-z

Chapter 4
Role of Nanotechnology in Plant Growth and Development

Hem C. Joshi, Richa Joshi, Prabha Pant, and Babita Patni

Abstract Nanotechnology showing a promise in current agriculture, for providing new methods to promote plant growth and development. These materials help with efficient nutrient uptake, improve resistance to environmental stress, and speed up growth processes. Because of their tiny size, they have unique qualities that allow precise delivery of nutrients, agrochemicals, and genetic material directly to plant cells. This precision reduces waste and lessens environmental harm. For example, nanofertilizers release nutrients slowly and in a controlled way, ensuring plants get what they need over time and minimizing loss through leaching. This results in better root growth, improved nutrient absorption, and stronger overall plant health. Additionally, nanotechnology has led to smart delivery approaches that react to environmental signals like pH and moisture, making sure plants receive inputs exactly when they are needed. This targeted approach not only boosts plant growth but also cuts unnecessary use of chemical fertilizers, which can harm the environment. Furthermore, nanoparticles can improve plant resistance to both biotic and abiotic stresses. They can carry growth-promoting substances or directly interact with plant cells to kickstart defence mechanisms. The combination of nanotechnology and plant biotechnology creates new possibilities for genetic manipulation, leading to more efficient gene transfer methods. This could result in crops with better traits.

Keywords Nanofertilizers · Biotic and Abiotic stresses · Nanoparticles · Agrochemicals · Plant Growth

H. C. Joshi (✉) · R. Joshi · P. Pant
Department of Botany, D.S.B. Campus, Kumaun University, Nainital, Uttarakhand, India

B. Patni
High Altitude Plant Physiology Research Centre, Hemvati Nandan Bahuguna Garhwal University, Srinagar, Garhwal, Uttarakhand, India

R. N. Pudake et al. (eds.), *Phyto Nanotoxicology*, Nanotechnology in Plant Sciences 7, https://doi.org/10.1007/978-3-032-24518-2_4

4.1 Introduction

Manipulating matter at the nanoscale (1–100 nm) gives materials novel properties compared to their bulk counterparts. This behavior is especially useful in plant sciences and agriculture. Incorporating nanoparticles in various shapes can improve crop yield and enhance resistance against diseases through targeted delivery of fertilizers and pesticides (Sandhya et al. 2014; Jiaoyang et al. 2018). Additionally, nano-biotechnology combines these nanoscale innovations into useful biosensors and new biofuels that promote sustainable agriculture (Ruiqiang and Rattan 2015). Quantum dots are one aspect of nanotechnology that can improve light harvesting for photosynthesis and maximize plant productivity. Thus, using nanotechnology in agriculture not only boosts crop yields but also helps reduce chemical inputs, which is key for achieving environmental sustainability (Debasis et al. 2012). This field continues to change, promising new agricultural advancements.

4.1.1 Overview of Nanotechnology

Nanotechnology deals with the design, study, production, and use of structures, devices, and systems that have precise control over their shape and size at the nanometer scale. A nanometer is a thousand times thinner as compared to a human hair; it is about the thickness of ten hydrogen atoms in a line. Materials at this scale show unique properties like greater strength, lighter weight, higher chemical reactivity, and improved electrical conductivity (Tatineni and Kulkarni 2024). These new properties have led to major progress in medicine, electronics, environmental science, and, more recently, agriculture.

In 1959, physicist Richard Feynman first introduced the idea of nanotechnology in his lecture "There's Plenty of Room at the Bottom." He elaborated on manipulating atoms and molecules (Sanjit et al. 2024). The field gained significant momentum in the 1980s when scientists discovered scanning tunneling microscopy (STM), fullerenes, and carbon nanotubes (Sanjit et al. 2024). In the following decades, improvements in fabrication and characterization techniques greatly expanded the possibilities for exploring nanoscale materials in different applications. This led to innovation in various industries.

4.1.2 Applications of Nanotechnology Across Sectors

This technology has quickly evolved into a transformative discipline influencing various fields, including medicine, electronics, energy, and environmental science. It plays an important role in enhancing efficiency, precision, and sustainability across these sectors, making it one of the most significant technological advancements of recent decades. In medicine science, nanotechnology has opened new chances for improving therapeutic approaches. The use of nanoparticles allows for targeted delivery of drugs, which increases bioavailability while minimizing systemic toxicity (Muhammad et al. 2024). Furthermore, nanomaterials have improved medical imaging techniques, supporting the early diagnosis of diseases through enhanced visualization capabilities (Brij et al. 2024).

The electronics and energy sectors have also benefited greatly from nanoscale innovations. Incorporating nanomaterials into electronic components has led to the development of smaller, faster, and more energy-efficient devices (Parisa et al. 2024). Similarly, advancements in nanotechnology have contributed to creating novel materials that enhance energy conversion and storage systems, paving the way for more sustainable and powerful energy solutions. Environmental science has witnessed significant progress through nanotechnology applications aimed at addressing pollution and water scarcity. Nanomaterials are being employed in water purification systems and in environmental remediation processes, providing effective solutions to contamination challenges (Parisa et al. 2024). Despite its tremendous potential, the use of nanomaterials has concerns about nanotoxicity and possible environmental risks, emphasizing the continued research on its safe and responsible application across all domains (Muhammad et al. 2024).

4.1.3 Introduction to Nanotechnology in Agriculture

The use of nanotechnology in agriculture is sometimes called “agricultural nanotechnology.” This is a new field of research and development (Sandhya et al. 2014; Jiaoyang et al. 2018). With the global population growing and environmental challenges increasing, there is a greater need for sustainable farming practices. These practices should improve productivity while reducing resource use and environmental harm. Promising solutions include better crop yields, improved stress tolerance, fewer chemical inputs, and support for plant health (Ruiqiang and Rattan 2015; Debasis et al. 2012). Nano-fertilizers can improve nutrient delivery and use without raising soil toxicity or increasing treatment frequency (Indu et al. 2024). Nano-pesticides allow for effective pest control with less environmental impact through targeted delivery systems (Qudrat et al. 2024). Real-time watch on plant health and soil conditions is now possible with nano-sensors. This technology enables precise management of inputs (Pamela et al. 2024).

Nanoparticles can enhance the physiological processes of plants to improve stress tolerance against abiotic factors (Hafiza et al. 2024). But environmental impacts after long exposure and the safety of nanoparticles are yet to be fully known, so regulatory frameworks need to be holistic (Pamela et al. 2024). So, despite nanotechnology’s great potential in agriculture, its challenges must be addressed for sustainability and safety.

4.2 Types of Nanoparticles Used in Agriculture

Nanoparticles are valuable instruments in agriculture today. They feature high surface areas, extreme reactivity, and unique interactions with biological systems, especially at the molecular scale. Classification can be conducted based on nanoparticle composition, structure, or origin. Common types for agricultural applications are metal and metal oxide, carbon-based, and polymeric NPs. Understanding the classification, synthesis, and characterization of these nanoparticles is important for the development of effective agricultural applications.

4.2.1 Classification of Nanoparticles

4.2.1.1 Metal and Metal Oxide Nanoparticles

Each nanoparticle type exhibits distinct properties contributing to crop health improvement, enhanced plant growth, and environmental resilience. Silver NPs are recognized for their potent antimicrobial effects, aiding in plant disease management. Zinc oxide nanoparticles, essential as micronutrients, help in enzyme activation and stress resistance. Titanium dioxide nanoparticles support photosynthesis and UV protection due to their photocatalytic properties. Iron oxide NPs play an important role in chlorophyll production and energy transfer, helping to correct iron deficiencies and boost plant yield. The integration of these nanomaterials in agriculture represents a promising approach toward sustainable and precision farming practices (Indu et al. 2024).

4.2.1.2 Carbon-Based Nanomaterials

Carbon nanotubes, graphene, and fullerenes are examples of carbon-based nanomaterials, possess unique structural and functional characteristics that make them highly suitable for agricultural applications (Kaur et al. 2024). Carbon nanotubes (CNTs) are tubular nanostructures composed of carbon atoms arranged in a hexagonal pattern. Owing to their excellent tensile strength, electrical conductivity, and large surface area, CNTs are being explored as carriers for nutrients and agrochemicals to enhance plant growth. They have also been reported to improve seed germination as well as stimulate root and shoot development. Graphene, consisting of a single atomic layer of carbon organized in a two-dimensional honeycomb lattice, and its oxidized counterpart, graphene oxide, offer remarkable surface area, mechanical robustness, and chemical stability. These properties have prompted investigations into their roles in promoting plant growth, improving soil quality, and enabling efficient agrochemical delivery systems. Fullerenes, spherical carbon molecules with notable reactivity and antioxidant capabilities, have attracted considerable attention for their likely application for protection of plants against oxidative stress and to strengthen their tolerance to harmful environmental conditions (Kaur et al. 2024).

4.2.1.3 Polymeric Nanoparticles

These are nanoscale carriers made from organic polymers that offer multiple benefits, such as biocompatibility, biodegradability, and the ability to encapsulate and transport active compounds in a controlled manner (Ahmed et al. 2024). Their versatility has made them highly valuable in agriculture, where precise delivery of fertilizers, pesticides, and plant growth regulators is increasingly important. Chitosan nanoparticles, derived from chitosan-a natural biopolymer obtained from chitin in crustacean exoskeletons-are commonly used as delivery systems for agrochemicals. These nanoparticles improve the stability and controlled release of pesticides, fertilizers, and growth-promoting substances, thereby increasing their effectiveness while reducing their ill effects on the environment (Ahmed et al. 2024).

Another widely researched polymer is poly (lactic-co-glycolic acid) (PLGA). PLGA NPs are designed to degrade gradually, releasing their cargo after exposure to specific environmental conditions. This controlled and sustained release mechanism helps enhance the efficiency of agrochemicals such as nutrients and pesticides, ensuring targeted action with minimal waste (Ahmed et al. 2024). Starch-based nanoparticles, produced from the natural polysaccharide starch, are another eco-friendly option. These biodegradable nanoparticles are highly effective for encapsulating fertilizers, herbicides, and insecticides, supporting improved agricultural productivity. Their controlled release properties reduce the need for repeated applications, thereby lowering costs and minimizing environmental contamination (Ahmed et al. 2024).

4.2.2 Synthesis and Characterization

The synthesis and characterization of nanoparticles are very important steps for ensuring efficacy and safety for agricultural applications. There are several methods used to produce nanoparticles, each with advantages depending on the properties and applications sought.

Nanoparticles can be produced using several synthesis techniques, each offering distinct benefits and certain limitations. Chemical synthesis remains one of the most widely used approaches for generating metal and metal oxide nanoparticles. Methods such as chemical reduction, sol–gel processing, and precipitation provide fine control over particle size, morphology, and composition. For example, silver nanoparticles are often obtained by reducing silver ions with reducing agents like sodium borohydride, enabling the fabrication of uniform and stable nanostructures (Patel et al. 2024).

Physical synthesis, on the other hand, relies on mechanical or energy-based processes to create nanoparticles. Techniques including ball milling, laser ablation, and physical vapor deposition are commonly employed, particularly for producing highly defined nanomaterials. Carbon-based nanostructures such as CNTs and graphene are frequently manufactured using these physical approaches (Patel et al. 2024). A greener alternative is biological or "green" synthesis, which utilizes microorganisms, enzymes, or plant extracts to fabricate nanoparticles. In this process, biomolecules act as reducing and stabilizing agents, allowing metal nanoparticles to be generated in an green manner. This technique has gained significant attention for its sustainability, cost-effectiveness, and reduced ecological impact (Sharma et al. 2024).

In addition, polymeric nanoparticles are often prepared through polymerization-based methods, including emulsion polymerization, nanoprecipitation, and interfacial polymerization. These approaches are particularly useful in agricultural applications, where precise control over release kinetics and targeted delivery of agrochemicals can significantly enhance efficiency and minimize environmental contamination (Patel et al. 2024). Although traditional chemical synthesis techniques deliver high purity and precise control, they often raise concerns regarding environmental safety. In contrast, green synthesis methods offer a more sustainable pathway, though challenges such as lower yield and reduced purity remain important considerations (Patel et al. 2024).

4.2.2.1 Characterization Techniques

The characterization of nanoparticles is vital in understanding their size, shape, surface characteristics, and chemical composition. Accurate analysis of these properties is essential to confirm the success of synthesis and to predict their behavior in various applications. Multiple advanced analytical techniques are employed to study the physical and chemical features of nanomaterials. Transmission Electron Microscopy (TEM) is one of the most widely used methods, offering high-resolution imaging that enables detailed observation of particle size, morphology, and structural features, particularly for metallic and metal oxide nanoparticles (Kumar et al. 2024). Scanning Electron Microscopy (SEM) is similarly valuable, especially for examining surface topography and composition, making it highly applicable to polymeric and carbon-based nanostructures (Verma et al. 2024).

X-Ray Diffraction (XRD) plays a crucial role in identifying crystalline phases, providing insights into crystallite size and phase composition, particularly in metal and metal oxide nanoparticles (Verma et al. 2024). For colloidal and suspension-based nanoparticles, Dynamic Light Scattering (DLS) is frequently utilized to determine particle size distribution and evaluate system stability by measuring the scattering of light within a suspension (Kumar et al. 2024).

Fourier Transform Infrared Spectroscopy (FTIR) is commonly applied to detect functional groups on nanoparticle surfaces, proving especially useful for functionalized and polymer-based nanostructures (Kumar et al. 2024; Verma et al.

2024). Additionally, Zeta Potential analysis gives important information about surface charge, an important characteristic governing colloidal stability and suitability for biological and environmental applications (Kumar et al. 2024; Verma et al. 2024). Despite the availability of these advanced tools, the characterization of complex nanomaterials remains challenging due to their structural diversity and varied synthesis methods. Therefore, continued innovation in analytical techniques is essential to close the gap between current capabilities and the level of detail required to ensure the safe and efficient use of nanoparticles across diverse fields (Singh and Mehta 2024).

4.3 Mechanisms of Nanoparticle Interaction with Plants

Nanoparticles interact with plants in diverse ways that alter their growth, development, and general health. Such mechanisms are vital in the development of effective agricultural applications. Studying these interactions is crucial for estimating both the benefits and potential risks of nanotechnology in agriculture. This includes how nanoparticles are absorbed by the roots or leaves, their translocation through vascular tissues such as the xylem and phloem, and their accumulation in different plant organs (Aracena Santos et al. 2024; Hafiza Zulfiqar et al. 2024).

4.3.1 Uptake and Translocation of Nanoparticles

4.3.1.1 Pathways for Nanoparticle Entry into Plants

Nanoparticles can enter plants through multiple routes, most commonly via roots, leaves, and stomata. The entry mechanism is affected by factors such as particle size, surface charge, chemical properties, and the physiological condition of the plant (Tripathi et al. 2020). For root uptake, three major pathways have been identified: apoplastic, symplastic, and endocytic. In the apoplastic transfer, NPs move along cell walls and intercellular spaces without crossing the plasma membrane. This passive transport is typically favourable for larger nanoparticles that are unable to penetrate cellular membranes. The symplastic pathway, on the other hand, involves the movement of nanoparticles through the cytoplasm, facilitated by plasmodesmata, which enables direct cell-to-cell transport of smaller particles once they have entered the cytoplasm. The third route, endocytosis, is an active and energy-driven process in which root cells internalize nanoparticles by forming vesicles, thereby enabling the uptake of particles that cannot diffuse across membranes passively (Shang et al. 2022). Foliar uptake is another major route by which nanoparticles can enter the plant system, particularly through stomatal and cuticular penetration. Stomata are tiny pores found mainly on the underside of leaves that facilitate gas exchange and transpiration. When nanoparticles are applied via foliar sprays, they may enter through these open stomata, a process influenced by nanoparticle size, surface charge, and environmental factors like humidity and light (Nair et al. 2011). Additionally, the cuticle-a waxy layer covering leaf surfaces-can serve as a route of entry for nanoparticles, especially those that are hydrophobic or capable of interacting with cuticle components. These particles may pass through the cuticle and enter the epidermal cells directly (Nair et al. 2011).

Nanoparticles can also be absorbed through seed uptake, particularly during seed coating and germination. When nanoparticles are applied to seed surfaces, they may penetrate the seed coat and reach the embryo. During germination, the developing radicle and cotyledons may absorb these particles, affect early plant development, and offer a route for targeted delivery of nutrients or protective agents (Zhao et al. 2012).

Once inside the plant, NPs are distributed and accumulated in various tissues through the vascular system, primarily the xylem and phloem. Water and minerals are transported through from roots to aerial parts. Nanoparticles absorbed by the roots can enter the xylem vessels and move upward via the transpiration stream. This movement is passive and influenced by the evaporation of water from the leaf surfaces. Characteristics such as nanoparticle size, shape, and surface properties affect mobility within the xylem. Smaller NPs exhibit higher mobility, whereas larger or aggregated particles may become immobilized in the xylem (Wang et al. 2013; Ma et al. 2010).

In contrast, the phloem transports photosynthates (e.g., sugars) from leaves to other parts such as roots, fruits, and storage organs. Nanoparticles can enter the phloem either through foliar uptake or after being transported to leaves via the xylem. From there, they are loaded into the sieve tubes and transported to various sink tissues. This phloem transport is slower and driven by pressure gradients established through sugar movement. Nanoparticles may accumulate in storage organs or actively growing tissues, depending on their interaction with phloem-associated cells (Rico et al. 2011; Corredor et al. 2009).

Nanoparticles may be localized in various organelles within plant cells depending on their physicochemical properties and their specific pathway of entry. For example, nanoparticles that enter cells via endocytosis may be directed to the vacuole, where they can be either stored or degraded, thus playing a role in cellular detoxification or nutrient management (Tripathi et al. 2021). The extent and location of nanoparticle accumulation in plant tissues or organs can result in beneficial or adverse effects on plant health. For instance, the targeted accumulation of ZnO nanoparticles in leaf tissues has improved zinc nutrition, contributing to enhanced plant growth and increased resistance to environmental stress. However, when nanoparticles accumulate excessively, they may become toxic, interfering with key cellular processes such as respiration, photosynthesis, and enzymatic activity, ultimately leading to physiological disorders or growth inhibition in plants (Shang et al. 2022).

4.3.1.2 Factors Influencing Nanoparticle Uptake and Translocation

The uptake and movement of nanoparticles within plants are governed by several interacting factors. Among the most important are the physicochemical characteristics of the nanoparticles themselves, including their size, shape, surface charge, solubility, and functional surface modifications. Generally, smaller nanoparticles are absorbed and transported more efficiently than larger ones. Furthermore, surface functionalization-particularly with biocompatible polymers-can enhance internalization while minimizing potential toxic effects (Tripathi et al. 2021). Plant-related factors also play a crucial role. Both the species and its developmental stage influence nanoparticle absorption and distribution. Variations among species and cultivars result in different capacities for nanoparticle uptake, while younger or actively growing plants often exhibit higher rates of absorption and translocation compared to older, fully matured plants (Singh et al. 2022).

Environmental conditions also have a pronounced effect. Factors like soil pH, temperature, humidity, and light intensity can influence nanoparticle solubility and availability. For instance, high humidity may enhance stomatal opening, thus facilitating foliar uptake, while acidic soil conditions can alter the mobility of metal oxide NPs in the rhizosphere (Tripathi et al. 2021). Furthermore, the concentration and dosage of nanoparticles applied are critical. Although higher concentrations may promote increased uptake, they also raise the risk of phytotoxic effects and oxidative stress (Sharma et al. 2012).

4.3.2 *Impact on Plant Physiology*

Nanoparticles can significantly influence plant physiology by interacting with cellular structures, metabolic pathways, photosynthetic processes, respiration, and nutrient uptake. These effects may be either beneficial or harmful, based on the NP's type, size, concentration, surface chemistry, plant species, and environmental conditions (Ma et al. 2010; Siddiqi and Husen 2017).

4.3.2.1 Effects on Cellular Structures and Metabolic Processes

One of the primary sites of interaction between NPs and plant cells is the cell membrane. The integrity of this membrane can be compromised by NPs such as silver (Ag NPs) and zinc oxide (ZnO-NPs), which can cause lipid peroxidation through the generation of reactive oxygen species (ROS), causing to more membrane permeability, ion leakage, and even cell death under high stress (Tripathi et al. 2017). However, certain polymeric nanoparticles have been shown to interact with membrane components in a way that enhances resistance to abiotic stress by reinforcing membrane stability (Khodakovskaya et al. 2012a).

Once inside the cell, nanoparticles may localize in organelles such as chloroplasts, mitochondria, and the nucleus. Titanium dioxide nanoparticles (TiO_2-NPs), for example, have been observed to enhance photosynthesis by increasing light absorption and ATP production in chloroplasts (Hong et al. 2005). Nevertheless, excessive accumulation can damage thylakoid membranes and inhibit photosynthetic efficiency. In mitochondria, nanoparticles may disrupt the electron transport chain (ETC), leading to decreased respiration and elevated ROS production, although low concentrations may sometimes stimulate mitochondrial function (Shah and Belozerova 2009). In the nucleus, gold nanoparticles (AuNPs) have shown promise in delivering genetic materials for transformation, but their unintended interactions can lead to chromosomal abnormalities or DNA damage (Wang et al. 2016).

One of the frequent outcomes of nanoparticle interactions with plant cellular structures is the increased production of reactive oxygen species (ROS). Although ROS-including hydrogen peroxide (H_2O_2), superoxide anions (O_2^-), and hydroxyl radicals (OH^-)-are integral to normal signaling and defense mechanisms, excessive accumulation can lead to oxidative damage of proteins, lipids, and nucleic acids, ultimately impairing cellular processes (Mittler 2002). Certain nanoparticles, such as zinc oxide (ZnO-NPs) and titanium dioxide (TiO_2-NPs), can further stimulate ROS formation through photochemical reactions (Navarro et al. 2008). To mitigate oxidative stress, plants rely on enzymatic antioxidants, including superoxide dismutase (SOD), catalase (CAT), and peroxidase (POD). However, when ROS generation surpasses the capacity of these defense systems, it results in cellular injury, metabolic disturbances, and inhibition of plant growth (Rico et al. 2011; Roy et al. 2022).

Nanoparticles also affect enzymatic activities crucial for metabolism, like Ag NPs have been found to inhibit nitrate reductase activity, impairing nitrogen assimilation and reducing biomass accumulation (García-Sánchez et al. 2015). On the contrary, some nanoparticles enhance the activity of enzymes like catalase and glutathione reductase, aiding the plant in coping with oxidative stress and improving resilience under adverse environmental conditions (Taran et al. 2014).

4.3.2.2 Influence on Photosynthesis, Respiration, and Nutrient Uptake

The physiological activities of photosynthesis, respiration, and nutrient uptake are essential for plant survival and productivity. Nanoparticles, depending on their composition, size, surface characteristics, and concentration, can influence these functions either positively or negatively (Siddiqi and Husen 2017; Rico et al. 2011).

Photosynthesis is particularly sensitive to nanoparticle interactions. TiO_2-NPs, enhance light absorption across ultraviolet to visible wavelengths, because of their photocatalytic properties. This leads to increased carbon fixation, sugar production, and biomass accumulation by improving light energy conversion in chloroplasts (Hong et al. 2005; Yang et al. 2006). Additionally, some nanoparticles, such as carbon nanotubes and fullerene derivatives, have been shown to increase chlorophyll content, thereby boosting photosynthetic efficiency by facilitating CO_2 assimilation and improving electron transport (Khodakovskaya et al. 2009; Tripathi et al. 2017). However, overexposure to metal oxide nanoparticles like CuO or Al_2O_3 can degrade chlorophyll pigments, disrupt chloroplast ultrastructure, and reduce the net photosynthetic rate (Zhao et al. 2012).

Nanoparticles may also affect the performance of **Photosystem I (PSI)** and **Photosystem II (PSII)**, the core protein complexes responsible for light energy conversion. Low concentrations of engineered nanoparticles can accelerate electron transport between PSII and PSI, thereby enhancing the photosynthetic rate (Giraldo et al. 2014). However, excessive nanoparticle exposure may disturb the redox balance and induce photoinhibition, resulting in a decline in photosystem efficiency and the production of ROS within chloroplasts (Ma et al. 2010).

Respiration, particularly mitochondrial respiration, is another physiological process influenced by nanoparticles. Certain nanomaterials like cerium oxide nanoparticles (CeO_2-NPs) exhibit antioxidant properties and can protect mitochondrial membranes from oxidative damage by scavenging excess ROS, leading to enhanced respiratory performance and ATP synthesis (Turgut et al. 2018). In contrast, toxic nanoparticles at elevated concentrations may disrupt the mitochondrial electron transport chain (ETC), inhibit key respiratory enzymes, and reduce ATP production (Shah and Belozerova 2009). This suppression of energy metabolism may lead to cellular energy deficits, impairing plant development and stress tolerance. Moreover, an excess of ROS generated during nanoparticle exposure may damage mitochondrial DNA and proteins, further hamper respiration and increasing metabolic costs (Mittler 2002; Roy et al. 2022).

Nutrient uptake is often improved when nanoparticles are used in optimal concentrations. Zinc oxide nanoparticles (ZnO-NPs), for instance, not only increase the bioavailability of zinc in soil but also facilitate its uptake by plants, leading to better enzyme activation and stress resistance (Raliya et al. 2015). Nanoparticles can also function as **nanofertilizers**, releasing essential nutrients in a controlled and targeted manner to improve nutrient use efficiency and minimize environmental losses (Liu and Lal 2015). These materials can also modulate root system architecture-promoting elongation and lateral root formation—which increases root surface area and enhances nutrient acquisition (Dimkpa et al. 2012).

Furthermore, some nanoparticles interact with root cells to upregulate nutrient transporter genes, thereby improving the uptake of key macronutrients like nitrogen, phosphorus, and potassium (Tripathi et al. 2017). However, when present in excessive quantities or if composed of toxic metals, such as silver or copper, nanoparticles can impair root cell membranes, inhibit transporter activities, and disrupt nutrient homeostasis. This may result in micronutrient deficiencies, oxidative stress, and reduced overall plant growth and productivity (Navarro et al. 2008; Rico et al. 2011).

4.4 Applications of Nanotechnology in Enhancing Plant Growth

Nanotechnology offers substantial potential for advancing modern agriculture by stimulating plant growth, protecting crops from both biotic and abiotic stresses, and strengthening plants' inherent stress response mechanisms (Fig. 4.1). Within this field, nanofertilizers have emerged as a particularly promising innovation aimed at addressing the shortcomings of traditional fertilizers. By enabling targeted and efficient nutrient delivery, nanofertilizers enhance nutrient use efficiency (NUE), reduce negative environmental impacts, and eventually contribute to higher crop yields and overall agricultural productivity (Liu and Lal 2015; De Rosa et al. 2010).

Nanofertilizers can be categorized based on their structural design and nutrient delivery mechanisms. A prominent strategy involves the development of nanoscale nutrient formulations, in which essential macro- and micronutrients such as nitrogen (N), phosphorus (P), potassium (K), zinc (Zn), and iron (Fe) are engineered into nanoparticles. Owing to their minute size and exceptionally high surface-area-to-volume ratio, these nano-formulated nutrients are readily absorbed through plant roots or foliage, thereby improving nutrient bioavailability and uptake efficiency. For instance, zinc oxide nanoparticles (ZnO-NPs) and iron oxide nanoparticles

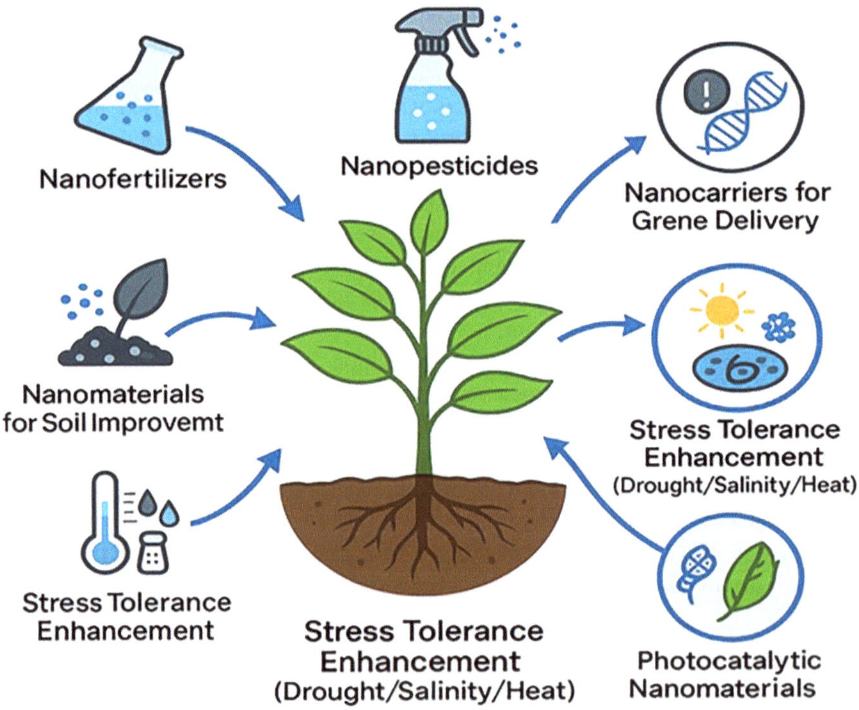

Fig. 4.1 Applications of nanotechnology in enhancing plant growth

(Fe_2O_3-NPs) have been reported to enhance the accumulation of Zn and Fe in major crops like wheat and maize, leading to improved plant growth and productivity (Raliya et al. 2015; Dimkpa et al. 2012).

Controlled-release nanofertilizers represent another innovative formulation where nutrients are embedded in a matrix or coated with biodegradable polymers that allow for slow and sustained nutrient release. This method ensures a time-dependent and need-based nutrient supply synchronized with plant developmental stages. It reduces the frequency of application and prevents nutrient deficits happened due to leaching, runoff, or volatilization, commonly seen in traditional fertilizers (Naderi and Danesh-Shahraki 2013; Liu and Lal 2015). These fertilizers not only enhance NUE but also reduce the risk of groundwater contamination.

Another important class of nanofertilizers involves nutrient encapsulation, where compounds such as urea are enclosed within nanocapsules composed of materials like chitosan, silica, or other biodegradable polymers. This protective encapsulation not only shields nutrients from premature degradation in the environment but also enables their controlled and sustained release, thereby enhancing fertilizer efficiency. Research has shown that urea-based encapsulated nanofertilizers can significantly decrease nitrogen losses through volatilization while improving nitrogen assimilation in plants, ultimately leading to greater biomass production and higher yields (Saharan et al. 2016; Subramanian et al. 2015).

Collectively, these innovative nanofertilizer platforms hold significant promise for transforming nutrient management in sustainable agriculture. By providing targeted, efficient, and responsive nutrient delivery tailored to plant physiological requirements, they offer the capacity to improve crop performance under both favorable and stress-prone conditions.

4.4.1 Advantages Over Conventional Fertilizers

Nanofertilizers offer a notable advantage over normal fertilizers by significantly increasing Nutrient Use Efficiency (NUE). Due to their nanoscale size and enhanced surface properties, nanofertilizers facilitate better penetration through plant tissues and more effective absorption at the cellular level. This improved delivery ensures that nutrients are utilized more efficiently by the plant, reducing losses due to leaching, volatilization, or immobilization in the soil. Additionally, many nanofertilizers possess controlled-release properties, which allow nutrients to be supplied gradually over time, matching the plant's growth requirements and thereby minimizing wastage (Liu and Lal 2015) (Fig. 4.2).

The environmental benefits of nanofertilizers are equally significant. Traditional fertilizers, when overused or improperly managed, often contribute to environmental degradation through processes such as nutrient runoff and leaching, which can cause eutrophication of nearby water bodies and deterioration of soil health. Nanofertilizers, on the other hand, are engineered for targeted delivery and slow release, requiring lower quantities to achieve the same or better outcomes. This

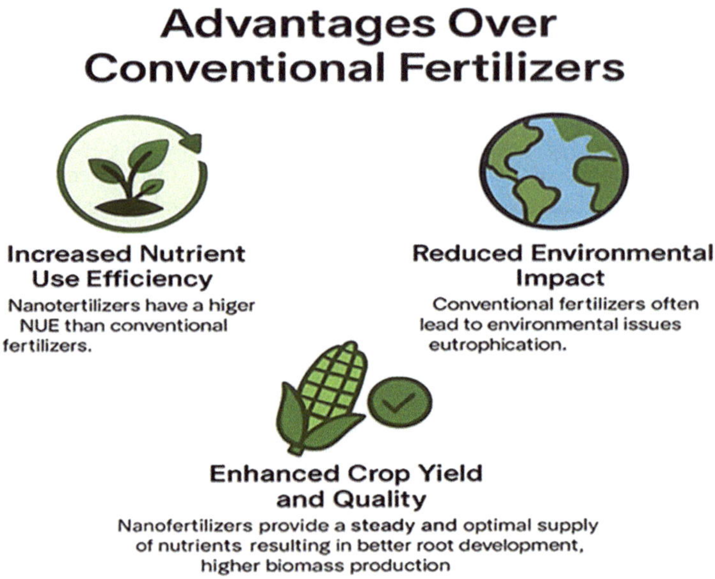

Fig. 4.2 Advantages of nanofertilizers over conventional fertilizers

targeted nutrient management reduces nutrient runoff, greenhouse gas emissions, and the potential for groundwater contamination, making nanofertilizers a more environmentally sustainable option (Dimkpa and Bindraban 2016).

Moreover, the application of nanofertilizers has been associated with enhanced crop yield and quality. By providing a consistent and adequate supply of essential nutrients throughout the plant's life cycle, nanofertilizers support vigorous root development, improved nutrient assimilation, and enhanced photosynthetic activity. This translates into increased biomass, better resistance to both abiotic and biotic stresses, and superior quality harvest. Studies across various crops have found that the use of nanofertilizers not only boosts yield but also improves attributes such as grain size, protein content, and micronutrient density in the edible parts of the plant (Ramesh et al. 2014; Prasad et al. 2017).

4.4.2 Growth Promotion and Stress Tolerance

Nanotechnology offers tremendous potential for promoting plant growth and strengthening tolerance to both abiotic and biotic stresses. Among its most impactful applications is the enhancement of drought resilience in crops. Hydrogel-based nanomaterials, for example, improve soil water retention and release moisture gradually, ensuring sustained hydration over extended periods. This slow release reduces irrigation frequency and helps plants endure water-limited conditions more effectively (Mishra et al. 2017). Beyond improving soil moisture availability, certain nanoparticles stimulate root elongation and branching, enabling plants to access deeper water sources. Silicon nanoparticles (Si-NPs) are particularly noteworthy, as they can activate stress-responsive genes associated with improved water-use efficiency and regulate transpiration by influencing stomatal activity, thereby reducing water loss (Rossi et al. 2019).

Nanoparticles are also valuable tools for enhancing salinity tolerance in plants. Elevated salt concentrations in soil impose osmotic stress and ion toxicity, leading to suppressed growth and reduced yields. Si-NPs have been shown to mitigate these adverse effects by increasing the activity of key antioxidant enzymes, including catalase, peroxidase, and superoxide dismutase, which neutralize ROS generated under salt stress (Suriyaprabha et al. 2012). Additionally, nanoparticles can facilitate selective ion transport, reduce sodium (Na^+) accumulation while promoting beneficial ions such as potassium (K^+). They also enhance the synthesis of osmoprotectants, such as proline and glycine betaine, crucial for maintaining osmotic balance and preserving cell turgor in saline environments (Khodakovskaya et al. 2012a, b). Collectively, these mechanisms support metabolic stability and sustained growth, enabling plants to perform better even under challenging high-salinity conditions.

4.5 Conclusion

The integration of nanotechnology into plant science and agriculture shows a promising frontier with far-reaching implications for sustainable crop production and global food security. This work has systematically explored how nanoparticles interact with plant systems, beginning with their uptake and translocation, which are governed by nanoparticle properties, plant species, developmental stage, environmental conditions, and concentration. These interactions influence key physiological and biochemical processes, including cellular structure integrity, metabolic pathways, photosynthesis, respiration, and nutrient assimilation.

The dual nature of nanoparticles as both growth-promoting agents and potential stressors highlights the importance of understanding their mode of action at the cellular and molecular levels. Beneficial effects, like increased enzymatic activity, antioxidant defense, chlorophyll content, and nutrient use efficiency, can significantly improve plant vigor and productivity when applied judiciously. However, concerns related to oxidative stress, phytotoxicity, and long-term ecological impacts necessitate careful risk assessment and regulatory oversight. Among the most promising applications is the development of nanofertilizers, which offer targeted and efficient nutrient delivery with reduced environmental impact. In addition, nanoparticles have shown significant potential in improving plant resilience to abiotic stresses like drought and salinity through enhanced water retention, ion regulation, and upregulation of stress-responsive genes.

While nanotechnology holds substantial promise for revolutionizing agriculture, its success depends on a balanced approach one that merges scientific innovation with environmental safety and socio-economic responsibility. Future efforts must focus on developing biodegradable and biocompatible nanomaterials, scaling up cost-effective delivery systems, and fostering interdisciplinary partnerships among researchers, policymakers, farmers, and industry stakeholders. As the field progresses, nanotechnology is expected to play a critical role in reshaping the agricultural landscape, contributing to increased crop yields, better resource management, and better resilience to climate challenges. This work underscores the need for continued research and informed application to ensure that nanotechnology not only meets the demands of modern agriculture but does so in a manner that is ecologically sound, economically viable, and socially acceptable.

References

Ahmed A, Khan MS, Javed R (2024) Polymeric nanoparticles for controlled agrochemical delivery. In: Nanotechnology for agricultural resilience, pp 89–112. Springer, Cham/Singapore

Aracena Santos T, Gómez R, Molina F (2024) Plant-nanoparticle interactions: uptake, translocation, and accumulation. In: Advances in nanobiotechnology in agriculture, pp 145–168. Springer, Cham

Brij M, Shalini R, Ayush S (2024) Nanotechnology-based innovations in medical diagnostics and imaging. In: Applications of nanobiotechnology in healthcare, pp 121–142. Springer, Cham

Corredor E, Testillano PS, Coronado MJ, González-Melendi P, Fernández-Pacheco R, Marquina C, Ibarra MR, de la Fuente JM, Rubiales D, Pérez-de-Luque A, Risueño MC (2009) Nanoparticle penetration and transport in living pumpkin plants: in situ subcellular identification. BMC Plant Biol 9:45

De Rosa MR, Monreal C, Schnitzer M, Walsh R, Sultan Y (2010) Nanotechnology in fertilizers. Nat Nanotechnol 5:91–94

Debasis M, Rajeev K, Nidhi S (2012) Nanotechnology in agriculture: a review. In: Advances in nanoscience and nanotechnology, pp 75–98

Dimkpa CO, Bindraban PS (2016) Fortification of micronutrients for efficient agronomic production: a review of nanofertilizer technologies. Agron Sustain Dev 29:1589–1600

Dimkpa CO, McLean JE, Latta DE, Manangón E, Britt DW, Johnson WP, Boyanov MI, Anderson AJ (2012) CuO and ZnO nanoparticles: phytotoxicity, metal speciation, and induction of oxidative stress in sand-grown wheat. J Nanopart Res 14:1125–1135

García-Sánchez S, Bernales I, Cristobal S (2015) Early plant responses to silver nanoparticles: impact on root development and auxin signaling pathways. In: Environmental science and technology, pp 5449–5457

Giraldo JP, Landry MP, Faltermeier SM, McNicholas TP, Iverson NM, Boghossian AA, Reuel NF, Hilmer AJ, Sen F, Brewster MM, Strano MS (2014) Plant nanobionics approach to augment photosynthesis and biochemical sensing. Nat Mater 13:400–408

Hafiza S, Anam R, Shahid M (2024) Enhancing crop stress tolerance using nanotechnology. In: Nanotechnological innovations in agriculture, pp 153–178

Hong F, Zhou J, Liu C, Yang F, Wu C, Zheng L, Yang P (2005) Effect of nano-TiO_2 on photochemical reaction of chloroplasts of spinach. Biol Trace Elem Res 105:269–279

Indu S, Ramesh G, Varsha M (2024) Role of metal and oxide nanoparticles in sustainable agriculture. In: Applications of nanomaterials in agri-biotechnology, pp 87–112. Springer, Singapore

Jiaoyang L, Meiying Z, Lian W (2018) Targeted delivery of agrochemicals using nanomaterials: challenges and future prospects. In: Nanotechnology and biosensors in agriculture, pp 99–118

Kaur M, Sharma V, Patel R (2024) Carbon-based nanomaterials: emerging tools in modern agriculture. In: Recent advances in nano-enabled agriculture, pp 211–234. Springer, Singapore

Khodakovskaya M, Dervishi E, Mahmood M, Xu Y, Li Z, Watanabe F, Biris AS (2009) Carbon nanotubes are able to penetrate plant seed coat and dramatically affect seed germination and plant growth. ACS Nano 3:3221–3227

Khodakovskaya MV, Silva KD, Nedosekin DA, Dervishi E, Biris AS, Shashkov EV, Galanzha EI, Zharov VP (2012a) Complex genetic, photothermal, and photoacoustic analysis of nanoparticle-plant interactions. Proc Natl Acad Sci 108:1028–1033

Khodakovskaya M, Dervishi E, Mahmood M, Xu Y, Li Z, Watanabe F, Biris AS (2012b) Carbon nanotubes induce growth enhancement of tobacco cells through the activation of stress-related genes. Small:273–282

Kumar V, Singh R, Arora N (2024) Advanced techniques for nanoparticle characterization in agricultural sciences. In: Nanomaterials and precision farming, pp 223–248. Springer, Cham

Liu R, Lal R (2015) Potentials of engineered nanoparticles as fertilizers for increasing agronomic productions. Sci Total Environ 514:131–139

Ma X, Geiser-Lee J, Deng Y, Kolmakov A (2010) Interactions between engineered nanoparticles (ENPs) and plants: phytotoxicity, uptake and accumulation. Sci Total Environ 408:3053–3061

Mishra V, Sharma R, Verma K (2017) Hydrogels in agriculture and their application. J Pharm Phytochem 2:85–90

Mittler R (2002) Oxidative stress, antioxidants and stress tolerance. Trends Plant Sci 7:405–410

Muhammad A, Fatima S, Irfan M (2024) Risk assessment and safe use of nanomaterials in environment and medicine. In: Environmental nanoscience and nanotechnology, pp 189–212. Springer, Cham

Naderi MR, Danesh-Shahraki A (2013) Nanofertilizers and their roles in sustainable agriculture. Int J Agric Crop Sci 5:2229–2232

Nair R, Varghese SH, Nair BG, Maekawa T, Yoshida Y, Kumar DS (2011) Nanoparticulate material delivery to plants. Plant Sci 179:154–163

Navarro E, Baun A, Behra R, Hartmann NB, Filser J, Miao A-J, Quigg A, Santschi PH, Sigg L (2008) Environmental behavior and ecotoxicity of engineered nanoparticles to algae, plants, and fungi: a review. Ecotoxicology 17:372–386

Pamela T, Jerome F, Anita B (2024) Nano-sensors for smart agriculture: potential and policy concerns. In: Smart agriculture and sustainable development, pp 55–78. Springer, Cham

Parisa K, Neeraj S, Kianoush F (2024) Advances in nanotechnology for electronics, energy, and environmental applications. In: Nanoscience for sustainable future, pp 33–61. Springer, Cham

Patel D, Sharma K, Mehra A (2024) Synthesis strategies for agricultural nanoparticles: from chemistry to green methods. In: Eco-friendly nanotechnology in agriculture, pp 65–98. Springer, Cham

Prasad R, Bhattacharyya A, Nguyen QD (2017) Nanotechnology in sustainable agriculture: recent developments, challenges, and perspectives. Front Microbiol 8:1014–1025

Qudrat U, Faizan A, Shazia B (2024) Nanopesticides: Advancements in pest management systems. In: Emerging trends in agrochemical nanotechnology, pp 123–144. Springer, Cham

Raliya R, Tarafdar JC, Biswas P (2015) Enhancing the mobilization of native phosphorus in the arid soil using ZnO nanoparticles synthesized by soil fungi. J Agric Food Chem 63:9128–9133

Ramesh M, Palanisamy K, Sharma N (2014) Nanofertilizer: a novel tool for sustainable agriculture. Proc Natl Acad Sci, India, Sect B Biol Sci 84:68–72

Rico CM, Majumdar S, Duarte-Gardea M, Peralta-Videa JR, Gardea-Torresdey JL (2011) Interaction of nanoparticles with edible plants and their possible implications in the food chain. J Agric Food Chem 59:3485–3498

Rossi L, Zhang W, Ma X (2019) Uptake, transformation, and physiological effects of metal-based nanoparticles in plants. Environ Sci Nano 6:2080–2092

Roy S, Das TK, Kole PC (2022) Nanoparticles in plants: uptake, transport and physiological activity in leaf and root. In: Protoplasma, pp 487–500

Ruiqiang L, Rattan L (2015) Nanobiotechnology in sustainable agriculture: role of biosensors and biofuels. In: Nanotechnology and plant sciences, pp 201–220

Saharan V, Sharma G, Yadav M, Choudhary MK, Sharma SS, Soni SL (2016) Synthesis and in vitro antifungal efficacy of Cu–chitosan nanoparticles against pathogenic fungi of tomato. Int J Biol Macromol 75:116–123

Sandhya R, Vishnu P, Harshita K (2014) Nanotechnology in agriculture: scope and current relevance. In: Nanoscience in food and agriculture, pp 11–34

Sanjit R, Harjeet K, Kavita B (2024) Evolution of nanotechnology: from Feynman's vision to agricultural revolutions. In: Foundations of nanotechnology, pp 1–20. Springer, Singapore

Shah V, Belozerova I (2009) Influence of metal nanoparticles on the soil microbial community and germination of lettuce seeds. Water Air Soil Pollut 197:143–148

Shang Y, Hasan MK, Ahammed GJ, Li M, Yin H, Zhou J (2022) Applications of nanotechnology in plant growth and crop protection: a review. Molecules 24:792

Sharma P, Bhatt D, Zaidi MGH, Saradhi PP, Khanna PK, Arora S (2012) Silver nanoparticle-mediated enhancement in growth and antioxidant status of *Brassica juncea*. Appl Biochem Biotechnol 167:2225–2233

Sharma S, Bansal M, Tripathi R (2024) Green synthesis of nanoparticles using plant extracts: agricultural applications and sustainability. In: Sustainable nanoscience in food and agriculture, pp 113–137. Springer, Singapore

Siddiqi KS, Husen A (2017) Plant response to engineered metal oxide nanoparticles. In: Nanomaterials in plants, algae and microorganisms: Concepts and controversies. Springer, Cham

Singh A, Kumar V, Pandey AK, Singh RP (2022) Plant–nanoparticle interactions: Uptake, translocation and impact on plant growth. In: Advances in plant nanotechnology. Springer, Singapore

Singh M, Mehta P (2024) Analytical challenges and future directions in nanoparticle characterization. In: Nanotechnology for sustainable agriculture, pp 271–295. Springer, Singapore

Subramanian KS, Manikandan A, Thirunavukkarasu M, Rahale CS (2015) Nano-fertilizers for balanced crop nutrition. In: Nanotechnologies in food and agriculture, pp 69–80

Suriyaprabha R, Karunakaran G, Yuvakkumar R, Rajendran V, Kannan N (2012) Silica nanoparticles for increased silica availability in maize (Zea mays. L) seeds under salinity stress. Curr Nanosci 8:676–682

Taran N, Storozhenko V, Svietlova N, Batsmanova L, Shvartau V, Kovalenko M (2014) Effect of zinc and copper nanoparticles on drought resistance of wheat seedlings. Nanoscale Res Lett 9:1–6. https://doi.org/10.1186/1556-276X-9-354

Tatineni S, Kulkarni V (2024) Fundamentals of nanotechnology: definitions and unique properties. In: Nanotechnology: principles and practices, pp 3–22. Springer, Singapore

Tripathi DK, Singh S, Singh VP, Prasad SM, Chauhan DK, Dubey NK (2017) Nanoparticles and their impact on photosynthesis in plants. In: Nanotechnology in plant sciences. Springer, Cham

Tripathi DK, Singh S, Singh VP, Prasad SM, Chauhan DK, Dubey NK (2020) Uptake, accumulation and toxicity of nanoparticles in plants. In: Nanomaterials in plants, algae and microorganisms: Concepts and controversies. Springer, Cham

Tripathi DK, Singh S, Singh VP, Prasad SM, Chauhan DK, Dubey NK (2021) Uptake, accumulation and toxicity of nanoparticles in plants: A review. In: Nanomaterials in plants, algae and microorganisms: Concepts and controversies. Springer, Cham

Turgut G, Koca FD, Yalçın S (2018) Cerium oxide nanoparticles and their role in alleviating oxidative stress in plants. In: Nanotechnology in plant sciences. Springer, Cham

Verma S, Kaushik R, Yadav D (2024) Characterization techniques for nano-agricultural materials: a practical guide. In: Handbook of nanotechnology in crop sciences, pp 197–222. Springer, Cham

Wang Z, Xie X, Zhao J, Liu X, Feng W, White JC, Xing B (2013) Xylem- and phloem-based transport of nanoparticles in plants. Environ Sci Technol 47:11757–11764. https://doi.org/10.1021/es403264k

Wang P, Lombi E, Zhao F-J, Kopittke PM (2016) Nanotechnology: A new opportunity in plant sciences. Trends in Plant Sci 21(8):699–712. https://doi.org/10.1016/j.tplants.2016.04.005

Yang F, Hong F, You W, Liu C, Gao F, Wu C, Yang P (2006) Influences of nano-anatase TiO_2 on the nitrogen metabolism of growing spinach. Biol Trace Elem Res 110:179–190. https://doi.org/10.1385/BTER:110:2:179

Zhao L, Sun Y, Hernandez-Viezcas JA, Hong J, Majumdar S, Niu G, Duarte-Gardea M, Peralta-Videa JR, Gardea-Torresdey JL (2012) Monitoring the environmental effects of CeO_2 and CuO nanoparticles on plant growth and physiology. Environ Sci Technol 46:11410–11418. https://doi.org/10.1021/es301479b

Zulfiqar H, Khan N, Yousaf M (2024) Mechanistic insights into nanoparticle interactions with plant systems. In: Nanotechnology applications in environmental and plant sciences, pp 179–202. Springer, Cham

Chapter 5
Physiological Processes for Uptake, Translocation, and Accumulation of Nanomaterials in Plants

Harshan Mahesh Kumar, Cannon Antony Fernandes, Vasantha Veerappa Lakshmaiah, and Praveen Nagella

Abstract Plants have the potential to uptake the nanoparticles (NPs) through roots or leaves. These nanoparticles can have a beneficial or detrimental effect on plants. Studies report that NPs can influence the physiological activities of the plants. They can improve the plant growth, nutritional uptake with improved crop yield. On the other hand, NPs can impose oxidative stress in plants and hinder their growth and photosynthesis. Nanoparticles are adsorbed and accumulated in plants; their specific effect depends on the type of NPs, their concentration, size, shape, environmental conditions, and plant species. Research towards a better understanding of nanoparticles impact on plants is required to optimise the conditions for its beneficial application of NPs in precision agriculture and to minimise their adverse effects on plants. Nano fertilizers and nano pesticides are widely used to mitigate the ill effects of chemicals used in the agricultural fields. Nanoparticles also serve as potential carriers of nutrients and their persistent release to improve their growth, metabolites, and crop yield. These NPs can also improve their immunity against pathogens as they boost the antioxidant properties, which play a significant role in plant immunity. This chapter focuses on the mechanism of nanoparticle uptake, translocation, and their effect on plants. This chapter also emphasises channelizing nanotechnology in agricultural science.

Keywords Apoplastic pathway · Symplastic pathway · Seed priming · Antioxidant · Hormonal regulation

H. M. Kumar · C. A. Fernandes · V. V. Lakshmaiah (✉) · P. Nagella
Department of Life Sciences, School of Sciences, Christ University, Bengaluru, India
e-mail: vasantha.vl@christuniversity.in

R. N. Pudake et al. (eds.), *Phyto Nanotoxicology*, Nanotechnology in Plant Sciences 7, https://doi.org/10.1007/978-3-032-24518-2_5

5.1 Introduction

From the inception of the idea by Richard Feynman in 1959, the field of nanotechnology has been consistently getting popular, showing a great deal of promise as a tool for sustainability in the twenty-first century. The application of nanotechnology has been observed in the production of energy, medicine, targeted drug delivery, catalysis, and various aspects in the field of agricultural sciences. According to the European Commission (EC, 2012), nanotechnology is one of the key enabling technologies that has potential for sustainable growth in any field. The term nanotechnology was coined by Norio Taniguchi in 1974, and can be defined as "Recognizing and utilizing the exclusive properties of the diverse matter at nanoscale dimensions (1–100 nm)". Nanotechnology is employed in the food industry with its implementation in food packaging materials to ensure its safety, quality, and shelf life (Mohammad et al. 2022). Nanotechnology is also used to detect microbial contaminations, enhance the bioavailability of nutrients, and improve mechanical barriers (Duncan 2011; Grumezescu and Holban 2018). Nanotechnology is also implemented extensively in the medical field, for example, in vitro diagnosis using nanoparticles as molecular imaging agents and input of cancer-based genetic changes and cell function changes (Haleem et al. 2023). Recent studies discuss the application of nanomaterials in the textile industry for achieving "smart textiles" (Shah et al. 2022). Researchers have found methods to chemically and physically integrate active nanomaterials that provide antimicrobial properties, such as silver (Ag), Cu_2O, chitosan, and others (Sun et al. 2019). Nanotechnology has also made its way into water treatment applications, including nanoscale filtration, adsorption capability of pollutants by nanoparticles, and contamination breakdown by nanoparticle catalysis (Tom 2021).

Nanoparticles (NPs), briefly, are particles at the scale of nanometres, which is one billionth of a meter (10^{-9}). The interesting and useful aspect of materials on this scale is their incredible properties that are otherwise not seen in macro materials. Thus, the principle of nanotechnology stems from the enhanced properties of nanomaterials. It enables manipulation of the matter in new structures with new or different properties, and its application in various fields is also through the structuring of devices at the nanoscale. We observe this effect in opaque materials becoming transparent at the nanoscale (copper), inert materials acting as efficient catalysts (platinum), insulators such as silicon becoming conductors, and so on. Gravity has little effect on nanomaterials, while electrostatic and covalent bonds show a greater influence (Moosa 2011). Nanoparticles also bear a property that allows for better adsorbent capacity with their increased surface area. For example, efficiency in removing contaminants, including heavy metals, inorganic and organic solutes, and toxins. It has proved to be a low-cost, quick treatment method (Tom 2021). Nanoparticles have been classified on numerous grounds, among which one of them is material-based, which includes (i) Carbon-based, which are NP's that contain carbon, (ii) Inorganic-based, which include metal and metal oxide NPs, (iii) Organic-based NPs include organic sourced NPs except for carbon-based NP's, and (iv) Composite-based NPs, which are multiphase NP's that contain either complex structures, or comprise nano materials in combination with larger materials (Jeevanandam et al. 2018). Apart from the nature of the material, NPs are classified based on their surface characteristics, size, morphology, uniformity, and agglomeration (Bowman 2017).

Sustainability is the grounds for majority of the research conducted today. Sustainable agriculture, in particular, is picking up pace quite rapidly, owing to the importance given to it during recent years as part of the 17 sustainable development goals of the United Nations. The ever-escalating world population, current effect on climate and environment, and increasing stress on water and energy demands all contribute towards the need for sustainability in agriculture and food security. The 2024 Revision of World Population Prospects was the latest revision of world population estimates by the Department of Economic and Social Affairs under the United Nations, in which demographers have estimated a growth of population in the coming decades, reaching as high as 10.3 billion by mid-2080 (UNDESA 2024). Having established the importance of advancing modern agriculture to the next level, nanotechnology shows tremendous scope, as discussed by various researchers (Arora et al. 2022). For instance, the use of chemical fertilizers has been rapidly decreasing, as their adverse effects, such as leaching out into water bodies, soil fertility damage, low nutrient uptake efficiency, and increased eutrophication of nearby water bodies, are being realised (Wilson et al. 2008; Zulfiqar et al. 2019; Kumari and Singh 2020). In response, nano-fertilizers and nano-biofertilizers have been recently explored, as they have benefited crop nutrition, reduced expenses, and caused less harm to the environment (Duhan et al. 2017). A recent analysis revealed a gain of about 18–29% by using nano-fertilizers in contrast to traditional fertilizers (Kah et al. 2018). The scope of nano-pesticides is also discussed widely, offering an efficient and sustainable disease control strategy (Usman et al. 2020). Other applications also include nano-biosensors to monitor physical and chemical conditions of soil, providing high sensitivity, rapid detection, and longer life (Scognamiglio 2013). Nanomaterials are also used in soil remediation, including techniques such as nano-assisted abiotic remediation, nano-assisted bioremediation, etc. (Usman et al. 2020).

This chapter focuses on the scope of nanotechnology, particularly in the area of plant uptake and their interaction, which can impact their growth and development, improve their defence and response to environmental stress. However, there are also harmful effects of these NPs on plant processes, which are influenced by the type, concentration, and plant species. There is a need for extensive research for a clearer understanding of the underlying principle through which plant-nanoparticle interactions take place. These fundamentals need to be explored to widen our knowledge of this contemporary technology for its application in many fields on a larger scale.

5.2 Mechanisms of Nanoparticle Uptake

To fully comprehend the working principle that governs the action of NPs, it is vital to understand the various mechanisms utilized in the uptake of NPs within plants. Generally, NPs are applied to plants as a spray and through soil application. The process by which NPs are adsorbed and transformed into plants and their effect on plant physiological processes is of significant research to mitigate their potential in the agriculture sector (Khan et al. 2022). A diagrammatic representation of the types of pathways involved in nanoparticle uptake is shown in Fig. 5.1.

5.2.1 Root Uptake of Nanoparticles

Multiple researchers have conducted experiments to show the accumulation of NPs through root adsorption. The uptake of NPs by the root system is by various means, like endocytosis, diffusion, and interaction with root exudates. The size, composition, and charge of these NPs can significantly influence their translocation. Initial interaction of NPs with the root system is adsorption, wherein the root system secretes mucus and organic acids, facilitating the positively charged nanoparticles to potentially penetrate plant roots. The nanoparticles are then greeted by many "root barriers"—cuticle, epidermis, cortex, endodermis, and Casparian strips. Nanoparticles less than 100 nm can penetrate through fine pores on the lateral roots. The size of the NPs is also an important characteristic, as it is the determining factor for entering through root tips, root cell wall pores, or plant stomata (Ali et al.

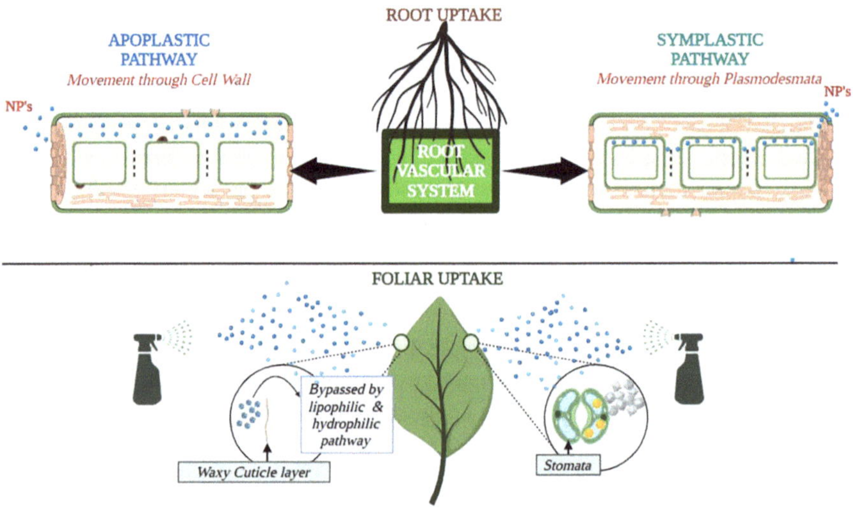

Fig. 5.1 Mechanisms involved in Nanoparticle uptake by plants

2021). NPs find their route to the xylem of the plant. NPs in the size range of 22–50 nm follow the apoplastic pathway. Root exudates can influence the assortment of NPs in the rhizosphere and determine the fate of NPs. These exudates can aggregate the NPs, leading to NPs and microbial interactions. This can significantly impact nanoparticle mobility and their availability to plants. The pivotal role of root exudates towards NPs uptake can be attributed to chemotactic response, change in soil matric physicochemical and aggregation dynamics (Zhou et al. 2011). NPs are observed to penetrate through most underdeveloped root tips, in the root system. After the epidermal region of the root is reached, there are two key pathways for the translocation of the NPs to various tissues of the plant—the apoplastic and symplastic pathways.

In the apoplastic pathway, NPs are found to diffuse into the spaces between the cell membrane and the cell wall through the cell wall pores. This transport occurs through the extracellular space outside the plasma membrane to xylem vessels. This kind of transport occurs radially across plant tissue and finally reaches the root central cylinder and vascular system. However, the transport of NPs to the xylem from the root system indicates breaking the barrier of the apoplastic pathway. This has been previously observed and documented in *Arabidopsis*, ryegrass, and cucumber (Avellan et al. 2017). Some NPs are hindered by the Casparian strips, where they can accumulate. This kind of transport in plant systems is through the non-living space to the vascular system of the plant (Khan et al. 2022).

The symplastic pathway represents a transport within cells where cell wall pores of adjacent cells (plasmodesmata) act as a transport channel. Alternatively, the NPs could just penetrate through the cell membrane to the cytoplasm, which is achieved by membrane channels, aquaporins, and endocytosis. The process of transport of NPs in plants occurs through 2–20 nm diameter of intercellular plasmodesmata (Luu and Maurel 2005).

5.2.2 Foliar Uptake of NPs

The most prevalent method of introducing NPs to plants in agriculture is simply by spraying them onto the leaf surfaces for absorption (Wang et al. 2023). This absorption occurs through foliar uptake pathways. The NPs face their first barrier to entry, which is the waxy cuticle layer. It generally overcomes this through either a lipophilic pathway (nonpolar solutes) or a hydrophilic pathway (polar solutes) (Avellan et al. 2019). It has been noted that, using these two pathways, NPs below a size of 4.8 nm can pass through. NPs suspended in water or air deposited on leaves are taken through stomata or penetrate through cuticle layers. Generally, modified nanoparticles of smaller size can easily penetrate the cuticle of leaves (Khan et al. 2022). On entry to mesophyll cells, the NPs can embark on an extracellular or plastid pathway. Further investigations into the principle of uptake of NPs of size 5 nm or greater are noteworthy.

An alternative to cuticular absorption of NPs is through stomatal openings. This is a familiar pathway, owing to the size of the stomatal openings, being approximately 25 μm in length and 3–10 μm in width (Eichert and Goldbach 2008). Plants such as *Cucurbita pepo*, *Lactuca sativa,* and *Arabidopsis thaliana* are among those in which stomatal uptake of NPs has been demonstrated (Larue et al. 2014).

5.3 Major Factors Affecting Nanoparticle Uptake

5.3.1 Properties of NPs That Influence Their Absorption in Plants

Various properties of NPs influence their absorptivity and translocation in plants. These include the size of the particle, chemical composition, surface modifications, and more. Among these, the size of the particles plays the key factor, expressing the highest influence in determining the uptake of NPs in plants.

5.3.2 Effect of Size

Prior works have revealed that inorganic-based NPs less than 50 nm in diameter utilizes the stomatal pathway to enter plant leaves (Ha et al. 2021). It has also been found that an increase in the size of the NPs corresponded to a decrease in its absorptivity into plants. A study conducted by Zhu et al. (2020), found ZnO particles of size 30 nm used the stomatal pathway to move through the epidermis of wheat leaves. They also found out how stomatal opening and closing had an effect on the transport of the NPs, proving that a decrease in stomatal diameter resulted in a 33.2% and 8.3% decrease in zinc concentration in the chloroplast and cytoplasm of wheat leaf cells, respectively (Zhu et al. 2020). In another study, Avellan et al. (2019) used coated gold NPs with a varying diameter (3, 10, 50 nm) and observed that the size did not affect their absorption capacity in *Triticum aestivum* leaves. Simultaneously, they also discovered that the uptake took place through disruption and/or diffusion through both the cuticle layer as well as the stomata (Avellan et al. 2019).

5.3.3 Effect of Surface Charge and Shape

Changes in the shape of the NPs correspond to changes in the surface area and contact points with the plant surface, which significantly influence the uptake of NPs. This was tested by Zhang et al. using rod-shaped and spherical gold nanoparticles using *Arabidopsis* sp. leaves (Zhang et al. 2022). Along with the shape, its surface charge also plays a crucial role, as demonstrated by the absorption experiments with graphene quantum dots on maize leaves (Sun et al. 2022) and zinc oxide NPs on wheat (Zhu et al. 2021). The experiments also showed that positively charged NPs were absorbed more effectively than negatively charged NPs. This is due to electrostatic attraction between the positively charged nanoparticles and negatively charged plant cell walls (Zhu et al. 2021).

5.3.4 Effect of Plant Species and Cultivar

Realising the importance of the density and size of the pores of stomata, their distribution and other characteristics of the leaves, it is easy to understand that the absorption of NPs varies greatly with different plant species. Monocots have a regular arrangement of stomatal openings, in contrast to dicots, which are more irregular (Wang et al. 2023). In terms of distribution, some species have stomata on both sides of the leaves, while some have them on just one side. Even when it is present on both sides, the number of pores as well as the ratio of openings on both sides varies greatly between species, the growth stages of the plant, and its life cycle also influence the NPs uptake. Research has been done comparing the absorption of CeO_2 nanoparticles in dicotyledonous pumpkin and monocotyledonous wheat. The results obtained by the experiment was in accordance to existing findings- dicots such as Pumpkin (*Cucurbita maxmima*) were more likely to absorb NPs than monocots such as Wheat (*Zea mays*). This difference in absorption might be attributed to their corresponding root structures, water uptake capacity as well as their sieving pore size (Schwabe et al. 2013).

5.3.5 Environmental Factors

External factors influence the foliar uptake of NPs since they affect stomatal activity. Abiotic factors such as CO_2 concentration, water levels, humidity and light intensity determine stomatal opening and closing (Gutschick 1999; Ainsworth and Rogers 2007; Xu and Zhou 2008) which consequently affects nanoparticle absorption, since NPs use stomatal openings as the point of entry. Increased CO_2 levels and water deficit environments will lead to smaller pore size of the stomata to reduce water loss, lowering NP uptake efficiency. Meanwhile, lower humidity and lower light intensity also causes closing of stomatal pores. The time of day chosen for foliar application of NPs must be in conjunction with the stomatal aperture cycle for better rates of uptake (Avellan et al. 2021).

Therefore, it is evident that there are a multitude of factors that affect the rate of uptake of any NPs in plants. The size of the NPs plays a crucial role, since the penetration of the NPs inside the plants is through minute pores or fine cuticular layers. Other characteristics of the NPs, such as their shape, surface characteristics, and surface charge, also determine the rate. The number of pores, their distribution, sizes, and arrangement vary greatly between different plants, and hence, the plant species also influences the rate of foliar uptake of NPs. Finally, environmental factors such as humidity, temperature, light levels, CO_2 concentration, and so on, lead to changes in stomatal activity, influencing the plant's characteristics, and thus indirectly affect the absorption rate of NPs as well.

5.4 Transport and Translocation of Nanoparticles in Plants

Once absorbed, the NPs translocate to various tissues of the plant, including newly developed seeds (Tripathi et al. 2017). The NPs that were introduced by foliar means migrate throughout the plant, covering shoots, fruits, roots, and all associated parts. This proliferation is capable only through the vascular tissues of the plant, that is, the xylem and phloem. Here again, the size of the NPs and their surface functional groups regulate the site of action and accumulation (Avellan et al. 2021).

5.4.1 Mechanism of Transport

Transport of NPs within plant cells are dependent on nanoparticle characteristics like size and surface properties as well as the plant cell characteristics like cell wall pore size. For the nanoparticles to produce its effect, the size of NP's must be suitable with respect to the cell wall pore size (Verano-Braga et al. 2014). Tube elements of the vascular tissues facilitate the movement of the NPs in the plants (Muller et al. 2014). Osmosis occurs towards the phloem when they have a relatively lower water potential and a higher sap concentration as compared to nearby xylem cells. This leads to an increased water pressure, which enables multidirectional flow to drive the sap towards all sides, including developing tissues. After which, sap unloading occurs at the sites while also unloading the NPs. This occurs either actively or passively (De Schepper et al. 2013). Alternatively, a reduction in sap concentration leads to water moving towards the xylem. This suggests that NPs can be exchanged between xylem and phloem (De Schepper et al. 2013).

The distribution of NPs throughout the plant occurs either through the symplastic or apoplastic pathway (Huang et al. 2022). NPs may also utilize protein channels present in the membranes, such as that used by Silicon based nano-courier formulation, which helped in transporting quercetin to increase plant stress tolerance (Guerriero et al. 2023). The novel study of Proteomics has revealed vast information regarding translocation of NPs in plants (Djanaguiraman et al. 2024). By studying the membrane proteins, it is possible to recognize its role in carrying NPs through the plant cells. The translocation mechanism has been demonstrated in recent research conducted using carbon nanotubes when applied as suspensions on French bean leaves. The NPs were found in both phloem sieve tubes as well as xylem vessels (Hasaneen et al. 2016).

When foliar application of NPs is performed in large scale agricultural setting, homogenous distribution of the nanomaterial throughout each plant is not attainable. As a result, inefficient uptake through stomata or leaf burns due to high accumulation at certain sites could occur, all of which affects the intended improvement of crops by NPs (Ghormade et al. 2011). Properties of both the NPs as well as the target plant may hinder its proper translocation. Thus, recent developments have been targeted at improving nanoparticle distribution and delivery. When ZnO nanoparticles were encapsulated by SiO_2 shells and applied, it was found that the micronutrient was delivered in a much more controlled and sustained manner (Gao et al. 2021). Similarly, Pepper plants (*Capsicum annum*) exposed to ZnO NPs coated with Zinc Phosphate shells were also utilized, and it was observed that the shell led to faster phloem loading, enabling better translocation of the nanomaterial. This also highlights the importance of phloem in the translocation of NPs (Rodrigues et al. 2024).

5.4.2 Accumulation of NPs

Translocation and accumulation are dependent on both the plant as well as NP characteristics. Earlier studies have demonstrated the accumulation of Au (gold) nanoparticles in the shoot of *Oryza sativa*, but their accumulation was not observed in the shoot of *Cucurbita pepo* and *Raphanus raphanistrum*. Additionally, there is a pattern observed with negatively charged Au NPs to accumulate in plant shoots while positively charged Au NPs accumulate in plant roots (Zhu et al. 2012). Therefore, the various vascular systems of the plants that transport the sap and water to all parts of the plant are also involved in the translocation of NPs and their subsequent accumulation in the seeds, fruits, shoots, and roots.

5.5 Effects of Nanoparticles on Plant Growth and Development

The uptake of NPs by the various mechanisms stated above has led to numerous effects in plant growth and development at various stages. These effects have been widely studied in recent years, and their stimulatory and inhibitory effects have been described.

5.5.1 Effect of Nanoparticles on Seed Germination and Vigour

The technique used to incorporate NPs into germinating seeds is known as seed priming, which is a rejuvenation technique involving controlled hydration at the beginning process of germination, that enables the penetration of the NPs into the tissue. Seed priming is seen to be a rapidly growing field of research, due to the sustainable mode of action in enhancing seed quality and its productivity (Das et al. 2022).

Over the recent years, multiple studies have been performed in the field of NP's and their effect on agricultural seed enhancement. The effect of numerous NPs, particularly in seed germination and vigour, has been studied, including Ag NPs, gold NPs, iron oxide NPs, ZnO NPs, copper NPs, carbon NPs, and silicon NPs, among many others (Table 5.1).

5.5.2 Stimulatory Effects of NPs on Seeds

Among the commonly observed changes, most indicated a stimulatory or enhancing effect in various aspects of seed growth and germination. Application of Au NPs

Table 5.1 Effects of various nanoparticles in seed priming of different commercial plant species

Plant species	NP Used	Effective concentration	Effect observed	References
Onion (*Allium cepa* L.)	AuNPs	5.4 μg/mL	Enhancement in seed germination and post germination development	Acharya et al. (2019)
Winged Bean (*Psophocarpus tetragonolobus* L.)	AgNPs	50 mg/L	Significant increase in germination and vigour.	Kumar et al. (2020)
Green gram (*Vigna radiata* L.)	AgNPs	20 ppm	Higher germination percentage	Chakraborty and Bordolui (2021)
Wheat (*Triticum aestivum* L.)	Iron oxide (IONPs)	400 ppm	Germination percentage and shoot length	Sundaria et al. (2019)
Black gram (*Vigna mungo* L.)	Zinc oxide (ZnO NPs); Copper (CuNPs)	600 mg/L; 300 mg/L	Improvements in seed quality	Raja et al. (2019)
Tomato (*Solanum lycopersicum* L.)	ZnO-NPs (under various stress levels)	750 ppm	Increase in antioxidative defense, improving seed vigor.	Ghosh et al. (2024)

resulted in a boost of seed germination, plant height, leaf parameters, and later developmental stages without any toxic effect (Acharya et al. 2019). Ag NPs have been noted to significantly increase germination percentage (88.33%) and vigour index (2848.44) in winged bean seeds (*Psophocarpus tetragonolobus* (L.) DC). It also improved antioxidant enzymes, including catalase, peroxidase, ascorbate peroxidase and superoxide dismutase. A higher concentration of soluble sugars, soluble protein, sucrose, and chlorophyll contents was also observed (Kumar et al. 2020). Iron oxide NPs and copper (Cu NPs) have also improved seed quality, germination, and vigour (Sundaria et al. 2019; Raja et al. 2019). Seed priming using ZnO NPs was shown to enhance antioxidative enzymatic and non-enzymatic antioxidative defences, amount of chlorophyll and osmotic adjustment. Consequently, it led to decreasing ROS accumulation as well as lipid peroxidation in cell membranes in tomato during juvenile seedling stages (Ghosh et al. 2024). The following figure represents the various stimulatory effects observed on seeds due to nanoparticle treatments (Fig. 5.2).

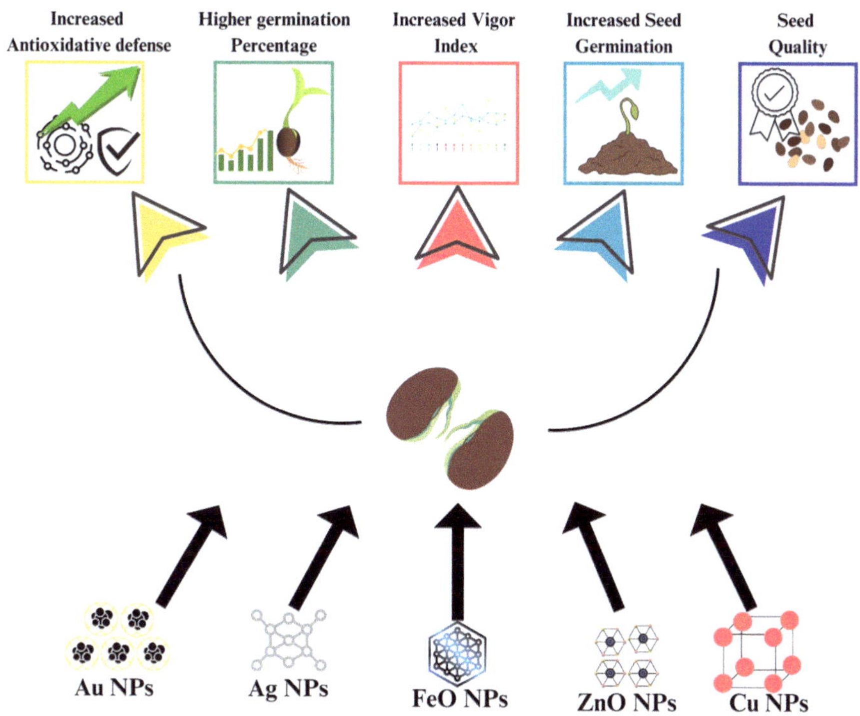

Fig. 5.2 Stimulatory effects of nanoparticles on seeds

5.5.3 Inhibitory Effects of NPs on Seeds

Seed priming with NPs can also show many inhibitory effects when compared to unprimed seeds, which depend on composition, concentration, and plant species. Among the few, it includes a reduction in pungency levels in Onion (*Allium cepa* L.) when treated with Au NPs (Acharya et al. 2019). ZnO NP treatment at higher concentrations negatively affected plant growth but significantly increased some other parameters, such as fruit quality in Habanero peppers (García-López et al. 2019). This inhibitory effect of NPs is attributed to the penetration and aggregation of NPs in seeds, resulting in toxic effects. It was also observed that under drought stress conditions, the germination rate and mean germination time shared a negative correlation with Si NP treatment seeds (Saeedeh et al. 2021).

5.5.4 Effect on Root Growth and Development

The utility of NPs has also been observed in root development of plants, particularly their morphology, branching, and length (Fig. 5.3). Variations were observed between untreated and treated date palm plants with nano silver at the rooting stage. More roots were observed in the treated plants grown *in vitro,* showing a higher root length at relatively lower concentrations (3 ml/L). Further increase in dose showed a decrease in root length and branching. Rooting percentage showed a

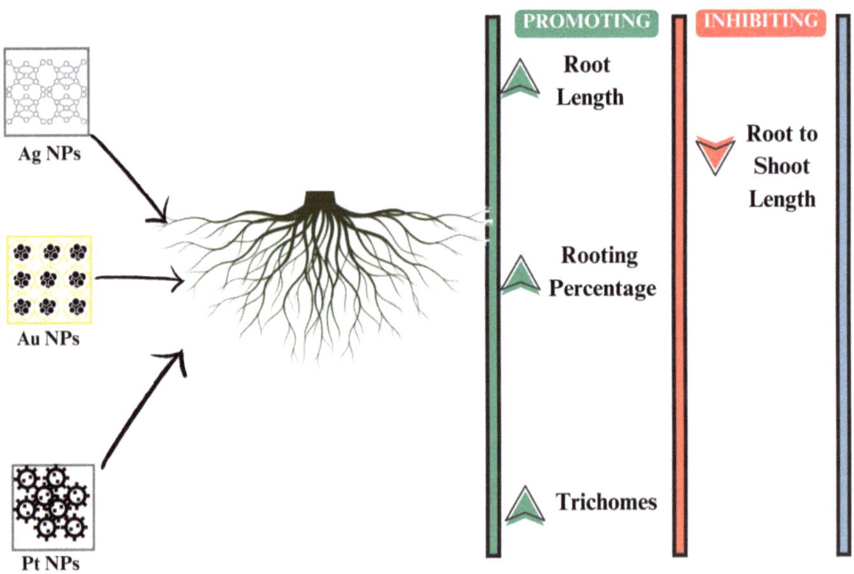

Fig. 5.3 Schematic representation of the effect of Nanoparticles on the root system of plants

steady increment with an increase in concentration of Ag NPs (Elsayh et al. 2022). The root development in Lavender (*Lavandula angustifolia* Mill.) has also been positively affected by silver and gold NPs. Rooting rates increased considerably in the presence of the NPs, and the roots were also found to be longer in the treated specimens. Lower concentrations of these NPs also influenced the development of more trichomes on the roots (Jadczak et al. 2019). When the effects of platinum NPs were tested on the Pea plants (*Pisum sativum* L.), the results showed an inhibitory response, with a 20% reduction in root to shoot lengths as compared to an untreated plant. Apart from this, the germination rate also decreased along with an increase in dormancy periods (Rahman et al. 2020).

5.5.5 *Effect on Shoot Development*

Like root development, NPs have a profound effect on the shoots. Comparison of effects of ZnO NPs and $ZnSO_4$ on the broccoli plant (*Brassica oleracea*) showed prominent positive effects on growth by the ZnO NPs, among which are higher leaf number, larger leaf area and taller plants were observed when they were treated with 800 mg/L ZnO NPs (Awan et al. 2021). Earlier works utilizing ZnO NPs have shown successful increase in biomass growth, protein synthesis, and tolerance to environmental stress in Cotton plant (*Gossypium hirsutum*) (Venkatachalam et al. 2017). In another study, when tomato seeds of Vengador and Rio Grande cultivars were treated with Ag NPs of varying concentrations, the biomass output was found to increase at 5 mg/L. But at higher concentrations, showed a reduction in biomass content with respect to the control (Guzmán-Báez et al. 2021) Table 5.2 summarises the positive and negative impacts of different NPs towards plant's response. Ag NPs act as a biostimulant due to their antioxidant regulating capability. In a study using *Arabidopsis* sp., Ag NPs induced gene expression that is responsible for cell proliferation, hormone pathways and metabolism. It also promoted ROS accumulation and root growth (Syu et al. 2014).But the reduction in biomass at higher concentrations may be attributed to the toxic effect of Ag NPs, reducing chlorophyll content, disrupting electron transfer to the reaction centres, as well as the oxygen evolution complex in PSII, all of which affect ATP and NADPH synthesis (Dewez et al. 2018).

5.5.6 *Fruiting and Reproductive Development*

Certain NPs also show effects on the reproductive development of commercial fruit-bearing plants. CeO_2 NPs showed a positive effect on different growth parameters of Strawberry (*Fragaria* × *ananassa* Duch.), including an increase in total ripe fruit number. The enhancement of yield is attributed to the higher pollen grain numbers, pollen germination rates and pollen tube elongation because of CeO_2 NP exposure

Table 5.2 Major nanoparticles and their impact on plant species

Nanomaterial used	Plant Species	Region of effect	Impact	References
Silver NPs (Ag)	Date palm (*Phoenix dactylifera L.*) Lavender (*Lavandula angustifolia* Mill.) Tomato (Vengador and Rio Grande cultivars) Passion fruit (*Passiflora edulis Sim* f. *edulis*)	Root Root and Shoot Root and Shoot Reproductive development	Positive at lower concentrations 3 mL/L, but further increase causes negative effect. Largely positive Positive at lower concentration; showed toxic effect at higher concentration Positive effect on flowering and fruiting	Elsayh et al. (2022), Jadczak et al. (2019), Guzmán-Báez et al. (2021), Phong et al. (2022)
Platinum NPs (Stabilised with PVP)	Pea (*Pisum sativum)*	Root, seed	Negative effect	Rahman et al. (2020)
Zinc NPs (Zn)	Broccoli *Brassica oleracea* var *italic*)	Shoot, Root, Leaves	Positive	Awan et al. (2021)
Cerium Oxide NPs	Strawberry (*Fragaria* × *ananassa* Duch.	Fruit	Positive	Dai et al. (2022)

(Dai et al. 2022). Supplementing Ag NPs also positively affected the flowering and fruiting of purple Passion fruit (*Passiflora edulis Sim* f. *edulis*) grown *in vitro*. 7 mg/L of Ag NPs showed the highest flowering rate. They also lowered the endogenous hormone production (Gibberellic acid, abscisic acid, melatonin) as compared with the control. Treatment of Ag NPs led to an increase in flower diameter and bloom rates (Phong et al. 2022). The effect of NPs on fruiting and development has been highlighted in Fig. 5.4.

5.6 Effects of Nanoparticles on Plant Physiological Processes

5.6.1 Oxidative Stress and Antioxidant Responses in Plants

NP's affect the physiology of plants through a multitude of approaches. A typical example of such action of NP's involves the reduction of reactive oxygen species (ROS) produced due to cadmium toxicity in plants, which is mitigated by Se NPs, apart from many other advantageous effects, as presented by Qi et al. These ROS,

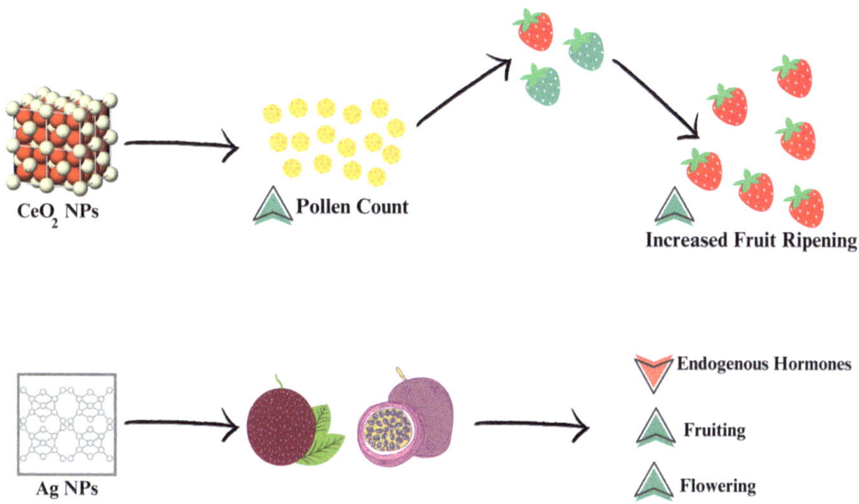

Fig. 5.4 Effect of CeO_2 and Ag NPs on the development of Strawberry and Passionfruit

which are usually produced in the form of radicals such as the hydroxyl radical (HO•), superoxide (O2•−), and singlet oxygen, are created during the reduction of molecular oxygen (Soenen et al. 2011). In the study, the presence of Se NPs was able to effectively reduce the adverse effects of ROS produced due to Cd toxicity in *Brassica napus* L. (Qi et al. 2021).

Moreover, NP's have been observed to have an effect of reducing antioxidant and non-enzymatic antioxidants produced by the plants as a defence mechanism against stressed conditions. These stresses may include displacement of essential elements within the biomass, or disruption of enzymes that are dependent on certain metals, all of which ultimately also leads to an increase of ROS in the plants (Chen et al. 2017). In response, the plants produce numerous kinds of enzymatic and non-enzymatic antioxidants, such as catalase, peroxidase, superoxide dismutase, and many others. A recent study revealed that coriander (*Coriandrum sativum* L.) exposed to exogenous lead (Pb) stress showed a significant increase in these antioxidant enzymes. Furthermore, application of silicon NPs appreciably decreased the activity of peroxidase and catalase enzymes (Fatemi et al. 2021). Similarly, Shah et al. noted that when maize hybrids were exposed to various concentrations of salinity stress, seed priming with titanium dioxide (TiO_2) using two methods—nano-priming and hydro-priming showed an effective change in the sodium oxide dismutase enzyme activity. The two methods of priming with TiO_2 also greatly influenced the increase in effect of phenylalanine ammonia lyase (PAL) enzyme action in its leaves, just as it influenced catalase activity. Particularly, the effective concentration of titanium dioxide for its effect to take place in *C. sativum* was at 60 ppm (Shah et al. 2021). Most crops are highly prone to oxidative stress, adversely affecting crop yield. Wheat is an important crop among them, which is widely studied in terms of sustainably promoting and protecting the growth of wheat crops. When wheat crops were treated with low concentrations of silver and gold NP's (20 mg/L and 10 mg/L, respectively), they showed a notable increase in the activity of the enzyme's catalase, ascorbate peroxidase, and guaiacol peroxidase. This is because silver (Ag) NPs could facilitate the plant to suppress the progressive death due to ROS produced from oxidative stress (Manaf et al. 2021).

5.6.2 Hormonal Regulation in Plants

Nanoparticles are considered to be excellent regulatory molecules in plants, which can modulate physiological activities, biochemical processes and hormonal regulations. Nanoparticles can ameliorate during exposure to biotic and abiotic stress by modulating the phytohormone signalling cascade. Plant growth regulators or hormones such as auxins, cytokinins and gibberellins are the majorly regulators of plant growth and development; however, abscisic acid, jasmonic acid and salicylic acids are the hormones associated with stress management and defence mechanisms. Studies report that NPs via synergistic and antagonistic action cross-talk in hormonal signalling for the biosynthesis of phytohormone, regulate the availability of the hormones or regulate the signalling process. Selenium NPs induced profuse root development and embryo-like structure in *Momordica charanti*a during in vitro organogenesis and callogenesis. This provides substantial proof on NPs role in phytohormone signalling. Ag NPs exposure to *Eruca sativa* had resulted in better root growth owing to increased biosynthesis of auxins (Tripathi et al. 2022; Vannini et al. 2013). Furthermore, AuNPs had a direct effect on the root architecture and overexpression of genes associated with auxin biosynthesis in *Arabidopsis,* while Ag NPs resulted in low auxin production and inhibited the gravitropism of the root (Sun et al. 2017). Recent studies have also explored the NP's and the various developmental cycles of the plant growth regulators, finding a plausible link between the two. The involvement of ZnO, SiO_2 and composite NPs in ethylene and cytokinin-regulated genes was recently evaluated by Azhar et al. (2021). The genes targeted by NPs were Acetyl-CoA synthetase, ACC oxidase, Ethylene Response 2 and Ethylene response Factor 1 for Ethylene hormone and Response regulators ARR7 and ARR15 for cytokinin. In the study, it was found that ethylene-based genes were expressed prominently in higher concentrations of ZnO NP treatment. Si NPs did not show much effect in the case of genes corresponding to ethylene, except ACC Oxidase 2. On the other hand, the genes regulating cytokinin were inhibited by ZnO NPs, and showed slight increase with respect to Si NP treatment. The composite NP treatment did not show a significant change (Azhar et al. 2021).

5.6.3 Nutrient Uptake and Transport

Improvement in plant nutrient acquisition has also been observed in many studies with the introduction of NPs. Nanoparticles assist in the adsorption and delivery of nutrients and when applied to soil, they can improve the soil fertility and microbial population in soil. NPs with its unique features such as low surface tension, high adsorption capacity, mobility, pH stability and slow, persistent release of nutrients have high potential application in agriculture. Nanofertilizers utilise carbon-based, metal-based, and porous NPs that are mixed with macro and micronutrients for effective delivery and distribution in plants (Wang et al. 2020). Application of zeo-urea nanofertilizer (N-NF) instead of synthetic urea to maize and star flower has been reported to grow faster with higher yield (Manikandan and Subramanian 2016). Similar results were observed when urea-modified zeolites were applied to soybeans over synthetic fertiliser (Liu and Lal 2014). Potato growth and yield were better when nano-N were utilised and reduced nitrate leaching. In a recent study by Zhao et al. carbon NPs and its effect on the growth parameters of *Zea mays* L. were investigated. Among the results, the experiment showed an incredible Nitrogen (N), Potassium (K) and Phosphate (P) uptake from soil, especially in the presence of fertilizers. The effective N uptake and K uptake was capped at 400 mg/kg and 200 mg/kg of carbon NP concentration, respectively. The mechanism that caused this increasing effect in the uptake may be due to more effective uptake by the aquaporins, rhizospheres and the general high sorption capability and surface area of the NPs. Ultimately, application of carbon NPs positively affected the plant growth as well as its capacity to accumulate essential nutrients (Zhao et al. 2021).

5.6.4 Photosynthesis in Plants

Nanoparticles based on their composition, size and plant species can have positive or negative effects on photosynthesis. Studies have reported that some NPs can reduce chlorophyll content, electron transport rate and damage to chloroplast. On the other hand, NPs can improve the chlorophyll content, activity of Rubisco enzyme, PSII and broadening of chloroplast for overall improved performance of photosynthesis. Single walled carbon nanotubes can show better absorption of light at UV, visible and infrared that can significantly improve photosynthesis. Similarly, application of multi walled carbon tubes to rice plants had enhanced its chlorophyll content. At low concentration of Ag NPs had improved the chlorophyll content and PSII efficiency (Ghorbanpour et al. 2021). The influence of NPs on plants has also been evaluated based on the photosynthesis rates. Iron oxide NPs (Fe_3O_4) have shown definitive increase in CO_2 assimilation in wheat plant in the presence of relatively high Iron oxide NP concentration (500 mg/L), which was estimated as P_n-R difference (Feng et al. 2022).

The mechanism of these nanoparticles in improving the photosynthesis rates have also been documented. Results obtained by Rai-Kalal et al. showed that ZnO NPs were able to stimulate an increased performance of the oxygen evolving complex of PSII in wheat plants. It had also enhanced the number of active reaction centres per chlorophyll molecules. Nano-primed seeds of wheat showed higher absorption, energy trapping efficiency and electron transport efficiency from the active reaction centres (Rai-Kalal and Jajoo 2021).

5.7 Future Perspectives of Nanotechnology in Plant Sciences

The concept of nanotechnology is closely linked to an eco-friendly, sustainable future. The newly emerging study on nanotechnology is an evidence for its value with respect to many fields, particularly in plant sciences and agriculture. Incorporation of NPs in fertilizers to create Nano-biofertilizers are currently being looked as feasible alternative to chemical fertilizers that can be utilized at more intensive rates due to the properties offered by the NPs such as its larger surface area and higher absorption rates of plants in taking up NPs. They also provide excellent protection for crops against damage. Silver, copper, zinc and silica-based nano-pesticides and nano-fungicides are the latest in the line of protectants, which show great promise in their application (Garg et al. 2023). The focus of development in agriculture is primarily to reduce costs, increase crop productivity and output to accommodate the ever-increasing demand of the growing human population. Agriculture has been continuously evolving, taking into account the profit margins of the farmers, land utilization, crop requirements, and sustainability in farming. A recent emerging model is precision agriculture, which utilizes various nanotechnology-based applications throughout the production and processing stages. Innovations like nanosensors for monitoring real-time soil and plant health, nanoscale carriers, nanofertilizers, nanoherbicides, and the like for improved nutrient utilization, and providing the ability for controlled and targeted release. Nanosensors in particular, have been recently used in applications such as environmental monitoring providing a valuable data for optimizing crop output (Yin et al. 2021). Precision agriculture involves reduction of wastes produced, maximizing yields and minimizing usage of inputs to only what is necessary. These aspects are managed by NPs as carriers, by a concept called as "Variable Rate Technology", which allows precise delivery of inputs, as well as protect them from degradation while increasing their efficacy, thereby limiting usage of resources (Yadav et al. 2023). Engineered nanoparticles have newly been implemented in alleviating crops under salt stress. These engineered nanoparticles are biocompatible, such as chitosan, silica, zinc, iron, copper, titanium, silver as well as silicon-based nanoparticles. These nanoparticles, once taken up by the plants, have proven to activate stress responsive genes, expanding the properties of plants that allows it to tolerate salt stress (Junedi et al.

2023). Synthesis of nanoparticles have also had massive developments, such as microbial nanoparticle synthesis and fungal based synthesis (Bhardwaj et al. 2023). Moreover, novel methods have been developed using NPs to biofortify wheat crops, thereby providing a solution for zinc-deficiency, which is a major concern today. The biofortifies zinc crops compensate for the inadequacy of zinc in most agricultural soils. Crops treated with ZnO NPs have shown significant increase in zinc content (Sánchez-Palacios et al. 2023). These nanomaterials have served as successful micronutrient delivery systems, proving in many ways that NPs have numerous applications, and more are yet to be discovered.

5.8 Conclusion

Nanotechnology holds great promise for an eco-friendly and sustainable future. The invaluable properties of nanoparticles have been applied in vast areas, with more applications to be discovered on the horizon. Nanoparticles possess both beneficial and harmful effects on plants however, research towards their application as agrochemicals is gaining more attention. In this chapter, the unique properties of NPs have been discussed, particularly those making it useful in agriculture and plant sciences. Qualities such as increased surface area, higher absorptivity and sustainable approaches in agriculture. Plants have proven themselves as good hosts to NPs, allowing nanoparticle uptake through various routes, specific to the property, site of action as well as function of the NPs. Both foliar as well as root uptake have been widely studied. Various factors such as nanoparticle size, environmental conditions and the species of plants all affect the rate of nanoparticle uptake. Effects of these NPs have been witnessed in countless developmental stages of commercially important plant species such as seed germination, root and shoot growth, reproductive maturation, and fruiting. The majority of nanoparticles show positive effects on growth at lower concentrations, while higher accumulation leads to adverse consequences. Overall, its benefits far outweigh its cons, making it a crucial part of agriculture research to explore its application for a better future. Within plants, NPs furnish protection from environmental and biological stresses, enhance hormonal control, improve nutrient uptake, transport, and enhance photosynthesis and respiratory pathways of plants. With new and emerging studies in nanoparticles opens a new avenue for the dynamic world of nanotechnology as a fundamental stepping stone for a greener and robust future.

References

Acharya P, Jayaprakasha GK, Crosby KM, Jifon JL, Patil BS (2019) Green-synthesized nanoparticles enhanced seedling growth, yield, and quality of onion (Allium cepa L.). ACS Sustain Chem Eng 7(17):14580–14590

Ainsworth EA, Rogers A (2007) The response of photosynthesis and stomatal conductance to rising [CO_2]: mechanisms and environmental interactions. Plant Cell Environ 30(3):258–270
Ali S, Mehmood A, Khan N (2021) Uptake, translocation, and consequences of nanomaterials on plant growth and stress adaptation. J Nanomater 2021(1):6677616
Arora S, Murmu G, Mukherjee K, Saha S, Maity D (2022) A comprehensive overview of nanotechnology in sustainable agriculture. J Biotechnol 355:21–41
Avellan A, Schwab F, Masion A, Chaurand P, Borschneck D, Vidal V et al (2017) Nanoparticle uptake in plants: gold nanomaterial localized in roots of Arabidopsis thaliana by X-ray computed nanotomography and hyperspectral imaging. Environ Sci Technol 51(15):8682–8691
Avellan A, Yun J, Zhang Y, Spielman-Sun E, Unrine JM, Thieme J et al (2019) Nanoparticle size and coating chemistry control foliar uptake pathways, translocation, and leaf-to-rhizosphere transport in wheat. ACS Nano 13(5):5291–5305
Avellan A, Yun J, Morais BP, Clement ET, Rodrigues SM, Lowry GV (2021) Critical review: role of inorganic nanoparticle properties on their foliar uptake and in planta translocation. Environ Sci Technol 55(20):13417–13431
Awan S, Shahzadi K, Javad S, Tariq A, Ahmad A, Ilyas S (2021) A preliminary study of influence of zinc oxide nanoparticles on growth parameters of Brassica oleracea var italic. J Saudi Soc Agric Sci 20(1):18–24
Azhar BJ, Noor A, Zulfiqar A, Zeenat A, Ahmad S, Chishti I et al (2021) Effect of ZnO, SiO2 and composite nanoparticles on *Arabidopsis thaliana* and involvement of ethylene and cytokinin signaling pathways. Pak J Bot 53(2):437–446
Bhardwaj I, Kumar V, Bhardwaj N, Verma R, Bhardwaj Y, Kumari T (2023) A glimpse into the performance and synthesis of microbial nanoparticles and its new advances in soil enrichment and plant nutrition: a review. Nanotechnol Environ Eng 8(4):943–964
Bowman DM (2017) More than a decade on: mapping today's regulatory and policy landscapes following the publication of nanoscience and nanotechnologies: opportunities and uncertainties. NanoEthics 11:169–186
Chakraborty A, Bordolui SK (2021) Impact of seed priming with Ag-nanoparticle and GA3 on germination and vigour in green gram. Int J Curr Microbiol App Sci 10(03):1499–1506
Chen Q, Zhang X, Liu Y, Wei J, Shen W, Shen Z, Cui J (2017) Hemin-mediated alleviation of zinc, lead and chromium toxicity is associated with elevated photosynthesis, antioxidative capacity; suppressed metal uptake and oxidative stress in rice seedlings. Plant Growth Regul 81:253–264
Dai Y, Li T, Wang Z, Xing B (2022) Physiological and proteomic analyses reveal the effect of CeO2 nanoparticles on strawberry reproductive system and fruit quality. Sci Total Environ 814: 152494
Das I, Gogoi B, Sharma B, Borah D (2022) Role of metal-nanoparticles in farming practices: an insight. 3 Biotech 12(11):294
De Schepper V, De Swaef T, Bauweraerts I, Steppe K (2013) Phloem transport: a review of mechanisms and controls. J Exp Bot 64(16):4839–4850
Dewez D, Goltsev V, Kalaji HM, Oukarroum A (2018) Inhibitory effects of silver nanoparticles on photosystem II performance in Lemna gibba probed by chlorophyll fluorescence. Curr Plant Biol 16:15–21
Djanaguiraman M, Anbazhagan V, Dhankher OP, Prasad PV (2024) Uptake, translocation, toxicity, and impact of nanoparticles on plant physiological processes. Plants 13(22):3137
Duhan JS, Kumar R, Kumar N, Kaur P, Nehra K, Duhan S (2017) Nanotechnology: the new perspective in precision agriculture. Biotechnol Rep 15:11–23
Duncan TV (2011) Applications of nanotechnology in food packaging and food safety: barrier materials, antimicrobials and sensors. J Colloid Interface Sci 363(1):1–24
Eichert T, Goldbach HE (2008) Equivalent pore radii of hydrophilic foliar uptake routes in stomatous and astomatous leaf surfaces–further evidence for a stomatal pathway. Physiol Plant 132(4):491–502

Elsayh SA, Arafa RN, Ali GA, Abdelaal WB, Sidky RA, Ragab TI (2022) Impact of silver nanoparticles on multiplication, rooting of shoots and biochemical analyses of date palm Hayani cv. by in vitro. Biocatal Agric Biotechnol 43:102400

European Commission (2012) Rural Development in the European Union. Statistical and Economic Information. Report 2012, European Union

Fatemi H, Esmaiel Pour B, Rizwan M (2021) Foliar application of silicon nanoparticles affected the growth, vitamin C, flavonoid, and antioxidant enzyme activities of coriander (*Coriandrum sativum* L.) plants grown in lead (Pb)-spiked soil. Environ Sci Pollut Res 28:1417–1425

Feng Y, Kreslavski VD, Shmarev AN, Ivanov AA, Zharmukhamedov SK, Kosobryukhov A et al (2022) Effects of iron oxide nanoparticles (Fe3O4) on growth, photosynthesis, antioxidant activity and distribution of mineral elements in wheat (*Triticum aestivum*) plants. Plants 11(14):1894

Gao X, Kundu A, Bueno V, Rahim AA, Ghoshal S (2021) Uptake and translocation of mesoporous SiO2-coated ZnO nanoparticles to Solanum lycopersicum following foliar application. Environ Sci Technol 55(20):13551–13560

García-López JI, Niño-Medina G, Olivares-Sáenz E, Lira-Saldivar RH, Barriga-Castro ED, Vázquez-Alvarado R et al (2019) Foliar application of zinc oxide nanoparticles and zinc sulfate boosts the content of bioactive compounds in habanero peppers. Plants 8(8):254

Garg D, Sridhar K, Stephen Inbaraj B, Chawla P, Tripathi M, Sharma M (2023) Nano-biofertilizer formulations for agriculture: a systematic review on recent advances and prospective applications. Bioengineering 10(9):1010

Ghorbanpour M, Movahedi A, Hatami M, Kariman K, Bovand F, Shahid MA (2021) Insights into nanoparticle-induced changes in plant photosynthesis. Photosynthetica 59(4):570–586

Ghormade V, Deshpande MV, Paknikar KM (2011) Perspectives for nano-biotechnology enabled protection and nutrition of plants. Biotechnol Adv 29(6):792–803

Ghosh T, Yadav SK, Choudhary R, Rao D, Sushma MK, Mandal A et al (2024) Effect of zinc oxide nanoparticle based seed priming for enhancing seed vigour and physio-biochemical quality of tomato seedlings under salinity stress. Russ J Plant Physiol 71(1):38

Grumezescu AM, Holban AM (eds) (2018) Impact of nanoscience in the food industry, vol 12. Academic Press, London

Guerriero G, Sutera FM, Hoffmann J, Leclercq CC, Planchon S, Berni R et al (2023) Nanoporous quercetin-loaded silicon-stabilized hybrid lipid nanoparticles alleviate salt stress in tomato plants. ACS Appl Nano Mater 6(5):3647–3660

Gutschick VP (1999) Biotic and abiotic consequences of differences in leaf structure. New Phytol 143(1):3–18

Guzmán-Báez GA, Trejo-Téllez LI, Ramírez-Olvera SM, Salinas-Ruíz J, Bello-Bello JJ, Alcántar-González G et al (2021) Silver nanoparticles increase nitrogen, phosphorus, and potassium concentrations in leaves and stimulate root length and number of roots in tomato seedlings in a hormetic manner. Dose-Response 19(4):15593258211044576

Ha N, Seo E, Kim S, Lee SJ (2021) Adsorption of nanoparticles suspended in a drop on a leaf surface of Perilla frutescens and their infiltration through stomatal pathway. Sci Rep 11(1): 11556

Haleem A, Javaid M, Singh RP, Rab S, Suman R (2023) Applications of nanotechnology in medical field: a brief review. Global Health J 7(2):70–77

Hasaneen MNAG, Abdel-aziz HMM, Omer AM (2016) Effect of foliar application of engineered nanomaterials: carbon nanotubes NPK and chitosan nanoparticles NPK fertilizer on the growth of French bean plant. Biochem Biotechnol Res 4(4):68–76

Huang D, Dang F, Huang Y, Chen N, Zhou D (2022) Uptake, translocation, and transformation of silver nanoparticles in plants. Environ Sci Nano 9(1):12–39

Jadczak P, Kulpa D, Bihun M, Przewodowski W (2019) Positive effect of AgNPs and AuNPs in in vitro cultures of Lavandula angustifolia Mill. Plant Cell Tissue Organ Cult 139:191–197

Jeevanandam J, Barhoum A, Chan YS, Dufresne A, Danquah MK (2018) Review on nanoparticles and nanostructured materials: history, sources, toxicity and regulations. Beilstein J Nanotechnol 9(1):1050–1074
Junedi MA, Mukhopadhyay R, Manjari KS (2023) Alleviating salinity stress in crop plants using new engineered nanoparticles (ENPs). Plant Stress 9:100184
Kah M, Kookana RS, Gogos A et al (2018) A critical evaluation of nanopesticides and nanofertilizers against their conventional analogues. Nat Nanotechnol 13:677–684
Khan I, Awan SA, Rizwan M, Hassan ZU, Akram MA, Tariq R et al (2022) Nanoparticle's uptake and translocation mechanisms in plants via seed priming, foliar treatment, and root exposure: a review. Environ Sci Pollut Res 29(60):89823–89833
Kumar VK, Muthukrishnan S, Rajalakshmi R (2020) Phytostimulatory effect of phytochemical fabricated nanosilver (AgNPs) on Psophocarpus tetragonolobus (L.) DC. seed germination: an insight from antioxidative enzyme activities and genetic similarity studies. Current. Plant Biol 23:100158
Kumari R, Singh DP (2020) Nano-biofertilizer: an emerging eco-friendly approach for sustainable agriculture. Proc Natl Acad Sci, India, Sect B Biol Sci 90(4):733–741
Larue C, Castillo-Michel H, Sobanska S, Cécillon L, Bureau S, Barthès V et al (2014) Foliar exposure of the crop Lactuca sativa to silver nanoparticles: evidence for internalization and changes in Ag speciation. J Hazard Mater 264:98–106
Liu R, Lal R (2014) Synthetic apatite nanoparticles as a phosphorus fertilizer for soybean (*Glycine max*). Sci Rep 4(1):5686
Luu DT, Maurel C (2005) Aquaporins in a challenging environment: molecular gears for adjusting plant water status. Plant Cell Environ 28(1):85–96
Manaf A, Wang X, Tariq F, Jhanzab HM, Bibi Y, Sher A et al (2021) Antioxidant enzyme activities correlated with growth parameters of wheat sprayed with silver and gold nanoparticle suspensions. Agronomy 11(8):1494
Manikandan A, Subramanian KS (2016) Evaluation of zeolite based nitrogen nano-fertilizers on maize growth, yield and quality on inceptisols and alfisols. Int J Plant Soil Sci 9(4):1–9
Mohammad ZH, Ahmad F, Ibrahim SA, Zaidi S (2022) Application of nanotechnology in different aspects of the food industry. Discov Food 2(1):12
Moosa AA (2011) Nanotechnology: principles and applications. Dar Dijlah, Amman Jordan
Muller O, Cohu CM, Stewart JJ, Protheroe JA, Demmig-Adams B, Adams WW III (2014) Association between photosynthesis and contrasting features of minor veins in leaves of summer annuals loading phloem via symplastic versus apoplastic routes. Physiol Plant 152(1):174–183
Phong TH, Hieu T, Tung HT, Mai NTN, Khai HD, Cuong DM et al (2022) Silver nanoparticles: a positive factor for in vitro flowering and fruiting of purple passion fruit (Passiflora edulis Sim f. edulis). Plant Cell Tissue Organ Cult 151(2):401–412
Qi WY, Li Q, Chen H, Liu J, Xing SF, Xu M et al (2021) Selenium nanoparticles ameliorate Brassica napus L. cadmium toxicity by inhibiting the respiratory burst and scavenging reactive oxygen species. J Hazard Mater 417:125900
Rahman MS, Chakraborty A, Mazumdar S, Nandi NC, Bhuiyan MNI, Alauddin SM et al (2020) Effects of poly (vinylpyrrolidone) protected platinum nanoparticles on seed germination and growth performance of Pisum sativum. Nano-Struct Nano-Objects 21:100408
Rai-Kalal P, Jajoo A (2021) Priming with zinc oxide nanoparticles improve germination and photosynthetic performance in wheat. Plant Physiol Biochem 160:341–351
Raja K, Sowmya R, Sudhagar R, Moorthy PS, Govindaraju K, Subramanian KS (2019) Biogenic ZnO and Cu nanoparticles to improve seed germination quality in blackgram (*Vigna mungo*). Mater Lett 235:164–167
Rodrigues S, Avellan A, Bland GD, Miranda MC, Larue C, Wagner M et al (2024) Effect of a zinc phosphate shell on the uptake and translocation of foliarly applied zno nanoparticles in pepper plants (Capsicum annuum). Environ Sci Technol 58(7):3213–3223

Saeedeh R, Mehrnaz H, Mansour G (2021) Silicon-nanoparticle mediated changes in seed germination and vigor index of marigold (*Calendula officinalis* L.) compared to silicate under PEG-induced drought stress. Gesunde Pflanzen 73(4):575–589
Sánchez-Palacios JT, Henry D, Penrose B, Bell R (2023) Formulation of zinc foliar sprays for wheat grain biofortification: a review of current applications and future perspectives. Front Plant Sci 14:1247600
Schwabe F, Schulin R, Limbach LK, Stark W, Bürge D, Nowack B (2013) Influence of two types of organic matter on interaction of CeO2 nanoparticles with plants in hydroponic culture. Chemosphere 91(4):512–520
Scognamiglio V (2013) Nanotechnology in glucose monitoring: advances and challenges in the last 10 years. Biosens Bioelectron 47:12–25
Shah T, Latif S, Saeed F, Ali I, Ullah S, Alsahli AA et al (2021) Seed priming with titanium dioxide nanoparticles enhances seed vigor, leaf water status, and antioxidant enzyme activities in maize (*Zea mays* L.) under salinity stress. J King Saud Univ Sci 33(1):101207
Shah MA, Pirzada BM, Price G, Shibiru AL, Qurashi A (2022) Applications of nanotechnology in smart textile industry: a critical review. J Adv Res 38:55–75
Soenen SJ, Rivera-Gil P, Montenegro JM, Parak WJ, De Smedt SC, Braeckmans K (2011) Cellular toxicity of inorganic nanoparticles: common aspects and guidelines for improved nanotoxicity evaluation. Nano Today 6(5):446–465
Sun J, Wang L, Li S, Yin L, Huang J, Chen C (2017) Toxicity of silver nanoparticles to *Arabidopsis*: inhibition of root gravitropism by interfering with auxin pathway. Environ Toxicol Chem 36(10):2773–2780
Sun D, Siddiqui MOR, Iqbal K (2019) Specialty testing techniques for smart textiles. In: Smart textile coatings and laminates. Woodhead Publishing, pp 99–116, Delhi, India
Sun H, Wang M, Wang J, Wang W (2022) Surface charge affects foliar uptake, transport and physiological effects of functionalized graphene quantum dots in plants. Sci Total Environ 812: 151506
Sundaria N, Singh M, Upreti P, Chauhan RP, Jaiswal JP, Kumar A (2019) Seed priming with iron oxide nanoparticles triggers iron acquisition and biofortification in wheat (Triticum aestivum L.) grains. J Plant Growth Regul 38:122–131
Syu YY, Hung JH, Chen JC, Chuang HW (2014) Impacts of size and shape of silver nanoparticles on Arabidopsis plant growth and gene expression. Plant Physiol Biochem 83:57–64
Tom AP (2021) Nanotechnology for sustainable water treatment—a review. Mater Today Proc 10: 16–78
Tripathi DK, Singh S, Singh VP, Prasad SM, Dubey NK, Chauhan DK (2017) Silicon nanoparticles more effectively alleviated UV-B stress than silicon in wheat (Triticum aestivum) seedlings. Plant Physiol Biochem 110:70–81
Tripathi D, Singh M, Pandey-Rai S (2022) Crosstalk of nanoparticles and phytohormones regulate plant growth and metabolism under abiotic and biotic stress. Plant Stress 6:100107
UNDESA (2024) World population prospects 2024
Usman M, Farooq M, Wakeel A, Nawaz A, Cheema SA, ur Rehman H et al (2020) Nanotechnology in agriculture: current status, challenges and future opportunities. Sci Total Environ 721:137778
Vannini C, Domingo G, Onelli E, Prinsi B, Marsoni M, Espen L, Bracale M (2013) Morphological and proteomic responses of *Eruca sativa* exposed to silver nanoparticles or silver nitrate. PLoS One 8(7):e68752
Venkatachalam P, Priyanka N, Manikandan K, Ganeshbabu I, Indiraarulselvi P, Geetha N et al (2017) Enhanced plant growth promoting role of phycomolecules coated zinc oxide nanoparticles with P supplementation in cotton (Gossypium hirsutum L.). Plant Physiol Biochem 110:118–127
Verano-Braga T, Miethling-Graff R, Wojdyla K, Rogowska-Wrzesinska A, Brewer JR, Erdmann H, Kjeldsen F (2014) Insights into the cellular response triggered by silver nanoparticles using quantitative proteomics. ACS Nano 8(3):2161–2175

Wang Z, Yue L, Dhankher OP, Xing B (2020) Nano-enabled improvements of growth and nutritional quality in food plants driven by rhizosphere processes. Environ Int 142:105831

Wang X, Xie H, Wang P, Yin H (2023) Nanoparticles in plants: uptake, transport and physiological activity in leaf and root. Materials 16(8):3097

Wilson MA, Tran NH, Milev AS, Kannangara GK, Volk H, Lu GM (2008) Nanomaterials in soils. Geoderma 146(1–2):291–302

Xu Z, Zhou G (2008) Responses of leaf stomatal density to water status and its relationship with photosynthesis in a grass. J Exp Bot 59(12):3317–3325

Yadav A, Yadav K, Ahmad R, Abd-Elsalam KA (2023) Emerging frontiers in nanotechnology for precision agriculture: advancements, hurdles and prospects. Agrochemicals 2(2):220–256

Yin H, Cao Y, Marelli B, Zeng X, Mason AJ, Cao C (2021) Soil sensors and plant wearables for smart and precision agriculture. Adv Mater 33:2007764

Zhang H, Goh NS, Wang JW, Pinals RL, González-Grandío E, Demirer GS et al (2022) Nanoparticle cellular internalization is not required for RNA delivery to mature plant leaves. Nat Nanotechnol 17(2):197–205

Zhao F, Xin X, Cao Y, Su D, Ji P, Zhu Z, He Z (2021) Use of carbon nanoparticles to improve soil fertility, crop growth and nutrient uptake by corn (*Zea mays* L.). Nanomaterials 11(10):2717

Zhou D, Jin S, Li L, Wang Y, Weng N (2011) Quantifying the adsorption and uptake of CuO nanoparticles by wheat root based on chemical extractions. J Environ Sci 23(11):1852–1857

Zhu ZJ, Wang H, Yan B, Zheng H, Jiang Y, Miranda OR et al (2012) Effect of surface charge on the uptake and distribution of gold nanoparticles in four plant species. Environ Sci Technol 46(22): 12391–12398

Zhu J, Li J, Shen Y, Liu S, Zeng N, Zhan X et al (2020) Mechanism of zinc oxide nanoparticle entry into wheat seedling leaves. Environ Sci Nano 7(12):3901–3913

Zhu J, Wang J, Zhan X, Li A, White JC, Gardea-Torresdey JL, Xing B (2021) Role of charge and size in the translocation and distribution of zinc oxide particles in wheat cells. ACS Sustain Chem Eng 9(34):11556–11564

Zulfiqar F, Navarro M, Ashraf M, Akram NA, Munné-Bosch S (2019) Nanofertilizer use for sustainable agriculture: advantages and limitations. Plant Sci 289:110270

Chapter 6
Application of Advanced Characterization Equipment for Nanotoxicity in Plants

Mathiazhakan Lavanya, Kuppusamy Selvam, and Muthugounder Subramanian Shivakumar

Abstract Nanotechnology is a rapidly evolving field characterized by materials in the scale 1 to 100 nm, which have unique physicochemical and biological properties. Despite these benefits of nanomaterials, there are increasing reports of toxicity of NMs plants and animals. Toxicity of NMs on humans and animals is widely studied, while there is a dearth of information regarding the toxicity of nanomaterials in plants and their cellular interactions. Studying Nanotoxicity in plants is vital, as plants are a source of food for all living organisms. Key factors influencing these interactions include size, shape, morphology, and surface-to-volume ratio, which are analyzed using various physicochemical characterization techniques. A few of the conventional methods such as seed germination, root, shoot elongation, determination of chlorophyll content, secondary metabolites, and antioxidant enzyme activities are used to assess the toxicity of NMs on plants. But these are not sufficient for understanding the cellular-level interaction with macromolecules. Recently, several histochemical, microscopical, and analytical methods have been developed for understanding the toxicity of NMs in living systems. These methods can fairly accurately predict the toxicity at a molecular and cellular level, which marks a significant advancement in nanotoxicity assessment. This chapter explores the latest developments in techniques and methods for the assessment of nanotoxicity in plants and their implications in the safety evaluation of food products.

Keywords Nanotechnology · Phytonanotoxicity · Photosynthesis · Toxicity assessment

M. Lavanya · K. Selvam
Department of Botany, Periyar University, Salem, Tamil Nadu, India

M. S. Shivakumar (✉)
Department of Biotechnology, Periyar University, Salem, Tamil Nadu, India

R. N. Pudake et al. (eds.), *Phyto Nanotoxicology*, Nanotechnology in Plant Sciences 7, https://doi.org/10.1007/978-3-032-24518-2_6

6.1 Introduction

In the twenty-first century, nanotechnology is considered an emerging technology across various research domains, including healthcare, energy, environmental protection, construction, agriculture, and food processing, owing to their unique physicochemical characteristics in their nanoscale (Malik et al. 2023). Nanotechnology is rapidly advancing and creating new opportunities across various fields. This progress stems from the development of nanomaterials, nanostructures, and nanosystems. Diverse research areas, ranging from organic chemistry to semiconductor physics, contribute to the fabrication of nanoparticles and nanostructured materials, enabling innovations from new materials design to atomic-scale matter manipulation. It is expected that nanomaterials and their products will become an integral part of everyday life soon. However, the continuous use of NMs has raised concerns about short and long-term toxicological effects on human health and the ecosystem. Several studies have been done for the assessment of nanotoxicity in animals and humans (Huang et al. 2021). These studies have shown that NMs produces a variety of sub-lethal toxic manifestations in animals, ranging from cytotoxicity, DNA damage, activation of stress genes, lowering of antioxidant capacity, and affecting the reproductive system (Bano et al. 2022). Very few studies have been undertaken for the assessment of the toxicity of NMs in plants.

NMs are used in plants as nanopesticides, nanofertilizers, growth stimulants, and as antibiotics for increasing crop productivity and for the control of diseases and insect damage (Bharti et al. 2025). NMs in agriculture has reduced the requirement of the aforesaid products for crop productivity and has contributed to a reduction in environmental footprint. But at the same time, NMs, due to their ability to penetrate the cell membrane (as in the case of metal nanoparticles), make it unpalatable and affect the nutritional value of the crop products. Consuming these low-nutritional and toxic foods is hazardous to the human and animal population, additionally can lead to biomagnification in the food chain. Hence, assessing the toxicity of NPs in plants is crucial. For toxicity assessment, conventional tools and protocols are inadequate and often require advanced and accurate characterization techniques for assessing their safety and understanding their interaction with biological systems. In this chapter, various advanced equipment and methods used for the characterization of phytotoxicity of nanomaterials and their toxicity assessment methods are discussed, along with the future development in this area.

6.2 Synthesis of Nanoparticles (NPs)

Top-down and bottom-up approaches are the two methods for synthesizing nanoparticles (Fig. 6.1). In the top-down method, large-sized materials are reduced to their nanoscale through thermal decomposition, ball milling, laser ablation, and lithographic techniques. On the contrary, the bottom-up approach uses nanoparticles

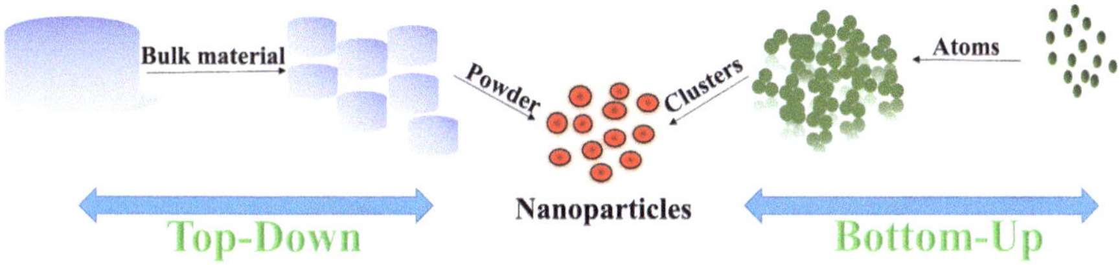

Fig. 6.1 Synthesis of nanoparticles using Top-down and Bottom-up approaches

formed by the self-assembly of smaller atoms or molecules, which include processes like pyrolysis, chemical vapor deposition, hydrothermal, sol-gel, and green synthesis methods (de Oliveira et al. 2020).

Chemical, physical, and biological methods are the three ways by which nanomaterials (NMs) can be synthesized (Dikshit et al. 2021). Physical synthesis is often chosen for producing NMs with high purity and uniform size and shapes. Physical methods include plasma chemical processes, vapor deposition, microwave irradiation, pulsed laser ablation, sono-chemical reduction, and gamma radiation (Harish et al. 2023). Chemical synthesis methods include polyol, microemulsions, thermal decomposition, ball milling, and electrochemical synthesis (Ogbezode et al. 2023). Physical and chemical synthesis is expensive, requiring high input cost and skilled manpower. The synthesis process also produces hazardous by-products (Ying et al. 2022).

Given these shortcomings, biologically synthesized nanoparticles are increasingly being used due to their biocompatibility and low toxicity, which minimize adverse effects on living organisms and biodiversity. This approach is rapid, simple, and cost-effective, utilizing plants, fungi, algae, yeast, lichen, and bacteria as reducing agents, thus eliminating the need for chemical additives (Zeebaree et al. 2023). Biological synthesis uses metals and metal oxides, which are reduced using plant-based metabolites for the synthesis of nanoparticles. Biologically synthesized NMs have significant applications in environmental and biomedical fields, including wastewater treatment, pollution control, pesticide formulation, disease diagnosis, drug delivery, food packaging and preservation (Dhaka et al. 2023; Mahajan et al. 2017).

6.3 Applications of Nanomaterials

Nanotechnology is a boon to the world because of its diverse range of applications in different fields such as pharmaceuticals, environment, agriculture, textiles, and cosmetics (Fig. 6.2).

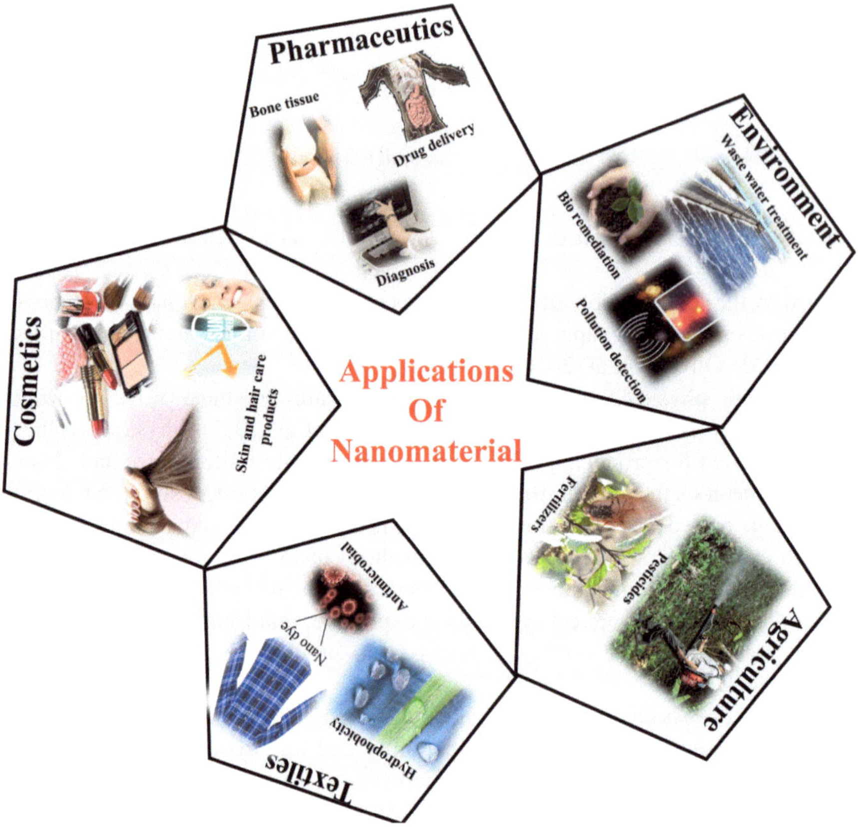

Fig. 6.2 Applications of nanomaterials

6.3.1 Pharmaceutical Applications

The pharmaceutical industry increasingly relies on nano drug delivery systems over conventional methods for their targeted delivery and controlled release at specific sites (Afzal et al. 2022). Unlike traditional systems, which suffer from poor absorption and adverse effects, nano systems enhance therapeutic efficacy and minimize side effects (Cheng et al. 2023). For example, recently Jahangirian and team (2017) reported that molybdenum disulfide nanosheets were used for tumor therapy. Lipid nanoparticles were used for drug delivery in nucleic acids (Mitchell et al. 2021). A study conducted by Azadpour et al. (2022) showed that folic acid-functionalized chitosan nanoparticles for co-delivery of imatinib and silver nanoparticles exhibit effective drug delivery on colon cancer.

Nanotechnology has been extensively used in tissue engineering for repairing, replacing, and improving the functioning of tissues and organs. These nano-engineered tissues and organs closely mimic the structure and physiology of their natural counterparts (Hasan et al. 2018). Carbon nanotubes, graphene oxide, metal nanoparticles, polymeric and composite-based nanomaterials, are used in the regeneration of bone tissue, skin tissue, neural tissues, and dental tissues (Zheng et al. 2021). Incorporation of hydroxyapatite nanoparticles with polycaprolactone is used in bone tissue engineering (Murugan and Parcha 2021). Studies using Cerium oxide have shown that it promotes osteogenesis differentiation and bone regeneration (Ren et al. 2022).

The integration of bioimaging and nanotechnology has transformed disease diagnosis and management by providing non-invasive, near-real-time feedback with high accuracy (Kumar and Mahajan 2024). Enhanced bioimaging techniques utilize nanomaterials to detect and monitor diseases with exceptional sensitivity and specificity. These advancements facilitate early disease detection, improving diagnostic accuracy and enabling timely medical interventions (Wang et al. 2024). Molecular imaging in disease diagnosis is achieved through the use of polymers, liposomes, ultra-small super-paramagnetic iron oxide nanoparticles, and gold nanoparticles.

6.3.2 Environmental Applications

Nanomaterials are increasingly being used for wastewater treatment and for sustainable improvement of water quality. Environmentally friendly manufacturing of nanomaterials avoids harmful by-products, making them efficient for water purification (Epelle et al. 2022; Bhattacharyya et al. 2025). Carbon nanotubes (CNTs), graphene oxides, silver, gold, zinc, copper, magnesium, and titanium oxide nanoparticles are extensively used due to their mechanical actions, including adsorption, degradation of contaminants, and removal of disinfectants. These properties make them highly effective in water purification and remediation processes (Jangid and Inbaraj 2021).

Soil remediation involves the removal of contaminants like hydrocarbons, heavy metals, and pesticides from soil. Advanced materials such as carbon nanotubes, bionanoparticles, and polymeric nanoparticles enhance this process by effectively degrading and removing pollutants, thereby improving the efficiency and sustainability of soil remediation efforts (Qian et al. 2020).

Gas sensors operate by converting gas concentrations into electrical signals and are valued for their affordability, rapid response, sensitivity, and selectivity. Additionally, they integrate with electronic systems without requiring extra wiring or calibration. These features make gas sensors indispensable in applications such as environmental monitoring, air quality assessment, vehicle exhaust analysis, medical diagnostics, and quality control in food and cosmetics (Khan et al. 2022).

6.3.3 Agricultural Application

Nanopesticides are finding increasing application in sustainable agriculture due to their efficiency, stability, and water solubility (Dangi and Verma 2021). Their nanoscale size enables controlled release and targeted delivery, reducing pesticide use and environmental harm (Tang et al. 2023; Preetha et al. 2024). Nano-fertilizers are being used, facilitating increased bioavailability for plants and for improving crop yield (Semenova et al. 2024). Mesoporous nanosilica-based fertilizers are highly stable and have high efficiency in being transported into plant cells. This extends the duration of fertilizer releases and reduces the input cost to the farmer (Liu et al. 2021; Basak et al. 2024). Nanotechnology also finds application in protecting the plant against abiotic stress like high temperatures, salinity, and drought (Jiang et al. 2021). Engineered nanomaterials can reduce oxidative damage and heavy metal toxicity, thereby enhance plant resilience and promote better growth under adverse conditions (Nawaz et al. 2023; Pudake et al. 2024).

6.3.4 Applications for Textiles

Nanotechnology is currently being used for producing durable fabrics with antibacterial properties (Makvandi et al. 2021). The integration of metal nanoparticles such as Ag, Zn, Cu, and carbon quantum dots has been instrumental in developing antibacterial fabrics (Bibi et al. 2024). Improved stain and water repellency of fabrics is achieved by integrating silica nanoparticles and hydrocarbon-based nanowhiskers. These nanomaterials form a peach-fuzz-like structure on the fabric's surface, creating high surface tension, preventing liquids from absorbing into the fibers (Yetisen et al. 2016).

6.3.5 UV Absorption

The depletion of the ozone layer has increased UV radiation exposure, leading to serious health issues like skin cancer and cataracts, as well as damaging textiles. To combat these effects, integrating nanomaterials such as TiO_2, ZnO, SiO_2, and Al_2O_3 into fabrics provides effective UV protection. These nanoparticles enhance fabric durability and functionality by absorbing and scattering UV rays, contributing to the development of UV-resistant clothing and extending material lifespan (Syduzzaman et al. 2023).

6.3.6 Applications in Cosmetics

Sunscreens containing nanomaterials like ZnO and TiO_2 effectively retain their UV filtration and absorption properties while eliminating the white, chalky appearance of traditional sunscreens. These nanoparticles not only provide seamless, transparent protection against harmful UV rays but also aid in repairing already damaged skin by reducing oxidative stress and promoting skin regeneration (Singh et al. 2020). Gold and silver nanomaterials penetrate deeply into the skin and help to reduce wrinkles, improve skin hydration, and promote a youthful appearance (Dhawan et al. 2020). Nanotechnology enhances shampoos and conditioners by improving nutrient delivery to the hair and scalp, boosting strength and shine. Nanoparticles also protect hair from environmental damage like UV light. Additionally, silver nanoparticles can effectively control dandruff, alleviating the social discomfort it may cause for individuals affected by the condition (Yukuyama et al. 2020).

6.4 Nanotoxicology

Nanomaterials are widely utilized in medicine, agriculture, cosmetics, and textiles. However, the increased nanomaterial use on medical and environmental fields has raised serious concerns about their safety in humans, animals, and plants. To address these concerns, researchers study the adverse effects of engineered nanomaterials or nanoparticles on living organisms through the field of nanotoxicology (Egbuna et al. 2021).

In the development of nanomaterials, toxicity is considered one of the primary challenges among chemists, engineers, and scientists. Accumulation of nanomaterials leads to acute or chronic toxicity in living organisms. The toxicity of nanomaterials (NMs) varies based on their physicochemical properties such as morphology, magnetic, electrical, optical, size, surface area, composition, crystalline structure, agglomeration, and concentration (Mazari et al. 2021). Nanoparticle toxicity was determined based on their shapes and size. A decrease in the size of NPs will result in a significant increase in the surface-to-volume ratio, which will lead to increased toxicity and reactivity of the NPs. Furthermore, the smaller size of NPs increases their ability to penetrate plant and animal tissues. Similarly, the shapes of the nanoparticles also determine their toxicity (Abbasi et al. 2023). Disposal of nanomaterials in the environment severely affects humans, animals, and plants (Fig. 6.3). Moreover, all living organisms directly or indirectly rely on plants for their food chain. The toxicity of nanomaterials (NMs) in different organisms, such as humans, animals, and plants, is discussed below.

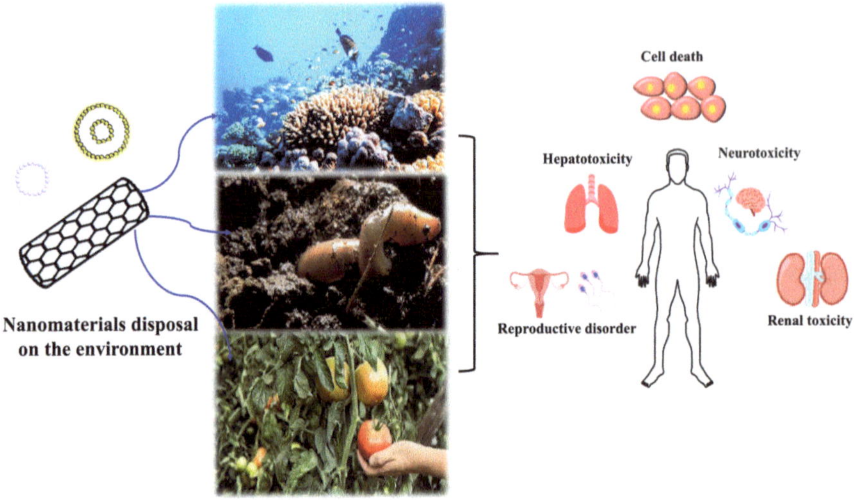

Fig. 6.3 Bioaccumulation of NMs causes toxicity in different organisms

6.4.1 *Toxicity on Humans*

Nanoparticles can enter the human body via inhalation, ingestion, or skin contact, potentially damaging cells, tissues, and organs. Accumulated NPs like CNTs, Ag, Au, TiO_2, ZnO, and SiO_2 can reach vital organs, causing nervous, endocrine, immune, reproductive, and cellular toxicity (Xuan et al. 2023; Pietroiusti et al. 2018). Gold NPs (5–80 nm) impair fertility and longevity via ROS generation (Yang et al. 2010). Small-sized TiO_2 NPs (<10 nm) and AgNPs induce immunotoxicity and cytotoxicity, respectively (Sajid et al. 2015). *In-vitro* methods such as microfluidic systems and cell viability assays can be used for NP toxicity assessment (Xuan et al. 2023).

6.4.2 *Toxicity on Aquatic Animals*

NPs pose serious risks to aquatic ecosystems due to their widespread usage and accumulation. They cause mechanical and physiological damage to aquatic organisms like fish and crustaceans (Jahan et al. 2017). In brine shrimp (Artemia), NPs induce DNA damage, apoptosis, and reduced hatching, indicating genotoxic and developmental toxicity. In daphnids, NPs such as titanium dioxide, fullerenes, cerium oxide, and diamond nanoparticles disrupt molting behavior and swimming speed, increasing mortality rates and affecting survival and reproduction. In tadpoles, NP accumulation in the gut impairs growth and reproduction. In rainbow trout, NPs accumulate on the gills, disrupting respiration and reducing locomotion, which compromises their survival in natural habitats (Bundschuh et al. 2016). According to Venâncio et al. (2022) deposition of polymethylmethacrylate nanoplastics on *Xenopus laevis* caused severe deformities; hence, the growth and survival rate ultimately decreased. Similarly, the polystyrene microplastics and nanoplastics affect the nutritional value, inhibit the feeding and physiological behaviors of *Chlamys farreri* (Sun et al. 2024). Nanoparticles such as Cd, As, Cu, Zn, Pb, and Ti cause adverse effects on humans during the high intake of *Cyprinus carpio* fish (Nikokherad et al. 2022).

6.4.3 Toxicity on Aquatic Plants

An aquatic ecosystem consists of both macrophytes and microphytes (phytoplankton, diatoms, and algae). These aquatic plants are vital components of aquatic ecosystems, which provide food and oxygen for animals (Taub 2018). The symbiotic relationship between the algae and microbes ensures nutrient cycling, water quality, and ecosystem stability (Saravanan et al. 2021). Additionally, some algae, such as Ulva, Nostoc, Chlorella, and Spirulina, are edible with high nutritional and medicinal value (Wang et al. 2019). But the release of NMs from households and industries interacts with the aquatic plants and reduces their photosynthetic activities, as well as affects their photosynthetic organelles (Lau et al. 2022). Exposure to Ag^+ ions from AgNPs induces toxic manifestations in duckweed due to overproduction of free radicals resulting in damage to cell membrane, proteins and DNA (Souza et al. 2021). Generation of ROS following exposure to AgNPs causes lethal effects on *Pseudokirchneriella subcapitata*, while in *Chlamydomonas reinhardtii*, AgNPs affect the photosystem II, thereby impacting the photosynthetic process (Sibiya et al. 2022). Accumulated polystyrene nanoplastics on *Spirulina platensis* result in reduced cell growth, dry weight, and decreased chlorophyll content. Contamination of polystyrene nanoplastics in aquatic environments leads to adverse impact on *Scenedesmus* sp and *Nitrobacter vulgaris* (Nagaraja et al. 2025). Exposure of TiO_2 NPs resulted in stunted growth in *Spirodela polyrrhiza* (Movafeghi et al. 2018).

6.4.4 Bioaccumulation of NMs in Food Crops and Green Leafy Vegetables

Carbohydrates are produced through photosynthesis in plants and are considered an essential nutrient for humans, animals, and microbes (Cleri and Cleri 2016). Plant roots absorb nanomaterials released by the pharmaceutical, cosmetic, food packaging industries, and agricultural applications. These absorbed nanomaterials are transported to various parts of the plant (Gowtham et al. 2024). Hence, morphology, physiology, and molecular activities are severely affected by these NMs exposures

(Tripathi et al. 2022). According to a report of Dev et al. (2018), exposure to Fe_3O_4, TiO_2, and carbon nanoparticles has adversely affected seed germination and root elongation in cucumbers. A study carried out by Biswas and Pal (2024) found that exposure to ZnO NPs adversely affected photosynthetic efficiency and oxidative stress in *Zea mays*, *Phaseolus vulgaris*, *Triticum aestivum*, and *Pisum sativam.* In rice plants, the interaction of AgNPs was reported to inhibit root development, cause tissue death, and lead to a significant decline in soluble carbohydrates (Tripathi et al. 2024).

The bioaccumulation of NPs on aquatic algae, crop products, and green leafy vegetables causes a severe health impact on humans and animals when consumed. Bioaccumulation of gold NPs on the leaves of tomatoes leads to adverse effects on human beings (Sharma et al. 2022). Mahmoud et al. (2021) reported that the toxicity effect of ZnO NPs on Indian river chicks did not show toxicity after 42 days of exposure. Similarly, ZnO NPs bioaccumulation leads to respiratory damage in Crab-eating macaques upon 14 days of exposure (Park et al. 2021).

6.4.5 Toxicity on Soil Organisms

The soil ecosystem is home to organisms, including bacteria, fungi, actinomycetes, invertebrates, and overground crops. Soil biota is vital for nutrient recycling and maintaining soil stability. In soil ecosystems, nanomaterials largely accumulate through industrial effluents and anthropogenic activities, which can adversely affect soil microfauna and reduce soil fertility (Xu et al. 2022). Nanoparticles such as Ag, TiO_2, ZnO, CeO_2, and Fe_3O_4 can disrupt the enzymatic activity and diversity of microbial communities. In order to maintain the functioning of the soil ecosystem, earthworms play an indispensable role because they are the largest animal group in terrestrial biomass. Toxicity of NPs affects the earthworm's community at the molecular and cellular level, as well as the reproduction, which leads to the mortality of earthworms (Chen et al. 2024). Ag ions released from AgNPs and $AgNO_3$NPs exhibit toxicity on *Folsomia candida* (Waalewijn-Kool et al. 2014). According to a study by Li et al. (2021b), the deposition of NMs on *Caenorhabditis elegans* causes neurotoxicity, nephrotoxicity, and reproductive toxicity. The exposure of TiO_2NPs (Abdel-Azeem et al. 2023) and ZnO NP (Abdel-Azeem and Osman 2021) leads to digestive gland alteration in *Helix aspersa*.

6.5 Uptake of Nanomaterials on Plants and Induction of Phytotoxicity

Nanomaterials (NMs) enter plants through root absorption, stomatal pores, hydathodes, and seed coats via applications of pesticides, fertilizers, and pollutants. Once inside, they move through xylem and phloem, potentially causing phytotoxic effects

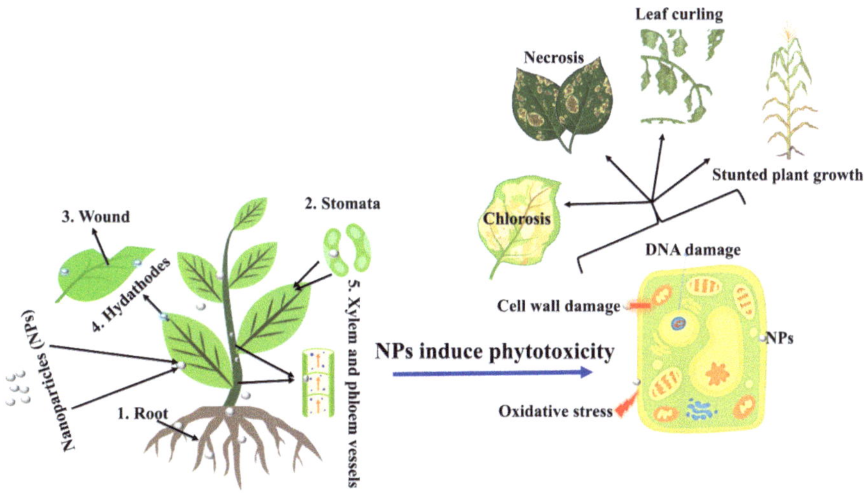

Fig. 6.4 Accumulated nanomaterials (NMs) induce phytotoxicity. The toxicity of NMs are entered through 1. Root absorption, 2. Stomatal aperture, 3. Wound, 4. Hydathode (Guttation) causes phytotoxicity

like leaf curling, chlorosis, necrosis, and stunted growth (Fig. 6.4) (Singh et al. 2022; Eid et al. 2025). These effects are visible morphologically, but internal toxicity must be assessed to understand cellular damage, photosynthetic efficiency, antioxidant responses, metabolic changes, and nutritional impact. Detection of NMs penetration and toxicity is achieved through morphological, anatomical, biochemical, and analytical techniques, which are discussed below.

6.6 Phytotoxicity Assessment of NMs Using Different Methods

6.6.1 Morphological Methods

6.6.1.1 Seed Germination, Shoot and Root Elongation

The accumulation of NMs during the application of nanopesticides or the disposal of industrial pollutants causes toxicity in plants. Nanoparticle toxicity can be assessed by morphological characteristics such as seed germination, germination index, root shoot elongation, and leaf surface area (Fig. 6.5). The toxicity of NMs is assessed by an *in vitro* method using plant growth media. Methods include seed soaking, hydroponics, and root dipping in NP solutions, allowing NP transfer from roots to shoots via xylem and phloem (Wang et al. 2020). NPs can also inhibit germination by penetrating seed coats and disrupting cell membranes (Verma et al. 2021). ZnONPs significantly inhibited germination and growth in cucumber, wheat, and

Fig. 6.5 Morphological assessment of nanomaterials toxicity on plants via seed germination, roots, shoot elongation, and leaf surface area measurements

green gram (Kumar et al. 2015). Green-synthesized AgNPs reduced germination, vigor index, and root/shoot length in mung beans (Anwar et al. 2021). However, halloysite nanotubes showed no effect on wheat seed germination (Chen et al. 2021).

Zhao et al. (2021) reported that yttrium oxide nanoparticles (Y_2O_3NPs) delayed rice seed germination at 50–100 mg/L and inhibited root elongation at 20 mg/L under hydroponic conditions. In *Coffea arabica*, ZnONPs increased biomass, while $ZnSO_4$ reduced it (Rossi et al. 2019). Andersen et al. (2016) found that TiO_2 affected 8 of 10 plant species, reducing germination, delaying cotyledon development, and shortening roots. CeO_2 had minimal effects but slightly altered root and cotyledon lengths in cucumber, lettuce, onion, ryegrass, and tomato. Milewska-Hendel et al. (2022) observed that negatively charged AuNPs (25 and 50 μg/mL) enhanced root diameter in *Hordeum vulgare* but reduced root hair formation at higher concentrations. These studies highlight that NP toxicity is determined by their concentration, plant species, and NP characteristics, significantly influencing plant growth and development.

6.6.2 *Anatomical Methods*

6.6.2.1 Microscopic Analysis

Microscopic techniques such as light microscopes, stereomicroscopes, confocal microscopes, and electron microscopes, namely SEM and TEM, are used to detect and analyze nanoparticle accumulation and morphological changes in plant tissues at various magnification levels (Fig. 6.6) (Sun et al. 2014). These techniques are used to visualize the interaction or accumulation of NMs within the plant cells and tissues. The effect of ZnONPs on the root of onion was observed under the light microscope, which showed damaged roots due to the toxicity of ZnONPs (Al-Khazali 2023). Hasan et al. (2021) reported that AgNPs at 100 ppm concentration produced cell death in *Lactuca sativa*. Similarly, TEM study shows polystyrene nanoplastics toxicity on leaves of *L. sativa,* accumulation of nanoplastics on leaf surface in the form of dark black spot and was associated with chloroplast structural damage (Lian et al. 2021). Liu et al. (2022) reported the translocation of nano-sized and micro-sized polystyrene in rice observed by confocal laser scanning microscopy, showing

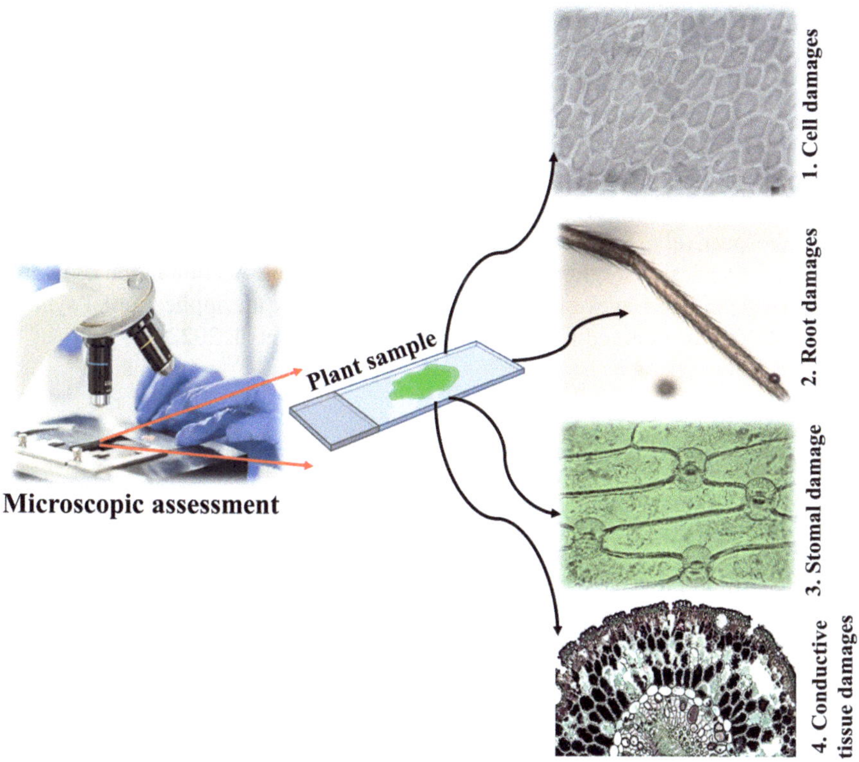

Fig. 6.6 Toxicity assessment of nanomaterials (NMs) using microscopic and histochemical assessment

accumulation of NPs in roots, stems, and leaves of rice. The toxicity effect of CuONPs on barley root and leaves, assessed using SEM, shows damaged roots, a decrease in root trichomes, and stomatal density. While the TEM analysis shows disturbed cell wall integrity of epidermis, endodermis, vacuolization, fragments disorganization in ER, mitochondrial swelling, and mitochondrial cristae destruction in barley crop (Rajput et al. 2018).

6.6.2.2 Histochemical Analysis

Histochemical studies are used to assess the chemical compositions or locations of plant metabolites such as phenols, flavonoids, tannins, alkaloids, lignins, starch and other antioxidant enzymes, etc. on the plant tissues or surfaces using some staining such as Toluidine Blue O (TBO), Iodine-Potassium Iodide (IKI), Sudan III, Congo red, Calcofluor White, Nitroblue tetrazolium (NBT), and 3′3′-diaminobenzidine (DAB) (Badria and Aboelmaaty 2019). Histochemical study on *Vigna radiata* using NBT and 3′-(p-hydroxyphenyl) fluorescein indicates increased ROS levels in roots in a dose-dependent manner. Phloroglucinol and hydrochloric acid (Ph-HCl) staining shows an increased level of lignification in roots. Findings of these studies confirm the increased level of ROS and lignin deposition as a result of CuONP application in *Vigna radiata* (Gopalakrishnan Nair et al. 2014). Increased concentration of AgNPs also results in increased production of SO- free radicals in roots of *Vigna radiate* as observed by NBT staining (Nair and Chung 2015). Green-synthesized ZnO NPs from *Agathosma betulina* were used to reduce the salt stress in sorghum plants. The histochemical study on control and ZnONPs-treated sorghum plants showed higher superoxide production and decreased superoxide production, respectively. These changes are noted based on the appearances of dark blue spots on the leaves of sorghum plants using NBT staining. Accumulation of hydrogen peroxide on the leaves of sorghum was prevented by the application of synthesized ZnONPs as observed by DAB staining (Rakgotho et al. 2022).

6.6.3 *Biochemical Methods*

6.6.3.1 Estimation of Chlorophyll Content

In plants, chlorophyll is due to its ability to fix sunlight into chemical energy. Reduced chlorophyll content can lead to chlorosis, resulting in reduced plant growth (Fig. 6.7) (Rafique et al. 2018). The accidental release of NMs results in their accumulation on the leaf surface and in chloroplasts, affecting the electron transport chain and light-harvesting complexes. Hence, the estimation of the amount of chlorophyll content in the plants is essential for determining the phytotoxicity of nanoparticles. High concentration of TiO_2NPs adversely reduces the chlorophyll content on *Zea mays* (Dağhan 2018). Wang et al. (2018) reported that the application

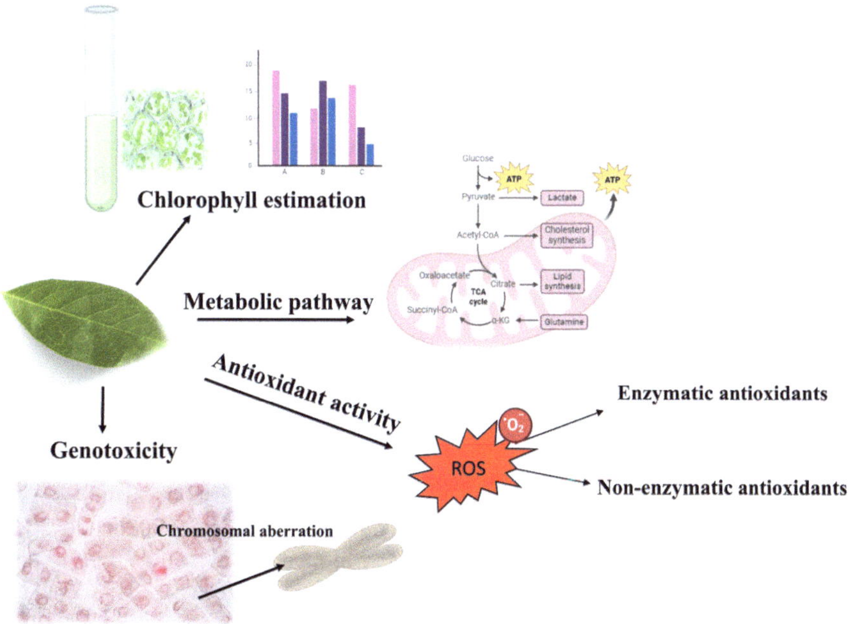

Fig. 6.7 Toxicity determination of nanomaterials (NMs) using biochemical methods such as chlorophyll a, b estimation, metabolic activity, antioxidant activity, genotoxicity, and cytotoxicity methods

of synthesized ZnONPs on tomato plants showed decreased chlorophyll a and b content, which subsequently led to a reduction in photosynthetic efficiency. Another study conducted by Răcuciu et al. (2022), magnetite NPs coated with aspartic acid exhibited high chlorophyll content in the maize crop. *Avicennia marina* mediated copper NPs revealed concentration-dependent chlorophyll content (Essa et al. 2021). In the presence of AgNPs, *Lemna gibba* showed a decrease in chlorophyll content (Dewez et al. 2018). Similarly, iron oxide NPs caused mortality in *Lemna minor* due to the decreased chlorophyll contents (Souza et al. 2019).

6.6.3.2 Assessment of Secondary Metabolites and Metabolic Pathways

Plants produce primary and secondary metabolites, which play an essential role in growth, reproduction, and protection from predators. Engineered NMs exposure on plants affects the production of both primary and secondary metabolites. These metabolic variations of plants can be assessed by qualitative and quantitative analysis (Večeřová et al. 2016). Shahhoseini et al. (2020) study shows that foliar applications of ZnO NPs on *Tanacetum parthenium* increased the essential oil titers compared to the control. Whereas the secondary metabolite Parthenolide levels were decreased. Kruszka et al. (2022) study shows that NPs such as Ag, Au, Cu, Pd, Ag and CuO NPs induced the changes in secondary metabolites in the cell suspension culture of *Hypericum perforatum*. The production of secondary metabolites in *Thymus daenensis* is influenced by single-walled carbon nanotubes in a concentration-dependent manner (Samadi et al. 2021). Exposure of AgNPs caused metabolic changes in shikimate-phenylpropanoid biosynthesis, tryptophan, and glucose metabolism pathways in *Arabidopsis thaliana* (Ke et al. 2018). Rose petals mediated CuONPs increase the amount of phytoconstituents, proline, and malondialdehyde content in *Solanum lycopersicum* and at the same time increase ROS levels in the plant (Solanki and Khan 2024).

6.6.3.3 Enzymatic and Nonenzymatic Antioxidants

During development, plants face environmental stresses like temperature, pH, metals, drought, and salinity. To cope, they activate an antioxidant defense system comprising enzymatic (ascorbate peroxidase, catalase, superoxide dismutase) and non-enzymatic (ascorbic acid, flavonoids) components (Juárez-Maldonado 2023). NMs affect these systems dose-dependently by inducing ROS and disrupting mitochondrial function (Hernández-Hernández et al. 2019; Pudake et al. 2019). ZnO NPs significantly increased catalase, peroxidase, and superoxide dismutase activity in tomato plants (Faizan et al. 2018). Konate et al. (2018) found that bulk iron oxide NPs boosted superoxide dismutase in roots, while nano-sized forms suppressed it in both roots and shoots.

6.6.3.4 Genotoxicity and Cytotoxicity Assessment

Genotoxicity and cytotoxicity assessment are used to study the effect of NMs toxicity disturbing the plants' genetic materials, such as genes, DNA, chromosomes, cell divisions, and their normal physiological functions (Badr et al. 2021). Genotoxicity and cytotoxicity effects of ZnONPs were assessed on the plants such as *Allium cepa*, *Nicotiana tabacum*, and *Vicia faba*. Comet assay shows significant DNA damage in the root of *A. cepa*, and *N. tabacum* (Ghosh et al. 2016). The genotoxicity and cytotoxicity effect of root tips of *Vicia faba* shows increased chromosomal aberrations, micronuclei, and reduced mitotic index at different concentrations of AgNPs exposure (Patlolla et al. 2012).

6.6.4 Analytical Methods

Analytical methods are used to evaluate the physicochemical properties of NMs, such as size, shape, crystalline structures, and their biocompatibility. These properties are used to determine the toxicity effects of accumulated NMs on the plant

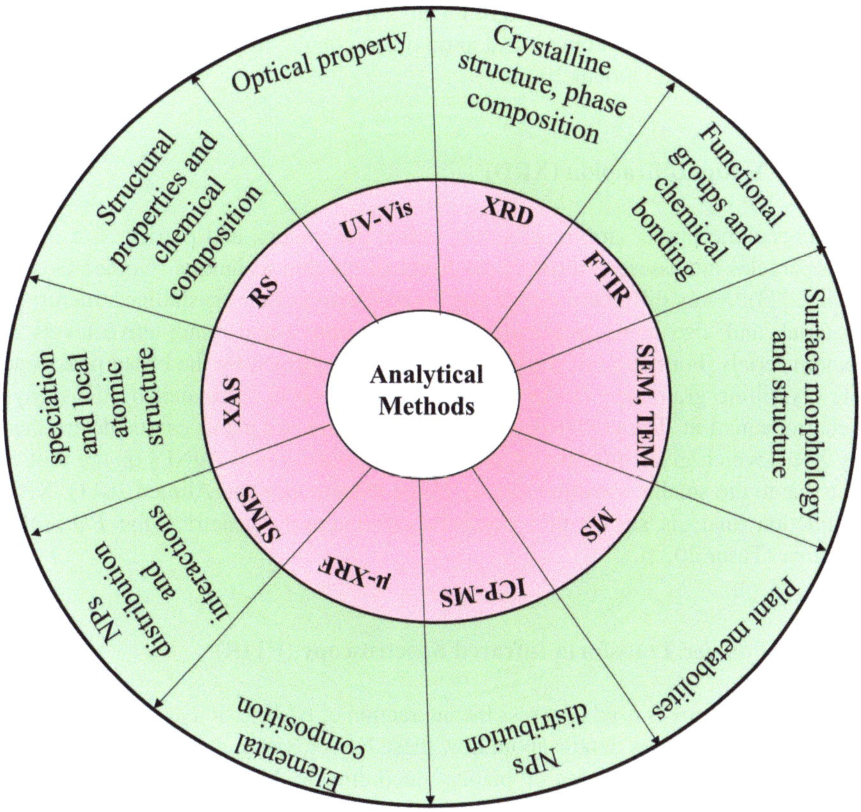

Fig. 6.8 Phytotoxicity assessment of nanomaterials (NMs) using different analytical methods

surface and their intercellular activities (Fig. 6.8). There are a few analytical methods discussed below in phytotoxicity assessment of nanomaterials.

6.6.4.1 UV-Visible Spectroscopy

UV-Vis spectroscopy is a simple, reliable, and widely used characterization technique for analyzing nano-sized materials such as transition metal ions, highly conjugated organic compounds, and biological macromolecules. It effectively assesses the size, shape, concentration, surface properties, and optics of metal nanoparticles in colloidal solution. The formation of different metallic nanoparticles from their respective metallic salts results in characteristic absorption peaks that can be monitored using UV-Vis spectroscopy (Vaseghi and Nematollahzadeh 2020). Surface plasmon resonance (SPR) is a key feature of metallic nanoparticles, like Au, Ag, and Cu, which depends on the size and shape of nanomaterials and has applications in catalytic, surface-enhanced Raman scattering, and biosensor studies

(Nurrohman and Chiu 2021). Additionally, a UV-Vis spectrometer is used to estimate the chlorophyll content and antioxidant content in plants at specific wavelengths (Feseha and Gholap 2020).

6.6.4.2 X-Ray Diffraction (XRD)

The crystalline phase, atomic structure, lattice parameters, and particle size of the nanoparticles are assessed through the X-ray diffraction technique (Gómez-Gómez et al. 2023). X-ray diffractometer of the XRD determines the crystalline structure of material, and also helps in identifying the qualitative and quantitative levels of nanomaterials. Formation of the particles can be confirmed by the broad peaks and the crystalline grain size of the nanomaterials, which is calculated by the Debye Scherrer equation. The phytotoxic effect of AgNPs on rice plants can be determined by XRD, which gives information about the accumulation of AgNPs on the root of rice due to the small crystalline size of 25 nm (Mazumdar and Ahmed 2011). XRD diffraction analysis shows that CuONPs can intensely penetrate the *Phaseolus vulgaris* (Tasar 2022).

6.6.4.3 Fourier Transform Infrared Spectroscopy (FTIR)

FTIR spectroscopy is used to assess the interaction of NMs toxicity on the plant cells and tissues. It provides details about how these NMs toxicity changes the functional groups or vibrational modes of plants. According to Morales et al. (2013), the toxicity of CeO_2 NPs on *Coriandrum sativum* estimated using FTIR analysis shows conformational and vibrational changes with the presence of lipids, carbohydrates, lignins, and amide in the NPs treated root and shoots. Similarly, CuONPs treated peanut leaves show changes in the biochemical constituents such as protein, chlorophyll, and carbohydrate (Suresh et al. 2016). AgNPs-treated rice roots show significant changes in intensity and shift position of peaks compared to untreated rice roots (Mazumdar and Ahmed 2011).

6.6.4.4 Electron Microscopy

The size, shape, and ultrastructure of nanoparticles affect plants. These are estimated by using a scanning electron microscope (SEM) and a transmission electron microscope (TEM). The microscopic images provide information about the phytotoxicity of NMs on plant cell walls, membranes, and organelles. Methods such as TEM-EDS are used for detecting the bioaccumulation of Prussian blue nanoparticles (PBNPs) in rice plants (Xu et al. 2024). PBNP-treated rice plants showed structural alterations such as distorted mitochondria, disrupted cristae, swollen chloroplasts, plasmolysis, and damaged vascular tissues, it was observed under TEM analysis. While SEM images exhibited the damaged root hairs and deformations of stomata. This may affect the absorption of nutrients and gaseous exchanges. SEM analysis on barley root exhibited cracked, broken, and bulged root morphology at the higher concentration of $MgZnAl_2O_4$. Additionally, HR-TEM provided an aggregation and cluster of NPs in roots. The toxicity of NPs was deposited on the periplasmic spaces, cell wall, cytoplasm, vacuole, and nucleus. These findings clearly showed that MgZnAl2O4 NPs toxicity adversely affects the growth and physiology of barley (Alghofaili et al. 2025).

6.6.4.5 Raman Spectroscopy (RS)

The vibrational and rotational patterns, as well as the electronic structures of nanomaterials, are effectively identified through Raman spectroscopy. Advanced techniques in Raman spectroscopy, such as stimulated Raman spectroscopy, surface-enhanced Raman spectroscopy, and resonance Raman spectroscopy, have further expanded its capabilities and are particularly valuable in applications like bioimaging and biofunctional analysis. Additionally, it was used to analyze the interaction of NPs with plants and assess their biochemical changes (Mudalige et al. 2019; Fernández-Galiana et al. 2024). The toxicity of Al_2O_3NPs on *Zea maize* was determined by confocal micro-Raman spectroscopy. Maize plants treated with NPs exhibited decreased biochemicals such as cellulose, carbohydrates, carotenoid, lignin, proteins, pectin, and aliphatic compounds compared to untreated maize. It occurred due to the toxicity of NPs (Tripathi et al. 2025).

6.6.4.6 Mass Spectroscopy (MS)

Liquid chromatography Mass spectrometry (LC-MS) and Gas chromatography Mass spectrometry (GC-MS) are the two main analytical techniques (Mazumdar et al. 2023). These are used to evaluate the presence of volatile and non-volatile compounds in the plants. Li et al. (2021a) found that in LC-MS analysis, the foliar applications of ZnO and SiO2 nanoparticles were showed increased levels of amino acids, flavonols, quercetin, rutin, and quercetin 3-O-glucoside in *Cucumis sativus*. Hence, the result of this study shows that NPs significantly influence both primary and secondary metabolites. According to a previous report, GC–MS analyses of cucumber fruits exposed to Cu NPs revealed significant changes in primary metabolite composition. These changes include reductions in sugars and amino acids, as well as significant changes in the profiles of organic acids and fatty acids (Zhao et al. 2017). Wu et al. (2017) reported that exposure of TiO_2 NPs on rice plants showed a significant increase in the concentration of sugars and organic acids. The concentrations of glucose-6-phosphate, glucose-1-phosphate, succinic acid and isocitric acid were increased at the exposure to the highest concentration of TiO_2 NPs (500 mg/L), but the highest concentration inhibited the metabolites such as sucrose, isomaltulose, and glyoxylic acid.

6.6.4.7 Inductively Coupled Plasma Mass Spectrometry (ICP-MS)

ICP-MS is an analytical technique used for quantifying the distribution or uptake of metal and metal oxide nanoparticles in plant cells and tissues. Single-particle inductively coupled plasma mass spectrometry (spICP-MS) analysis shows the fractionation of AgNPs on soybean callus culture (da Silva and Arruda 2023). Wheat seeds grown in a medium containing different concentrations of ZnONPs, show an increased amount of Zn content in the NPs treated seeds in a concentration-dependent manner, measured using ICP-MS (Alsuwayyid et al. 2022). Similarly, accumulation of AgNPs was observed on the root and shoot of *Arabidopsis thaliana* (Bao et al. 2016). Physiological effects of accumulated Y/FITC-doped SiNPs on the leaves of sweet basil were observed using laser ablation inductively coupled plasma mass spectroscopy (LA-ICP-MS) (Ko et al. 2019).

6.6.4.8 Micro X-Ray Fluorescence (μ-XRF)

Accumulation and distribution of nanoparticles on the plant tissues or surfaces and their interactions are determined by μ-XRF. In *Zea mays*, elemental distribution of CeO_2NPs was determined by μ-XRF mapping, showing distribution of Cu, K, Mn, and Zn in the kernels into cobs, but Ca and Fe were found in other parts of the kernels. Whereas ZnONPs-treated plants showed a reduced amount of Zn and Ca. Finding of this study suggests, CeO2 and ZnO NPs compromise the yield and nutritional quality of corn, which leads to serious implications for food safety (Zhao et al. 2015). Similarly, a cucumber plant grown under different concentrations of CeO_2 and ZnO NPs supplemented soils, assessed using μ-XRF, shows uptake of Ce was observed in the leaf and fruit of the cucumber. μ-XRF study shows the presence of K, Ca in the lamina and veins, while accumulation of ZnO NPs was observed in the root, leaf, stem, and fruits (Zhao et al. 2013). Distribution of Lanthanum oxide nanoparticles (La_2O_3) on the leaves of *Pfaffia glomerata* was determined by μ-XRF, which showed the presence of accumulated La in the entire leaf of *P. glomerata* (Neves et al. 2019).

6.6.4.9 Secondary Ion Mass Spectrometry (SIMS)

Secondary Ion Mass Spectrometry is used to detect trace elements, isotopic composition, and molecular fragments within plant tissues (McPhail 2006). Leaves of *Lactuca sativa* assessed using ToF-SIMS showed the presence of Ag^+ ions on the surface of the leaf cuticle (Larue et al. 2014). ToF-SIMS analysis of AgNPs and CeO_2NPs effect on *Triticum aestivum*, *Brassica napus*, and *Hordeum vulgare* showed the deposition of Ag and Ce into ions and particulates (Wagener et al. 2019).

6.6.4.10 X-Ray Absorption Spectroscopy (XAS)

XAS is used in phytotoxicity assessment of nanoparticles. It provides detailed information about the speciation, oxidation state, and local atomic structure of nanoparticles in plants. There are two types of XAS namely X-ray Absorption near Edge Structure and Extended X-ray Absorption Fine Structure (Chandrakasan et al. 2021). A study conducted by Duran et al. (2017) reported the deposition of Cu NPs in the seed coat, embryo, and the radicle of *Phaseolus vulgaris* seeds. Similarly, XAS results show that seeds of *Parkinsonia florida*, *Prosopis juliflora-velutina,* and *Salsola tragus* grown in ZnO NPs solution show biotransformed roots with the presence of zinc ions (De La Rosa et al. 2011). A similar study conducted by López-Moreno et al. (2010) demonstrated the accumulation of nanoceria on the plant tissues such as *Medicago sativa*, *Zea mays*, *Cucumis sativus,* and *Lycopersicon esculentum.*

6.7 Future of Nanomaterials

Nanotechnology has transformed healthcare by enabling sustained drug release, targeted delivery, organ replacement, and advancements in medical devices. Similarly nano nano-based drugs will become smarter with targeted delivery of drugs to the affected cells and organs. This reduces the need for higher doses and is target specific. Green-synthesized nanomaterials using biological materials are popular for their cost-effectiveness and biocompatibility (Lavanya et al. 2024; Alqarni et al. 2022). Furthermore, wireless nanochips and devices aid in early disease diagnosis, including heart attacks and blood sugar prediction (Kaushik et al. 2022). The portability and efficiency of these nanosensors make them invaluable tools for disease detection and management, offering promising solutions to improve human health (Ma et al. 2024).

In agriculture, the use of nanourea has already started and is expected to be widely used in agriculture. The use of nanourea has improved the yield and reduced the fertilizers used in the field. Nevertheless, uncontrolled release of urea and fertilizers can also pollute the environment, soil acidification, leaching, and contamination of groundwater (Ashitha et al. 2021). In this regard, further research needs to be conducted on the fate of nanourea and nanofertilizers in terrestrial and aquatic ecosystems to produce an environmentally sustainable solution.

Pesticides have been the mainstay of agriculture for the protection of crops from diseases and insect pests (Mubeen et al. 2023). Pesticides, though effective on insect pests, also result in environmental contamination and toxic effects on non-target organisms (Ofuya et al. 2023). Nanotechnology has reduced the amount of active ingredients used in pesticides and has been effective in reducing the pesticide input in the agroecosystems. Though there are reports of toxic effects of nanopesticides, further research in developing smart pesticides that are stable and release active ingredients in a need-based manner is underway. Further biopolymer-based pesticides incorporating plant or microbial-based bioactive principles are being tested to further reduce the environmental footprint of chemical pesticides. Looking ahead, the development of smart field systems equipped with wireless sensors will enable eco-friendly pesticide applications, allowing for targeted release in response to pest invasions (Kookana et al. 2014).

6.8 Conclusion

Today, with the development of nanotechnology and its use in several industries, it has gained attention among scientists and researchers. In nanotechnology, the synthesis of nanoparticles using green resources is highly appreciated by the World Health Organization (WHO) for being eco-friendly, non-hazardous, and cost-effective. With increasing use of NMs, the problem of nanotoxicity in animals, plants, and the environment has to be addressed, as the current regulatory measures regarding the toxicity tests for conventional industrial products are not sufficient. Newer tests should be specifically designed for the assessment of the toxicity of nanomaterials. These should include long-term and short-term studies encompassing all the environmental niches. Especially the marine and aquatic ecosystems to be focused more as these are fragile. Further advanced instruments and protocols should be made available to all the stakeholders, and a strict regulatory framework/measure should be enforced. Biosensors need to be developed for quick identification/toxicity assessment of NMs, with the advanced instruments, NMs can be easily assessed for their toxicity and aid in developing safe NMs for human use and environmental applications.

References

Abbasi R, Shineh G, Mobaraki M, Doughty S, Tayebi L (2023) Structural parameters of nanoparticles affecting their toxicity for biomedical applications: a review. J Nanopart Res 25(3):43

Abdel-Azeem HH, Osman GY (2021) Oxidative stress and histopathological effect of zinc oxide nanoparticles on the garden snail Helix aspersa. Environ Sci Pollut Res 28:9913–9920

Abdel-Azeem HH, Osman GY, Mohamed AH (2023) Potential toxic effects of titanium oxide (TiO2) nanoparticles on the biological, biochemical, and histological aspects of the land snail Helix aspersa. Environ Sci Pollut Res 30(32):78127–78138

Afzal O, Altamimi AS, Nadeem MS, Alzarea SI, Almalki WH, Tariq A, Mubeen B, Murtaza BN, Iftikhar S, Riaz N, Kazmi I (2022) Nanoparticles in drug delivery: from history to therapeutic applications. Nanomaterials 12(24):4494

Alghofaili F, Tombuloglu H, Almessiere MA, Tombuloglu G, Akhtar S, Turumtay EA, Baykal A, Turumtay H (2025) Phytotoxicity and growth enhancement properties of magnesium and zinc co-doped aluminum oxide nanoparticles on barley (Hordeum vulgare L.). Environ Sci Pollut Res 32:1–21

Al-Khazali ZK (2023) Phytotoxic nature of different concentrations of zinc oxide nanoparticles on Allium Cepa L. Root Meristem cells. J Nanostruct 13(4):1168–1183
Alqarni LS, Alghamdi MD, Alshahrani AA, Nassar AM (2022) Green nanotechnology: recent research on bioresource-based nanoparticle synthesis and applications. J Chem 2022(1): 4030999
Alsuwayyid AA, Alslimah AS, Perveen K, Bukhari NA, Al-Humaid LA (2022) Effect of zinc oxide nanoparticles on Triticum aestivum L. and bioaccumulation assessment using ICP-MS and SEM analysis. J King Saud Univ Sci 34(4):1–8
Andersen CP, King G, Plocher M, Storm M, Pokhrel LR, Johnson MG, Rygiewicz PT (2016) Germination and early plant development of ten plant species exposed to titanium dioxide and cerium oxide nanoparticles. Environ Toxicol Chem 35(9):2223–2229
Anwar N, Mehmood A, Ahmad KS, Hussain K (2021) Biosynthesized silver nanoparticles induce phytotoxicity in Vigna radiata L. Physiol Mol Biol Plants 27(9):2115–2126
Ashitha A, Rakhimol KR, Mathew J (2021) Fate of the conventional fertilizers in the environment. Controlled release fertilizers for sustainable agriculture. pp 25–39. Academic Press, Cambridge, London
Azadpour A, Hajrasouliha S, Khaleghi S (2022) Green synthesized-silver nanoparticles coated with targeted chitosan nanoparticles for smart drug delivery. J Drug Delivery Sci Technol 74:103554
Badr A, El-Shazly HH, Mohamed HI (2021) Plant responses to induced genotoxicity and oxidative stress by chemicals. In: Induced genotoxicity and oxidative stress in plants, pp 103-–131. Springer, Singapore
Badria FA, Aboelmaaty WS (2019) Plant histochemistry: a versatile and indispensible tool in localization of gene expression, enzymes, cytokines, secondary metabolites and detection of plants infection and pollution. Acta Sci Pharm Sci 3(7):88–100
Bano I, Skalickova S, Arbab S, Urbankova L, Horky P (2022) Toxicological effects of nanoselenium in animals. J Anim Sci Biotechnol 13(1):72
Bao D, Oh ZG, Chen Z (2016) Characterization of silver nanoparticles internalized by Arabidopsis plants using single particle ICP-MS analysis. Front Plant Sci 7:32
Basak S, Bhattacharyya P, Pudake RN, Lokhande PE, Rednam U, Chakrabarti S (2024) Metal-organic framework as nanocarriers for agricultural applications: a review. Front Nanotechnol 6: 1385981
Bharti A, Hooda V, Singh A, Pudake RN, Jain U, Chauhan N (2025) Advancing sustainable crop yield enhancement through environmentally friendly nanoparticle synthesis: a comprehensive review. J Soil Sci Plant Nutr 25:1–32
Bhattacharyya P, Pudake RN, Chakrabarti S (2025) A silver functionalized ZIF-8 derived heterostructure with excellent visible-light-driven photocatalytic removal of antibiotics. Front Nanotechnol 7:1477247
Bibi A, Afza G, Afzal Z, Farid M, Sumrra SH, Hanif MA, Jinadasa BKKK, Zubair M (2024) Synthetic vs. natural antimicrobial agents for safer textiles: a comparative review. RSC Adv 14(42):30688–30706
Biswas A, Pal S (2024) Plant-nano interactions: a new insight of nano-phytotoxicity. Plant Physiol Biochem 210:108646
Bundschuh M, Seitz F, Rosenfeldt RR, Schulz R (2016) Effects of nanoparticles in fresh waters: risks, mechanisms and interactions. Freshw Biol 61(12):2185–2196
Chandrakasan G, Ayala MT, Trejo JFG, Marcus G, Carroll DL (2021) Mapping and distribution of speciation changes of metals from nanoparticles in environmental matrices using synchrotron radiation techniques. Environ Nanotechnol Monitor Manag 16:100491
Chen L, Guo Z, Lao B, Li C, Zhu J, Yu R, Liu M (2021) Phytotoxicity of halloysite nanotubes using wheat as a model: seed germination and growth. Environ Sci Nano 8(10):3015–3027
Chen H, Yang L, Zhao S, Xu H, Zhang Z (2024) Long-term toxic effects of iron-based metal-organic framework nanopesticides on earthworm-soil microorganism interactions in the soil environment. Sci Total Environ 917:170146

Cheng X, Xie Q, Sun Y (2023) Advances in nanomaterial-based targeted drug delivery systems. Front Bioeng Biotechnol 11:1177151

Cleri F, Cleri F (2016) Energy production and storage for life. In: The Physics of living systems, pp 113–158. Cham, Springer Int. Publ

da Silva ABS, Arruda MAZ (2023) Exploring single-particle ICP-MS as an important tool for the characterization and quantification of silver nanoparticles in a soybean cell culture. Spectrochim Acta B At Spectrosc 203:106663

Dağhan HATİCE (2018) Effects of TiO_2 nanoparticles on maize (Zea mays L.) growth, chlorophyll content and nutrient uptake. Appl Ecol Environ Res 16:1–11

Dangi K, Verma AK (2021) Efficient & eco-friendly smart nano-pesticides: emerging prospects for agriculture. Mater Today Proc 45:3819–3824

De La Rosa G, Lopez-Moreno ML, Hernandez-Viezcas JA, Montes MO, Peralta-Videa J, Gardea-Torresdey J (2011) Toxicity and biotransformation of ZnO nanoparticles in the desert plants Prosopis juliflora-velutina, Salsola tragus and Parkinsonia florida. Int J Nanotechnol 8(6–7): 492–506

de Oliveira PF, Torresi RM, Emmerling F, Camargo PH (2020) Challenges and opportunities in the bottom-up mechanochemical synthesis of noble metal nanoparticles. J Mater Chem A 8(32): 16114–16141

Dev A, Srivastava AK, Karmakar S (2018) Nanomaterial toxicity for plants. Environ Chem Lett 16: 85–100

Dewez D, Goltsev V, Kalaji HM, Oukarroum A (2018) Inhibitory effects of silver nanoparticles on photosystem II performance in Lemna gibba probed by chlorophyll fluorescence. Curr Plant Biol 16:15–21

Dhaka A, Mali SC, Sharma S, Trivedi R (2023) A review on biological synthesis of silver nanoparticles and their potential applications. Results Chem 6:101108

Dhawan S, Sharma P, Nanda S (2020) Cosmetic nanoformulations and their intended use. In: Nanocosmetics, pp 141–169. Elsevier, Amsterdam, Netherlands

Dikshit PK, Kumar J, Das AK, Sadhu S, Sharma S, Singh S, Gupta PK, Kim BS (2021) Green synthesis of metallic nanoparticles: applications and limitations. Catalysts 11(8):902

Duran NM, Savassa SM, Lima RGD, de Almeida E, Linhares FS, van Gestel CA, Pereira de Carvalho HW (2017) X-ray spectroscopy uncovering the effects of Cu based nanoparticle concentration and structure on Phaseolus vulgaris germination and seedling development. J Agric Food Chem 65(36):7874–7884

Egbuna C, Parmar VK, Jeevanandam J, Ezzat SM, Patrick-Iwuanyanwu KC, Adetunji CO, Khan J, Onyeike EN, Uche CZ, Akram M, Ibrahim MS (2021) Toxicity of nanoparticles in biomedical application: nanotoxicology. J Toxicol 2021(1):9954443

Eid NA, Ibrahim AM, Elsharawy AA, Salem KF (2025) Nanofertilizers for plant viral disease management. In: Nanofertilizers for sustainable agriculture: assessing impacts on health, environment, and economy. Springer Nature Switzerland, Cham, pp 149–187

Epelle EI, Okoye PU, Roddy S, Gunes B, Okolie JA (2022) Advances in the applications of nanomaterials for wastewater treatment. Environments 9(11):141

Essa HL, Abdelfattah MS, Marzouk AS, Shedeed Z, Guirguis HA, El-Sayed MM (2021) Biogenic copper nanoparticles from Avicennia marina leaves: impact on seed germination, detoxification enzymes, chlorophyll content and uptake by wheat seedlings. PLoS One 16(4):0249764

Faizan M, Faraz A, Yusuf M, Khan ST, Hayat S (2018) Zinc oxide nanoparticle-mediated changes in photosynthetic efficiency and antioxidant system of tomato plants. Photosynthetica 56:678–686

Fernández-Galiana Á, Bibikova O, Vilms Pedersen S, Stevens MM (2024) Fundamentals and applications of Raman-based techniques for the design and development of active biomedical materials. Adv Mater 36(43):2210807

Feseha A, Gholap AV (2020) Determination of chlorophylls and carotenes content in some vegetables of ethiopiaby using UV-VIS spectroscopy. Ann Food Sci Technol 21(4):413

Ghosh M, Jana A, Sinha S, Jothiramajayam M, Nag A, Chakraborty A, Mukherjee A, Mukherjee A (2016) Effects of ZnO nanoparticles in plants: cytotoxicity, genotoxicity, deregulation of antioxidant defenses, and cell-cycle arrest. Mutat Res Genet Toxicol Environ Mutagen 807: 25–32

Gómez-Gómez AL, Martínez-Ayala AL, Moguel-Concha DDR, Borges-Martínez JE, Perea-Flores MDJ, Dávila-Ortiz G (2023) Relationship of nanomaterials' structure based on their application in the food industry: physicochemical and techno-functional characteristics. Appl Sci 13(12): 7167

Gopalakrishnan Nair PM, Kim SH, Chung IM (2014) Copper oxide nanoparticle toxicity in mung bean (Vigna radiata L.) seedlings: physiological and molecular level responses of in vitro grown plants. Acta Physiol Plant 36:2947–2958

Gowtham HG, Shilpa N, Singh SB, Aiyaz M, Abhilash MR, Nataraj K, Amruthesh KN, Ansari MA, Alomary MN, Murali M (2024) Toxicological effects of nanoparticles in plants: mechanisms involved at morphological, physiological, biochemical and molecular levels. Plant Physiol Biochem 210:108604

Harish V, Ansari MM, Tewari D, Yadav AB, Sharma N, Bawarig S, García-Betancourt ML, Karatutlu A, Bechelany M, Barhoum A (2023) Cutting-edge advances in tailoring size, shape, and functionality of nanoparticles and nanostructures: a review. J Taiwan Inst Chem Eng 149: 105010

Hasan A, Morshed M, Memic A, Hassan S, Webster TJ, Marei HES (2018) Nanoparticles in tissue engineering: applications, challenges and prospects. Int J Nanomedicine 13:5637–5655

Hasan M, Mehmood K, Mustafa G, Zafar A, Tariq T, Hassan SG, Loomba S, Zia M, Mazher A, Mahmood N, Shu X (2021) Phytotoxic evaluation of phytosynthesized silver nanoparticles on lettuce. Coatings 11(2):225

Hernández-Hernández H, Quiterio-Gutiérrez T, Cadenas-Pliego G, Ortega-Ortiz H, Hernández-Fuentes AD, Cabrera de la Fuente M, Valdés-Reyna J, Juárez-Maldonado A (2019) Impact of selenium and copper nanoparticles on yield, antioxidant system, and fruit quality of tomato plants. Plants 8(10):355

Huang HJ, Lee YH, Hsu YH, Liao CT, Lin YF, Chiu HW (2021) Current strategies in assessment of nanotoxicity: alternatives to in vivo animal testing. Int J Mol Sci 22(8):4216

Jahan S, Yusoff IB, Alias YB, Bakar AFBA (2017) Reviews of the toxicity behavior of five potential engineered nanomaterials (ENMs) into the aquatic ecosystem. Toxicol Rep 4:211–220

Jahangirian H, Lemraski EG, Webster TJ, Rafiee-Moghaddam R, Abdollahi Y (2017) A review of drug delivery systems based on nanotechnology and green chemistry: green nanomedicine. Int J Nanomedicine 12:2957–2978

Jangid P, Inbaraj MP (2021) Applications of nanomaterials in wastewater treatment. Mater Today Proc 43:2877–2881

Jiang M, Song Y, Kanwar MK, Ahammed GJ, Shao S, Zhou J (2021) Phytonanotechnology applications in modern agriculture. J Nanobiotechnol 19:1–20

Juárez-Maldonado A (2023) Influence of nanomaterials on non-enzymatic antioxidant defense activities in plants. In: Nanomaterial interactions with plant cellular mechanisms and macromolecules and agricultural implications. Springer International Publishing, Cham, pp 273–298

Kaushik S, Soni V, Skotti E (eds) (2022) Nanosensors for futuristic smart and intelligent healthcare systems. CRC Press, Boca Raton, pp 1–108

Ke M, Qu Q, Peijnenburg WJGM, Li X, Zhang M, Zhang Z, Lu T, Pan X, Qian H (2018) Phytotoxic effects of silver nanoparticles and silver ions to Arabidopsis thaliana as revealed by analysis of molecular responses and of metabolic pathways. Sci Total Environ 644:1070–1079

Khan Y, Sadia H, Ali Shah SZ, Khan MN, Shah AA, Ullah N, Ullah MF, Bibi H, Bafakeeh OT, Khedher NB, Eldin SM (2022) Classification, synthetic, and characterization approaches to nanoparticles, and their applications in various fields of nanotechnology: a review. Catalysts 12(11):1386

Ko JA, Furuta N, Lim HB (2019) New approach for mapping and physiological test of silica nanoparticles accumulated in sweet basil (Ocimum basilicum) by LA-ICP-MS. Anal Chim Acta 1069:8–35

Konate A, Wang Y, He X, Adeel M, Zhang P, Ma Y, Ding Y, Zhang J, Yang J, Kizito S, Rui Y (2018) Comparative effects of nano and bulk-Fe3O4 on the growth of cucumber (Cucumis sativus). Ecotoxicol Environ Saf 165:547–554

Kookana RS, Boxall AB, Reeves PT, Ashauer R, Beulke S, Chaudhry Q, Cornelis G, Fernandes TF, Gan J, Kah M, Lynch I (2014) Nanopesticides: guiding principles for regulatory evaluation of environmental risks. J Agric Food Chem 62(19):4227–4240

Kruszka D, Selvakesavan RK, Kachlicki P, Franklin G (2022) Untargeted metabolomics analysis reveals the elicitation of important secondary metabolites upon treatment with various metal and metal oxide nanoparticles in Hypericum perforatum L. cell suspension cultures. Ind Crop Prod 178:114561

Kumar PPP, Mahajan R (2024) Gold polymer nanomaterials: a promising approach for enhanced biomolecular imaging. Nanotheranostics 8(1):64

Kumar S, Patra AK, Datta SC, Rosin KG, Purakayastha TJ (2015) Phytotoxicity of nanoparticles to seed germination of plants. Int J Adv Res 3(3):854–865

Larue C, Castillo-Michel H, Sobanska S, Cécillon L, Bureau S, Barthès V, Ouerdane L, Carrière M, Sarret G (2014) Foliar exposure of the crop Lactuca sativa to silver nanoparticles: evidence for internalization and changes in Ag speciation. J Hazard Mater 264:98–106

Lau ZL, Low SS, Ezeigwe ER, Chew KW, Chai WS, Bhatnagar A, Yap YJ, Show PL (2022) A review on the diverse interactions between microalgae and nanomaterials: growth variation, photosynthetic performance and toxicity. Bioresour Technol 351:127048

Lavanya M, Selvam K, Prakash P, Shivaswamy MS (2024) Antioxidant, antimicrobial, α-amylase and α-glucosidase inhibitory activities of methanolic leaf extract of Breynia vitis-idaea using in vitro and in silico techniques. Process Biochem 136:156–168

Li S, Liu J, Wang Y, Gao Y, Zhang Z, Xu J, Xing G (2021a) Comparative physiological and metabolomic analyses revealed that foliar spraying with zinc oxide and silica nanoparticles modulates metabolite profiles in cucumber (Cucumis sativus L.). Food Energy Secur 10(1):e269

Li Y, Zhong L, Zhang L, Shen X, Kong L, Wu T (2021b) Research advances on the adverse effects of nanomaterials in a model organism, Caenorhabditis elegans. Environ Toxicol Chem 40(9): 2406–2424

Lian J, Liu W, Meng L, Wu J, Chao L, Zeb A, Sun Y (2021) Foliar-applied polystyrene nanoplastics (PSNPs) reduce the growth and nutritional quality of lettuce (Lactuca sativa L.). Environ Pollut 280:116978

Liu C, Zhou H, Zhou J (2021) The applications of nanotechnology in crop production. Molecules 26(23):7070

Liu Y, Guo R, Zhang S, Sun Y, Wang F (2022) Uptake and translocation of nano/microplastics by rice seedlings: evidence from a hydroponic experiment. J Hazard Mater 421:126700

López-Moreno ML, de la Rosa G, Hernández-Viezcas JA, Peralta-Videa JR, Gardea-Torresdey JL (2010) X-ray absorption spectroscopy (XAS) corroboration of the uptake and storage of CeO2 nanoparticles and assessment of their differential toxicity in four edible plant species. J Agric Food Chem 58(6):3689–3693

Ma X, Tian Y, Yang R, Wang H, Allahou LW, Chang J, Williams G, Knowles JC, Poma A (2024) Nanotechnology in healthcare, and its safety and environmental risks. J Nanobiotechnol 22(1): 715

Mahajan R, Bhadwal AS, Kumar N, Madhusudanan M, Pudake RN, Tripathi RM (2017) Green synthesis of highly stable carbon nanodots and their photocatalytic performance. IET Nanobiotechnol 11:360–364

Mahmoud MA, Yahia D, Abdel-Magiud DS, Darwish MH, Abd-Elkareem M, Mahmoud UT (2021) Broiler welfare is preserved by long-term low-dose oral exposure to zinc oxide nanoparticles: preliminary study. Nanotoxicology 15(5):605–620

Makvandi P, Iftekhar S, Pizzetti F, Zarepour A, Zare EN, Ashrafizadeh M, Agarwal T, Padil VV, Mohammadinejad R, Sillanpaa M, Maiti TK (2021) Functionalization of polymers and nanomaterials for water treatment, food packaging, textile and biomedical applications: a review. Environ Chem Lett 19:583–611
Malik S, Muhammad K, Waheed Y (2023) Nanotechnology: a revolution in modern industry. Molecules 28(2):661
Mazari SA, Ali E, Abro R, Khan FSA, Ahmed I, Ahmed M, Nizamuddin S, Siddiqui TH, Hossain N, Mubarak NM, Shah A (2021) Nanomaterials: applications, waste-handling, environmental toxicities, and future challenges–a review. J Environ Chem Eng 9(2):105028
Mazumdar H, Ahmed GU (2011) Phytotoxicity effect of silver nanoparticles on Oryza sativa. Int J ChemTech Res 3(3):1494–1500
Mazumdar S, Singhal S, Tyagi J, Mishra A, Joshi NC, Pudake RN, Farooq M, Varma A (2023) Understanding of belowground biochemical communication in millets through metabolomics. In: Millet Rhizosphere. Springer Nature, Singapore, pp 259–274
McPhail DS (2006) Applications of secondary ion mass spectrometry (SIMS) in materials science. J Mater Sci 41:873–903
Milewska-Hendel A, Gepfert W, Zubko M, Kurczyńska E (2022) Morphological, histological and ultrastructural changes in Hordeum vulgare (L.) roots that have been exposed to negatively charged gold nanoparticles. Appl Sci 12(7):3265
Mitchell MJ, Billingsley MM, Haley RM, Wechsler ME, Peppas NA, Langer R (2021) Engineering precision nanoparticles for drug delivery. Nat Rev Drug Discov 20(2):101–124
Morales MI, Rico CM, Hernandez-Viezcas JA, Nunez JE, Barrios AC, Tafoya A, Flores-Marges JP, Peralta-Videa JR, Gardea-Torresdey JL (2013) Toxicity assessment of cerium oxide nanoparticles in cilantro (Coriandrum sativum L.) plants grown in organic soil. J Agric Food Chem 61(26):6224–6230
Movafeghi A, Khataee A, Abedi M, Tarrahi R, Dadpour M, Vafaei F (2018) Effects of TiO2 nanoparticles on the aquatic plant Spirodela polyrrhiza: evaluation of growth parameters, pigment contents and antioxidant enzyme activities. J Environ Sci 64:130–138
Mubeen I, Mfarrej MFB, Razaq Z, Iqbal S, Naqvi SAH, Hakim F, Mosa WF, Moustafa M, Fang Y, Li B (2023) Nanopesticides in comparison with agrochemicals: outlook and future prospects for sustainable agriculture. Plant Physiol Biochem 198:107670
Mudalige T, Qu H, Van Haute D, Ansar SM, Paredes A, Ingle T (2019) Characterization of nanomaterials: tools and challenges. In: Nanomaterials for food applications, pp 313–353. Elsevier, Amsterdam, Netherlands
Murugan S, Parcha SR (2021) Fabrication techniques involved in developing the composite scaffolds PCL/HA nanoparticles for bone tissue engineering applications. J Mater Sci Mater Med 32(8):93
Nagaraja SS, Gouda Y, Miguez D, Muralidaran Y, Romanholo Ferreira LF, Américo-Pinheiro JHP, Mulla SI, Mishra P (2025) Distinctive toxic repercussions of polystyrene nano plastic towards aquatic non target species Nitrobacter vulgaris, Scenedesmus sp and Daphnia magna. Ecotoxicology 34(1):61–75
Nair PMG, Chung IM (2015) Physiological and molecular level studies on the toxicity of silver nanoparticles in germinating seedlings of mung bean (Vigna radiata L.). Acta Physiol Plant 37: 1–11
Nawaz A, Rehman HU, Usman M, Wakeel A, Shahid MS, Alam S, Sanaullah M, Atiq M, Farooq M (2023) Nanobiotechnology in crop stress management: an overview of novel applications. Discov Nano 18(1):74
Neves VM, Heidrich GM, Rodrigues ES, Enders MS, Muller EI, Nicoloso FT, Carvalho HWPD, Dressler VL (2019) La2O3 nanoparticles: study of uptake and distribution in Pfaffia glomerata (spreng.) pedersen by LA-ICP-MS and μ-XRF. Environ Sci Technol 53(18):10827–10834
Nikokherad H, Esmaili-Sari A, Moradi AM, Bahramifar N, Mostafavi PG (2022) Bioaccumulation capacity of Chlorella vulgaris and Spirulina platensis exposed to silver nanoparticles and silver nitrate: bio-and health risk assessment approach. Algal Res 64:102671

Nurrohman DT, Chiu NF (2021) A review of graphene-based surface plasmon resonance and surface-enhanced raman scattering biosensors: current status and future prospects. Nanomaterials 11(1):216
Ofuya TI, Okunlola AI, Mbata GN (2023) A review of insect pest management in vegetable crop production in Nigeria. Insects 14(2):111
Ogbezode JE, Ezealigo US, Bello A, Anye VC, Onwualu AP (2023) A narrative review of the synthesis, characterization, and applications of iron oxide nanoparticles. Discov Nano 18(1):125
Park J, Choi I, Lee SH, Song J, Lim HJ (2021) Repeated intratracheal instillation of zinc oxide nanoparticles induced pulmonary damage and a systemic inflammatory response in cynomolgus monkeys. Nanotoxicology 15(5):621–635
Patlolla AK, Berry A, May L, Tchounwou PB (2012) Genotoxicity of silver nanoparticles in Vicia faba: a pilot study on the environmental monitoring of nanoparticles. Int J Environ Res Public Health 9(5):1649–1662
Pietroiusti A, Stockmann-Juvala H, Lucaroni F, Savolainen K (2018) Nanomaterial exposure, toxicity, and impact on human health. Wiley Interdiscip Rev Nanomed Nanobiotechnol 10(5): e1513
Preetha JSY, Sriram D, Premasudha P, Pudake RN, Anand AV, Arun M (2024) Recent advancements and applications of nano-agrochemicals against plant fungal diseases. In: Nanoparticles in plant biotic stress management, pp 403–418. Springer Nature, Singapore
Pudake RN, Mittal J, Tripathi RM, Tyagi J, Mohanta TK (2019) Biochemical responses of maize seedlings exposed to SnNPs. Micro Nano Lett 14:645–649
Pudake RN, Tripathi RM, Gill SS (2024) Nanotechnology for abiotic stress tolerance and management in crop plants. Academic Press, Elsevier, London
Qian Y, Qin C, Chen M, Lin S (2020) Nanotechnology in soil remediation-applications-vs. implications. Ecotoxicol Environ Saf 201:110815
Răcuciu M, Tecucianu A, Oancea S (2022) Impact of magnetite nanoparticles coated with aspartic acid on the growth, antioxidant enzymes activity and chlorophyll content of maize. Antioxidants 11(6):1193
Rafique R, Zahra Z, Virk N, Shahid M, Pinelli E, Park TJ, Kallerhoff J, Arshad M (2018) Dose-dependent physiological responses of Triticum aestivum L. to soil applied TiO_2 nanoparticles: alterations in chlorophyll content, H2O2 production, and genotoxicity. Agric Ecosyst Environ 255:95–101
Rajput V, Minkina T, Fedorenko A, Sushkova S, Mandzhieva S, Lysenko V, Duplii N, Fedorenko G, Dvadnenko K, Ghazaryan K (2018) Toxicity of copper oxide nanoparticles on spring barley (Hordeum sativum distichum). Sci Total Environ 645:1103–1113
Rakgotho T, Ndou N, Mulaudzi T, Iwuoha E, Mayedwa N, Ajayi RF (2022) Green-synthesized zinc oxide nanoparticles mitigate salt stress in Sorghum bicolor. Agriculture 12(5):597
Ren S, Zhou Y, Zheng K, Xu X, Yang J, Wang X, Miao L, Wei H, Xu Y (2022) Cerium oxide nanoparticles loaded with nanofibrous membranes promote bone regeneration for periodontal tissue engineering. Bioact Mater 7:242–253
Rossi L, Fedenia LN, Sharifan H, Ma X, Lombardini L (2019) Effects of foliar application of zinc sulfate and zinc nanoparticles in coffee (Coffea arabica L.) plants. Plant Physiol Biochem 135: 60–166
Sajid M, Ilyas M, Basheer C, Tariq M, Daud M, Baig N, Shehzad F (2015) Impact of nanoparticles on human and environment: review of toxicity factors exposures, control strategies, and future prospects. Environ Sci Pollut Res 22:4122–4143
Samadi S, Saharkhiz MJ, Azizi M, Samiei L, Karami A, Ghorbanpour M (2021) Single-wall carbon nano tubes (SWCNTs) penetrate Thymus daenensis Celak. plant cells and increase secondary metabolite accumulation in vitro. Ind Crop Prod 165:113424
Saravanan A, Kumar PS, Varjani S, Jeevanantham S, Yaashikaa PR, Thamarai P, Abirami B, George CS (2021) A review on algal-bacterial symbiotic systems for effective treatment of wastewater. Chemosphere 271:129540

Semenova NA, Burmistrov DE, Shumeyko SA, Gudkov SV (2024) Fertilizers based on nanoparticles as sources of macro-and microelements for plant crop growth: a review. Agronomy 14(8):1646
Shahhoseini R, Azizi M, Asili J, Moshtaghi N, Samiei L (2020) Effects of zinc oxide nanoelicitors on yield, secondary metabolites, zinc and iron absorption of Feverfew (Tanacetum parthenium (L.) Schultz Bip.). Acta Physiol Plant 42:1–18
Sharma S, Shree B, Sharma A, Irfan M, Kumar P (2022) Nanoparticle-based toxicity in perishable vegetable crops: molecular insights, impact on human health and mitigation strategies for sustainable cultivation. Environ Res 212:113168
Sibiya A, Jeyavani J, Santhanam P, Preetham E, Freitas R, Vaseeharan B (2022) Comparative evaluation on the toxic effect of silver (Ag) and zinc oxide (ZnO) nanoparticles on different trophic levels in aquatic ecosystems: a review. J Appl Toxicol 42(12):1890–1900
Singh S, Pandey SK, Vishwakarma N (2020) Functional nanomaterials for the cosmetics industry. In: Handbook of functionalized nanomaterials for industrial applications, pp 717–730. Elsevier, Amsterdam, Netherlands
Singh S, Kandhol N, Pandey S, Singh VP, Tripathi DK, Chauhan DK (2022) Nitric oxide overcomes copper and copper oxide nanoparticle-induced toxicity in Sorghum vulgare seedlings through regulation of ROS and proline metabolism. Funct Plant Biol 50:183
Solanki B, Khan MS (2024) Phytotoxic impact of copper oxide nanoparticles fabricated from rose petals (Rosa indica) on germination, biological growth, and phytochemicals of tomato (Solanum lycopersicum). S Afr J Bot 172:125–139
Souza LRR, Bernardes LE, Barbetta MFS, da Veiga MAMS (2019) Iron oxide nanoparticle phytotoxicity to the aquatic plant Lemna minor: effect on reactive oxygen species (ROS) production and chlorophyll a/chlorophyll b ratio. Environ Sci Pollut Res 26:24121–24131
Souza LR, Corrêa TZ, Bruni AT, da Veiga MA (2021) The effects of solubility of silver nanoparticles, accumulation, and toxicity to the aquatic plant Lemna minor. Environ Sci Pollut Res 28:16720–16733
Sun D, Hussain HI, Yi Z, Siegele R, Cresswell T, Kong L, Cahill DM (2014) Uptake and cellular distribution, in four plant species, of fluorescently labeled mesoporous silica nanoparticles. Plant Cell Rep 33:1389–1402
Sun Y, Zhao X, Sui Q, Sun X, Zhu L, Booth AM, Chen B, Qu K, Xia B (2024) Polystyrene nanoplastics affected the nutritional quality of Chlamys farreri through disturbing the function of gills and physiological metabolism: comparison with microplastics. Sci Total Environ 910: 168457
Suresh S, Karthikeyan S, Jayamoorthy K (2016) FTIR and multivariate analysis to study the effect of bulk and nano copper oxide on peanut plant leaves. J Sci Adv Mater Dev 1(3):343–350
Syduzzaman M, Hassan A, Anik HR, Akter M, Islam MR (2023) Nanotechnology for high-performance textiles: a promising frontier for innovation. ChemNanoMat 9(9):e202300205
Tang Y, Zhao W, Zhu G, Tan Z, Huang L, Zhang P, Gao L, Rui Y (2023) Nano-pesticides and fertilizers: solutions for global food security. Nanomaterials 14(1):90
Tasar N (2022) Mitotic effects of copper oxide nanoparticle on root development and root tip cells of Phaseolus vulgaris L. seeds. Microsc Res Tech 85(12):3895–3907
Taub FB (2018) Standardized aquatic microcosm—development and testing. In: Aquaticecotoxicology. pp 47–92, CRC Press, Boca Raton, USA
Tripathi D, Singh M, Pandey-Rai S (2022) Crosstalk of nanoparticles and phytohormones regulate plant growth and metabolism under abiotic and biotic stress. Plant Stress 6:100107
Tripathi S, Mahra S, Sharma S, Mathew S, Sharma S (2024) Interaction of silver nanoparticles with plants: a focus on the phytotoxicity, underlying mechanism, and alleviation strategies. Plant Neurobiol 9:100082
Tripathi A, Awasthi A, Baran C, Uttam KN (2025) Rapid investigation of the alteration in biochemical profile of maize seedlings treated with aluminium oxide nanoparticles using spectroscopic techniques. Spectrochim Acta A Mol Biomol Spectrosc 330:125639

Vaseghi Z, Nematollahzadeh A (2020) Nanomaterials: types, synthesis, and characterization. In: Green synthesis of nanomaterials for bioenergy applications, pp 23–82. Hoboken, New Jersey
Večeřová K, Večeřa Z, Dočekal B, Oravec M, Pompeiano A, Tříska J, Urban O (2016) Changes of primary and secondary metabolites in barley plants exposed to CdO nanoparticles. Environ Pollut 218:207–218
Venâncio C, Melnic I, Tamayo-Belda M, Oliveira M, Martins MA, Lopes I (2022) Polymethylmethacrylate nanoplastics can cause developmental malformations in early life stages of Xenopus laevis. Sci Total Environ 806:150491
Verma DK, Patel S, Kushwah KS (2021) Effects of nanoparticles on seed germination, growth, phytotoxicity and crop improvement. Agric Rev 42(1):1–11
Waalewijn-Kool PL, Klein K, Forniés RM, van Gestel CA (2014) Bioaccumulation and toxicity of silver nanoparticles and silver nitrate to the soil arthropod Folsomia candida. Ecotoxicology 23: 1629–1637
Wagener S, Jungnickel H, Dommershausen N, Fischer T, Laux P, Luch A (2019) Determination of nanoparticle uptake, distribution, and characterization in plant root tissue after realistic long-term exposure to sewage sludge using information from mass spectrometry. Environ Sci Technol 53(9):5416–5426
Wang XP, Li QQ, Pei ZM, Wang SC (2018) Effects of zinc oxide nanoparticles on the growth, photosynthetic traits, and antioxidative enzymes in tomato plants. Biol Plant 62:801–808
Wang J, Zhang M, Fang Z (2019) Recent development in efficient processing technology for edible algae: a review. Trends Food Sci Technol 88:251–259
Wang W, Liu J, Ren Y, Zhang L, Xue Y, Zhang L, He J (2020) Phytotoxicity assessment of copper oxide nanoparticles on the germination, early seedling growth, and physiological responses in Oryza sativa L. Bull Environ Contam Toxicol 104:770–777
Wang B, Hu S, Teng Y, Chen J, Wang H, Xu Y, Wang K, Xu J, Cheng Y, Gao X (2024) Current advance of nanotechnology in diagnosis and treatment for malignant tumors. Signal Transduct Target Ther 9(1):200
Wu B, Zhu L, Le XC (2017) Metabolomics analysis of TiO2 nanoparticles induced toxicological effects on rice (Oryza sativa L.). Environ Pollut 230:302–310
Xu Z, Long X, Jia Y, Zhao D, Pan X (2022) Occurrence, transport, and toxicity of nanomaterials in soil ecosystems: a review. Environ Chem Lett 20(6):3943–3969
Xu K, Lan Y, Xing C, Rao Y, Zhu E, Yan J, Wang S, Wu XL, Cai M, Duan X (2024) Phytotoxicity of Prussian blue nanoparticles to rice and the related defence mechanisms: in vivo observations and physiological and biochemical analysis. Chemosphere 368:143724
Xuan L, Ju Z, Skonieczna M, Zhou PK, Huang R (2023) Nanoparticles-induced potential toxicity on human health: applications, toxicity mechanisms, and evaluation models. MedComm 4(4): e327
Yetisen AK, Qu H, Manbachi A, Butt H, Dokmeci MR, Hinestroza JP, Skorobogatiy M, Khademhosseini A, Yun SH (2016) Nanotechnology in textiles. ACS Nano 10(3):3042–3068
Ying S, Guan Z, Ofoegbu PC, Clubb P, Rico C, He F, Hong J (2022) Green synthesis of nanoparticles: current developments and limitations. Environ Technol Innov 26:102336
Yukuyama MN, De Araújo GLB, Bou-Chacra NA (2020) Nanomaterials for hair care applications. In: Nanocosmetics, pp 205–225. Elsevier, London
Zeebaree AYS, Zebari OIH, Zeebaree SYS, Thanoon RD, Yaseen YN (2023) Current contributions of novel nanoparticles-based colorimetric sensors for detection of SCN ions in different aqueous models. A review. Arab J Chem 16(12):105297
Zhao L, Sun Y, Hernandez-Viezcas JA, Servin AD, Hong J, Niu G, Peralta-Videa JR, Duarte-Gardea M, Gardea-Torresdey JL (2013) Influence of CeO2 and ZnO nanoparticles on cucumber physiological markers and bioaccumulation of Ce and Zn: a life cycle study. J Agric Food Chem 61(49):11945–11951
Zhao L, Sun Y, Hernandez-Viezcas JA, Hong J, Majumdar S, Niu G, Duarte-Gardea M, Peralta-Videa JR, Gardea-Torresdey JL (2015) Monitoring the environmental effects of CeO2 and ZnO

nanoparticles through the life cycle of corn (Zea mays) plants and in situ μ-XRF mapping of nutrients in kernels. Environ Sci Technol 49(5):2921–2928

Zhao L, Hu J, Huang Y, Wang H, Adeleye A, Ortiz C, Keller AA (2017) 1H NMR and GC–MS based metabolomics reveal nano-Cu altered cucumber (Cucumis sativus) fruit nutritional supply. Plant Physiol Biochem 110:138–146

Zhao X, Zhang W, He Y, Wang L, Li W, Yang L, Xing G (2021) Phytotoxicity of Y2O3 nanoparticles and Y3+ ions on rice seedlings under hydroponic culture. Chemosphere 263: 127943

Zheng X, Zhang P, Fu Z, Meng S, Dai L, Yang H (2021) Applications of nanomaterials in tissue engineering. RSC Adv 11(31):19041–19058

Yang Y, Qu Y, Lü X (2010) Global gene expression analysis of the effects of gold nanoparticles on human dermal fibroblasts. J Biomed Nanotechnol 6(3):234–246

Chapter 7
Analytical Methods to Characterize Nanomaterials' Interaction with Plant Systems

Silvia Carlo, Luca Pagano, Nelson Marmiroli, and Marta Marmiroli

Abstract The increasing use of engineered nanomaterials (ENMs) in agricultural practices offers promising benefits but also raises significant concerns about their interactions with plant systems and potential environmental impacts. This chapter provides an in-depth overview of advanced analytical methodologies that elucidate the localization, chemical speciation, and quantification of ENMs within plant tissues. Key techniques include electron and fluorescence microscopy, synchrotron-based X-ray spectroscopies, mass spectrometry, and single-cell analytical tools; together, these enable the investigation of nanoparticle uptake, translocation, biotransformation, and associated physiological responses. The integration of omics approaches and systems biology further provides a holistic framework for correlating the physical behavior of ENMs with molecular and cellular plant responses. The combined analytical strategies discussed here play a pivotal role in addressing the complexity of nanobiointeractions and in advancing the development of safe and effective nano-enabled agricultural technologies.

Keywords Analytical methods · Localization · Speciation · Quantification · Biotransformation · Nanobiointeractions

7.1 Introduction

Integrating nanotechnology into agriculture has emerged as a transformative approach, offering innovative solutions for crop enhancement, pest control, and environmental remediation. Engineered nanomaterials (ENMs), characterized by their nanoscale dimensions and unique physicochemical properties, have been

S. Carlo · M. Marmiroli (✉)
Department of Chemistry, Life Sciences and Environmental Sustainability, University of Parma, Parma, Italy
e-mail: marta.marmiroli@unipr.it

L. Pagano · N. Marmiroli
Consorzio Interuniversitario Nazionale per le Scienze Ambientali (CINSA), University of Parma, Parma, Italy

R. N. Pudake et al. (eds.), *Phyto Nanotoxicology*, Nanotechnology in Plant Sciences 7, https://doi.org/10.1007/978-3-032-24518-2_7

developed to enhance agricultural productivity and sustainability. Their application as nanofertilizers, nanopesticides, and nanosensors is expanding rapidly (Adisa et al. 2019; Bharti et al. 2025). Indeed, their small size and high surface area facilitate efficient nutrient delivery and targeted pest control, reducing the need for excessive chemical inputs compared to their bulk counterparts (Pagano et al. 2017; Preetha et al. 2024). Interestingly, ENMs can serve directly as nanopesticides and nanofertilizers or can be functionalized for use as nanocarriers, where active ingredients are encapsulated for direct delivery to target sites. This mechanism lowers the overall dosage required, minimizes environmental dispersion and contaminant build-up, and reduces costs. Comparative studies show that nanocarriers can match or exceed the efficacy of conventional formulations while often exhibiting lower toxicity (Kah et al. 2018). The same rationale applies to nanofertilizers: in China, up to 50% of applied nitrogen is lost through volatilization and another 5–10% through leaching. Nanometric formulations, whether composed directly of micro- and macronutrients or acting as carriers, alleviate these losses and mitigate associated environmental risks (Kah et al. 2018; Basak et al. 2024).

However, employing these materials in agriculture raises critical concerns about their interactions with plants and the broader environment. Once introduced into the soil, nanoparticles can be absorbed by plant roots and translocated to aerial parts, potentially accumulating in different tissues (Rico et al. 2011; Zhang and Su 2024). This highlights the need for a thorough understanding of their behavior, fate, and potential risks. Additionally, the processes of nanoparticle internalization and biotransformation within plants are intricate and not yet fully understood. Factors such as nanoparticle dissolution, aggregation, and interactions with plant biomolecules can alter their chemical form, affecting behavior and toxicity levels (Muzammil et al. 2023). Moreover, the accumulation of nanomaterials in edible plant parts raises concerns about their potential entry into the food chain and impacts on human health (Iavicoli et al. 2017). Investigating the molecular and physiological mechanisms of nanoparticle uptake, translocation, and biotransformation in plants is crucial to shed light on their effects on plant health and development.

Nanoparticles can enter the plant system through several routes, strictly dependent on the mode of application, whether foliar spray or soil amendment, and on species-specific traits (Biswas and Pal 2024). After foliar deposition, ENMs penetrate leaf tissues via both lipophilic and hydrophilic pathways, including direct entry through stomata and small wounds. Uptake efficiency depends strongly on particle size and surface chemistry, underscoring the need to identify precisely which nanomaterials cross each barrier. Once inside, ENMs bind to macromolecules such as proteins, RNAs, and photosynthates and are mobilized through the phloem from shoots to roots, often undergoing biotransformation that facilitates vascular transport (Hong et al. 2021). Because these processes remain poorly resolved, advanced analytical approaches such as synchrotron-based spectroscopies and high-resolution microscopies are essential for localizing ENMs *in planta* and determining their chemical speciation after uptake. For example, synchrotron-based X-ray fluorescence microscopy has been applied to investigate the systemic effects and improvements in food quality in winter wheat (*Triticum aestivum* L.) treated with foliar sprays of ZnO and $ZnSO_4$ nanoparticles. This study revealed that Zn-based nanoparticles significantly increased the zinc content in the grains, with the spatial distribution of Zn closely linked to the specific type of nanomaterial applied. Additionally, X-ray absorption spectroscopy indicated that, once internalized and accumulated in the grain, zinc undergoes biotransformation and is predominantly present as zinc phosphate, regardless of the nanoparticle source, providing important insights into the safety of the resulting edible products (Zhang et al. 2017).

Although the soil application of nanomaterials has been studied more extensively than foliar spraying, soil itself is a heterogeneous matrix whose physical, chemical, and biological complexity can alter nanoparticle form and reactivity. Interactions with soil microbiota and root exudates add further layers of abiotic and biotic transformation before ENMs even contact the root surface (Guo et al. 2021; Hortas 2025; Zhang et al. 2023). Techniques able to map both distribution and speciation at the root–soil interface are therefore critical for evaluating nanoparticle stability and bioavailability. Micro X-ray fluorescence (μXRF) analysis has demonstrated that cerium oxide nanoparticles (CeO_2 NPs), when applied as a soil amendment to barley (*Hordeum vulgare* L.), are predominantly retained in the soil and on the root surface, primarily in particulate form. Within the rhizosphere, cerium mainly occurs at the soil–root interface, where specific hotspots reveal substantial reduction of Ce(IV) to Ce(III). Interestingly, near these redox-active regions, a limited but detectable internalization of cerium is observed, primarily in the form of CeO_2 nanoparticles. These findings suggest that the soil microenvironment plays a crucial role in the partial reduction of Ce(IV) to Ce(III), enhancing cerium bioavailability and uptake by the plant (Rico et al. 2018).

Regardless of soil conditions, the diversity of root architectures makes it challenging to gain mechanistic insight into root uptake. Root surfaces carry a net negative charge due to the exudation of organic acids, which favours the adsorption of positively charged ENMs (Galindo-Castañeda et al. 2024; Sun et al. 2023). Internalization may proceed via both symplastic and apoplastic pathways, with subsequent biotransformation enabling xylem loading or accumulation in intra- or intercellular compartments (Lv et al. 2019). Because ENM behavior within roots varies with both plant species and particle characteristics, each plant–nanomaterial combination must be examined individually to resolve its specific uptake, translocation, and transformation dynamics.

Another critical aspect to consider when working with ENMs is that their interaction with plants can trigger a broad spectrum of responses at molecular, cytological, and physiological levels. One of the primary concerns is the internalization and subsequent accumulation of nanoparticles within plant tissues, which can cause physical damage at the cellular level, including membrane disruption and possible damage to cellular organelles and structures, compromising the structural integrity and functional capacity of cells (Pagano et al. 2018a). The imbalance induced by nanoparticle internalization may interfere with energy metabolism and lead to excessive production of reactive oxygen species (ROS), generating oxidative stress that alters the intracellular biochemical environment and potentially damages lipids, proteins, and nucleic acids, resulting in cytotoxic effects. Given these potential impacts, it is crucial to monitor specific biochemical and physiological biomarkers to assess plant responses to ENM exposure. Key indicators include membrane lipid peroxidation levels, which reflect oxidative damage; proline content, which serves as an osmoprotectant and stress marker; and parameters related to cellular respiration. Equally important are physiological parameters such as chlorophyll concentration, photosynthetic efficiency, and stomatal conductance, which are highly sensitive to environmental stress and are critical for determining the overall health and functionality of the plant (Singh et al. 2025).

Both the biochemical and physiological imbalances induced by the interaction between ENMs and plants, as well as the direct effects of nanoparticle internalization, can significantly impact the integrity of the genome, including both nuclear and organellar DNA. Furthermore, these interactions may influence the regulation of gene expression, potentially leading to substantial alterations in cellular function and plant development. Investigating how gene expression is modulated following nanoparticle exposure provides critical insights into the systemic effects of ENMs on plant biology, enabling a more refined understanding of the adaptive strategies employed by plants. Adopting a systems biology approach that investigates specific molecular pathways across different tissues allows the construction of a comprehensive picture of the actions and fate of nanoparticles within the plant, or at the interface if internalization does not occur, and enables discernment of the modes of action of different nanomaterials (Munir et al. 2023; Oliveira et al. 2023). Integrating these findings with complementary methodologies that assess nanoparticle structure, localization, and biotransformation further enhances our ability to characterize their behavior. Such correlations can reveal how specific nanoparticles exert their effects, whether these effects are consistent across plant species, and to what extent species-specific variability exists.

From a genomic perspective, it has been reported that certain nanoparticles can exhibit genotoxic effects, raising concerns about potential DNA damage (Gowtham et al. 2024; Muzammil et al. 2023). This aspect is particularly critical, as genotoxicity poses risks not only to plant health but also to environmental safety. Therefore, rigorous investigation into the genotoxic potential of ENMs is imperative to accurately assess the risks associated with their environmental release. It is essential to distinguish between reversible modulations in gene expression and actual DNA damage, and, in the latter case, to determine the nature and extent of such damage. Such comprehensive analyses are fundamental to ensure that nanomaterials designed for environmental or agricultural use are safe and do not pose long-term risks to plant genomes or broader ecosystems. This aspect is crucial for the development of standardized, safe, and sustainable applications of nanotechnology in agriculture and provides the foundation for addressing the even more complex issue of the ecosystem-wide impact of nanomaterials.

Assessing the environmental risks of nanomaterials in agriculture is complex due to their dynamic behaviors, transformations, and interactions with different biological systems, which makes traditional risk assessment frameworks less effective as they may not fully account for the unique features of nanomaterials (Holden et al. 2016). Designing nanoparticles with controlled release properties, biodegradability,

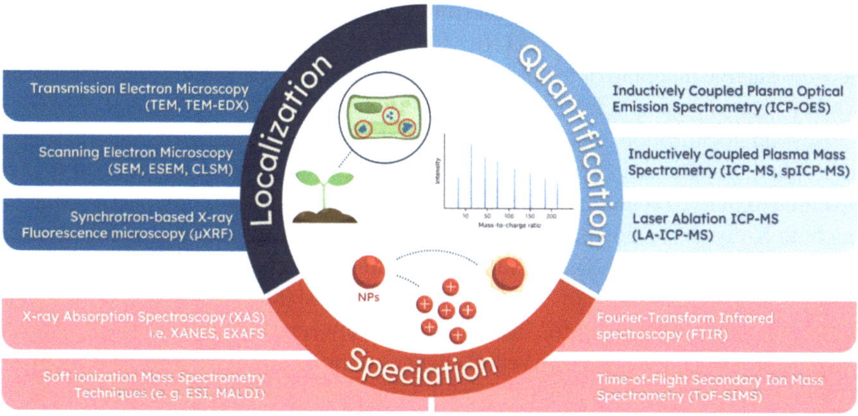

Fig. 7.1 Analytical approaches used to investigate plant-nanoparticle interaction, grouped by purpose: Localization (TEM, SEM, μXRF) to determine NPs distribution within tissues and cells; Quantification (ICP-OES, ICP-MS, LA-ICP-MS) to measure elemental content and NPs concentrations; Speciation (XANES, EXAFS, FTIR, ToF-SIMS, ESI, MALDI) to reveal the chemical forms and molecular interactions of elements and NPs in plant systems

and minimal toxicity will enhance their safety and efficacy in agricultural applications. The use of environmentally friendly materials for the production of ENMs could enable the application of nanotechnology in agriculture, allowing its benefits to be realized while reducing the potential risks related to their dispersion and accumulation in the environment (Balusamy et al. 2023; Wahab et al. 2024). A holistic understanding of nanomaterial–plant interactions is essential for the responsible application of nanotechnology in agriculture. This includes considering the entire lifecycle of ENMs, from synthesis and application to degradation and potential accumulation in the environment. Interdisciplinary collaboration among scientists, regulators, and stakeholders is vital to developing comprehensive guidelines and policies that ensure the safe and sustainable use of nanotechnology in agriculture. Given these considerations, a diverse range of analytical methodologies has been developed or adapted to address the specific challenges related to plant–nanomaterial systems, focusing on three key aspects: localization, speciation, and quantification, which underlie mechanisms of nanobiotransformation (Fig. 7.1).

Localization techniques, such as transmission electron microscopy (TEM) and scanning electron microscopy (SEM), and environmental scanning electron microscope (ESEM) offer direct visualization of nanoparticles at the cellular and subcellular levels (Sembada and Lenggoro 2024; Zhao et al. 2022). These imaging methods, often coupled with energy-dispersive X-ray spectroscopy (EDX), allow for the identification of the anatomical compartments in which ENMs accumulate and to verify whether internalization has occurred (Mourdikoudis et al. 2018; de Paiva et al. 2024). Confocal laser scanning microscopy (CLSM), particularly when nanoparticles are fluorescently labeled, enables three-dimensional imaging and tracking of ENM movement in living tissues (Zhao et al. 2022). For elemental localization at high spatial resolution (μms), synchrotron-based X-ray fluorescence microscopy (XRF) and micro-XRF (μXRF) represent powerful tools, as they allow non-destructive mapping of metal-based nanoparticles across tissues and within cells while preserving sample integrity (Castillo-Michel et al. 2017; Zhao et al. 2022).

Beyond localization, understanding the chemical speciation of nanomaterials after plant uptake is essential for assessing their transformation, mobility, and potential toxicity. Techniques such as X-ray absorption spectroscopy (XAS), including both X-ray absorption near-edge structure (XANES) and extended X-ray absorption fine structure (EXAFS), provide detailed information on the oxidation state and coordination environment of elements within ENMs (Mourdikoudis et al. 2018). These methods are critical for discriminating between nanoparticulate forms and dissolved ions, and for detecting redox changes or possible biotransformation caused by interaction with biological molecules within the plant. Additionally, Fourier-transform infrared spectroscopy (FTIR) can investigate the interactions between ENMs and plant biomolecules by identifying changes in vibrational modes associated with proteins, lipids, and polysaccharides, revealing potential binding or structural modification events (de Paiva et al. 2024).

Accurate quantification of ENMs in plant tissues is equally important to establish dose–response relationships and to evaluate uptake efficiency and translocation dynamics. Inductively coupled plasma mass spectrometry (ICP-MS) and inductively coupled plasma optical emission spectrometry (ICP-OES) enable quantification of concentrations of metal-based nanoparticles in complex biological matrices (Naozuka et al. 2024; Zhang et al. 2024a, b). For size-dependent quantification and particle characterization, single-particle ICP-MS (spICP-MS) has emerged as a key technique, providing information on particle size distribution, particle number concentration, and aggregation state in plant extracts (Mourdikoudis et al. 2018). In more complex biological samples, such as plant extracts, electrospray ionization (ESI) and matrix-assisted laser desorption/ionization (MALDI) introduce intact biomolecule–nanoparticle complexes into the mass spectrometer, preserving molecular interactions that are otherwise lost during ionization processes (Mourdikoudis et al. 2018; Zhao et al. 2022). Laser ablation techniques, such as LA-ICP-MS, combine spatially resolved sampling with ICP-based detection, making them ideal for mapping nanoparticle distribution in tissues (Tanaka and Ogra 2024; Zhang et al. 2024a, b; Zhao et al. 2022).

The use of mass analyzers, such as triple quadrupole (QQQ) and time-of-flight (ToF) mass spectrometers, enhances the selectivity and resolution of both quantitative and qualitative analyses. QQQ-MS is particularly effective for targeted quantification and multi-elemental analysis with high sensitivity, especially when coupled with collision/reaction cell technology for interference removal (Yang et al. 2002). Furthermore, techniques such as ToF-SIMS (time-of-flight secondary ion mass spectrometry) provide spatially resolved chemical imaging at the submicron level (Mourdikoudis et al. 2018), allowing investigation of potential ENM biotransformation inside plant tissues as well as identification of transformation products and adsorption patterns on plant surfaces. A rapidly advancing frontier is single-cell analysis, which characterizes nanoparticle uptake, distribution, and transformation at

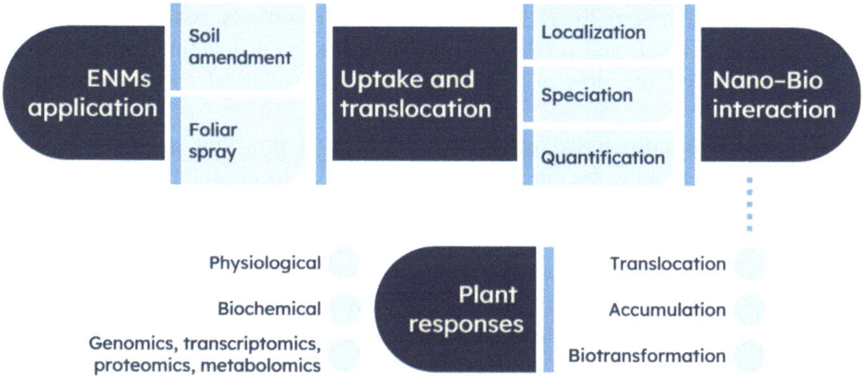

Fig. 7.2 Schematic representation of the implication of nanomaterials usage in plant systems: principal characterization approaches to reveal nanoparticles' behavior in plants; biological response derived from the biointeraction of nanomaterials in pristine or biotransformed form

the individual cell level. This approach elucidates cell-to-cell variability in nanoparticle interactions and identifies target cell types that may be more susceptible to accumulation or damage. Emerging techniques, such as single-cell ICP-MS and microfluidics-integrated ESI-MS, deliver ultra-trace elemental and molecular information from single cells (Yu et al. 2020), representing a paradigm shift in the study of biological responses to ENMs with high resolution.

Finally, integrating biochemical and omics approaches, such as transcriptomics, proteomics, and metabolomics, with these analytical tools offers a systems-level perspective on how nanoparticle exposure alters plant function at the molecular level. When combined with precise localization and speciation data, these approaches link the presence and transformation state of ENMs with specific physiological or gene expression responses, facilitating a mechanistic understanding of their action. Here, the authors will explore these techniques methodologically, outlining their principles and representative applications in the study of ENM–plant interactions (Fig. 7.2).

7.2 Analytical Techniques

7.2.1 ENMs Localization

The localization of nanomaterials within plant tissues is critical for understanding their internalization, accumulation, and translocation pathways, as well as elucidating their interactions with biological systems. A range of analytical techniques can address this purpose, each offering specific advantages depending on the physicochemical properties of the nanomaterial, the level of spatial resolution required, and the biological scale of interest, whether at the tissue, cellular, or subcellular level

(Sembada and Lenggoro 2024). Microscopy-based techniques, such as TEM, SEM, ESEM, EDX, CLSM, and μXRF, provide high-resolution imaging and elemental analysis, enabling the precise visualization of nanomaterials within biological structures.

Among microscopy-based techniques, TEM is a powerful and widely used method for investigating the internalization and precise localization of ENMs within plant cells. TEM enables the acquisition of ultra-high-resolution images at the nanometer level by transmitting a focused beam of high-energy electrons through ultra-thin sections of biological specimens (Zhao et al. 2022). As electrons pass through the sample in a vacuum column, electromagnetic lenses focus the beam into a narrow path, which interacts with the sample before reaching the detector to form a detailed image. The final image arises from variations in the number of electrons transmitted through the sample. Denser regions of the specimen allow fewer electrons to pass, resulting in darker areas, while less dense regions permit more electrons to pass and thus appear lighter (Franken et al. 2020). One key advantage of TEM is its capacity to resolve nanoscale structures within cellular compartments, allowing accurate localization of nanoparticles in specific organelles such as vacuoles, chloroplasts, mitochondria, and the cytoplasm (Ďúranová et al. 2024). This spatial resolution is critical for understanding the pathways of nanoparticle uptake, their potential interactions with cellular machinery, and any associated physiological responses within plant cells (Malatesta 2021). The use of S/TEM-EDX in leaf sections of zucchini treated with CuO NPs demonstrated the presence of Cu aggregates in the cells, both in nanometric and non-nanometric form (Pagano et al. 2016). These findings show that CuO nanoparticles, once applied, can be absorbed by the roots, translocated to the aerial parts, and internalized within cellular compartments. This observation was supported by elemental analysis via ICP-MS and gene expression profiling, which together revealed that, although nanoparticles accumulated to a greater extent in plant tissues compared to their bulk and ionic. The potential use of nanoparticles as delivery systems highlights the need to identify the most effective formulations to ensure targeted cargo delivery. Gold engineered nanomaterials (Au ENMs) are commonly employed for delivering nucleic acids into plant cells, although their size and shape can vary. Using TEM in combination with complementary analytical techniques, researchers investigated the localization of DNA-functionalized gold nanoparticles (AuNPs) within *Nicotiana benthamiana* leaves, differentiating based on particle size (5–20 nm) and morphology (spherical and rod-shaped). The study revealed that the internalization of AuNPs and their associated cargo depends on nanoparticle shape and size, with smaller, rod-shaped particles exhibiting higher and more sustained association with plant cell walls (Zhang et al. 2021). However, TEM has limitations, including the requirement for extensive sample preparation, which may introduce artefacts, and the inability to provide elemental composition information without coupling with additional techniques.

When TEM, SEM, and ESEM are coupled with EDX, additional insights into the sample's elemental composition can be obtained. EDX detects the characteristic X-rays emitted when the electron beam displaces inner-shell electrons in the atoms of the sample, leading to electronic transitions as higher-energy electrons fill the vacancies (Marmiroli et al. 2020; Scimeca et al. 2018; Zhao et al. 2022). Each element emits X-rays at specific energies, allowing both qualitative and quantitative elemental analysis (Scimeca et al. 2018). This is particularly useful for identifying and mapping the presence of nanomaterials, especially metal-based nanoparticles, within or on the surface of plant tissues. The SEM-EDX combination thus provides complementary information: TEM, SEM, and ESEM offer high-resolution imaging of organ architecture and cell morphology, while EDX delivers localized elemental data.

In a study on European beech (*Fagus sylvatica* L.) exposed to silver nanoparticles (Ag-NPs) via droplet application, SEM coupled with EDX confirmed the presence of silver in nanoparticulate form, both as individual particles and aggregates. SEM imaging of the leaf surface revealed nanoparticle accumulation around and on the stomatal openings, while cross-sections of the leaves showed that Ag was internalized only in its nanoparticulate form, not as Ag ions. Within the tissue, Ag-NPs adhered to the cell walls of mesophyll and palisade cells, and the combination of SEM and EDX analyses confirmed that the observed particles and aggregates were composed of silver, substantiating both their adhesion to the leaf surface and their stomatal internalization (Ballikaya et al. 2023). This evidence highlights that SEM-EDX provides valuable surface information related to both localization and identification of ENMs in plant tissues, but it is limited in its ability to probe internal structures and may not detect nanoparticles that have penetrated beyond the surface layers. The penetration depth of both SEM and ESEM is in the range of micrometers (Marmiroli et al. 2020).

Understanding the dynamics of nanomaterial interactions in living specimens, as well as tracking non-metallic ENMs, presents a significant challenge that conventional TEM and SEM techniques alone cannot fully address. The use of fluorescently labelled ENMs represents a promising strategy to overcome these limitations and enable more effective visualization and tracking within biological systems. These fluorescent tags allow visualization of nanomaterials behavior in real time, providing crucial insights into their uptake, translocation, and intracellular distribution, and giving information that may not be accessible through conventional electron-based imaging techniques, especially for non-metallic or low-contrast nanoparticles (Proença et al. 2022). Confocal laser scanning microscopy (CLSM) is a highly effective technique for such applications, as it utilizes point illumination from laser beams to sequentially scan fluorescently labelled specimens, collecting emitted fluorescence through a pinhole aperture that eliminates out-of-focus light. This optical sectioning capability allows the generation of sharp, high-resolution images at specific depths within the sample, enabling the acquisition of thin optical slices without the need for physical sectioning. These individual slices can then be digitally reconstructed to form detailed three-dimensional representations of the specimen's internal architecture in a comprehensive 3D-Tomography (Elliott 2020). Technically, CLSM relies on the excitation of fluorophores conjugated to ENMs or to specific cellular structures, which emit light at characteristic wavelengths when stimulated by selective laser illumination. By using multiple fluorescent dyes with distinct excitation/emission profiles, CLSM can simultaneously visualize different components or pathways within the same specimen, offering multiparametric analysis (Li et al. 2020; Păvăloiu et al. 2020). This method is particularly valuable for studying ENM interactions *in vivo* or *in situ* within plant tissues, as it allows real-time monitoring of nanoparticle dynamics under physiologically relevant conditions (Li et al. 2020; Păvăloiu et al. 2020). Moreover, its non-destructive nature and compatibility with live-cell imaging make CLSM a powerful tool for elucidating the mechanisms of nanomaterial uptake, intracellular trafficking, and potential accumulation in specific organelles or tissue compartments.

Specifically, the assessment of non-metallic nanoparticles, such as nano- and microplastics, remains a significant challenge due to the difficulty in determining their localization within plant tissues and elucidating their modes of action. To address this problem, researchers used CLSM to visualize polystyrene (PS) beads of varying sizes, which were fluorescently labelled using Nile Blue or 4-chloro-7-nitro-1,2,3-benzoxadiazole via a swelling method, applied to wheat (*Triticum aestivum* L.) plants grown in Hoagland solution. The study revealed a clear concentration-dependent fluorescence signal at the root level, indicating potential microplastic accumulation. In contrast, fluorescence within the vascular tissues and epidermis was minimal for 2.0 μm beads and nearly absent for 5.0 and 7.0 μm beads, while 0.2 μm PS beads formed aggregates predominantly within the xylem and along the cortical cell walls, suggesting translocation via the apoplastic pathway through intercellular spaces. These findings underscore the potential of CLSM to provide valuable insights into the mechanisms of internalization and translocation of fluorescently labelled nanomaterials within plant tissues (Li et al. 2020). Although CLSM enables the tracking of organic nanomaterials, it presents certain limitations, including limited penetration depth and the potential for photobleaching of fluorescent labels, which may affect its effectiveness in long-term (hours) imaging studies.

Furthermore, while SEM and TEM techniques provide high spatial resolution and detailed structural information, they require extensive sample preparation, which can compromise native morphology (Fischer et al. 2012; Keevend et al. 2020). Exposure to a high-energy electron beam can also induce thermal and radiation damage, leading to alterations in the chemical and structural integrity of biological specimens, particularly in hydrated or delicate plant tissues (Egerton et al. 2004). To overcome these limitations and enable non-destructive analysis with minimal sample alteration, synchrotron-based techniques offer a valuable alternative. These methods use highly focused and tunable photon beams, rather than electrons, to probe the sample, thereby avoiding damage to the specimen structure. The use of synchrotron radiation allows for high-resolution, element-specific imaging while better preserving the native state of the specimen (Kopittke et al. 2018).

Among these techniques, μXRF represents a powerful tool for elemental mapping. A micro-focused X-ray beam is directed to the sample, causing the emission of characteristic secondary fluorescent X-rays from the constituent elements that are detected and analyzed to produce spatially resolved maps of elemental distribution (Marmiroli et al. 2021; Sala et al. 2024). The technique is highly sensitive and capable of detecting a broad range of elements. Servin and team demonstrated the translocation mechanisms and potential excretion pathways of TiO_2 NPs in hydroponically grown cucumber plants using μXRF. They observed that Ti was transported from the roots to the leaf trichomes, where it accumulated, suggesting a potential role of trichomes as sinks or excretory sites for TiO_2 NPs. Furthermore, μXRF mapping revealed high concentrations of Ti in the phloem cell walls, indicating that the nanoparticles successfully entered the plant's vascular transport system (Servin et al. 2012). The ability of the μXRF technique to provide high-resolution, semi-quantitative *in situ* elemental analysis while preserving the structural integrity of biological samples makes it an indispensable tool for accurately interpreting the behaviour of nanomaterials within complex biological matrices (Kopittke et al. 2018; Sala et al. 2024).

7.2.2 Elemental Speciation and Structural Interaction from ENMs Exposure

Synchrotron-based techniques could offer a deeper understanding of nanomaterial speciation once applied in plant systems. Both EXAFS and XANES are components of X-ray absorption spectroscopy (XAS), a synchrotron-based technique used to determine the local structural and chemical environment of specific elements within a sample (Chandrakasan et al. 2021; Yano and Yachandra 2010). These methods enable the identification of the chemical form (including oxidation state, coordination geometry, and bonding environment) of elements, particularly metals. This is achieved by directing a narrow-focused, high-energy X-ray beam of variable wavelength to the sample, inducing photoelectric absorption when the energy exceeds the binding energy of core electrons in an atom (usually K or L shells) (Baker et al. 2017). The absorption coefficient μ(E) increases sharply at this point, producing an absorption edge. Fine oscillations in the absorption spectrum beyond this edge carry structural information and are categorized into XANES (0–50 eV above the edge) and EXAFS (50–1000 eV above the edge) (Chandrakasan et al. 2021). Each element has specific edge energies, enabling precise identification of the element. The EXAFS technique makes it possible to identify the valence and the nearest neighboring atoms of a metallic target atom (scatterer) and determine their distance from the scattering atom.

With EXAFS, a core photoelectron is ejected and behaves like a spherical wave propagating outward that is scattered by neighboring atoms: the interference between the outgoing and scattered waves modulates the absorption coefficient (Yano and Yachandra 2010). This interference produces oscillations in the absorption spectrum above the edge that are analyzed to extract radial distances (R) to neighbouring atoms, coordination numbers (N) that represent how many neighbours surround the absorbing atom, types of neighboring atoms (Z), disorder, and thermal motion (Yano and Yachandra 2010). EXAFS is used to identify complexation of ENMs with ligands such as phytochelatins, organic acids, or amino acids, to differentiate between adsorbed, precipitated, or bio transformed species, to understand the local coordination environment of metals, and to distinguish ENMs from dissolved ions or secondary mineral phases (Baker et al. 2017). As an example, researchers investigated the potential phytotoxicity and biotransformation of cadmium sulfide (CdS) quantum dots (QDs) in the model plant *Arabidopsis thaliana* using XAS, which enabled analysis of the local atomic environment of cadmium in plant tissues following exposure to either CdS QDs or ionic cadmium ($CdSO_4$). EXAFS analysis provided detailed insights into Cd–S and Cd–O bond distances, revealing variation in Cd–O coordination between treatments. These findings indicate that CdS QDs undergo biotransformation dynamics upon uptake by the plant root system, even if the original CdS structure was not entirely disrupted, as evidenced by a higher proportion of Cd–S compared to Cd–O coordination. The use of EXAFS in this study sheds light on the environmental fate of CdS QDs, offering valuable data for their risk assessment and potential ecotoxicological impact (Marmiroli et al. 2020).

XANES arises due to transitions of core electrons to unoccupied electronic states, especially those close to the Fermi level. The spectral features in this region (pre-edge and near-edge features) are sensitive to coordination geometry, electronic structure, and oxidation state, since higher oxidation states generally shift the edge to higher energies (Chandrakasan et al. 2021). Edge energy position gives the oxidation state, pre-edge peaks indicate the presence of bound or discrete states, line intensity relates to the density of unoccupied states, and symmetry refers to that of the absorbing atom. XANES is used for determining the valence states of metal-based ENMs, assessing biotransformation of ENMs within plant tissues, and tracking chemical changes due to interactions with root exudates or cellular components (Geoghegan et al. 2022). In a study on soybean (*Glycine max*), μ-XANES combined with μ-XRF was employed to investigate the accumulation and potential biotransformation of ZnO and CeO_2 NPs within the roots, vascular system, and pods. The analysis demonstrated that both Ce and Zn could translocate from the roots to aerial tissues and then accumulate in the pods, but their chemical speciation varied depending on the nanoparticle applied. In the case of Ce, approximately 79% of the Ce detected in the grains retained its original CeO_2 NPs structure, indicating limited biotransformation, while only a small percentage of Ce(IV) was reduced to Ce(III). Conversely, ZnO NPs exhibited different behavior: although μ-XRF confirmed zinc translocation and accumulation in pods, μ-XANES analysis revealed that Zn was no longer present in its original ZnO NPs form. Instead, zinc had precipitated, and the resulting species exhibited spectral features closely resembling those of zinc–organic acid complexes, particularly Zn–citrate. This suggests transformation into Zn–O coordinated forms within the soybean grains, indicating that Zn–citrate may represent a major Zn species following ZnO NPs exposure and thus providing critical insights into the implications for food safety associated with the agricultural use of these nanoparticles (Hernández-Viezcas et al. 2013).

In light of this scientific evidence, EXAFS and XANES together provide element-specific characterization of metal-based nanoparticles in complex biological matrices. However, these techniques require complex sample preparation. In addition to characterize organic compounds, biomolecules, and surface modifications of ENMs in biological matrices, the use of a technique such as FTIR (Fourier Transformed Infra-Red) spectroscopy can be a valuable resource (Pasieczna-Patkowska et al. 2025). FTIR is a vibrational spectroscopic technique that provides information about the functional groups and molecular bonding environment in a sample by detecting the absorption of infrared (IR) radiation by molecular vibrations (Eid 2021). The specific wavelengths absorbed correspond to the vibrational modes of chemical bonds: when IR radiation interacts with a molecule, certain frequencies are absorbed if the energy matches the energy difference between vibrational energy levels (Dendisová et al. 2018). The resulting spectrum represents absorbance or transmittance versus wavenumber and reveals characteristic vibrational fingerprints of molecular bonds (Dendisová et al. 2018). The raw output is mathematically converted into a frequency-domain spectrum via a Fourier transformation, yielding high-resolution spectra rapidly and with enhanced signal-to-noise ratio. Through FTIR, it is possible to detect surface coatings or modifications on ENMs that influence their reactivity and transport, thereby highlighting their interactions with plant tissues and potential binding to molecular structures within cells (Husen and Iqbal 2019). In the context of green-synthesis approaches that employ biological extracts, there is a need for accurate assessment of potential residual bioorganic compounds on nanoparticle surfaces, as these may influence their behavior, stability, and safety in planta (Marmiroli et al. 2024; Vora et al. 2020). FTIR can therefore assist in identifying surface-associated functional groups and stabilizing agents, offering critical insights into nanoparticle stability, biocompatibility, and suitability for agricultural or environmental applications (Pasieczna-Patkowska et al. 2025). Furthermore, FTIR has proven to be a valuable tool in detecting the subtle effects of CeO_2 NPs in cereal crops such as rice, wheat, and barley: while nano-Ce often induces only low levels of physiological toxicity, FTIR has revealed compositional modifications within the root xylem, indicating biochemical alterations that are not evident through conventional physiological measurements (Rico et al. 2015). Such findings highlight the power of FTIR-based approaches in uncovering hidden or early-stage impacts of nanoparticle exposure in plants.

Another noteworthy high-resolution analytical technique for nanomaterials characterization in plant tissues is ToF-SIMS, which provides qualitative and semi-quantitative chemical information about the outermost molecular layers of a sample (Barut and Fletcher 2023). ToF-SIMS is particularly suited for imaging surface chemistry, molecular fragments, and isotopic composition at high lateral and mass resolution, revealing nanomaterial interactions with biological systems (Jia et al. 2023). The technique involves directing a pulsed primary ion beam onto the surface of a solid sample under ultra-high vacuum (UHV) conditions. The energetic impact of these ions causes sputtering of the sample surface, leading to the ejection of secondary ions, atoms, molecules, or fragments (Barut and Fletcher 2023). The ejected secondary ions are accelerated into a time-of-flight mass analyzer: since all ions have the same kinetic energy, their flight time to the detector depends on their mass-to-charge ratio (m/z) (Jia et al. 2023). The resulting mass spectrum provides information on elemental, isotopic, and molecular composition. ToF-SIMS is uniquely capable of mapping the spatial distribution of metallic nanoparticles and associated organic ligands within plant tissues (Singh et al. 2020), allowing co-localization studies of ENMs with biomolecules, detection of molecular fragments indicative of ENM surface modifications or their interaction products, and differentiation between intact nanoparticles and dissolved or transformed forms through mass spectral signatures (Jia et al. 2023). Differences in molecular ion patterns can be used to monitor biochemical perturbations, while detection of stable isotope-labelled nanomaterials allows tracing of uptake pathways or metabolic incorporation, providing insights into ENM biotransformation and long-distance transport (Singh et al. 2020). For example, ToF-SIMS imaging on *Phaseolus vulgaris* enables monitoring of the spatial distribution and transport pathways of elements such as Na, K, and Rb within the xylem with high sensitivity, providing quantitative insights into nutrient movement at both cellular and tissue levels (Metzner et al. 2008).

In this context, soft ionization mass spectrometry techniques such as ESI and MALDI are valuable tools for the qualitative and quantitative analysis of complex molecular species associated with ENMs in plant systems (Ho et al. 2003). ESI is a soft ionization technique that facilitates the analysis of polar and thermally labile molecules, such as peptides, proteins, organic acids, and various plant metabolites (Steckel and Schlosser 2019). In ESI, a high voltage is applied to a liquid sample, typically introduced via a capillary, resulting in the generation of fine aerosols of charged droplets. As the solvent evaporates, the surface charge density of the droplets increases until Coulombic repulsion causes them to undergo successive fission, ultimately producing gas-phase ions of the analyte (Ho et al. 2003). These ions are then directed into a mass analyzer, such as a quadrupole, ToF, or Orbitrap, where their mass-to-charge (m/z) ratios are determined (Ho et al. 2003; Steckel and Schlosser 2019). One of the key strengths of ESI is its compatibility with liquid chromatography, enabling the separation of complex mixtures before ionization (Ho et al. 2003). For example, ESI-MS can identify small molecules or secondary metabolites that bind to nanoparticle surfaces, investigate changes in metabolic profiles upon ENM exposure, or detect oxidation or reduction products generated by catalytic activity at nanoparticle interfaces. In the context of ENM speciation, ESI is also instrumental in identifying metal–ligand complexes, such as those involving phytochelatins, glutathione, or organic acids that chelate metal ions released from nanoparticles (Mendoza-Cózatl et al. 2008; Połeć-Pawlak et al. 2007).

MALDI offers a complementary ionization approach, particularly well-suited for analyzing larger biomolecules, such as proteins, oligosaccharides, or polymeric structures, with minimal fragmentation (Kaspar et al. 2011). In MALDI, the analyte is co-crystallized with a low molecular weight matrix, usually an organic acid that strongly absorbs UV light, on a metal plate. A pulsed laser beam is then directed at the sample, causing the matrix to absorb the energy and facilitate the desorption and ionization of the analyte into the gas phase (Zhang et al. 2024a). The resulting ions, typically singly charged, are accelerated into a mass analyzer (commonly ToF), producing a spectrum of m/z values (Selby et al. 2024). MALDI is particularly advantageous when analyzing nanoparticle-bound biomolecules, such as the protein corona that may form on ENMs in plant fluids or extracts. Moreover, MALDI imaging mass spectrometry (MALDI-IMS) extends the utility of the technique by enabling the spatial mapping of molecules directly within tissue sections, allowing for visualization of the distribution of specific proteins, lipids, or metabolites in plant tissues following ENM exposure, without requiring homogenization or extraction steps that might interfere with localization (Zhang et al. 2024a; Vats et al. 2024). This enabled the spatial resolution of oligosaccharide and lipid distributions across distinct tissue compartments in corn kernels and germinated wheat sections mounted on a platinum nanomaterial-coated matrix, thereby demonstrating the capability of MALDI-IMS to preserve anatomical context while capturing detailed molecular localization (Shen et al. 2023).

7.2.3 ENMs Quantification

In the field of plant nanotoxicology and nanomaterial bio-interaction studies, accurately determining the transport and bioaccumulation of ENMs requires highly sensitive and specific analytical techniques to quantify them within biological tissues. Among these, ICP-MS, ICP-OES, and spICP-MS have emerged as indispensable tools, each offering unique advantages depending on the analytical objective (Zhang et al. 2024a, b). ICP-MS is characterized by its exceptional sensitivity and broad elemental coverage, making it particularly valuable for detecting trace and ultra-trace concentrations of metals and metalloids in complex biological matrices (Naozuka et al. 2024). In this technique, the sample is typically digested into a liquid form using strong acids such as HNO_3 and H_2O_2 and introduced into a high-temperature argon plasma, where it is rapidly vaporized, atomized, and subsequently ionized (Pagano et al. 2017). The resulting ions are extracted through a vacuum interface into a mass analyzer, most commonly a quadrupole, where they are separated according to their mass-to-charge (m/z) ratios and finally quantified by a detector such as an electron multiplier (Yu et al. 2020). Although this approach does not distinguish between ionic and particulate forms of the metal, it provides essential data on the degree of uptake and translocation and the concentrations of each element, particularly when coupled with proper sample digestion and fractionation protocols. In more advanced configurations, ICP-MS can be hyphenated with separation techniques such as high-performance liquid chromatography (HPLC) or field-flow fractionation (FFF), enabling speciation studies to distinguish between dissolved species and nanoparticulate fractions (Naozuka et al. 2024).

ICP-OES operates on a similar principle in terms of sample introduction and plasma generation but differs fundamentally in its detection mechanism. Here, the excited atoms and ions in the plasma emit light at element-specific wavelengths as they return to lower energy states; this emission is captured and dispersed by a spectrometer, allowing the identification and quantification of elements based on their characteristic optical signatures (Douvris et al. 2023). While ICP-OES does not reach the sensitivity levels of ICP-MS, it offers excellent robustness and tolerance for complex matrices, which is advantageous when dealing with plant digests that may contain high levels of organic or inorganic matter (Zhang et al. 2024a, b). ICP-OES is often used for routine quantification of elemental concentrations in plant samples following ENM exposure, particularly when concentrations are expected in the ppm or high ppb range (Pagano et al. 2016, 2017).

To overcome the inability of conventional ICP-MS to differentiate between dissolved ions and particulate matter, spICP-MS has been developed as a powerful technique for characterizing nanoparticles in suspension with the use of extreme dilution of the sample that allows individual nanoparticles to enter the plasma sequentially (Zhang et al. 2024a, b). Each nanoparticle, upon vaporization and ionization, produces a discrete ion cloud that is detected as a sharp, transient signal or "pulse" (Laborda et al. 2016). By analyzing these pulses over time in time-resolved mode, spICP-MS enables the determination of several critical particle parameters, including particle size derived from signal intensity, particle number concentration from pulse frequency, and size distribution (Naozuka et al. 2024). Additionally, it allows for simultaneous quantification of the dissolved fraction, thus offering a direct means of evaluating nanoparticle dissolution, transformation, or aggregation in biological systems (Zhang et al. 2024a, b). In plant studies, spICP-MS is particularly valuable for confirming whether intact ENMs are internalized and transported within tissues. For instance, this technique has been effectively employed to investigate the co-administration of metallic nanoparticles in lettuce (*Lactuca sativa*), providing detailed insights into tissue-specific distribution patterns and confirming the internalization of the nanomaterials (Wu et al. 2023). Furthermore, spICP-MS enables assessment of trophic transfer to insects and soil microfauna. Wu and colleagues quantified the degree of nanoparticle biomagnification in snails, thereby offering critical data for evaluating environmental risks associated with ENM exposure across trophic levels (Wu et al. 2023). However, successful application of spICP-MS requires careful sample preparation to preserve the particulate nature of nanomaterials, often involving enzymatic digestion or mild chemical extraction rather than total acid digestion (Naozuka et al. 2024).

While traditional ICP-MS offers unparalleled sensitivity for elemental quantification, it requires liquid samples and consequently loses spatial information. To overcome this limitation, LA-ICP-MS combines the elemental sensitivity of ICP-MS with the capability for direct, *in situ* micro-scale spatial analysis of solid samples (Zhang et al. 2024a, b). LA-ICP-MS operates by directing a finely focused pulsed laser beam to the surface of a solid specimen such as a thin section of plant tissue, resulting in the localized ablation of material in the form of a plume of aerosolized particles (Limbeck et al. 2015). These particles are then transported by an inert carrier gas, usually argon or helium, into the inductively coupled plasma torch, where they are vaporized, atomized, and ionized under extreme temperatures. The ions generated in the plasma are then analyzed by a mass spectrometer, providing elemental composition data with detection limits comparable to those of conventional ICP-MS, while the preservation of the spatial integrity of the sample is guaranteed, enabling the generation of high-resolution elemental maps across tissue sections (Doble et al. 2021). The laser spot size can be adjusted from a few micrometers to tens of micrometers, allowing targeting of specific anatomical features such as vascular bundles, epidermis, or cellular compartments, which enables differentiation of elemental distributions at sub-organ or cellular resolution and facilitates the study of nanoparticle sequestration within specific compartments (Gao et al. 2023). Additionally, by rastering the laser across the sample surface in a controlled pattern, two-dimensional elemental distribution images can be reconstructed, revealing the precise localization of ENMs and their transformed species within plant tissues, thus providing insights into plant transport pathways and potential bioaccumulation (Cui et al. 2023).

The advantage of LA-ICP-MS in analyzing multiple elements simultaneously proves beneficial when assessing the speciation and transformation of ENMs, such as nanoparticle dissolution, precipitation, or interactions with endogenous plant elements. Despite its strengths, LA-ICP-MS requires careful optimization of laser parameters and sample preparation protocols to minimize matrix effects, avoid thermal damage, and preserve tissue morphology (Doble et al. 2021). Furthermore, quantitative analysis can be challenging due to the lack of suitable matrix-matched standards, although recent advances in calibration approaches and software have greatly improved the quantitative reliability of LA-ICP-MS in biological tissues. For instance, LA-ICP-MS has enabled the investigation of the effects of ZnO NPs on rice (*Oryza sativa*), revealing their accumulation specifically within the vascular tissues of the rhizodermis and procambium in roots. This spatial localization supports the hypothesis that intact nanoparticles can persist within plant tissues and that particle size plays a pivotal role in mediating phytotoxicity (Li et al. 2023). In another study, LA-ICP-MS allowed direct visualization of foliar-applied mesoporous silica-encapsulated ZnO nanoparticles within the phloem cells of plant leaves, providing clear evidence that nanoparticles approximately 70 nm in size can reach the phloem. Notably, the study demonstrated that translocation efficiency significantly improved when nanoparticles were applied to the abaxial (lower) leaf surface, underscoring the critical role of application strategy in modulating nanoparticle uptake and systemic movement within the plant (Gao et al. 2023). Collectively, these analytical techniques offer complementary strengths in detecting, quantifying, and spatially resolving nanomaterial behavior within plant systems, such as uptake pathways, distribution patterns, and potential transformation processes, contributing to a deeper understanding of their bioavailability, mobility, and ecological implications in agricultural and environmental contexts.

7.3 Functional Strategies in Assessing Response Mechanisms

Analytical evidence, even if methodologically robust, maintains a substantial bias in the comprehension of biological phenomena. The analytical characterization of physicochemical forms, localization, and quantitation must be coupled with physiological and/or molecular information, which may derive from genetic or "-omic" tests. Such integration can reduce the complexity of biological phenomena by providing a more robust understanding. A systems biology approach, as a holistic perspective that considers multiple viewpoints, is key to unlocking the potential of analytical methodologies as tools for deepening the understanding of interactions between nanomaterials and the plant–soil system (Gechev and Petrov 2022).

Within the plant–soil system, soil physicochemical characteristics may significantly influence nanomaterial aggregation and dissolution, thereby modifying bioavailability as well as uptake, translocation, and accumulation in tissues and organs. Light and temperature can induce structural changes, for example, during foliar spray amendment (Hong et al. 2021). Once within plant tissues, biotransformation processes affect particle stability, altering the behavior in terms of biomolecular interactions, leading to modulation of stress response mechanisms (Ma et al. 2018). Nanomaterials are subject to various processes that result in partial dissolution or structural modifications, during weathering in soil, trophic transfer, and translocation within plant tissues (Servin et al. 2017). These interactions are dynamic and alter the original pristine particle structure, potentially triggering ion release as well as restructuring or destructuring the particle (Pagano et al. 2023). Nanomaterial biotransformation may depend on interactions with biological molecules which stabilize the outer surface, including peptides (such as amino acids and glutathione), secondary metabolites, fatty acids, and components of cell membranes (Marmiroli et al. 2020). Particle properties, which include size, charge, stability, and dissolution, may affect biotransformation processes, promoting functionalization with proteins (for example, corona proteins) or enzymatic modification within the cytoplasm and organelles (Ma et al. 2018; Marmiroli et al. 2020). When internalized by cells, nanomaterials may maintain their crystal structure or may be destructured and converted into less complex structures by chelation or biological modification, reducing toxicity and their risk of accumulation and translocation (Wang et al. 2022). These post-uptake alterations involve specific parameters such as the nature of the ligand atoms as well as the bond distance with other atoms.

Investigating nanobiotransformation dynamics helps clarify the physicochemical forms of ENMs (such as nanocrystal structures) within tissues and characterizes structural variations within newly biotransformed complexes, including identification of biomolecules interacting with the nanomaterial (Castillo-Michel et al. 2017; Marmiroli et al. 2020). It is also important to consider how genetic diversity among different plants and cultivars may influence uptake and translocation (Deng et al. 2020). Among the most informative insights into ENM biotransformation and localization of metal-based nanostructured compounds within plants, are those obtained using synchrotron-based imaging, elemental speciation, and identification of atomic neighbours. From this perspective, μ-XRF can provide information on the presence and localization of specific elements within tissues, while EXAFS and XANES spectroscopy can reveal the valence state, the elemental coordination environment, and the molecular species present in the tissue sample (Castillo-Michel et al. 2017). Importantly, these powerful methodologies offer the possibility to address important environmental issues, including the chemical activities of element speciation in complex matrices, trace element cycles, and environmental pollutants.

Particular attention should be given to genetics, epigenetic mechanisms, and descriptive tools such as omics techniques (Marmiroli et al. 2022; Pagano et al. 2022). Genetic and epigenetic descriptors of interactions between pristine or biotransformed ENMs and biomolecules can partially or fully depict the biological behavior of physicochemical forms present within tissues. For example, these descriptors may relate to effects on genome replication, with particular consideration of organellar genomes (plastid or mitochondrial DNA). Pagano et al. (2022) described how ENMs modulate organelle functionality and organelle genome replication. Observed differences in organellar DNA (orgDNA) stoichiometry represent a potential morpho-functional adaptive response to ENM exposure, triggered by bioenergetic redox balance modifications. The consequent reduction in photosynthetic activity or cellular respiration rate may be attributed to ENMs and to the relative dissolution rate of metal salts and metal ion toxicity, resulting from biotransformation processes.

On the other hand, ENMs can be considered alongside other abiotic environmental stressors such as drought, high or low temperature, salinity, or heavy metals, which may potentially cause extensive damage to plants, including nuclear DNA and organelle DNA. In this scenario, microRNAs (miRNAs) play a fundamental regulatory role in major stress response pathways (Pagano et al. 2021). Plants have evolved complex mechanisms to sense and respond to environmental changes, in which miRNAs have critical early regulatory roles in either encompassing or attenuating the stress effect. Relevant miRNAs have been confirmed in many species, both monocot and dicot, highlighting the phylogenetic conservation of pathways involved in abiotic stress response (Morris and Mattick 2014). Interestingly, in the case of nanomaterials, miRNA modulation in tissues can be closely linked to the physicochemical forms available in different tissues, where stress response and miRNA modulation converge along translocation routes from roots to shoots. Conversely, considering ENM physicochemical stability, which mainly influences translocation and dissolution tendency, the degree of similarity in general miRNA modulation is preserved (Morris and Mattick 2014). Due to the high level of conservation of the response to environmental abiotic stressors, elucidation of miRNA modulation in ENM response, mostly related to oxidative stress response and adaptive mutations, provides new functional tools for early monitoring plant exposure, effects and response.

In terms of functional descriptive tools offered by omics disciplines, several studies have been published in recent years across plant species (Ma et al. 2018; Ruotolo et al. 2018). By coupling molecular and analytical techniques, it is possible to identify pivotal players in responses to ENM exposure, characterizing particle behavior in specific organs and tissues in either conserved or species-specific ways (Pagano et al. 2018a, b).

From a genomic perspective, the study of detected mutants has historically been a powerful instrument for identifying key plant functions. Integrating functional genomics with analytical techniques has led to significant advances in mechanistically understanding responses to ENM exposure. Marmiroli et al. (2014) demonstrated through Ac/Ds mutants *of Arabidopsis thaliana* that detoxification mechanisms for CdS quantum dots (QDs) and ionic cadmium differ substantially in both physiological effects and stress response effectors. Tolerant mutants (e.g., *atnp01* and *atnp02*) were susceptible only to ionic cadmium, whereas *cad* mutants, altered in glutathione-mediated mechanisms, exhibited sensitivity to ionic cadmium but physiological behavior similar to wild-type accessions. These findings were corroborated analytically: total Cd content was quantitatively similar across mutants, yet physiological effects (e.g., reduced photosynthetic activity, cellular respiration) were markedly attenuated in tolerant lines. Synchrotron-based EXAFS studies revealed that Cd from QDs binds to multiple molecular species in plant cells, creating varied local environments. Partial degradation of QD structure was observed, with Cd retaining some sulphur bonds, suggesting reduced mobility/bioavailability (Marmiroli et al. 2020). At transcriptomic, proteomic, and metabolomic levels, advances in high-resolution platforms have enabled detailed study of gene, pathway, and network modulation at genetic, physical, and co-expression levels (Majumdar et al. 2019; Marmiroli et al. 2021). This information not only clarifies basic research implications but also aids in developing biomarkers for exposure, effect, and susceptibility. Cross-comparisons between ENM types and their bulk counterparts have identified both commonly modulated biomarkers and ENM-specific responses (Pagano et al. 2017). From this perspective, the tight bind between analytical measurement and biological evidence reduces bias in interpreting complex biological phenomena and serves as a functional tool to shed light on more realistic conditions (e.g. field studies) and scenarios of plant-soil-nanomaterials interactions.

7.4 Conclusion

From an analytical perspective, integrating advanced quantification and localization techniques has been instrumental in revealing key aspects of ENM uptake, distribution, accumulation, and transformation within plant systems. Spatially resolved techniques, in particular, enable researchers to localize nanoparticles in specific anatomical regions (e.g., root epidermis, xylem, phloem), while mass spectrometric approaches facilitate identification of associated biomolecules, transformation products, and coordination environments (Fig. 7.3). Such analytical advancements play a pivotal role in interpreting the complex interplay between nanomaterials and plant biology, helping to bridge the gap between exposure and effect.

New information gained from the multitude of published studies is pushing forward the research to the resolution of new challenges and raising questions from both technological and methodological perspectives. The complexity of the ENM-plant interaction underlies numerous uncertainties in different physical, chemical, biological, and broader ecological aspects: from the relevance of concentration considered, in terms of effectiveness of the treatment and differences in acute or

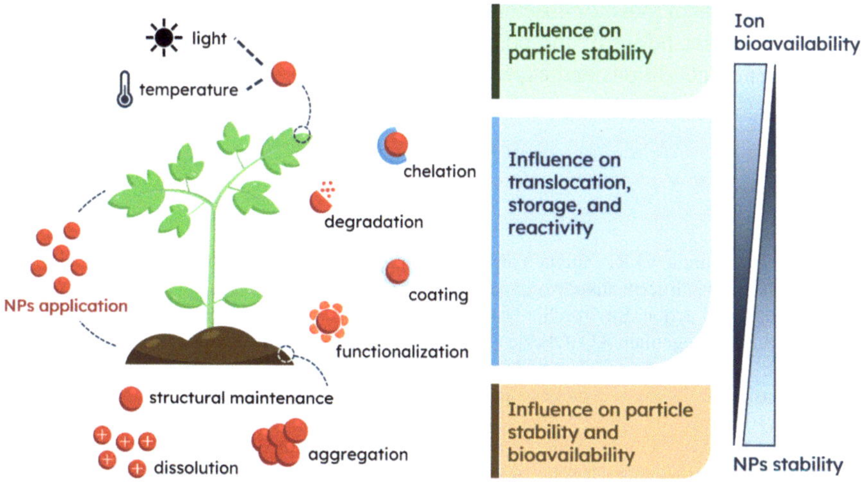

Fig. 7.3 Schematic representation of the factors influencing nanoparticle behavior and bioavailability in plants: physicochemical transformations associated with the application method (soil or foliar), and biotransformation processes following uptake and translocation, highlighting how plant-related mechanisms modulate nanoparticle fate. These interactions affect particle stability and ion bioavailability, with reduced stability generally corresponding to increased ion availability

chronic exposure, to the adoption of novel materials whose composition do not allow a direct assessment of the dose/effect ratio (Pagano and Marmiroli 2022). Other factors include interaction with the growth medium, whether *in vitro* (solid or hydroponic experimental conditions) or *on field* experiments, depending on soil composition (Pagano et al. 2018b); interaction with soil microbiota (Khan et al. 2021); interaction with biomolecules (*e.g.* corona protein, chelation, functionalization) or cellular structures (Pagano et al. 2023). All these potential issues may be only partially addressed by the present analytical techniques. It is essential to consider that, in certain cases, concentrations necessary to be observed (especially when considering localization and speciation) require being comprised in the ppm range, which becomes quite unrealistic, especially for realistic scenario exposure experiments. All these considerations can be applied only to metal or metal oxide-based nanomaterials, while becoming largely inapplicable to carbon-based nanomaterials, nanocomposites, or, more recently, nanoplastics (Ayanda et al. 2024; di Luca et al. 2024).

From a methodological perspective, several advances in quantification have already been achieved, although it is necessary to balance particle observation (in terms of purity, concentration, redox state, etc.) with sample destructiveness, limits of detection, resolution, elements determined, and potential interferences during sample preparation and analytical quantification (Djingova et al. 2013; Silva et al. 2022). Assessing exposure (physiologically and biochemically) and molecular effects of novel nanocomposites and nanoplastics becomes functional, from an applicative point of view, for developing, improving, and supporting available analytical tools for dose/effect assessment and biotransformation. This also increases understanding of the complexity of nanobiointeractions under different environmental conditions and exposure durations.

References

Adisa IO, Pullagurala VLR, Peralta-Videa JR, Dimkpa CO, Elmer WH, Gardea-Torresdey JL, White JC (2019) Recent advances in nano-enabled fertilizers and pesticides: a critical review of mechanisms of action. Environ Sci Nano 6(7):2002–2030. https://doi.org/10.1039/c9en00265k

Ayanda OS, Mmuoegbulam AO, Okezie O, Iya NID, Mohammed SE, James PH, Muhammad AB, Unimke AA, Alim SA, Yahaya SM, Ojo A, Adaramoye TO, Ekundayo SK, Abdullahi A, Badamasi H (2024) Recent progress in carbon-based nanomaterials: critical review. J Nanopart Res 26:106. https://doi.org/10.1007/s11051-024-06006-2

Baker ML, Mara MW, Yan JJ, Hodgson KO, Hedman B, Solomon EI (2017) K- and L-edge X-ray Absorption Spectroscopy (XAS) and Resonant Inelastic X-ray Scattering (RIXS) determination of Differential Orbital Covalency (DOC) of transition metal sites. Coord Chem Rev 345:182–208. https://doi.org/10.1016/j.ccr.2017.02.004

Ballikaya P, Mateos JM, Brunner I, Kaech A, Cherubini P (2023) Detection of silver nanoparticles inside leaf of European beech (Fagus sylvatica L.). Front Environ Sci 10. https://doi.org/10.3389/fenvs.2022.1107005

Balusamy SR, Joshi AS, Perumalsamy H, Mijakovic I, Singh P (2023) Advancing sustainable agriculture: a critical review of smart and eco-friendly nanomaterial applications. J Nanobiotechnol 21(1):372. https://doi.org/10.1186/s12951-023-02135-3

Barut I, Fletcher JS (2023) Cell and tissue imaging by secondary ion mass spectrometry. Biointerphases 18:061202. https://doi.org/10.1116/6.0003140

Basak S, Bhattacharyya P, Pudake RN, Lokhande PE, Rednam U, Chakrabarti S (2024) Metal-organic framework as nanocarriers for agricultural applications: a review. Front Nanotechnol 6: 1385981

Bharti A, Hooda V, Singh A, Pudake RN, Jain U, Chauhan N (2025) Advancing sustainable crop yield enhancement through environmentally friendly nanoparticle synthesis: a comprehensive review. J Soil Sci Plant Nutr 25:1–32

Biswas A, Pal S (2024) Plant-nano interactions: a new insight of nano-phytotoxicity. Plant Physiol Biochem 210:108646. https://doi.org/10.1016/j.plaphy.2024.108646

Castillo-Michel HA, Larue C, Pradas del Real AE, Cotte M, Sarret G (2017) Practical review on the use of synchrotron based micro- and nano- X-ray fluorescence mapping and X-ray absorption spectroscopy to investigate the interactions between plants and engineered nanomaterials. Plant Physiol Biochem 110:13–32. https://doi.org/10.1016/j.plaphy.2016.07.018

Chandrakasan G, Ayala MT, García Trejo JF, Marcus G, Carroll D (2021) Mapping and distribution of speciation changes of metals from nanoparticles in environmental matrices using synchrotron radiation techniques. Environ Nanotechnol Monitor Manag 16:100491. https://doi.org/10.1016/j.enmm.2021.100491

Cui Z, He M, Chen B, Hu B (2023) In-situ elemental quantitative imaging in plant leaves by LA-ICP-MS with matrix-matching external calibration. Anal Chim Acta 1275:341588. https://doi.org/10.1016/j.aca.2023.341588

de Paiva PSK, da Silva PM, Rangel Miguel TAB, Grillo R, de Souza FAG, de Castro ME (2024) Nanoparticles and plants: a focus on analytical characterization techniques. Plant Sci 348: 112225. https://doi.org/10.1016/j.plantsci.2024.112225

Dendisová M, Jeništová A, Parchaňská-Kokaislová A, Matějka P, Prokopec V, Švecová M (2018) The use of infrared spectroscopic techniques to characterize nanomaterials and nanostructures: a review. Anal Chim Acta 1031:1–14. https://doi.org/10.1016/j.aca.2018.05.046

Deng C, Wang Y, Cota-Ruíz K, Reyes A, Sun Y, Peralta-Videa JR, Hernández-Viezcas JA, Turley RS, Niu G, Li C, Gardea-Torresdey JL (2020) Bok Choy (Brassica rapa) grown in copper oxide nanoparticles-amended soils exhibits toxicity in a phenotype-dependent manner: translocation, biodistribution and nutritional disturbance. J Hazard Mater 398:122978. https://doi.org/10.1016/j.jhazmat.2020.122978

di Luca C, Garcia J, Munoz M, Hernando-Pérez M, de Pedro ZM, Casas JA (2024) Strategies for the quantification and characterization of nanoplastics in AOPs research. Chem Eng J. https://doi.org/10.1016/j.cej.2024.152490

Djingova R, Mihaylova V, Lyubomirova V, Tsalev DL (2013) Multielement analytical spectroscopy in plant ionomics research. Appl Spectrosc Rev 48(5):384–424. https://doi.org/10.1080/05704928.2012.703153

Doble P, de Vega RG, Bishop D, Hare DJ, Clases D (2021) Laser ablation-inductively coupled plasma-mass spectrometry imaging in biology. Chem Rev 121(19):11769–11822. https://doi.org/10.1021/acs.chemrev.0c01219

Douvris C, Trey V, Bussan DD, Bartzas G, Thomas RH (2023) How ICP-OES changed the face of trace element analysis: review of the global application landscape. Sci Total Environ 905: 167242–167242. https://doi.org/10.1016/j.scitotenv.2023.167242

Ďúranová H, Kšiňan S, Kuželová L, Šimora V, Ďurišová L, Olexíková L, Ernst D, Kolenčík M (2024) Nanoparticle-plant interactions: physico-chemical characteristics, application strategies, and transmission electron microscopy-based ultrastructural insights, with a focus on stereological research. Chemosphere. https://doi.org/10.1016/j.chemosphere.2024.142772

Egerton RF, Li P, Malac M (2004) Radiation damage in the TEM and SEM. Micron 35(6):399–409. https://doi.org/10.1016/j.micron.2004.02.003

Eid MM (2021) Characterization of nanoparticles by FTIR and FTIR-microscopy. In: Handbook of consumer nanoproducts, pp 1–30. https://doi.org/10.1007/978-981-15-6453-6_89-1

Elliott AD (2020) Confocal microscopy: principles and modern practices. Curr Protoc Immunol 92(1). https://doi.org/10.1002/cpcy.68

Fischer ER, Hansen BT, Nair V, Hoyt FH, Dorward DW (2012) Scanning electron microscopy. Curr Protoc Microbiol Chapter 2:Unit 2B.2. https://doi.org/10.1002/9780471729259.mc02b02s25

Franken LE, Grünewald K, Boekema EJ, Stuart MCA (2020) A technical introduction to transmission electron microscopy for soft-matter: imaging, possibilities, choices, and technical developments. Small 16(14):1906198. https://doi.org/10.1002/smll.201906198

Galindo-Castañeda T, Hartmann M, Lynch JP (2024) Location: root architecture structures rhizosphere microbial associations. J Exp Bot 75(2):594–604. https://doi.org/10.1093/jxb/erad421

Gao X, Kundu A, Persson DP, Szameitat A, Minutello F, Husted S, Ghoshal S (2023) Application of ZnO nanoparticles encapsulated in mesoporous silica on the abaxial side of a Solanum lycopersicum leaf enhances Zn uptake and translocation via the phloem. Environ Sci Technol 57(51):21704–21714. https://doi.org/10.1021/acs.est.3c06424

Gechev T, Petrov V (2022) Plant systems biology in 2022 and beyond. Int J Mol Sci 23(8): 4159–4159. https://doi.org/10.3390/ijms23084159

Geoghegan BL, Liu Y, Peredkov S, Dechert S, Meyer F, DeBeer S, Cutsail GE (2022) Combining valence-to-core X-ray emission and Cu K-edge X-ray absorption spectroscopies to experimentally assess oxidation state in organometallic Cu(I)/(II)/(III) complexes. J Am Chem Soc. https://doi.org/10.1021/jacs.1c09505

Gowtham HG, Shilpa N, Singh SB, Aiyaz M, Abhilash MR, Nataraj K, Amruthesh KN, Ansari MA, Alomary MN, Murali M (2024) Toxicological effects of nanoparticles in plants: mechanisms involved at morphological, physiological, biochemical and molecular levels. Plant Physiol Biochem 210:108604. https://doi.org/10.1016/j.plaphy.2024.108604

Guo H, Han F, Shang H, Xiong S, Huynh M, Thistle L, Meng L, He L, Xing B (2021) New insight into naturally formed nanosilver particles: role of plant root exudates. Environ Sci Nano 8(6): 1580–1592. https://doi.org/10.1039/d0en01188f

Hernández-Viezcas JÁ, Castillo-Michel H, Andrews JC, Cotte M, Rico C, Peralta-Videa JR, Ge Y, Priester JH, Holden PA, Gardea-Torresdey JL (2013) In situ synchrotron X-ray fluorescence mapping and speciation of CeO_2 and ZnO nanoparticles in soil cultivated soybean (Glycine max). ACS Nano 7(2):1415–1423. https://doi.org/10.1021/nn305196q

Ho CS, Lam CWK, Chan MHM, Cheung RCK, Law LK, Lit LCW, Suen MWM, Tai HL (2003) Electrospray ionisation mass spectrometry: principles and clinical applications. Clin Biochem Rev 24(1):3–12. PMID: 18568044; PMCID: PMC1853331

Holden PA, Gardea-Torresdey JL, Klaessig F, Turco RF, Mortimer M, Hund-Rinke K, Cohen Hubal EA, Avery D, Barceló D, Behra R, Cohen Y, Deydier-Stephan L, Ferguson PL, Fernandes TF, Herr Harthorn B, Henderson WM, Hoke RA, Hristozov D, Johnston JM et al (2016) Considerations of environmentally relevant test conditions for improved evaluation of ecological hazards of engineered nanomaterials. Environ Sci Technol 50(12):6124–6145. https://doi.org/10.1021/acs.est.6b00608

Hong J, Wang C, Wagner DC, Gardea-Torresdey JL, He F, Rico CM (2021) Foliar application of nanoparticles: mechanisms of absorption, transfer, and multiple impacts. Environ Sci Nano 8(5): 1196–1210. https://doi.org/10.1039/D0EN01129K

Hortas L (2025) Ecotoxicological effects of nanoparticles on soil microbiota and plant growth. J Ecol Toxicol 9(1):1. https://doi.org/10.4172/jety.1000259

Husen A, Iqbal M (2019) Nanomaterials and plant potential: an overview. Nanomater Plant Potential 3–29. https://doi.org/10.1007/978-3-030-05569-1_1

Iavicoli I, Leso V, Beezhold DH, Shvedova AA (2017) Nanotechnology in agriculture: opportunities, toxicological implications, and occupational risks. Toxicol Appl Pharmacol 329:96–111. https://doi.org/10.1016/j.taap.2017.05.025

Jia F, Zhao X, Zhao Y (2023) Advancements in ToF-SIMS imaging for life sciences. Front Chem 11. https://doi.org/10.3389/fchem.2023.1237408

Kah M, Kookana RS, Gogos A, Bucheli TD (2018) A critical evaluation of nanopesticides and nanofertilizers against their conventional analogues. Nat Nanotechnol 13(8):677–684. https://doi.org/10.1038/s41565-018-0131-1

Kaspar S, Peukert M, Svatoš A, Matros A, Mock H (2011) MALDI-imaging mass spectrometry—an emerging technique in plant biology. Proteomics 11(9):1840–1850. https://doi.org/10.1002/pmic.201000756

Keevend K, Coenen T, Herrmann IK (2020) Correlative cathodoluminescence electron microscopy bioimaging: towards single protein labelling with ultrastructural context. Nanoscale 12:15588–15603. https://doi.org/10.1039/D0NR02563A

Khan ST, Adil SF, Shaik MR, Alkhathlan HZ, Khan M (2021) Engineered nanomaterials in soil: their impact on soil microbiome and plant health. Plants 11(1):109–109. https://doi.org/10.3390/plants11010109

Kopittke PM, Punshon T, Paterson DJ, Tappero RV, Wang P, Blamey FPC, van der Ent A, Lombi E (2018) Synchrotron-based X-ray fluorescence microscopy as a technique for imaging of elements in plants. Plant Physiol 178(2):507–523. https://doi.org/10.1104/pp.18.00759

Laborda F, Bolea E, Jiménez-Lamana J (2016) Single particle inductively coupled plasma mass spectrometry for the analysis of inorganic engineered nanoparticles in environmental samples. Trends Environ Anal Chem 9:15–23. https://doi.org/10.1016/j.teac.2016.02.001

Li L, Luo Y, Peijnenburg WJGM, Li R, Yang J, Zhou Q (2020) Confocal measurement of microplastics uptake by plants. MethodsX 7:100750. https://doi.org/10.1016/j.mex.2019.11.023

Li Z, Yan W, Li Y, Xiao Y, Shi Y, Zhang X, Lei J, Min K, Pan Y, Chen X, Liu Q, Jiang G (2023) Particle size determines the phytotoxicity of ZnO nanoparticles in rice (Oryza sativa L.) revealed by spatial imaging techniques. Environ Sci Technol 57(36):13356–13365. https://doi.org/10.1021/acs.est.3c03821

Limbeck A, Galler P, Bonta M, Bauer G, Nischkauer W, Vanhaecke F (2015) Recent advances in quantitative LA-ICP-MS analysis: challenges and solutions in the life sciences and environmental chemistry. Anal Bioanal Chem 407(22):6593–6617. https://doi.org/10.1007/s00216-015-8858-0

Lv J, Christie P, Zhang S (2019) Uptake, translocation, and transformation of metal-based nanoparticles in plants: recent advances and methodological challenges. Environ Sci Nano 6(1):41–59. https://doi.org/10.1039/c8en00645h

Ma C, White JC, Zhao J, Zhao Q, Xing B (2018) Uptake of engineered nanoparticles by food crops: characterization, mechanisms, and implications. Ann Rev Food Sci Technol (New in 2010) 9(1):129–153. https://doi.org/10.1146/annurev-food-030117-012657

Majumdar S, Pagano L, Wohlschlegel JA, Villani M, Zappettini A, White JC, Keller AA (2019) Proteomic, gene and metabolite characterization reveal the uptake and toxicity mechanisms of cadmium sulfide quantum dots in soybean plants. Environ Sci Nano 6(10):3010–3026. https://doi.org/10.1039/c9en00599d

Malatesta M (2021) Transmission electron microscopy as a powerful tool to investigate the interaction of nanoparticles with subcellular structures. Int J Mol Sci 22(23):12789. https://doi.org/10.3390/ijms222312789

Marmiroli M, Pagano L, Savo Sardaro ML, Villani M, Marmiroli N (2014) Genome-wide approach in Arabidopsis thaliana to assess the toxicity of cadmium sulfide quantum dots. Environ Sci Technol 48(10):5902–5909. https://doi.org/10.1021/es404958r

Marmiroli M, Lepore GO, Pagano L, d'Acapito F, Gianoncelli A, Villani M, Lazzarini L, White JC, Marmiroli N (2020) The fate of CdS quantum dots in plants as revealed by extended X-ray absorption fine structure (EXAFS) analysis. Environ Sci Nano 7(4):1150–1162. https://doi.org/10.1039/c9en01433k

Marmiroli M, Pagano L, Rossi R, De La Torre-Roche R, Lepore GO, Ruotolo R, Gariani G, Bonanni V, Pollastri S, Puri A, Gianoncelli A, Aquilanti G, d'Acapito F, White JC, Marmiroli N (2021) Copper oxide nanomaterial fate in plant tissue: nanoscale impacts on reproductive tissues. Environ Sci Technol 55(15):10769–10783. https://doi.org/10.1021/acs.est.1c01123

Marmiroli M, Marmiroli N, Pagano L (2022) Nanomaterials induced genotoxicity in plant: methods and strategies. Nanomaterials 12(10):1658–1658. https://doi.org/10.3390/nano12101658

Marmiroli M, Villani M, Scarponi P, Carlo S, Pagano L, Sinisi V, Lazzarini L, Pavlicevic M, Marmiroli N (2024) Green synthesis of CuO nanoparticles from macroalgae Ulva lactuca and Gracilaria verrucosa. Nanomaterials. https://doi.org/10.3390/nano14131157

Mendoza-Cózatl DG, Butko E, Springer F, Torpey JW, Komives EA, Kehr J, Schroeder JI (2008) Identification of high levels of phytochelatins, glutathione and cadmium in the phloem sap of Brassica napus. A role for thiol-peptides in the long-distance transport of cadmium and the effect of cadmium on iron translocation. Plant J 54(2):249–259. https://doi.org/10.1111/j.1365-313x.2008.03410.x

Metzner R, Schneider H, Breuer U, Schroeder WH (2008) Imaging nutrient distributions in plant tissue using time-of-flight secondary ion mass spectrometry and scanning electron microscopy. Plant Physiol 147(4):1774–1787. https://doi.org/10.1104/pp.107.109215

Morris KV, Mattick JS (2014) The rise of regulatory RNA. Nat Rev Genet 15(6):423–437. https://doi.org/10.1038/nrg3722

Mourdikoudis S, Pallares RM, Thanh NTK (2018) Characterization techniques for nanoparticles: comparison and complementarity upon studying nanoparticle properties. Nanoscale 10(27): 12871–12934. https://doi.org/10.1039/C8NR02278J

Munir N, Gulzar W, Abideen Z, Hasanuzzaman M, El-Keblawy A, Zhao F (2023) Plant–nanoparticle interactions: transcriptomic and proteomic insights. Agronomy 13(8):2112. https://doi.org/10.3390/agronomy13082112

Muzammil S, Ashraf A, Siddique MH, Aslam B, Rasul I, Abbas R, Afzal M, Faisal M, Hayat S (2023) A review on toxicity of nanomaterials in agriculture: current scenario and future prospects. Sci Prog 106(4):00368504231221672. https://doi.org/10.1177/00368504231221672

Naozuka J, Oliveira AP, Nomura CS (2024) Evaluation of the effect of nanoparticles on the cultivation of edible plants by ICP-MS: a review. Anal Bioanal Chem 416(11):2605–2623. https://doi.org/10.1007/s00216-023-05076-w

Oliveira HC, Seabra AB, Kondak S, Adedokun OP, Kolbert Z (2023) Multilevel approach to plant–nanomaterial relationships: from cells to living ecosystems. J Exp Bot 74(12):3406–3424. https://doi.org/10.1093/jxb/erad107

Pagano L, Marmiroli M (2022) Future questions and approaches in plant–nanoparticle research, pp 239–251. https://doi.org/10.1016/b978-0-323-85032-2.00009-9

Pagano L, Servin AD, De La Torre-Roche R, Mukherjee A, Majumdar S, Hawthorne J, Marmiroli M, Maestri E, Marra RE, Isch SM, Dhankher OP, White JC, Marmiroli N (2016) Molecular response of crop plants to engineered nanomaterials. Environ Sci Technol 50(13): 7198–7207. https://doi.org/10.1021/acs.est.6b01816

Pagano L, Pasquali F, Majumdar S, De La Torre-Roche R, Zuverza-Mena N, Villani M, Zappettini A, Marra RE, Isch SM, Marmiroli M, Maestri E, Dhankher OP, White JC, Marmiroli N (2017) Exposure of Cucurbita pepo to binary combinations of engineered nanomaterials: physiological and molecular response. Environ Sci Nano 4(7):1579–1590. https://doi.org/10.1039/c7en00219j

Pagano L, Maestri E, Caldara M, White JC, Marmiroli N, Marmiroli M (2018a) Engineered nanomaterial activity at the organelle level: impacts on the chloroplasts and mitochondria. ACS Sustain Chem Eng 6(10):12562–12579. https://doi.org/10.1021/acssuschemeng.8b02046

Pagano L, Maestri E, White JC, Marmiroli N, Marmiroli M (2018b) Quantum dots exposure in plants: minimizing the adverse response. Curr Opin Environ Sci Health 6:71–76. https://doi.org/10.1016/j.coesh.2018.09.001

Pagano L, Rossi R, Paesano L, Marmiroli N, Marmiroli M (2021) miRNA regulation and stress adaptation in plants. Environ Exp Bot 184:104369. https://doi.org/10.1016/j.envexpbot.2020.104369

Pagano L, Marmiroli M, Villani M, Magnani J, Rossi R, Zappettini A, White JC, Marmiroli N (2022) Engineered nanomaterial exposure affects organelle genetic material replication in Arabidopsis thaliana. ACS Nano. https://doi.org/10.1021/acsnano.1c08367

Pagano L, Rossi R, White JC, Marmiroli N, Marmiroli M (2023) Nanomaterials biotransformation: in planta mechanisms of action. Environ Pollut 318:120834–120834. https://doi.org/10.1016/j.envpol.2022.120834

Pasieczna-Patkowska S, Cichy M, Flieger J (2025) Application of Fourier Transform Infrared (FTIR) spectroscopy in characterization of green synthesized nanoparticles. Molecules. https://doi.org/10.3390/molecules30030684

Păvăloiu RD, Sha'at F, Sha'at M, Nechifor G (2020) Intracellular uptake study of polymeric nanoparticles loaded with cardiovascular drugs using confocal laser scanning microscopy. Chem Proc 3(1):140. https://doi.org/10.3390/ecsoc-24-08427

Połeć-Pawlak K, Ruzik R, Lipiec E (2007) Investigation of Cd(II), Pb(II) and Cu(I) complexation by glutathione and its component amino acids by ESI-MS and size exclusion chromatography coupled to ICP-MS and ESI-MS. Talanta 72(4):1564–1572. https://doi.org/10.1016/j.talanta.2007.02.008

Preetha JSY, Sriram D, Premasudha P, Pudake RN, Anand AV, Arun M (2024) Recent advancements and applications of nano-agrochemicals against plant fungal diseases. In: Nanoparticles in plant biotic stress management. Springer, pp 403–418

Proença PLF, Carvalho LB, Campos EVR, Fraceto LF (2022) Fluorescent labeling as a strategy to evaluate uptake and transport of polymeric nanoparticles in plants. Adv Colloid Interf Sci:102695–102695. https://doi.org/10.1016/j.cis.2022.102695

Rico CM, Majumdar S, Duarte-Gardea M, Peralta-Videa JR, Gardea-Torresdey JL (2011) Interaction of nanoparticles with edible plants and their possible implications in the food chain. J Agric Food Chem 59(8):3485–3498. https://doi.org/10.1021/jf104517j

Rico CM, Peralta-Videa JR, Gardea-Torresdey JL (2015) Differential effects of cerium oxide nanoparticles on rice, wheat, and barley roots: a fourier transform infrared (FT-IR) microspectroscopy study. Appl Spectrosc 69(2):287–295. https://doi.org/10.1366/14-07495

Rico CM, Johnson MG, Marcus MA (2018) Cerium oxide nanoparticles transformation at the root–soil interface of barley (Hordeum vulgare L.). Environ Sci Nano 5(8):1807–1812. https://doi.org/10.1039/C8EN00316E

Ruotolo R, Maestri E, Pagano L, Marmiroli M, White JC, Marmiroli N (2018) Plant response to metal-containing engineered nanomaterials: an omics-based perspective. Environ Sci Technol 52(5):2451–2467. https://doi.org/10.1021/acs.est.7b04121

Sala S, Rengefors K, Kiventerä J, Patanen M, Gefors L, Werdinius C, Winge S, Broberg K, Kalbfleisch S, Sigfridsson Clauss K (2024) Applications of X-ray fluorescence microscopy with synchrotron radiation: from biology to materials science. Radiat Phys Chem. https://doi.org/10.1016/j.radphyschem.2024.112491

Scimeca M, Bischetti S, Lamsira HK, Bonfiglio R, Bonanno E (2018) Energy Dispersive X-ray (EDX) microanalysis: a powerful tool in biomedical research and diagnosis. Eur J Histochem 62(1):2841–2841. https://doi.org/10.4081/ejh.2018.2841

Selby KG, Hubecky EM, Zerda-Pinto V, Korte CE, Bressendorff GA, Tucker K (2024) Mass spectrometry imaging for environmental sciences: a review of current and future applications. Trends Environ Anal Chem. https://doi.org/10.1016/j.teac.2024.e00232

Sembada AA, Lenggoro IW (2024) Transport of nanoparticles into plants and their detection methods. Nanomaterials 14(2):131. https://doi.org/10.3390/nano14020131

Servin AD, Castillo-Michel H, Hernández-Viezcas JA, Corral-Díaz B, Peralta-Videa JR, Gardea-Torresdey JL (2012) Synchrotron micro-XRF and micro-XANES confirmation of the uptake and translocation of TiO2 nanoparticles in cucumber (Cucumis sativus) plants. Environ Sci Technol 46(14):7637–7643. https://doi.org/10.1021/es300955b

Servin AD, Pagano L, Castillo-Michel H, De La Torre-Roche R, Hawthorne J, Hernández-Viezcas JA, Loredo-Portales R, Majumdar S, Gardea-Torresday J, Dhankher OP, White JC (2017) Weathering in soil increases nanoparticle CuO bioaccumulation within a terrestrial food chain. Nanotoxicology 11(1):98–111. https://doi.org/10.1080/17435390.2016.1277274

Shen Y-L, Zhuang S-J, Yang F, Gong C, Xu X (2023) Prefabricated platinum nanomaterial matrix for MALDI-MS imaging of oligosaccharides and lipids in plant tissues. Front Plant Sci 14. https://doi.org/10.3389/fpls.2023.1105374

Silva ALP, Silva SAM, Duarte A, Barceló D, Rocha-Santos T (2022) Analytical methodologies used for screening micro(nano)plastics in (eco)toxicity tests. Green Anal Chem 3:100037–100037. https://doi.org/10.1016/j.greeac.2022.100037

Singh AV, Jungnickel H, Leibrock L, Tentschert J, Reichardt P, Katz A, Laux P, Luch A (2020) ToF-SIMS 3D imaging unveils important insights on the cellular microenvironment during biomineralization of gold nanostructures. Sci Rep 10(1):261–261. https://doi.org/10.1038/s41598-019-57136-w

Singh KM, Jha AB, Dubey RS, Sharma P (2025) Nanoparticle-mediated mitigation of salt stress-induced oxidative damage in plants: insights into signaling, gene expression, and antioxidant mechanisms. Environ Sci Nano 12:2983–3017. https://doi.org/10.1039/D5EN00174A

Steckel A, Schlosser G (2019) An organic chemist's guide to electrospray mass spectrometric structure elucidation. Molecules 24(3):611. https://doi.org/10.3390/molecules24030611

Sun X-D, Ma J-Y, Feng L-J, Duan J-L, Xie X-M, Zhang X-H, Kong X, Ding Z (2023) Magnetite nanoparticle coating chemistry regulates root uptake pathways and iron chlorosis in plants. Proc Natl Acad Sci U S A 120(27):e2304306120

Tanaka Y, Ogra Y (2024) Single-cell analysis of elemental contents by laser ablation–inductively coupled plasma mass spectrometry. J Am Soc Mass Spectrom 35(9):2082–2089. https://doi.org/10.1021/jasms.4c00136

Vats M, Cillero-Pastor B, Cuypers E, Heeren R (2024) Mass spectrometry imaging in plants, microbes, and food: a review. Analysis. https://doi.org/10.1039/d4an00644e

Vora RN, Joshi AN, Joshi N (2020) Green synthesis and characterization of gold nanoparticles using Mucuna monosperma. J Nanosci Technol 6(3):901–904. https://doi.org/10.30799/jnst.309.20060301

Wahab A, Muhammad M, Ullah S, Abdi G, Shah GM, Zaman W, Ayaz A (2024) Agriculture and environmental management through nanotechnology: eco-friendly nanomaterial synthesis for soil-plant systems, food safety, and sustainability. Sci Total Environ 926:171862. https://doi.org/10.1016/j.scitotenv.2024.171862

Wang Y, Deng C, Elmer WH, Dimkpa CO, Sharma S, Navarro G, Wang Z, LaReau J, Steven BT, Wang Z, Zhao L, Li C, Dhankher OP, Gardea-Torresdey JL, Xing B, White JC (2022) Therapeutic delivery of nanoscale sulfur to suppress disease in tomatoes: in vitro imaging and orthogonal mechanistic investigation. ACS Nano 16(7):11204–11217. https://doi.org/10.1021/acsnano.2c04073

Wu J, Sun J, Bosker T, Vijver MG, Peijnenburg WJGM (2023) Toxicokinetics and particle number-based trophic transfer of a metallic nanoparticle mixture in a terrestrial food chain. Environ Sci Technol 57(7):2792–2803. https://doi.org/10.1021/acs.est.2c07660

Yang L, Amad M, Winnik WM, Schoen AE, Schweingruber H, Mylchreest I, Rudewicz PJ (2002) Investigation of an enhanced resolution triple quadrupole mass spectrometer for high-throughput liquid chromatography/tandem mass spectrometry assays. Rapid Commun Mass Spectrom 16(21):2060–2066. https://doi.org/10.1002/rcm.824

Yano J, Yachandra VKX (2010) X-ray absorption spectroscopy. Photosynth Res 102:241–254. https://doi.org/10.1007/s11120-009-9473-8

Yu X, He M, Chen B, Hu B (2020) Recent advances in single-cell analysis by inductively coupled plasma-mass spectrometry: a review. Anal Chim Acta 1137:191–207. https://doi.org/10.1016/j.aca.2020.07.041

Zhang H-Y, Su W-H (2024) Classification, uptake, translocation, and detection methods of nanoparticles in crop plants: a review. Environ Sci Nano 11:1847–1870

Zhang T, Sun H, Lv Z, Cui L, Mao H, Kopittke PM (2017) Using synchrotron-based approaches to examine the foliar application of ZnSO4 and ZnO nanoparticles for field-grown winter wheat. J Agric Food Chem 66(11):2572–2579. https://doi.org/10.1021/acs.jafc.7b04153

Zhang H, Goh NS, Wang JF, Wang JW, Pinals RL, González-Grandío E, Demirer GS, Butrus S, Fakra SC, Del Rio FA, Zhai R, Zhao B, Park S-J, Landry MP (2021) Nanoparticle cellular internalization is not required for RNA delivery to mature plant leaves. Nat Nanotechnol 1–9. https://doi.org/10.1038/s41565-021-01018-8

Zhang W, Meng J, Huang Y, Sarkar B, Singh BP, Zhou X, Gao J, Teng Y, Wang H, Chen W (2023) Effects of soil grain size and solution chemistry on the transport of biochar nanoparticles. Front Environ Sci 10:1114940. https://doi.org/10.3389/fenvs.2022.1114940

Zhang H, Lu KH, Ebbini M, Huang P, Lu H, Li L (2024a) Mass spectrometry imaging for spatially resolved multi-omics molecular mapping. Npj Imaging 2:20. https://doi.org/10.1038/s44303-024-00025-3

Zhang K, Liu Q, Wang Y, Liu X, Zhou X, Yan B (2024b) Advances and challenges in tracking interactions between plants and metal-based nanoparticles. Nanomaterials 14(23):1939. https://doi.org/10.3390/nano14231939

Zhao B, Luo Z, Zhang H, Zhang H (2022) Imaging tools for plant nanobiotechnology. Front Genome Edit 4:1029944. https://doi.org/10.3389/fgeed.2022.1029944

Chapter 8
Molecular Mechanism of Nanoparticle Toxicity Regulation in Plants

Sumira Malik, Jutishna Bora, Aditi, Shalini Mehta, Sweta Sinha, Archna Dashmana, and Smita Lata

Abstract The swift emergence of nanotechnology in agriculture has brought considerable benefits through engineered nanoparticles (ENPs), which have increased crop production, improved nutrient absorption, and reduced pest resistance. With enhanced exposure to engineered nanoparticles comes the potential for toxicity, more accurately referred to as nanotoxicity, associated with the reactions of plants exposed to nanomaterials. Knowing how plants detect, respond, and mitigate their exposure to nanoparticle-induced stress is needed for their responsible and sustainable utilization in agriculture. The exposure of plants to many nanoparticles can induce oxidative stress because of more reactive oxygen species (ROS) accumulation. This imbalance of ROS damages the cellular structures composed of lipids, proteins, and nucleic acids, ultimately inhibiting growth and altering plant physiology. To balance excess oxidative stress, plants also activate subsequent detoxification mechanisms to reduce ROS levels. Antioxidant enzymes like catalase, superoxide dismutase (SOD), and peroxidases are the part detoxification mechanisms. Molecular regulation, beyond antioxidant responses, also includes signal transduction pathways controlled by mitogen-activated protein kinases (MAPKs), stress-responsive transcription factors including WRKY, DREB, NAC, and hormonal signaling networks involving abscisic acid, ethylene, jasmonic acid, and

Sumira Malik and Jutishna Bora contributed equally with all other contributors.

S. Malik
Amity Institute of Biotechnology, Amity University Jharkhand, Ranchi, Jharkhand, India

University Center for Research & Development (UCRD) Chandigarh University, Mohali, Punjab, India

J. Bora · Aditi · S. Sinha · S. Lata (✉)
Amity Institute of Biotechnology, Amity University Jharkhand, Ranchi, Jharkhand, India
e-mail: slata@rnc.amity.edu

S. Mehta
Ranchi Women's College, Ranchi University, Ranchi, Jharkhand, India

A. Dashmana
Himalayan School of Biosciences, Swami Rama University, Dehradun, Uttarakhand, India

R. N. Pudake et al. (eds.), *Phyto Nanotoxicology*, Nanotechnology in Plant Sciences 7, https://doi.org/10.1007/978-3-032-24518-2_8

salicylic acid to modulate the expression of defence-related genes, increase cell wall integrity, and increase repair processes. This chapter discusses and summarizes all these molecular processes happening in plants after the exposure to the nanomaterials.

Keywords Nanoparticles · Nanotoxicity · Plants · Oxidative stress · Signal transduction · Transcription factors · Hormonal regulation · Antioxidants · Molecular mechanism

8.1 Introduction

Nanoparticles are extremely small particles, typically measured below 100 nanometres, which is one-billionth of a meter (Altammar 2023). Their unique characteristics differ markedly from those of the same materials in bulk form. This is mainly because of their diminutive size, which causes the surface area to increase, which results in an increase in their surface reactivity and introducing quantum effects. The increased surface area relative to volume significantly boosts their reactivity. These particles exhibit distinctive optical, electrical, and magnetic properties that can be adjusted by changing their size and shape. Quantum effects emerge because of the confinement of electrons within nanoparticles, contributing to their unique behaviours. Nanoparticles are defined by the International Organization for Standardization (ISO) as materials having exterior dimensions within the nanoscale range. The dimensions of nanoparticles in the longest and shortest axes do not vary substantially. When the difference between these dimensions exceeds threefold, terms like nanofibers or nanoplates are preferred over nanoparticles. Nanoparticles come in various shapes, sizes, and structural forms. The nanomaterials may be of spherical, cylindrical, conical, tubular, hollow core, twisted, or asymmetrical shapes. Typically, they are between one to one hundred nanometers in size. When the size drops below 1 nm, the label "atom clusters" is often used. In terms of structure, nanoparticles can be amorphous or crystalline, including one or more crystal solids. They could be agglomerated clusters or solitary particles (Dolez 2015).

8.1.1 Commonly Studied Nanoparticles in Plant Science

8.1.1.1 Titanium Dioxide (TiO_2) NPs

Several studies have been done to estimate the effects of TiO_2 NPs on microorganisms, planktons, fish, mice, and rats, but studies on effect of these NPs on plant remains are inconclusive. TiO_2 NPs improved the germination of canola and stimulated the radicle and plumule development (Mahmoodzadeh et al. 2013). According to the report of Jaberzadeh et al. (2013), under the water stress, the

TiO_2 NPs enhanced growth and other parameters in wheat plants. It controls nitrogen metabolism enzyme activity, including glutamate dehydrogenase, nitrate reductase, glutamic-pyruvic transaminase and glutamine synthase, which assisted the plants to absorb nitrate. It also helps in conversion of inorganic nitrogen to organic counterpart as present protein, which may enhance the increment of fresh and dry biomass of plant (Yang et al. 2006).

TiO_2 NPs are a very good photocatalytic agent that triggers an oxidation-reduction reaction under the exposure to light (Athanasekou et al. 2018). It significantly enhances the viability and chlorophyll content of aged seeds. It activates Ribulose 1 5-bisphosphate carboxylase (Rubisco) and enhance photosynthesis, thus boosting the growth and development of plants (Yang et al. 2006). It also increases the light absorbance, accelerates light energy transportation and conversion, preserves chloroplasts against ageing, and extends the photosynthetic period of the chloroplasts (Ze et al. 2011). This could be because TiO_2 NPs shield the chloroplast against excess light by enhancing the activity antioxidant enzymes (Hong et al. 2005).

8.1.1.2 SiO_2 NPs

Beneficial properties of SiO_2 NPs in terms of plant growth promotion have been reported. In less concentrations it enhanced the germination (Suriyaprabha et al. 2012). This study in maize reported enhanced seed germination through superior nutrients supply, and better pH and conductivity in the germination medium. Bao-shan and team (Bao-Shan et al. 2004) supplied nano-SiO_2 to Changbai larch (*Larix olgensis*) seedlings and identified that it increased root growth parameters, and chlorophyll content. It has also enhanced the germination rate under abiotic stress. In tomato (Siddiqui and Al-Whaibi 2014) and squash (Haghighi et al. 2012), the germination of seeds and the antioxidant-promoting effect of nano-SiO_2 under the stress of NaCl. In another study, Shah and Belozerova (2009) evaluated nano-SiO_2, and found that it plays an important role in lettuce seeds germination. The application of SiO_2 and TiO_2 NPs together enhances seed germination of soybean through its ability to boost nitrate reductase (Lu ChangMei et al. 2002); and by promoting the water and nutrients absorption and utilization (Zheng et al. 2005). In salinity stress, these NPs enhanced the fresh and dry biomass, chlorophyll and proline content in leaves. This helped the plants to achieve tolerance to abiotic stress (Goswami et al. 2022; Haghighi et al. 2012; Li Bo et al. 2012; Siddiqui and Al-Whaibi 2014). In rice, Wang et al. (2014) conducted study with QDs, and silica coated QDs. They found that the silica coated with QDs spurred significant growth in the root of the rice plant. This growth and development of plants happened because of improvement in parameters like gas exchange, stomatal conductance, chlorophyll fluorescence, photosynthetic and transpiration rate, electron transport rate and photochemical quench (Siddiqui and Al-Whaibi 2014; Xie YinFeng et al. 2011).

8.1.1.3 Zinc Oxide (*ZnO)* NPs

More studies indicate that ZnO NPs boost the growth and health of plants. But with an increased concentration, it reduced the seed germination. Application of NPs on germination is dependent not only on the concentrations of NPs but also on the plant under study. de la Rosa et al. (2013) used ZnO NPs with different doses to the seeds of cucumber, alfalfa and tomato and found that NP treatment only induced germination in cucumber seeds. According to Raliya and Tarafdar (2013), these NPs had a significant effect on plant growth and rhizospheric microbial parameters in cluster bean rhizosphere. Uptake ZnO NPs by roots of *Vigna radiata* and *Cicer arietinum* was confirmed by microscopy and inductive coupled plasma/atomic emission spectroscopy (Mahajan et al. 2011). When ZnO NPs were added to tissue culture media, it enhanced somatic embryogenesis, shoots, and regeneration of plantlets in addition to catalyzing proline production, catalase, superoxide dismutase, and peroxidase activity (Helaly et al. 2014).

8.1.1.4 Manganese (Mn) NPs

Interesting findings were also observed for hydroponically grown mung bean plants with Mn or $MnSO_4$ NPs; and at lower doses it produced substantial increases in root and shoot biomass (10–100%), while similar doses of the bulk Mn produced some toxicological effects. The findings of this study suggestedthat the toxicity was due to a large initial release of manganese ions from the salt form, while the nanoparticulate forms allowed for a more controlled and slow release of nano metal. Additionally, the nanoparticle treatment resulted in improved chlorophyll content, oxygen evolution, carotene phosphorylation, and nitrogen metabolism in plants (Pradhan et al. 2014).

8.1.2 Physical and Chemical Properties Influencing Plant Interaction

The physical and chemical behavior of nanoparticles (NPs) is important for defining the NPs and plants interaction. These properties will affect the uptake of NPs by plants, the way NPs distribute in plant tissues, and the potential impact on its growth and development. Table 8.1 gives an illustration of nanoparticles' toxicity in different plants.

Table 8.1 Nanoparticle toxicity in different plants

Nanoparticles	Plants	Toxicty	Refrences
CuO	*Orzya sativa* (Rice), *Zea mays*	Strong root growth inhibition (as much as 97 per cent)	Muzammil et al. (2023)
ZnO	*Glycine max* (Soybean)	Retarded root growth decreased cell viability, augmented oxidative stress, and changed transcription.	Muzammil et al. (2023)
Al_2O_3	*Sinapis alba*	Reduced seed germination and increase in oxidative stress markers	Muzammil et al. (2023)
TiO_2	*Lactuca sativa*	Decrease in root growth, biomass and antioxidant defense concentrations	Muzammil et al. (2023)
Polystyrene NPs (plastic)	*Allium sepa* (Onion)	Reduce seed germination and root growth	Wang et al. (2023)

8.1.2.1 Key Factors Affecting Leaves

The effect of metal-based NP's size on its uptake in plants has been evaluated in various investigations. According to the results of studies, it has been discovered that nanoparticles lesser than 50 nm, may penetrate the plant leaf through the stomata-based pathway (Ha et al. 2021). With the increase in size of the particles, the leaves became less able to absorb nanoparticles. Nanoparticles have been found in the foliar uptake of many studies. In one study with wheat leaves, fluorescein isothiocyanate (FITC) labelled ZnO nanoparticles of 30 nm was used (Zhu et al. 2020). Using confocal microscopy, it was discovered that ZnO NPs primarily penetrate the epidermis of wheat leaves via the stomatal route and then moved to the chloroplasts. The study also examined the impact of stomatal opening and closing on ZnO NPs absorption; and found that when the stomatal diameter is reduced, the Zn in chloroplast and cytoplasm fell by 33.2% and 8.3, respectively. Avelian and team (Avellan et al. 2019) employed Au NPs of varying diameter (3, 10, 50 nm) with a coating added to target wheat leaves; and found that all NPs could enter leaf. Meanwhile, their findings suggested that the absorption was through disruption and/or diffusion by the cuticle layer as well as stomata. In a different study, scientists established that MgO nanoparticles with a size of 2735 nm were absorbed in leaves of watermelon (*Citrullus lanatus*). Along with metal-based nanoparticles, scientists also found that leaves may absorb nanoparticles composed of natural materials, silica, and polymers. Their study findings revealed that the nature of nanoparticles had an immense influence on the dangerous size that could be assimilated by the leaves. Through TEM, studies have established that the SiO_2 NPs (54 nm) was capable of entering the leaves of *A. thaliana* through the stomatal route (El-Mahrouk et al. 2024). Zhao and team (Zhao et al. 2018) incubated leaves of cucumber with FITC-labelled mesoporous silica nanoparticles (MSNs) and discovered that the leaves were able to absorb the FITC-labelled MSNs. Scientists, by using TEM, discovered that polystyrene nanoplastic measuring 93.6 nm was able to penetrate the leaf barrier of lettuce and then enter the phloem (Lian et al. 2021). More recently, scientists found that SiO_2 nanoparticles made with chitosan and having a mean diameter of 166 nm were taken up in rice leaf and were located inside the cells (Jia-Yi et al. 2022).

Besides the size of NPs, the shape and charge on the surface also affect the entry of nanoparticles into plant tissues. Nanoparticles of varying shapes possess dissimilar interfacial characteristics that bring about alteration in its surface area and resultant contact angle with the plant's external surfaces, which ultimately influence NP absorption (Sun et al. 2021a). Zhang and team (Zhang et al. 2022) compared same size rod-shaped and spherical gold nanoparticles, they found rod-shaped nanoparticles were more easily absorbed and internalised by *Arabidopsis*. Also, plant leaves can absorb both positive and negatively charged nanoparticles. There is prior research on the impact of the NP's surface charge on the absorption of graphene quantum dots (GQDs) in maize leaf, and the ZnO NPs in wheat (Zhu et al. 2021). Findings have indicated that both positively and negatively charged GQDs were absorbed through stomata in maize. By using confocal microscopy, similar results were reported in case of a negative and positive charged ZnO NPs in wheat. Simultaneously, it revealed that positively charged nanoparticles adsorbed the leaves more strongly than negatively charged ones, the latter being mostly because of the electrostatic interaction of positively charged NPs with negative charge on the plant cell walls (Zhu et al. 2021).

8.1.2.2 Key Factors Affecting Roots

Nanoparticle size is believed to be the most critical element in its root absorption. Past research has found that plant roots can take up nanomaterials of less than 10 nm in diameter. It was confirmed by the studies on the absorption of gold (3.5 nm) and CeO_2 (8 ± 1 nm), NPs in the root of *Vicia faba* L. and maize, respectively (Hu et al. 2022; Zhao et al. 2012). Also, in case of TiO_2 nanoparticles (36–140 nm) in wheat roots it was size dependent process. The larger the particle size, the smaller the overall amount of absorption. Whereas NPs larger in terms of particle size than 140 nm cannot be absorbed (Larue et al. 2012). Due to the size constraint, it is widely assumed that for metal NPs greater than 100 nm, it is hard to get absorbed in plants via roots (Banerjee et al. 2019). Interestingly, silicon and natural polymer-based nanoparticles with sizes greater than 100 nm can be absorbed. As an example, *Arabidopsis* root was found to be penetrated by Si NPs having a particle size of 200 nm after 6 weeks of treatment (Sun et al. 2020). Translocation of zein NPs (~135 nm) into sugarcane roots was demonstrated with the help of confocal and transmission electron microscopy (Prasad et al. 2017).

As discussed earlier, the plant uptake of NPs is also determined by its surface charge (Sun et al. 2020). There is negative surface charge the plant root cell walls, and positively charged NPs are prone to settle due to electrostatic interactions, but they cannot enter the root tissue (Bosker et al. 2019). The reversible addition chain transfer-polymerization was used to prepare nanoparticles of various sizes (20–100 nm) and varied surface charges by Parkinson et al. (2022). They discovered that the uncharged and the negatively charged nanoparticles (22 nm) were capable of being absorbed by the root of *A. thaliana*. Later they translocated to the xylem of the root and was proven using confocal microscopy. Comparatively, the positively charged NPs are only able to build up in the root epidermis and not in the root tissues (Parkinson et al. 2022).

8.2 Mechanisms of NPs Uptake and Transport in Plants

Knowledge of the uptake of nanoparticles in plants is important to apply them with full potential in agriculture. Their path from outside environment to inside plant tissues is a complex process that is dependent on the characteristics of the nanoparticles, and furthermore on the plant species. Nanoparticles can carry particular materials, compounds or genetic data into plants; they can be used as vectors to deliver their payloads to their concentration sites. Precise delivery of NPs can be beneficial for the accurate and efficient use in plants. Applications of nanomaterials in agriculture include the administration of fertilizer or nutrients, genetic material, and pesticides or herbicides (Fig. 8.1). The nutrients, including micronutrients like iron, zinc and other trace elements, can be delivered to plant cells through nanoparticles (Ramírez-Ortiz et al. 2020). Tombuloglu and team (Tombuloglu et al. 2020) were able to synthesize micronutrient NPs composites and apply them to *Hordeum vulgare* L. and showed how these micronutrients could be incorporated successfully into the plant tissue. The movement of these NPs greatly enhanced the quantity of micronutrients elements in plant tissues. In particular, the concentrations of Fe, Zn, and Cu were increased by a factor of about 5, 3 and 18 times, respectively, when compared with the control (Tombuloglu et al. 2020). This method is referred to

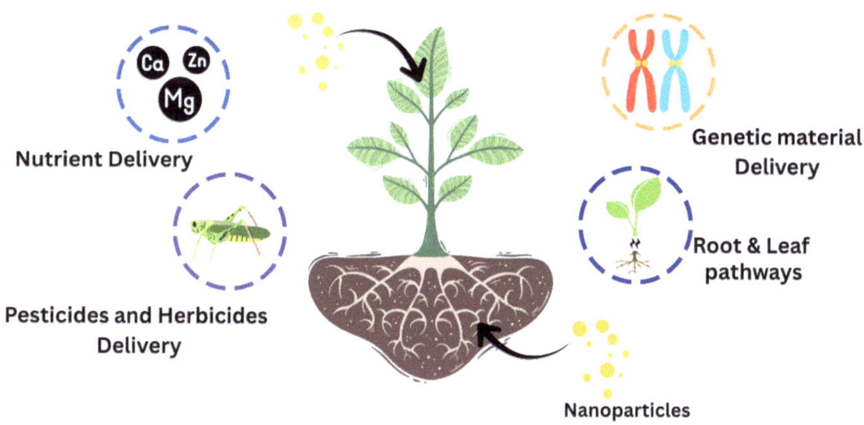

Fig. 8.1 Nanoparticles that are employed to deliver nutrients (fertilizer), genetic material, and pesticides into plants

as nanotechnology based nutrient delivery and is intended to increase the nutrient absorption and consumption of plants.

Nanoparticles can find their way into plant cells by a number of different paths, including endocytosis, diffusion through cell wall, or direct uptake via ion channels and transporters (Hong et al. 2021). After the nanoparticles gets entry into the cells, they release the nutrients contained within them. This precise release will make the nutrients available in the plant at the rate that is equivalent to the needs of the plant (Tombuloglu et al. 2020). Nanoparticle-mediated nutrient delivery can facilitate nutrient uptake and utilize nutrients more effectively, which means that fewer traditional fertilizers are required and fewer nutrient runoffs and leaching contribute to the eco-harmful effect.

8.2.1 Root uptake and Transport Through the Vascular System

Nanoparticles initially intermingle with the roots of the plant by adsorbing on the root surface. Roots may also produce chemicals like mucus or organic acids, that makes root surface negatively charges. This makes it easy to collect and absorb positively charged nanoparticles (Lv et al. 2019). The lateral roots can be developed to form new nanoparticle adsorption interfaces to allow the nanoparticles a potential route to the root column (Peng et al. 2015) (Fig. 8.2). The structure and the role of the root epidermis resemble to that of the plant leaf epidermis. But the epidermis on root

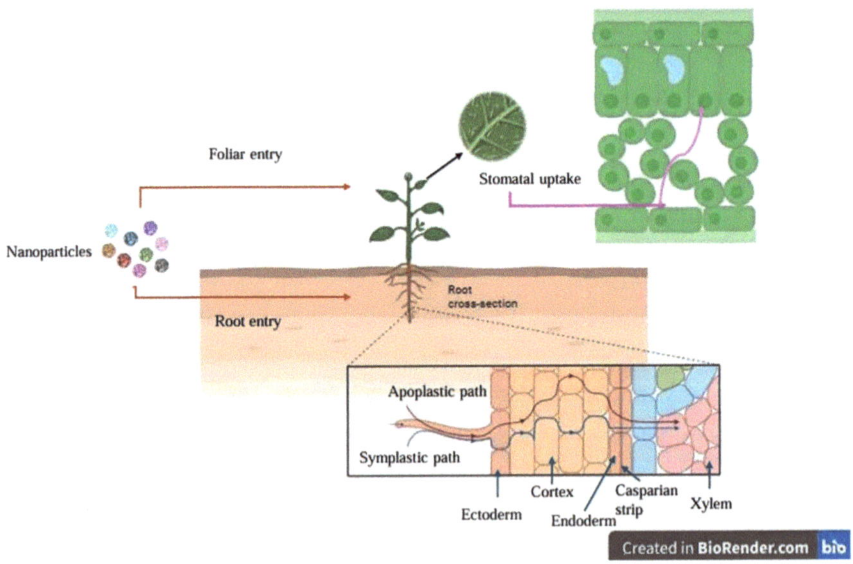

Fig. 8.2 Different pathways of the Nanoparticles' Entry

tip and root hair haven't well developed. In this area, once the nanoparticles are in place, they have immediate contact with the root epidermis and penetrate it (Rajput et al. 2020). The semi-permeability of the relative response of the root cell wall epidermal cells. Small pores in the cell wall of the root may not allow the travel of large particles (Peng et al. 2015). Without the exodermis, nanoparticles can travel within xylem of the root (Peng et al. 2015). Research has revealed that nanoparticles may also enter the plasma membrane of the epidermis to create new holes, via which larger-diameter nanoparticles will enter (Lv et al. 2019). The nanoparticles can find various ways in the plant to enter the cell, such as the ion pathway, endocytosis, binding to cell membrane proteins, or a sort of physical damage to the membrane. In some studies, plant roots took up nanoparticles and entered plant cells mainly by hydrophilic pathways (Li et al. 2020). Endocytosis is also another important pathway through which NPs are taken in plant cells. The NPs are taken into cells as a result of the infolding of the plant cell plasma membrane. The research has revealed that endocytosis of the nanoparticles can occur in plant protoplasts with sizes smaller than 1 μm; therefore, there is no particle size selectivity in endocytosis taken up nanoparticles (Ma and Yan 2018). Liu et al. (2009) postulated that the root cells of *Catharanthus roseus* absorb carbon-based NPs and carbon nanotubes by means of endocytosis. Furthermore, the plants may absorb nanoparticles through the blend with transport or channel proteins that are present on the outer epidermis (Grillo et al. 2021).

8.2.2 Foliar uptake and Interaction with Plant Cuticles

In agricultural use, nanoparticles are commonly applied onto the leaf surface where they settle and then enter the plants via the cuticle or stomata. The important components of the waxy cuticle of the epidermis of leaves are cutin and pectin. It helps prevent water loss to the environment when plants grow and act as a main natural shield against the entrance of nanoparticles into leaves (Yang et al. 2015). Nevertheless, the surface of the waxy stratum corneum contains two different channels (Avellan et al. 2019); one of them is a hydrophilic channel, the other is a lipophilic channel. Both the channels have varying diameters, that is from 0.6 nm to 4.8 nm (Avellan et al. 2019). Hydrophilic nanoparticles having a diameter of lower than 4.8 nm can diffuse into the hydrophilic channels (Banerjee et al. 2019). Lipophilic nanoparticles can pass through the diffusion and infiltration process of lipophilic channels found in the leaf cuticle (Bussieres 2014). Hu and team (Hu et al. 2022) reported that smaller carbon dots (<2 nm) may enter in cotton leaves through the cuticular pathway by using the modern confocal fluorescence microscopy. Nevertheless, absorption of nanoparticles by plants through the epidermis is limited by the pore channels size within the cuticle. Reports on some research studies have shown that when NPs are deposited on the leaf surface, they might build up in the leaf epidermis and vascular tissue; and further could be translocated to other plant tissues. It has been hypothesized that an alternative route of exposure to nanoparticles is possible, i.e. the stomatal pathway (Fig. 8.2). The leaf stomata are important to the functionality of water and gas interchange in plants. Stomata size is usually 10–100 μm, and their number and size differ among the different plant species. The size exclusion limit (SEL) of stomatal aperture to nanoparticle diffusion has not been determined due to the actual its SEL is still being investigated due to the geometry and physiology of stomata (Ha et al. 2021).

8.2.3 *Pathways of Intracellular Transport and Storage*

In plant systems, the uptake and internal movement of nanoparticles (NPs) are critical for their application in agriculture, especially for targeted delivery of nutrients, pesticides, or genetic material. After reaching inside the plant cells, NPs follow specific intracellular pathways to move toward their target location or be stored. The study of the routes of intracellular transport and storage of nanoparticles (NPs) in plants is currently a point of interest in recent studies, due to its importance in enhancing nanobiotechnology use in agriculture. This movement within the plant tissues is complicated, as the main routes are the apoplastic, symplastic, and vacuolar ones that have a different role in the transport and storage of NPs (Wang et al. 2023).

8.2.3.1 Apoplastic Pathway

The apoplastic pathway indicates movement of nanoparticles across a biological system along cell walls and extracellular spaces, skipping the cytoplasm completely. Movement along the apoplastic pathway is heavily reliant on the size, charge, and chemical properties of the nanoparticles' surface. The smallest neutrally charged nanoparticles will diffuse through the apoplast more easily than larger or highly charged particles (Schwab et al. 2016). Movement that follows an apoplastic pathway happens exterior to the plasma membrane, largely via the cell walls and intercellular spaces, such as xylem vessels. This pathway enables the nanoparticles to passively migrate through diffusion or by the force of transpiration, and to get to the vascular tissues without breaching cell membranes at the start. In recent experiments, NPs smaller than 50 nm in size mostly use this route when transporting over long distances, particularly in the root and leaves (Sembada and Lenggoro 2024). The Casparian strip, however, forms a selective barrier, which does not allow free movement of larger nanoparticles into the stele, thereby regulating the internal movement inside the plant (Larue et al. 2012).

8.2.3.2 Symplastic Pathway

The symplastic pathway is characterized by the direct passage of NPs into the cytoplasm of the neighboring cells through the plasmodesmata or small channels that bind cells together. To use this route, NPs are required to penetrate through the plasma membrane, which may happen through endocytosis or the opening up of pores, thereby enabling internalization in the cytoplasm. Rico et al. (2011) have shown that nanomaterials less than 20 nm in size can penetrate cells via endocytosis and avoid membrane obstructions to allow distribution within the cell. After entry, NPs can travel freely within the cytosol of the plant and access different organelles and tissues. This pathway is very significant in the specific transport of the compound directly into the cells, such as the nucleus or chloroplasts, where some biochemical processes occur (Zhou et al. 2024).

8.2.3.3 Vacuolar Sequestration

The vacuolar mechanism is also important in the storage and detoxification of NPs. Once cells are entered, NPs may be compartmentalised in vacuoles, or large storage organelles, which control cellular homeostasis and act as depositories of potentially toxic material. New studies indicate NPs such as CeO_2 and graphene oxide preferentially accumulate in the vacuoles as a site of detoxification and storage. Sequestration can be a process that occurs within the tonoplast (vacuolar membrane), although this is frequently facilitated by the action of particular transporter proteins or by fusion with endosomal structures (Wei et al. 2024). This reservoir assists in absorbing external stressors, including drought, pathogen invasion, by regulating cellular reactions and homeostasis (Khan et al. 2023). Sequestration most probably takes place in the tonoplast (vacuolar membrane), but it probably takes place with the help of certain transporters or due to fusion with endosomal bodies (Wei et al. 2024). In addition, the NP vacuolar sequestration is regulated dynamically under the influence of external stimuli, as well as internal cellular requirements. In stressful environments such as drought, it is possible to increase NPs osmolyte concentration and antioxidant activity through vacuole-specific targeting and thus helps mitigate stress. Movement into vacuoles is generally aided by a series of autophagy or vessels or vesicular transport pathways, which are now coming to be recognized as key routes of intracellular storage and detoxification of nanomaterials. These routes highlight the capacity of the plant to handle and process NPs effectively with the least level of toxicity (Wu et al. 2022).

8.3 Molecular Mechanisms of Nanoparticle-Induced Toxicity in Plants

Molecular toxicity of nanoparticles in plants is also not negligible, in addition to numerous positive impacts. It can harm plants in diverse manners, e.g., causing oxidative stress of plants, causing physical harm to plants, e.g. stomatal closure because of nanoparticle aggregation (Gao et al. 2023).

8.3.1 Reactive Oxygen Species (ROS) Generation and Oxidative Stress

Nanoparticles can interfere with several plant metabolism processes, like the uptake of micronutrients (Liu and Lal 2015), oxidative patterns (Kumari et al. 2024), and gene regulation (Nair and Chung 2015). It has been hinted that in the case of nanoparticles at supra-physiological levels, they trigger the making of ROS and results the oxidative bursts (Zia-ur-Rehman et al. 2023). It has been implied that even the entry of nanoparticles into plant cells disturbs the electron transport chains in mitochondria and chloroplasts, triggering an oxidative burst because ROS are generated (Djanaguiraman et al. 2024). The effects of ROS generated during the reaction between the nanoparticle and any of the cell elements include changes in protein structure or conformation, lipid peroxidation, and damage to nucleic acids (Dilnawaz et al. 2023). Also, the released ROS may result into apoptosis or necrosis that would ultimately result in cell death (Al et al. 2023). Nevertheless, nanoparticles generated by ROS can also be involved in signaling that can lead to stress tolerance. The quantity of ROS in a plant cell have control whether it will be in a signaling or in a damaging mode. Another similarity between the plant and animal cells is that the antioxidant defense systems are well-developed and made up of enzymatic and non-enzymatic molecules that serve to contain the levels of ROS. Peroxidase (POX), superoxide dismutase (SOD), catalase (CAT), ascorbate peroxidase (APX), and glutathione peroxidase (GPX) are the major antioxidant enzymes that participate in the scavenging of ROS. Along with these enzymes, the non-enzymatic pathway through ascorbate, glutathione, carotenoids, tocopherols and phenols (Djanaguiraman et al. 2024). Ma et al. (2013) shown clear evidence of the action of glutathione in compensating for the adverse impact of Ag NPs. It has been shown through reports that there is increased activity of antioxidant enzymes when exposed to nanoparticles. This shows that the antioxidant defensive system of plants is regulated by nanoparticles. Besides the antioxidant defense mechanism, nanoparticles in the regulation of hormones produced in plants also exist (Vinković et al. 2017). On these grounds, it is evident that the genetic mechanism that is related to antioxidant defense and hormone production are expressed by nanomaterials.

8.3.2 Damage of Cellular Organelles

Due to the peculiarities of their nature, nanoparticles (NPs) are generally more common in agricultural use, and their interactions with plant cells pose risks of intoxication. After entering the plant cell, NPs translocate to the cytoplasm and even interact with valuable organelles such as chloroplasts, mitochondria, vacuoles, and the nucleus (Tripathi et al. 2023). Cellular dysregulation is often brought about by their internalization of NPs (Fig. 8.3). The ability of NPs to cause cellular damage by oxidative stress via overaccumulation of reactive oxygen species (ROS) is one of the

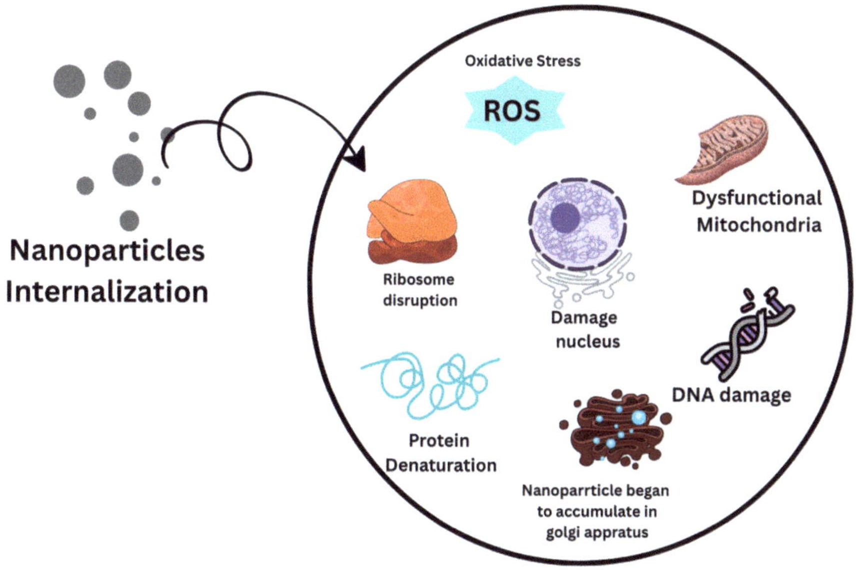

Fig. 8.3 Cellular organelles are damaged by nanoparticles

core issues of the profession. As an illustration, the mitochondria and chloroplasts may lead to excessive ROS accumulation by nanoparticles, such as zinc oxide (ZnO) and silver (Ag), which damage the corresponding membrane, and, as a result, change cellular energy production and photosynthesis (Poddar et al. 2020). Damage to thylakoid membranes followed by degradation of chlorophyll pigments is coupled with destruction of photosynthetic activity, and eventual alteration of plant productivity and growth (Khan et al. 2023). Damage to mitochondria, which is manifested by a change in shape, swelling and loss of cristae, interferes with the normal operation of the electron transport sequence and the production of ATP. Also, damage to the vacuoles may interfere with nutrient accumulation and ionic homeostasis, which is fundamental to cellular functions (Djanaguiraman et al. 2024). Nanomaterials can also disrupt the nuclear envelope, causing strands of DNA in the nucleus to be broken or mutated, causing changes in the gene expression and cell cycle regulation, and affecting the future development of plants (Heikal and Şuţan 2021). The morphological changes of the cells that cause cellular distress (including plasmolysis or higher membrane permeability) can be accompanied by changes in intracellular organelles. It has been confirmed in multiple studies that the buildup of nanomaterials disturbs cellular structures and biochemical pathways. Even though nanoparticles may enhance nutrient use efficiency and avoid pest damage, the stressor between the beneficial applications and phytotoxicity remains a topic of research. The insight into the relations between nanoparticles and plants is important to the safety of low-risk and safer nanomaterials and use in agriculture to implement nanotechnology in crop production in a more efficient way.

8.3.3 DNA damage and Genotoxicity

In toxicology, genotoxicity refers to damage affecting genetic material, while cytotoxicity relates to injury at the cellular processes. The study of toxic effects on genetic material has gained significant attention, particularly because the genome of germ cells governs an organism's heritable traits. Research into genome damage has contributed to the establishment of genetic toxicology, a specialized field focused on identifying and analyzing genetic toxicity (Lynch et al. 2011). Certain plant species serve as valuable genetic assay systems, aiding in the detection of mutagens and the development of analytical techniques. These plant-based systems have aided in our understanding of mutagenesis. Several investigations have reported the genotoxic effects of engineered nanomaterials (ENMs) in plants. For instance, Panda et al. (2011) employed the comet assay to analyze DNA damage in response to Ag NPs, observing a dose-dependent increase in genomic instability linked to ROS formation. Similarly, Faisal et al. (2013) applied the same tool to assess the impact of nickel oxide nanoparticles (NiO NPs) on *Solanum lycopersicum* seedlings, reporting significant DNA damage along with increased rates of apoptosis (21.8%) and necrosis (24.0%). Ciğerci et al. (2015) studied the effects of indium tin oxide (ITO, In_2O_3/SnO_2, 90/10%) nanoparticles (1–100 mg L^{-1}) on *Allium cepa* root meristematic cells, revealing alterations in the cell cycle, evidenced by an increased number of cells progressing into mitosis compared to untreated controls. They also studied the physiological and molecular impacts of Ag NPs (1–10 mg L^{-1}) on *Brassica rapa* species, finding that turnip cells experienced dose-dependent DNA damage. Further validated ZnO NPs-induced chromosomal aberrations in *A. cepa* using the comet assay, highlighting substantial DNA fragmentation. These findings underscore the importance of assessing nanoparticle-induced genotoxicity in plants, as this damage may have broad implications for plant health, genetic stability, and ecosystem balance. Figure 8.4 illustrates the ROS generated from the exposure to nanoparticles, and oxidative stress plays a critical role in damaging essential organelles and therefore affecting plant health at the cellular level.

8.4 Plant Defence Mechanisms Against Nanoparticles

Plant cells that encounter nanoparticles (NPs) often experience over-production of ROS, causing oxidative stress that can affect cellular components like lipids, proteins and nucleic acids. As discussed earlier, the plant response to this oxidative stress is the activation of a complicated antioxidant system consisting of both enzymatic and non-enzymatic mechanisms. The antioxidant system regulates redox homeostasis and reduces oxidative damage (Tripathi et al. 2017). The enzymatic antioxidants primarily detoxify ROS. The primary enzymes of interest include SOD, which catalyzes the dismutation of superoxide radical to hydrogen peroxide (H_2O_2). The CAT enzyme decomposes $H_2O_2$to water and oxygen. Also, the

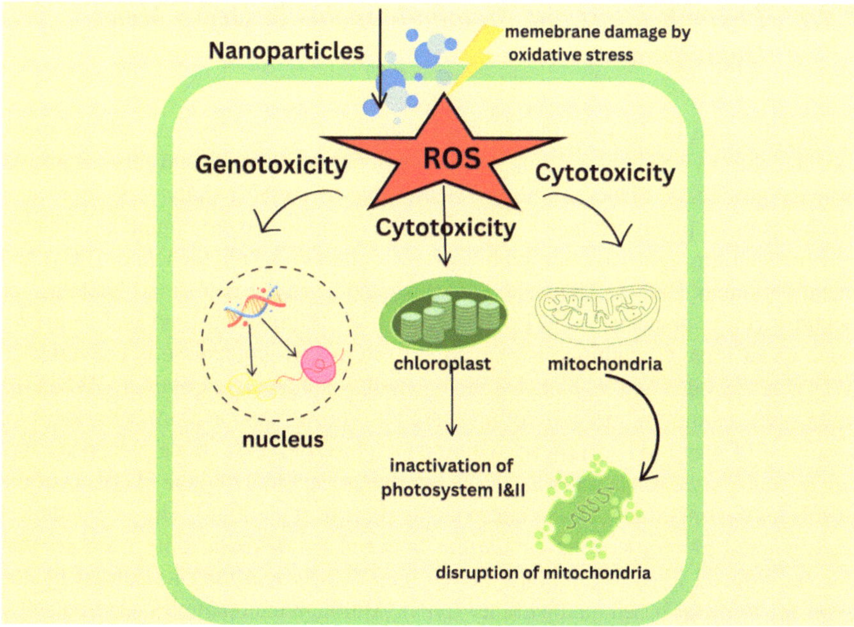

Fig. 8.4 The image shows the toxic influences of nanomaterials on plant cells. It shows how NPs damage cellular function through oxidative stress. Once inside the cell, nanoparticles disrupt the plasma membrane and cause excessive production of ROS. These ROS are a key intermediary to both genotoxicity and cytotoxicity

ascorbate peroxidase (APX) exploits ascorbate as a substrate to detoxify H_2O_2. The interplay of these three enzymes is essential in preventing the build-up of ROS in the plant tissues through defence from oxidative stress induced by NPs. In addition to enzymatic defences, plants also rely on non-enzymatic antioxidants, such as ascorbic acid (vitamin C), glutathione, tocopherols (vitamin E), and flavonoids. These molecules directly scavenge ROS and stabilize free radicals, acting as a secondary line of defence. For instance, glutathione not only neutralizes ROS but also participates in the regeneration of other antioxidants and in redox signalling pathways that mediate stress responses (Hernandez-Viezcas et al. 2011). Research has found that the antioxidant response in plants changes based on the type, size, and concentration of nanoparticles. For example, exposure to AgNPs in *Zea mays* led to a considerable increases in SOD, CAT, and APX activity, suggesting a compensatory mechanism to mitigate oxidative damage (Sharma et al. 2012). Similarly, TiO_2 NPs have been found to induce ROS generation in *A. thaliana*, triggering a robust antioxidant defense response (Ma et al. 2013).

8.4.1 Phytochelatins and Metallothioneins in Heavy Metal Detoxification

Metal NPs, due to their distinctive physicochemical properties, can release metal ions into the environment, causing a serious risk to human health and ecosystems. As primary producers, plants are highly susceptible to metal toxicity. The molecular mechanisms in plants for heavy metal stresses have developed complex responses and proceed through two main pathways *viz.* synthesis of phytochelatin (PCs) and metallothionein (MTs). They are well-designed cysteine-rich small proteins, or peptides (PCs) or proteins (MTs) that have high affinities to bind and sequester toxic metal ions. Glutathione is converted to PCs, and glutathione is converted to MTs gene encoded and regulated at the transcriptional level, respectively. When the metals, e.g. cadmium, mercury, or lead, are introduced into the plant cells, the thiol groups of PCs and MTs will react with the metal ions to form stable complexes that can be transported to vacuoles and stored in the process to avoid the ill effects of the metals on the important cellular components (Faizan et al. 2024). The recent studies have pointed to very striking similarities between the detoxification processes that can be induced by the traditional heavy metal stress and the one caused by the exposure to metal-based nanoparticles (NPs). Although nanoparticles can frequently exhibit distinct physical and chemical properties as compared to bulk metals, when they dissolve, they may give rise to ions that cause analogous effects, such as heavy metals, by triggering comparable stress reactions. Further, the quantity of nanoparticles or ions in the rhizosphere and within the tissues of the plant results in the upregulation of the PC and MT genes, and biosynthesis amplification is a vital initial defense response. Proteomic analysis has shown that nano-ZnO or nano-TiO_2 causes exposure to metal-containing nanoparticles to induce antioxidant and metal-binding pathways, PC synthase activity, and increased levels of the MT protein to prevent oxidative damage and promote detoxification (Cobbett and Goldsbrough 2002). Remarkably enough, the response of plant metabolism to the interaction of nanoparticles can influence the magnitude and the specificity of PC and MT response. As an example, it is the sulfur-containing PCs that are especially concerned with metal ions released by nanoparticles binding and neutralizing them, and the protection of oxidative stress by the MTs is a more general process. More environmental factors, like silicon or secondary elements, could also be synergistic to the detoxification caused by nanoparticles by facilitating a higher concentration of PCs and MTs and tolerance. Recent research has also proposed that new nanomaterials like NO-releasing nanocarriers are able to reduce metal toxicity as well as stimulate endogenous plant defence by those molecular ligands (Seregin and Kozhevnikova 2023).

8.4.2 Gene and Transcription Factor Expression in Response to Stress

Genes and transcription factors respond to stress, which is paramount to offering protection to cells from a variety of stressors, including nanoparticle (NP) exposure. Genes that code for antioxidant enzymes that neutralize ROS are examples of stress-responsive genes (Zia-ur-Rehman et al. 2023). Nanoparticles can change the genetic composition of plants to enable them to endure different stresses. Nanomaterials can control, amplify, or inhibit the processes that govern and metabolic functions in plants (Abideen et al. 2022). Indicatively, exposure of selenium nanoparticles to plants triggered the effects of DNA methylation that normally trigger epigenetic modifications (Abideen et al. 2022). Gene expression studies in *Arabidopsis thaliana* saw a response to ZnO NPs in changes of gene expression. It eventually caused the downgrade of the biomass in plants. Meanwhile, nanoparticles in connection with cellular organization, biosynthetic pathways, and oxidative phosphorylation were deregulated (Landa et al. 2012). Munir et al. (2021) reported that nanomaterials can alter plant metabolic pathways, disrupt DNA, damage membranes, photosynthesis, suppress diseases and stress resistance. A study in*A. thaliana* revealed that exposure to Ag NP has been linked to enhanced gene expression that are involved in repairing oxidative or DNA damage and water homeostasis (Qian et al. 2013; Nair and Chung 2015). The metabolic genes of glutathione involved in the mitigation of abiotic stress in plants were also regulated under the condition of subjecting *A. thaliana* to CeO_2 and indium oxide NPs (Ma et al. 2013). A second experiment done in the same plant revealed that exposure to TiO_2 induced the expression of the light-harvesting complex II (LHCII) gene that eventually led to an increase in the light-harvesting capacity of the chloroplast (Ze et al. 2011). Moreover, ZnO nanoparticle exposure induced the glycolysis metabolic pathway in the maize that triggered the phosphogluco-isomerases and phosphorylases. ZnO nanoparticles were also involved in improving drought stress and enhancing the rate of photosynthesis by enhancing the uptake of carbon by plants (Sun et al. 2021b). Salinity-related genes, which included dehydration-responsive element binding protein (DREB), cytosolic ascorbate peroxidase (cAPX), manganese superoxide dismutase (MnSOD) and Glutathione-S-transferase (GST), were found to be enhanced by exposure to silicon nanoparticles in the strawberry plants. (Moradi et al. 2022). When using silver nanoparticles and sodium chloride in wheat plants, SOD, APX, GR levels were upregulated by 44 -, 82-, and 89-%, respectively. (Wahid et al. 2020). The genes increased expression helped plants in the abiotic stress tolerance. Figure 8.5 illustrates some of the effects of nanoparticles on the regulation of genes in plants.

Similarly, nanoparticles such as lanthanum oxide (La_2O_3) can control some gene expression in plants. As an illustration, aquaporin (AQP) genes were downregulated by La_2O_3 NP exposure in root and shoot maize plants. A decrease in the expression of aquaporin genes is a sign that the growth of plants has ceased, since aquaporins are the genes that encode the channel proteins that help to conduct water between

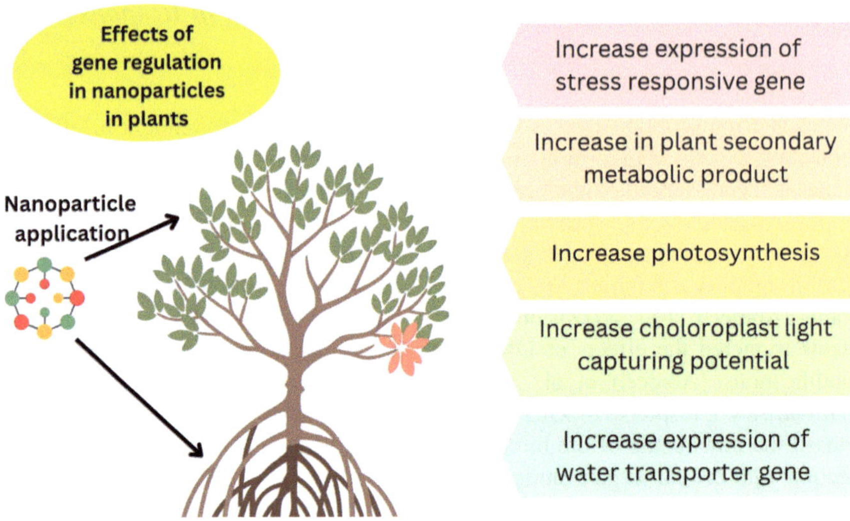

Fig. 8.5 Effects of nanoparticles on gene regulation in plants

Table 8.2 Gene expression affected by nanoparticles in different plants

Nanoparticles	Plants	NPs regulated gene	Expression level	References
AgNPs	*Eruca sativa*	JAL	Increase	Vannini et al. (2013)
Cerium oxide	*Arabidopsis thaliana*	Glutathione	Increase	Ma et al. (2013)
Graphene	Tomato	GPX, PAL and CAT	Decrease	Soraki et al. (2021)
Graphene oxide	*Lepidium sativum*	DPPH and PAL	Increase	Golkar et al. (2021)
Selenium	Wheat	Water transporter gene	Increase	Soliman et al. (2023)
Silicon	Strawberry	DREB, cAPX, MnSOD and GST	Increase	Moradi et al. (2022)

cells (Yue et al. 2017). Recent research has also seen nanoparticles as stimulators of plant growth in agriculture, together with the enhanced increase of plant secondary metabolites. Besides enhancing the synthesis of flavonoids and glutathione in tomato by graphene/Cu nanocomposites, the genes of the GPX, PAL (phenylalanine ammonialyase), and CAT were also induced by nanocomposites (Soraki et al. 2021). The processes of scavenging ROS in plant cells are significant in all 3 of the enzymes. Table 8.2 lists the potential impacts on gene expression of various nanoparticles.

8.4.3 Hormonal Modulation in Response to Exposure to Nanoparticles

Phytohormones are metabolites that regulate important biochemical reactions and physiological developments, and response to biotic and abiotic stress in plant. Auxins, gibberellins, cytokinins, ethylene, abscisic acid (ABA), and brassinosteroids all interact flexibly, orchestrating important physiological processes such as cell division, differentiation, elongation, and organogenesis (Mukherjee et al. 2022). Phytohormonal crosstalk help plants to develop and grow in environments with varying conditions (Gilroy and Breen 2022; Kapoor et al. 2021). Yet, the impacts of climate change and global warming are compounded with other complex biotic and abiotic stressors to reduce plants' ability to adapt and respond (Zandalinas et al. 2024). In Fig. 8.6 shows the hormonal modulation to nanoparticle exposure. Inflammation-related genes are involved in signaling pathways, whereas regulatory mediators of the immune response to NP-induced damage are encoded by cytokines and other inflammatory mediators. To alleviate these issues, recent studies have looked to novel approaches, including adjustments to hormonal signalling. Nanoparticles may lower the negative effects of stress, although on the flip side, plants' responses to NPs may be stimulating or depressing and will be different for every plant based on the actual NP used (Gao et al. 2023; Joudeh and Linke 2022). There has been extensive study of metal (Ag, Au, Cu, Se) and metal oxide nanoparticles (SiO_2, ZnO, Fe_3O_4, TiO_2), because they have strong potential and are environmentally benign in agriculture (Alabdallah and Hasan 2021; Rasheed et al. 2022). Plants possess exquisite mechanisms for absorbing and translocating NPs that can enter from soil through roots or from the air through leaves

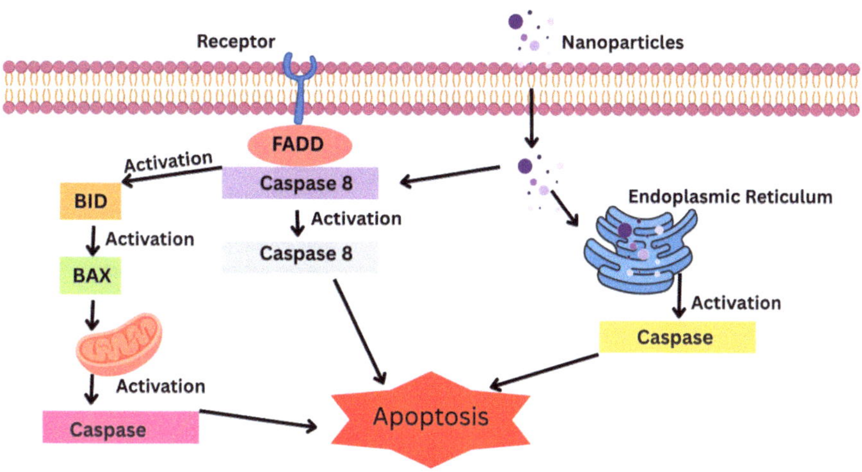

Fig. 8.6 Hormonal modulation in response to nanoparticle exposure, resulting in apoptosis via caspase activation and other mechanisms

(Kurczyńska et al. 2021). After they are internalized, NPs could interact with phytohormone biosynthesis pathways, receptor binding and signal transduction. These NP signaling interactions are likely responsible for changes to plant growth and development, their responses to environmental stress by affecting hormone-receptor interactions, altering protein transporters, or translocating hormones (Sabagh EL et al. 2022). NPs also serve as effective and potentially controlled delivery tools for phytohormones, increasing their bioavailability, function and specifically, when plants are stressed (Khayri Al et al. 2023). Although there are challenges, including the numerous unknowns associated with interactions of NPs with phytohormone and inconsistency in response across plant systems and NP types (Tripathi and Goshisht 2022).

8.5 Environmental and Biological Factors Influencing Nanoparticle Toxicity

Presently, majority of the research focuses on the adverse effects of anthropogenic NPs on plants that are also dependent on the chemical properties, size and reactivity of the NPs. The NP's opponents to the growth of plants can be blamed mainly on the fact that NPs are overly numerous (Khan et al. 2024). NPs are produced in the soil ecosystem and, in particular, the farmland where they may get in and accumulated by plants (Rajput et al. 2018). NP build-ups in plants usually modify physiological mechanisms, leading to low photosynthesis and transpiration rate, which affects plant growth (Rajput et al. 2018) (Fig. 8.7). It has been reported in literature that plastic NPs can also affect the length of roots and stems by modifying the germination rates and have an effect on the plant growth by modifying the photosynthetic processes, causing oxidative damage, altering the action of antioxidant enzymes, and

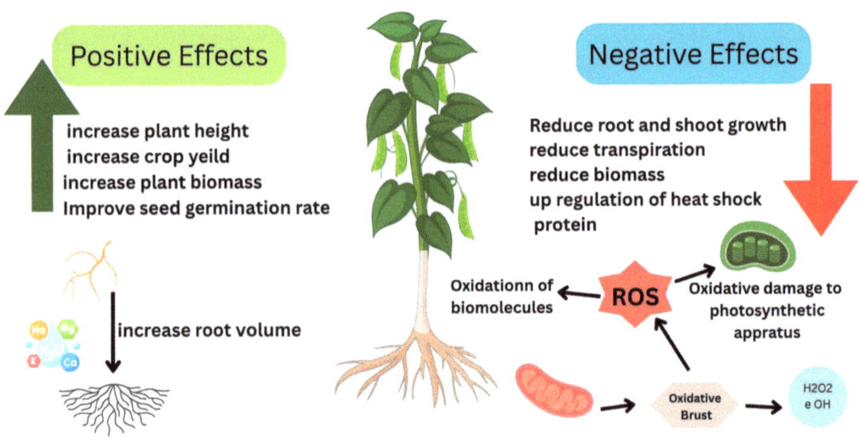

Fig. 8.7 Positive and negative effects of nanoparticles in plants

changing the nutrients of edible crops (Siddique et al. 2018). Giorgetti et al. (2020) compared the rate of germination and root length of the onion following 72 h of exposure to plastic NPs. The reported results found that root length was reduced in 0.1 and 1.0 g/L dose, which was in line with the inhibitory action of NPs on the growth of the root at higher doses (Giorgetti et al. 2020). The process can be supported by the fact that the particles are adsorbed onto the peripheral tissue of the roots and prevent the absorption of moisture and, accordingly, the decrease in hydraulic conductivity and the decrease in transpiration of the plant (Kalčíková et al. 2017). A study examined the effects of 500 μM and 1000 μM of Ag NPs on the growth of cucumber seedlings. The findings showed that total biomass, chlorophyll pigments and carotenoid, photosynthesis efficiency, and micronutrients like Zn and Fe declined during the treatment (Tripathi et al. 2017). In *A. thaliana*, scientists determined the toxicity of ZnO NPs on and observed that higher doses of the particle (50, 100, and 200 mg L − 1) had an inhibitory effect on the leaf size, chlorosis, and lateral roots of the plants, which was probably determined by the inhibition of nutrient uptake by roots leading to the overall reduced concentration of the micronutrients (Nair and Chung 2015). To be more precise, root elongation of *A. thaliana* after treating with CDs (at 3 nm) at 125, 250, 500 and 1000 mg L-1 concentrations, was reduced by about 25, 40, 56 and 61 percent, respectively, which points to the fact that the phytotoxicity of CDs was dose-dependent (Chen et al. 2020).

8.6 Application of Nanoparticles in Agriculture

The nanoparticles are used in different agricultural applications; and China is ranked as the top in the research and product development (Yata et al. 2018). In agriculture, it is applicable through making nanofertilizers, nanoherbicides, nano fungicides, and nanosensors (Fig. 8.8). Nanofertilizers enhance healthy growth of the crops by aiding the required uptake of the micronutrients required to ensure healthy development of the plant. The nanofertilizer can be prepared as Zn, Si, and TiO_2 (López-Valdez et al. 2018), Cu NPs (VandeVoort and Arai 2018), and even the polymeric NPs in the form of dendrimers as nanocarriers (Kah et al. 2018). Nanopesticides produce protective measures against biotic stress. Their primary use is in encapsulating pesticides to be released under protection, enhancing the selectivity of the pesticides and stability of the active ingredient compound (Pascoli et al. 2018).

8.6.1 Nanofertilizers

Fertilizers are natural or manmade materials that provide the chemical components that are essential for plant growth and productivity; and also enhances the natural fertility of soils by ameliorating the shortages of micronutrients (Meena et al. 2016). ZnO NPs have specifically shown to possess usefulness in surmounting zinc

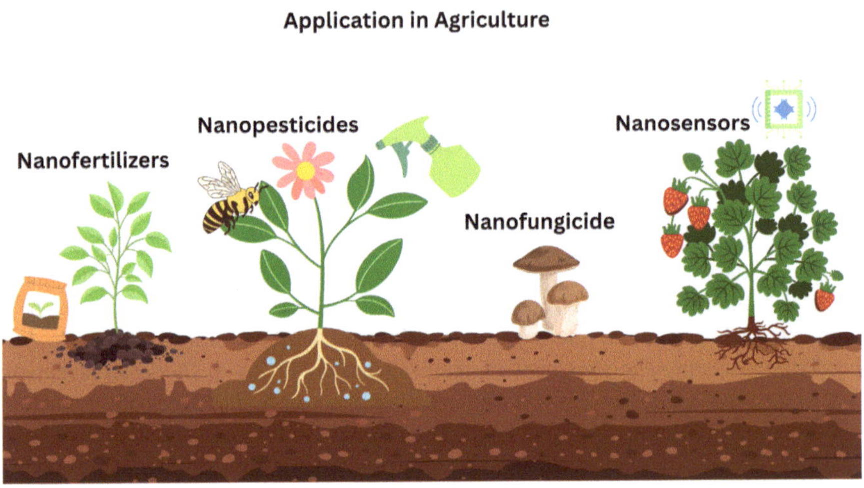

Fig. 8.8 Nanoparticle application in agriculture

deficiency, particularly in rice, owing to foliar treatment using NPs. Besides, ZnO is also capable of improving plant growth and yield data and improving the function of the dehydrogenase enzyme (Bala et al. 2019). Nanocomposites may be used as a delivery vehicle for nutrients like nitrogen. Madusanka et al. (2017) encapsulated urea-modified hydroxyapatite (HAP) NPs intercalated within the clay montmorillonite, and this enhanced yield in rice (*Oryza sativa*) in addition to controlling nitrogen availability. Bioengineered NPs like vesicles made of biological membrane, were also used as a source of zinc used to fertilize the foliage of *Brassica oleracea* and were reported to be very effective and easily enter the protoplast of the plant (Rios et al. 2019).

8.6.2 Nanoherbicides

Novel nanoformulations are made to minimize the ill effects of herbicides on the external environment, not only by prolonging and extending controlled release, but also providing protection from environmental degradation, such as UV degradation (Li et al. 2007). It also increase selectivity by protecting non-target plant species and microbes and beneficial insects. Polymeric NPs offer the greatest potential for encapsulating herbicides for slow diffusion and release (Abigail and Chidambaram 2017). Chitosan/tripolyphosphate NPs incorporated a popular nonselective herbicide -paraquat (Grillo et al. 2014). The chitosan/tripolyphosphate NPs incorporated paraquat nanoformulation was found to have a lower toxicity than the original paraquat, while maintaining the effectiveness of paraquat after formulation, and offered true promise to protect various plant species after herbal treatments. Conversely, photocatalytic NPs have been targeted at degrading herbicides. ZnO NPs have the potential to form the degradation of herbicide diquat under the sun (Shibin et al. 2015) and can degrade 70 to 90 per cent of herbicide (Páez et al. 2019). The easily decomposable imazethapyr herbicide exhibited a higher rate of degradation under solar irradiation along with TiO_2, which demonstrated physisorption on the photocatalyst surface (Ishiki et al. 2005). In another study, the WO_3–TiO_2 nanocomposite reported complete degradation of imazapyr herbicide under UV in 2 hours (Ismail et al. 2016).

8.6.3 Nanopesticides

The chemical stability, solubility, bioavailability, photodecomposition, and adsorption of the pesticide to the soil are all factors that lead to the toxicity of the pesticide (Kumar et al. 2019); the ultimate role of nanotools is to modify those characteristics by developing nanocarriers that will enable its gradual release (Jampílek and Kráľová 2017). Some of the advantages of the usage of nanostructured compounds by plants include higher yield, durability and higher nutrient content. Conversely, the impact of such materials in the long-term perspective remains to be understood in full, as such nano-compounds may act as nutrients and pest defense (insects, plants, bacteria) as well as cause stress to the other species in the ecosystem, which may result in an ecological hazard (Bourguet et al. 2016). Polymer-based nanoformulations are the most attractive formulation of pesticides (Kumar et al. 2019).

8.7 Conclusion

In this chapter, the absorption and transportation of NPs in plants and its toxicity on plant bioprocesses are reviewed. Their uptake into the plant leaf occurs using the pores on the plant cell wall and stomata, and in primary roots, through root cell wall pores, and needlestick spots. They enter by both the symplastic and apoplastic pathways in cells and transmitted between the various tissues by the xylem and phloem. The absorption and transportation process depends on its physical attributes, which include size, chemical composition, surface charge, and surface modification. Moreover, they exert varied impacts on seed germination and plant growth. The various classes of nanoparticles have been shown to cause both low and high concentrations of plant growth and injuries, respectively. The genotoxicity of nanoparticles may also include chromosomal aberration and micronuclei, that alter the gene expression of plants, and this toxicity may be caused by various mechanisms, including its toxicity. Although the impact of NPs on plants is not controversial, few studies have given evidence on the mechanism of plant toxicity caused by nanoparticles. The molecular mechanisms of nanoparticle toxicity on plant cells are based on the relationship between the nanoparticles and cellular barriers, the induction of oxidative stress, and the disruption of the gene and metabolic network. On entry, nanoparticles appear to produce ROS, altering the properties of proteins, damaging DNA and lipids, and causing a redox imbalance in the plant cells. Plants also overcome this toxicity by increasing antioxidant enzymes (SOD, CAT, APX, GPX) and stress-related gene expression and modifying secondary metabolite signalling and hormone synthesis. These regulatory mechanisms play an important role in cellular homeostasis and resilience to nanoparticle stress. Moreover, nanoparticles influence the key metabolic processes (photosynthesis, the assimilation of carbon dioxide and nitrogen) and can lead to physiological inhibition, nutrient disequilibrium, membrane dysfunction and the consumption of energy, in some cases, causing DNA damage and genotoxicity. We can conclude that the toxicity of nanomaterials differs with the chemical composition, size, shape, and concentration. Also, it is further regulated in a species-specific regulatory network and ecological contexts. Therefore, it is crucial to understand such molecular regulation pathways because it is necessary to create safe and sustainable applications of nanotechnology in agriculture and reduce ecological risks.

References

Abideen Z, Hanif M, Munir N, Nielsen BL (2022) Impact of nanomaterials on the regulation of gene expression and metabolomics of plants under salt stress. Plants 11(5):691

Abigail EA, Chidambaram R (2017) Nanostructured materials–fabrication to applications. In: IntechOpen, Rijeka, pp 207–212

Alabdallah NM, Hasan MM (2021) Plant-based green synthesis of silver nanoparticles and its effective role in abiotic stress tolerance in crop plants. Saudi J Biol Sci 28(10):5631–5639

Al-Khayri JM, Rashmi R, Surya Ulhas R, Sudheer WN, Banadka A, Nagella P, ... Almaghasla MI (2023) The role of nanoparticles in response of plants to abiotic stress at physiological, biochemical, and molecular levels. Plants 12(2):292

Altammar KA (2023) A review on nanoparticles: characteristics, synthesis, applications, and challenges. Front Microbiol 14:1155622

Athanasekou CP, Likodimos V, Falaras P (2018) Recent developments of TiO2 photocatalysis involving advanced oxidation and reduction reactions in water. J Environ Chem Eng 6:7386–7394

Avellan A, Yun J, Zhang Y, Spielman-Sun E, Unrine JM, Thieme J et al (2019) Nanoparticle size and coating chemistry control foliar uptake pathways, translocation, and leaf-to-rhizosphere transport in wheat. ACS Nano 13(5):5291–5305

Bala R, Kalia A, Dhaliwal SS (2019) Evaluation of efficacy of ZnO nanoparticles as remedial zinc nanofertilizer for rice. J Soil Sci Plant Nutr 19(2):379–389

Banerjee K, Pramanik P, Maity A, Joshi DC, Wani SH, Krishnan P (2019) Methods of using nanomaterials to plant systems and their delivery to plants (mode of entry, uptake, translocation, accumulation, biotransformation and barriers). In: Advances in phytonanotechnology. Academic, pp 123–152

Bao-Shan L, Shao-Qi D, Chun-Hui L, Li-Jun F, Shu-Chun Q, Min Y (2004) Effect of TMS (nanostructured silicon dioxide) on growth of Changbai larch seedlings. J For Res 15(2): 138–140

Bosker T, Bouwman LJ, Brun NR, Behrens P, Vijver MG (2019) Microplastics accumulate on pores in seed capsule and delay germination and root growth of the terrestrial vascular plant Lepidium sativum. Chemosphere 226:774–781

Bourguet D, Guillemaud T, Lichtfouse E (2016) Sustainable agriculture reviews. Vol 7, ISSN 2210-4410. Springer Dordrecht Heidelberg, London New York.

Bussieres P (2014) Estimating the number and size of phloem sieve plate pores using longitudinal views and geometric reconstruction. Sci Rep 4(1):4929

Chen J, Wu L, Lu M, Lu S, Li Z, Ding W (2020) Comparative study on the fungicidal activity of metallic MgO nanoparticles and macroscale MgO against soilborne fungal phytopathogens. Front Microbiol 11:365

Ciğerci İH, Liman R, Özgül E, Konuk M (2015) Genotoxicity of indium tin oxide by Allium and Comet tests. Cytotechnology 67(1):157–163

Cobbett C, Goldsbrough P (2002) Phytochelatins and metallothioneins: roles in heavy metal detoxification and homeostasis. Annu Rev Plant Biol 53:159–182. https://doi.org/10.1146/annurev.arplant.53.100301.135154

de la Rosa G, López-Moreno ML, de Haro D, Botez CE, Peralta-Videa JR, Gardea-Torresdey JL (2013) Effects of ZnO nanoparticles in alfalfa, tomato, and cucumber at the germination stage: root development and X-ray absorption spectroscopy studies. Pure Appl Chem 4th 85(12): 2161–2174

Dilnawaz F, Misra AN, Apostolova E (2023) Involvement of nanoparticles in mitigating plant's abiotic stress. Plant Stress 10:100280

Djanaguiraman M, Priyanka AS, Haripriya S, Kalarani MK, Umapathi M (2024) Nanoselenium improves drought tolerance of sorghum via reduced transpiration rate and osmolytic accumulation. S Afr J Bot 172:93–108

Dolez PI (2015) Nanomaterials definitions, classifications, and applications. In: Nanoengineering. Elsevier, pp 3–40

El-Mahrouk ESM, Atef EAM, Gabr MK, Aly MA, Mohamed AE, Eisa EA, Gururani MA (2024) Response of Salvia officinalis to zinc and silicon nanoparticles and pollen extract as alternates to traditional fertilizers. Front Plant Sci 15:1469691

EL Sabagh A, Islam MS, Hossain A, Iqbal MA, Mubeen M, Waleed M, ... Abdelhamid MT (2022) Phytohormones as growth regulators during abiotic stress tolerance in plants. Front Agron 4: 765068

Faisal M, Saquib Q, Alatar AA, Al-Khedhairy AA, Hegazy AK, Musarrat J (2013) Phytotoxic hazards of NiO-nanoparticles in tomato: a study on mechanism of cell death. J Hazard Mater 250:318–332

Faizan M, Alam P, Hussain A, Karabulut F, Tonny SH, Cheng SH et al (2024) Phytochelatins: key regulator against heavy metal toxicity in plants. Plant Stress 11:100355

Gao M, Chang J, Wang Z, Zhang H, Wang T (2023) Advances in transport and toxicity of nanoparticles in plants. J Nanobiotechnol 21(1):75

Gilroy E, Breen S (2022) Interplay between phytohormone signalling pathways in plant defence–other than salicylic acid and jasmonic acid. Essays Biochem 66(5):657–671

Giorgetti L, Spanò C, Muccifora S, Bottega S, Barbieri F, Bellani L, Castiglione MR (2020) Exploring the interaction between polystyrene nanoplastics and Allium cepa during germination: internalization in root cells, induction of toxicity and oxidative stress. Plant Physiol Biochem 149:170–177

Golkar P, Bakhtiari MA, Bazarganipour M (2021) The effects of nanographene oxide on the morpho-biochemical traits and antioxidant activity of Lepidium sativum L. under in vitro salinity stress. Sci Hortic 288:110301

Goswami P, Mathur J, Srivastava N (2022) Silica nanoparticles as novel sustainable approach for plant growth and crop protection. Heliyon 8:e09908

Grillo R, Pereira AE, Nishisaka CS, De Lima R, Oehlke K, Greiner R, Fraceto LF (2014) Chitosan/tripolyphosphate nanoparticles loaded with paraquat herbicide: an environmentally safer alternative for weed control. J Hazard Mater 278:163–171

Grillo R, Mattos BD, Antunes DR, Forini MM, Monikh FA, Rojas OJ (2021) Foliage adhesion and interactions with particulate delivery systems for plant nanobionics and intelligent agriculture. Nano Today 37:101078

Ha N, Seo E, Kim S, Lee SJ (2021) Adsorption of nanoparticles suspended in a drop on a leaf surface of Perilla frutescens and their infiltration through stomatal pathway. Sci Rep 11(1): 11556

Haghighi M, Afifipour Z, Mozafarian M (2012) The effect of N-Si on tomato seed germination under salinity levels. J Biol Environ Sci 6(16)

Heikal YM, Şuţan NA (2021) Mechanisms of genotoxicity and oxidative stress induced by engineered nanoparticles in plants. In: Induced genotoxicity and oxidative stress in plants. Springer Singapore, Singapore, pp 151–197

Helaly MN, El-Metwally MA, El-Hoseiny H, Omar SA, El-Sheery NI (2014) Effect of nanoparticles on biological contamination of'in vitro'cultures and organogenic regeneration of banana. Aust J Crop Sci 8(4):612–624

Hernandez-Viezcas JA, Castillo-Michel H, Servin AD, Peralta-Videa JR, Gardea-Torresdey JL (2011) Spectroscopic verification of zinc absorption and distribution in the desert plant Prosopis juliflora-velutina (velvet mesquite) treated with ZnO nanoparticles. Chem Eng J 170(2-3): 346–352

Hong F, Zhou J, Liu C, Yang F, Wu C, Zheng L, Yang P (2005) Effect of nano-TiO2 on photochemical reaction of chloroplasts of spinach. Biol Trace Elem Res 105(1):269–279

Hong J, Wang C, Wagner DC, Gardea-Torresdey JL, He F, Rico CM (2021) Foliar application of nanoparticles: mechanisms of absorption, transfer, and multiple impacts. Environ Sci Nano 8(5): 1196–1210

Hu D, Bai Y, Li L, Ai M, Jin J, Song K, Liu D (2022) Phytotoxicity assessment study of gold nanocluster on broad bean (Vicia Faba L.) seedling. Environ Pollut Bioavailab 34(1):284–296

Ishiki RR, Ishiki HM, Takashima K (2005) Photocatalytic degradation of imazethapyr herbicide at TiO2/H2O interface. Chemosphere 58(10):1461–1469

Ismail AA, Abdelfattah I, Helal A, Al-Sayari SA, Robben L, Bahnemann DW (2016) Ease synthesis of mesoporous WO3–TiO2 nanocomposites with enhanced photocatalytic performance for photodegradation of herbicide imazapyr under visible light and UV illumination. J Hazard Mater 307:43–54

Jaberzadeh A, Moaveni P, Moghadam HRT, Zahedi H (2013) Influence of bulk and nanoparticles titanium foliar application on some agronomic traits, seed gluten and starch contents of wheat subjected to water deficit stress. Notulae Botanicae Horti Agrobotanici Cluj-Napoca 41(1): 201–207

Jampílek J, Kráľová K (2017) Nanopesticides: preparation, targeting, and controlled release. In: New pesticides and soil sensors. Academic, pp 81–127

Jia-Yi Y, Meng-Qiang S, Zhi-Liang C, Yu-Tang X, Hang W, Jian-Qiang Z et al (2022) Effect of foliage applied chitosan-based silicon nanoparticles on arsenic uptake and translocation in rice (Oryza sativa L.). J Hazard Mater 433:128781

Joudeh N, Linke D (2022) Nanoparticle classification, physicochemical properties, characterization, and applications: a comprehensive review for biologists. J Nanobiotechnol 20(1):262

Kah M, Kookana RS, Gogos A, Bucheli TD (2018) A critical evaluation of nanopesticides and nanofertilizers against their conventional analogues. Nat Nanotechnol 13(8):677–684

Kalčíková G, Gotvajn AŽ, Kladnik A, Jemec A (2017) Impact of polyethylene microbeads on the floating freshwater plant duckweed Lemna minor. Environ Pollut 230:1108–1115

Kapoor RT, Salvadori MR, Rafatullah M, Siddiqui MR, Khan MA, Alshareef SA (2021) Exploration of microbial factories for synthesis of nanoparticles–a sustainable approach for bioremediation of environmental contaminants. Front Microbiol 12:658294

Khan S, Zahoor M, Khan RS, Ikram M, Islam NU (2023) The impact of silver nanoparticles on the growth of plants: the agriculture applications. Heliyon 9(6)

Khan A, Farhan A, Maqbool F, Maqsood N, Qayyum W, Haider A et al (2024) Exploring the transporters and mechanisms of arsenic detoxification in plants and potential role of nanoparticles in alleviating arsenic stress. Plant Growth Regul 104(1):95–119
Kumar S, Nehra M, Dilbaghi N, Marrazza G, Hassan AA, Kim KH (2019) Nano-based smart pesticide formulations: emerging opportunities for agriculture. J Control Release 294:131–153
Kumari A, Gupta AK, Sharma S, Jadon VS, Sharma V, Chun SC, Sivanesan I (2024) Nanoparticles as a tool for alleviating plant stress: mechanisms, implications, and challenges. Plants 13(11): 1528
Kurczyńska E, Godel-Jędrychowska K, Sala K, Milewska-Hendel A (2021) Nanoparticles—plant interaction: what we know, where we are? Appl Sci 11(12):5473
Landa P, Vankova R, Andrlova J, Hodek J, Marsik P, Storchova H et al (2012) Nanoparticle-specific changes in Arabidopsis thaliana gene expression after exposure to ZnO, TiO2, and fullerene soot. J Hazard Mater 241:55–62
Larue C, Veronesi G, Flank AM, Surble S, Herlin-Boime N, Carrière M (2012) Comparative uptake and impact of TiO2 nanoparticles in wheat and rapeseed. J Toxic Environ Health A 75(13–15): 722–734
Li Bo LB, Tao GongSheng TG, Xie YinFeng XY, Cai XianLei CX (2012) Physiological effects under the condition of spraying nano-SiO2 onto the Indocalamus barbatus McClure leaves. J Nanjing Forestry Univ 36(04):161
Li ZZ, Chen JF, Liu F, Liu AQ, Wang Q, Sun HY, Wen LX (2007) Study of UV-shielding properties of novel porous hollow silica nanoparticle carriers for avermectin. Pest Manag Sci Formerly Pesticide Sci 63(3):241–246
Li L, Luo Y, Li R, Zhou Q, Peijnenburg WJ, Yin N et al (2020) Effective uptake of submicrometre plastics by crop plants via a crack-entry mode. Nat Sustainability 3(11):929–937
Lian J, Liu W, Meng L, Wu J, Chao L, Zeb A, Sun Y (2021) Foliar-applied polystyrene nanoplastics (PSNPs) reduce the growth and nutritional quality of lettuce (Lactuca sativa L.). Environ Pollut 280:116978
Liu R, Lal R (2015) Potentials of engineered nanoparticles as fertilizers for increasing agronomic productions. Sci Total Environ 514:131–139
Liu Q, Chen B, Wang Q, Shi X, Xiao Z, Lin J, Fang X (2009) Carbon nanotubes as molecular transporters for walled plant cells. Nano Lett 9(3):1007–1010
López-Valdez F, Miranda-Arámbula M, Ríos-Cortés AM, Fernández-Luqueño F, De-La-Luz V (2018) Agricultural nanobiotechnology. Modern Agric Sustain Future
Lu ChangMei LC, Zhang ChaoYing ZC, Wen JunQiang WJ, Wu GuoRong WG, Tao MingXuan TM (2002) Research of the effect of nanometer materials on germination and growth enhancement of Glycine max and its mechanism
Lv J, Christie P, Zhang S (2019) Uptake, translocation, and transformation of metal-based nanoparticles in plants: recent advances and methodological challenges. Environ Sci Nano 6(1):41–59
Lynch AM, Sasaki JC, Elespuru R, Jacobson-Kram D, Thybaud V, De Boeck M, Aardema MJ, Aubrecht J, Benz RD, Dertinger SD (2011) New and emerging technologies for genetic toxicity testing. Environ Mol Mutagen 52:205–223
Ma X, Yan J (2018) Plant uptake and accumulation of engineered metallic nanoparticles from lab to field conditions. Curr Opin Environ Sci Health 6:16–20
Ma C, Chhikara S, Xing B, Musante C, White JC, Dhankher OP (2013) Physiological and molecular response of Arabidopsis thaliana (L.) to nanoparticle cerium and indium oxide exposure. ACS Sustain Chem Eng 1(7):768–778
Madusanka N, Sandaruwan C, Kottegoda N, Sirisena D, Munaweera I, De Alwis A et al (2017) Urea–hydroxyapatite-montmorillonite nanohybrid composites as slow-release nitrogen compositions. Appl Clay Sci 150:303–308
Mahajan P, Dhoke SK, Khanna AS (2011) Effect of nano-ZnO particle suspension on growth of mung (Vigna radiata) and gram (Cicer arietinum) seedlings using plant agar method. Journal of Nanotechnology 2011(1):696535

Mahmoodzadeh H, Nabavi M, Kashefi H (2013) Effect of nanoscale titanium dioxide particles on the germination and growth of canola (Brassica napus)
Meena VS, Maurya BR, Verma JP, Meena RS (eds) (2016) Potassium solubilizing microorganisms for sustainable agriculture, vol 331. Springer, New Delhi
Moradi P, Vafaee Y, Mozafari AA, Tahir NAR (2022) Silicon nanoparticles and methyl jasmonate improve physiological response and increase expression of stress-related genes in strawberry cv. Paros under salinity stress. Silicon 14(16):10559–10569
Mukherjee AA, Singh RK, Mishra R, Bag S (2022) Application of blockchain technology for sustainability development in agricultural supply chain: justification framework. Oper Manag Res 15(1):46–61
Munir N, Hanif M, Dias DA, Abideen Z (2021) The role of halophytic nanoparticles towards the remediation of degraded and saline agricultural lands. Environ Sci Pollut Res 28(43): 60383–60405
Muzammil S, Ashraf A, Siddique MH, Aslam B, Rasul I, Abbas R, Afzal M, Faisal M, Hayat S (2023) A review on toxicity of nanomaterials in agriculture: current scenario and future prospects. Sci Prog 106(4):368504231221672. https://doi.org/10.1177/00368504231221672
Nair PMG, Chung IM (2015) Study on the correlation between copper oxide nanoparticles induced growth suppression and enhanced lignification in Indian mustard (Brassica juncea L.). Ecotoxicol Environ Saf 113:302–313
Páez MR, Ochoa-Muñoz Y, Rodriguez-Páez JE (2019) Efficient removal of a glyphosate-based herbicide from water using ZnO nanoparticles (ZnO-NPs). Biocatal Agric Biotechnol 22: 101434
Panda KK, Achary VMM, Krishnaveni R, Padhi BK, Sarangi SN, Sahu SN, Panda BB (2011) In vitro biosynthesis and genotoxicity bioassay of silver nanoparticles using plants. Toxicol In vitro 25(5):1097–1105
Parkinson SJ, Tungsirisurp S, Joshi C, Richmond BL, Gifford ML, Sikder A et al (2022) Polymer nanoparticles pass the plant interface. Nat Commun 13(1):7385
Pascoli M, Lopes-Oliveira PJ, Fraceto LF, Seabra AB, Oliveira HC (2018) State of the art of polymeric nanoparticles as carrier systems with agricultural applications: a minireview. Energy Ecol Environ 3(3):137–148
Peng C, Duan D, Xu C, Chen Y, Sun L, Zhang H et al (2015) Translocation and biotransformation of CuO nanoparticles in rice (Oryza sativa L.) plants. Environ Pollut 197:99–107
Péret B, Larrieu A, Bennett MJ (2009) Lateral root emergence: a difficult birth. J Exp Bot 60(13): 3637–3643
Poddar K, Sarkar D, Sarkar A (2020) Nanoparticles on photosynthesis of plants: effects and role. In: Green nanoparticles: synthesis and biomedical applications. Springer International Publishing, Cham, pp 273–287
Pradhan S, Patra P, Mitra S, Dey KK, Jain S, Sarkar S et al (2014) Manganese nanoparticles: impact on non-nodulated plant as a potent enhancer in nitrogen metabolism and toxicity study both in vivo and in vitro. J Agric Food Chem 62(35):8777–8785
Prasad A, Astete CE, Bodoki AE, Windham M, Bodoki E, Sabliov CM (2017) Zein nanoparticles uptake and translocation in hydroponically grown sugar cane plants. J Agric Food Chem 66(26): 6544–6551
Qian H, Peng X, Han X, Ren J, Sun L, Fu Z (2013) Comparison of the toxicity of silver nanoparticles and silver ions on the growth of terrestrial plant model Arabidopsis thaliana. J Environ Sci 25(9):1947–1956
Rajput VD, Minkina TM, Behal A, Sushkova SN, Mandzhieva S, Singh R et al (2018) Effects of zinc-oxide nanoparticles on soil, plants, animals and soil organisms: a review. Environ Nanotechnol Monit Manag 9:76–84
Rajput V, Minkina T, Mazarji M, Shende S, Sushkova S, Mandzhieva S et al (2020) Accumulation of nanoparticles in the soil-plant systems and their effects on human health. Ann Agric Sci 65(2):137–143

Raliya R, Tarafdar JC (2013) ZnO nanoparticle biosynthesis and its effect on phosphorous-mobilizing enzyme secretion and gum contents in Clusterbean (Cyamopsis tetragonoloba L.). Agric Res 2(1):48–57

Ramírez-Ortiz J, Castro-Quintero D, Lerma-Córdoba C, Yela-Ceballos F, Escobar-Córdoba F (2020) Mental health consequences of the COVID-19 pandemic associated with social isolation. Colomb J Anesthesiol 48(4)

Rasheed A, Li H, Tahir MM, Mahmood A, Nawaz M, Shah AN et al (2022) The role of nanoparticles in plant biochemical, physiological, and molecular responses under drought stress: a review. Front Plant Sci 13:976179

Rico CM, Majumdar S, Duarte-Gardea M, Peralta-Videa JR, Gardea-Torresdey JL (2011) Interaction of nanoparticles with edible plants and their possible implications in the food chain. J Agric Food Chem 59(8):3485–3498

Rios JJ, Garcia-Ibañez P, Carvajal M (2019) The use of biovesicles to improve the efficiency of Zn foliar fertilization. Colloids Surf B: Biointerfaces 173:899–905

Schwab F, Zhai G, Kern M, Turner A, Schnoor JL, Wiesner MR (2016) Barriers, pathways and processes for uptake, translocation and accumulation of nanomaterials in plants–critical review. Nanotoxicology 10(3):257–278

Sembada AA, Lenggoro IW (2024) Transport of nanoparticles into plants and their detection methods. Nano 14(2):131

Seregin IV, Kozhevnikova AD (2023) Phytochelatins: sulfur-containing metal(loid)-chelating ligands in plants. Int J Mol Sci 24(3):2430. https://doi.org/10.3390/ijms24032430

Shah V, Belozerova I (2009) Influence of metal nanoparticles on the soil microbial community and germination of lettuce seeds. Water Air Soil Pollut 197(1):143–148

Sharma P, Jha AB, Dubey RS, Pessarakli M (2012) Reactive oxygen species, oxidative damage, and antioxidative defense mechanism in plants under stressful conditions. J Bot 2012(1):217037

Shibin OM, Yesodharan S, Yesodharan EP (2015) Sunlight induced photocatalytic degradation of herbicide diquat in water in presence of ZnO. J Environ Chem Eng 3(2):1107–1116

Siddique A, Kandpal G, Kumar P (2018) Proline accumulation and its defensive role under diverse stress condition in plants: an overview. J Pure Appl Microbiol 12(3):1655–1659

Siddiqui MH, Al-Whaibi MH (2014) Role of nano-SiO2 in germination of tomato (Lycopersicum esculentum seeds Mill.). Saudi J Biol Sci 21(1):13–17

Soliman MH, Alnusairi GS, Khan AA, Alnusaire TS, Fakhr MA, Abdulmajeed AM et al (2023) Biochar and selenium nanoparticles induce water transporter genes for sustaining carbon assimilation and grain production in salt-stressed wheat. J Plant Growth Regul 42(3):1522–1543

Soraki RK, Gerami M, Ramezani M (2021) Effect of graphene/metal nanocomposites on the key genes involved in rosmarinic acid biosynthesis pathway and its accumulation in Melissa officinalis. BMC Plant Biol 21(1):260

Sun XD, Yuan XZ, Jia Y, Feng LJ, Zhu FP, Dong SS et al (2020) Differentially charged nanoplastics demonstrate distinct accumulation in Arabidopsis thaliana. Nat Nanotechnol 15(9):755–760

Sun H, Lei C, Xu J, Li R (2021a) Foliar uptake and leaf-to-root translocation of nanoplastics with different coating charge in maize plants. J Hazard Mater 416:125854

Sun L, Song F, Zhu X, Liu S, Liu F, Wang Y, Li X (2021b) Nano-ZnO alleviates drought stress via modulating the plant water use and carbohydrate metabolism in maize. Arch Agron Soil Sci 67(2):245–259

Suriyaprabha R, Karunakaran G, Yuvakkumar R, Rajendran V, Kannan N (2012) Silica nanoparticles for increased silica availability in maize (Zea mays. L) seeds under hydroponic conditions. Curr Nanosci 8(6):902–908

Tombuloglu H, Ercan I, Alshammari T, Tombuloglu G, Slimani Y, Almessiere M, Baykal A (2020) Incorporation of micro-nutrients (nickel, copper, zinc, and iron) into plant body through nanoparticles. J Soil Sci Plant Nutr 20(4):1872–1881

Tripathi N, Goshisht MK (2022) Recent advances and mechanistic insights into antibacterial activity, antibiofilm activity, and cytotoxicity of silver nanoparticles. ACS Appl Bio Mater 5(4):1391–1463

Tripathi A, Liu S, Singh PK, Kumar N, Pandey AC, Tripathi DK et al (2017) Differential phytotoxic responses of silver nitrate (AgNO3) and silver nanoparticle (AgNps) in Cucumis sativus L. Plant Gene 11:255–264

Tripathi S, Mahra S, Tiwari K, Rana S, Tripathi DK, Sharma S, Sahi S (2023) Recent advances and perspectives of nanomaterials in agricultural management and associated environmental risk: a review. Nanomaterials 13(10):1604

VandeVoort AR, Arai Y (2018) Macroscopic observation of soil nitrification kinetics impacted by copper nanoparticles: implications for micronutrient nanofertilizer. Nanomaterials 8(11):927

Vannini C, Domingo G, Onelli E, Prinsi B, Marsoni M, Espen L, Bracale M (2013) Morphological and proteomic responses of Eruca sativa exposed to silver nanoparticles or silver nitrate. PLoS One 8(7):e68752

Vinković T, Novák O, Strnad M, Goessler W, Jurašin DD, Parađiković N, Vrček IV (2017) Cytokinin response in pepper plants (Capsicum annuum L.) exposed to silver nanoparticles. Environ Res 156:10–18

Wahid I, Kumari S, Ahmad R, Hussain SJ, Alamri S, Siddiqui MH, Khan MIR (2020) Silver nanoparticle regulates salt tolerance in wheat through changes in ABA concentration, ion homeostasis, and defense systems. Biomolecules 10(11):1506

Wang A, Zheng Y, Peng F (2014) Thickness-controllable silica coating of CdTe QDs by reverse microemulsion method for the application in the growth of rice. J Spectrosc 2014(1):169245

Wang X, Xie H, Wang P, Yin H (2023) Nanoparticles in plants: uptake, transport and physiological activity in leaf and root. Materials 16(8):3097

Wei L, Liu J, Jiang G (2024) Nanoparticle-specific transformations dictate nanoparticle effects associated with plants and implications for nanotechnology use in agriculture. Nat Commun 15(1):7389

Wu D, Zhang J, Du W, Yin Y, Guo H (2022) Toxicity mechanism of cerium oxide nanoparticles on cyanobacteria Microcystis aeruginosa and their ecological risks. Environ Sci Pollut Res 29(23): 34010–34018

Xie YinFeng XY, Li Bo LB, Zhang QianQian ZQ, Zhang ChunXia ZC, Lu KouPing LK, Tao GongSheng TG (2011) Effects of nano-TiO2 on photosynthetic characteristics of Indocalamus barbatus

Yang F, Hong F, You W, Liu C, Gao F, Wu C, Yang P (2006) Influence of nano-anatase TiO2 on the nitrogen metabolism of growing spinach. Biol Trace Elem Res 110(2):179–190

Yang C, Powell CA, Duan Y, Shatters R, Zhang M (2015) Antimicrobial nanoemulsion formulation with improved penetration of foliar spray through citrus leaf cuticles to control citrus huanglongbing. PLoS One 10(7):e0133826

Yata VK, Tiwari BC, Ahmad I (2018) Nanoscience in food and agriculture: research, industries and patents. Environ Chem Lett 16(1):79–84

Yue L, Ma C, Zhan X, White JC, Xing B (2017) Molecular mechanisms of maize seedling response to La 2 O 3 NP exposure: water uptake, aquaporin gene expression and signal transduction. Environ Sci Nano 4(4):843–855

Zandalinas SI, Peláez-Vico MÁ, Sinha R, Pascual LS, Mittler R (2024) The impact of multifactorial stress combination on plants, crops, and ecosystems: how should we prepare for what comes next? Plant J 117(6):1800–1814

Ze Y, Liu C, Wang L, Hong M, Hong F (2011) The regulation of TiO2 nanoparticles on the expression of light-harvesting complex II and photosynthesis of chloroplasts of Arabidopsis thaliana. Biol Trace Elem Res 143(2):1131–1141

Zhang H, Goh NS, Wang JW, Pinals RL, González-Grandío E, Demirer GS et al (2022) Nanoparticle cellular internalization is not required for RNA delivery to mature plant leaves. Nat Nanotechnol 17(2):197–205

Zhao L, Peralta-Videa JR, Varela-Ramirez A, Castillo-Michel H, Li C, Zhang J et al (2012) Effect of surface coating and organic matter on the uptake of CeO2 NPs by corn plants grown in soil: insight into the uptake mechanism. J Hazard Mater 225:131–138

Zhao P, Cao L, Ma D, Zhou Z, Huang Q, Pan C (2018) Translocation, distribution and degradation of prochloraz-loaded mesoporous silica nanoparticles in cucumber plants. Nanoscale 10(4): 1798–1806

Zheng L, Hong F, Lu S, Liu C (2005) Effect of nano-TiO2 on strength of naturally aged seeds and growth of spinach. Biol Trace Elem Res 104(1):83–91

Zhou H, Chen J, Zhang X, Chen J, Chen J, Jia S et al (2024) Exploring the application of graphene oxide-based nanomaterials in the repair of osteoporotic fractures. Nano 14(6):553

Zhu J, Li J, Shen Y, Liu S, Zeng N, Zhan X et al (2020) Mechanism of zinc oxide nanoparticle entry into wheat seedling leaves. Environ Sci Nano 7(12):3901–3913

Zhu J, Wang J, Zhan X, Li A, White JC, Gardea-Torresdey JL, Xing B (2021) Role of charge and size in the translocation and distribution of zinc oxide particles in wheat cells. ACS Sustain Chem Eng 9(34):11556–11564

Zia-ur-Rehman M, Anayatullah S, Irfan E, Hussain SM, Rizwan M, Sohail MI et al (2023) Nanoparticles assisted regulation of oxidative stress and antioxidant enzyme system in plants under salt stress: a review. Chemosphere 314:137649

Chapter 9
Omics Approaches to Examine the Nanomaterial's Impact on Biochemical, Physiological, and Molecular Responses in Plants

Payal Chakraborty, Sougta Bhattacharjee, Saheb Pal, and Asha Kumari

Abstract The increasing usage of nanoparticles in agricultural and environmental applications has sparked worries about their potential consequences on plant health. It is necessary to understand the biochemical, physiological, and molecular interactions that nanomaterials have with plants to evaluate the advantages and disadvantages of these materials. Techniques in omics, including transcriptomics, proteomics, metabolomics, and genomics, provide strong instruments for investigating the intricate, multifaceted reactions of plants to nanomaterials. The molecular pathways, changes in gene expression, protein alterations, and metabolic shifts brought on by exposure to nanomaterials may all be thoroughly analysed thanks to these high-throughput approaches. While proteomics and metabolomics offer insights into protein function and metabolic reprogramming under nanomaterial stress, genomic and transcriptome investigations aid in estimating gene expression changes linked to stress response, growth regulation, and defence mechanisms. Proteomics finds proteins implicated in stress tolerance, such as antioxidant enzymes, chaperones, and stress-related proteins, while genomic and transcriptome investigations identify important stress-responsive genes and signalling networks activated in response to nanomaterials. Metabolomics also emphasizes how nanomaterials might alter metabolic profiles, including the build-up of antioxidants, osmolytes, and secondary metabolites that help reduce stress. By combining these omics methods, researchers can comprehensively understand how nanomaterials affect plant biology. This information is crucial for creating safer nanomaterials, reducing any threats to plant health and ecosystems, and creating sustainable farming methods.

Keywords Nanoparticles · Transcriptomics · Proteomics · Metabolomics · Plant biology

---

P. Chakraborty
Amity Institute of Organic Agriculture, Amity University, Noida, Uttar Pradesh, India

S. Bhattacharjee · S. Pal · A. Kumari (✉)
ICAR-Indian Agricultural Research Institute, Gauria karma, Hazaribagh, Jharkhand, India
e-mail: asha.kumari@icar.org.in

R. N. Pudake et al. (eds.), *Phyto Nanotoxicology*, Nanotechnology in Plant Sciences 7, https://doi.org/10.1007/978-3-032-24518-2_9

9.1 Introduction

By the end of 2050, global population might reach 9.6 billion people. For the food security this growing population, it will necessitate an increase in food supply by a 70–100% yield (Alabdallah and Hasan 2021). Significant yield losses are caused by numerous factors, including increased abiotic stressors, decreased fertile land, excessive fertilizer and pesticide use, and global warming and climate change (Hasan et al. 2021). Global food supply is therefore seriously threatened by the decline in crop output; and to guarantee it, the right action needs to be taken to eradicate the detrimental effects of abiotic stress on crops. The food, medicine, and agricultural sectors frequently use nanotechnology (NT), which has become a promising field (Alabdallah and Hasan 2021). Nanomaterials are substances with structural elements at the nanoscale (1–100 nanometers). Their small size led to in distinct physical, chemical, and biological characteristics. They can be categorized as follows: Carbon-based nanomaterials (e.g., graphene, carbon nanotubes), Metal-based nanomaterials (e.g., gold, silver nanoparticles), Dendrimers (branched synthetic polymers), Quantum dots (semiconductor nanocrystals exhibiting distinctive optical properties), among others. Nanomaterials are extensively utilized in health, electronics, energy storage, and environmental applications.

Because of their non-threatening applications in the agricultural industry, several nanoparticles (NPs), such as titanium dioxide (TiO_2), iron oxide (Fe_3O_4), zinc oxide (ZnO), silicon oxide (SiO_2), copper (Cu-NPs), and selenium (Se-NPs), have attracted a lot of attention lately (Rai and Gupta 2024). These materials are fabricated using various methods, including physical, chemical, and green procedures (Akhtar et al. 2022). Although NPs have beneficial effects on plant growth and development, the results can differ depending on the crop species, size, concentration, exposure duration of NPs. In many cases, NPs increased antioxidant activity and resulted in shielding of plants from oxidative damage (Ahmed et al. 2021). By reducing the hydrogen peroxide (H_2O_2) and malondialdehyde (MDA) and maintaining the photosynthetic mechanism, NPs can lessen the harmful consequences of drought (Ahmed et al. 2021; Pudake 2024). Additionally, under stress conditions, NPs can boost electron transmission and light absorption in chloroplasts, progressing photosynthetic efficiency and plant growth. They can also enter the plant chloroplast and play a role in the photo system-II (PS-II) reaction (Maity et al. 2018). Concerns have been raised in relation to the toxicological and environmental effects of NPs, in addition to their widespread use in various products and commercial applications. Plants that use them excessively experience oxidative stress and physiological abnormalities, which lower their antioxidant activities and gas exchange properties (Wang et al. 2018). Additionally, NPs disrupt the processes of root growth and cell division and lower the mitotic index (Kumari et al. 2011). By changing the properties of growth medium and soil microbial activity, NPs also produce indirect toxicity (Ge et al. 2014). The concentrations, characteristics, and application techniques of NPs determine how biologically they function. NPs are applied to crops using a variety of methods, such as soil application, foliar sprays, and seed treatments

Table 9.1 Influence of nanoparticles (NPs) on various biochemical, molecular, and physiological processes in plants

Aspect	Impact	References
Biochemical	**Enhanced nutrient uptake:** NPs can increase the expression of aquaporins, improving water and nutrient absorption. **Antioxidant activity:** Exposure to NPs may activate plant defense systems, leading to increased quantity of reactive oxygen species (ROS) and enhanced antioxidant responses.	Rasheed et al. (2022) Pérez-Labrada et al. (2020)
Molecular	**Gene Expression modulation:** Regulate stress related genes expression, along with genes related to growth and development. **DNA integrity:** Certain NPs have been associated with DNA damage, including chromosomal aberrations.	Al-Khayri et al. (2023) Djanaguiraman et al. (2024)
Physiological	**Growth promotion:** Some NPs, like Ag and Au, have been found to have beneficial impact on plant growth parameters, like root and shoot length, leaf surface area, and total biomass. **Stress tolerance:** NPs can improve plant resistance to abiotic stresses like drought and salinity by modifying physiological processes. **Photosynthetic efficiency:** NPs may affect photosynthesis by influencing chlorophyll content and photosynthetic activity, although effects can vary depending on NP type and concentration.	Thabet and Alqudah (2024)

(Mahdy et al. 2020). NPs enter plant tissue through the root junction and injured areas, where they enter through cell walls and membranes (Tripathi et al. 2017b). According to Sharif and team (Sharif et al. 2013), NPs can also enter cell cytoplasm, stomata, trichrome, hydathodes, and leaf cuticles. According to Zhang and Monteiro-Riviere (2009), NPs connect with a variety of organelles in the cytoplasm of cells, disrupting several metabolic processes. NPs are also found to be absorbed into the seed, through the parenchymatic intercellular gaps in the seed coat. (Tripathi et al. 2017b). Plant species, age, organ, and NPs characteristics like size and origin, stability, and its interactions with roots, soil, and soil microbes all play important role in NPs accumulation, translocation, and uptake (Ali et al. 2021; Mittal et al. 2020). They have influenced various biochemical, molecular, and physiological processes in plants (Jha and Pudake 2016; Pudake et al. 2019); and Table 9.1 provide the summary. The use of metal NPs has a vast scope in modulating the growth of crop plants to improve productivity and ensure food supply for the increasing population (Francis et al. 2024). The applied nanoparticles interact with the plants at molecular levels, affecting various physiological, biochemical, and genetic processes like sensing, signaling of transcription, processing of the transcripts, translation, as well as post-translational modification of the protein (Ali et al. 2023) (Table 9.2). However, the level of such interactions depends upon the composition, type, size, surface charge, concentration, and the plant species it applies to. Upon application, the metal nanoparticles can activate one or more definite signaling mechanism involved in nutrient uptake, stress responses as well as hormonal regulation (Mustafa

Table 9.2 Effect of metal nanoparticles on various biochemical, molecular, and physiological processes in plants

Sr. No	Metal nanoparticle used	Key findings	Reference
1	Aluminum oxide	Rapid damage to some RNA molecules leading to potential changes in the expression of genes	Burklew et al. (2012)
2	Cerium oxide	Alteration of gene expression levels involved in cellular ion transport, antioxidant defense mechanisms, and stress signaling pathways. In salt stress it act as a scavenger of ROS	Khan et al. (2022); Preetha et al. (2023)
3	Copper	Increase in production of defence-related signaling biomolecules like salicylic and jasmonic acid in plants	Shang et al. (2020)
4	Gold	Interaction with plant-receptor legend affecting signaling pathways	Francis et al. (2024)
		Functionalization with auxin hormone analogs to activate various signaling pathways to modulate plant growth	Ferrari et al. (2021)
		Alteration in the RNA processing mechanism leads to variability in the transcriptome	Tiwari et al. (2016)
5	Iron oxide	Protein phosphorylation through methylation, acetylation and glycosylation	El-Moneim et al. (2021); Shah et al. (2022)
6	Silver	Enhancement of transcription of stress-responsive genes and triggering of signaling cascades that are responsible for abiotic stress tolerance	Khan et al. (2023)
		Modulation of protein synthesis	PeharecStefani'c et al. (2019)
		Increase in synthesis of proteins like CDK-2, 1,6-biphosphate aldolase, and protochlorophyllide oxidoreductase in roots	Francis et al. (2024)
		Modification of the gene expression level of cellular metabolism processes like IAA-8, RD22, and NCED3	Tripathi et al. (2017a); Ghosh and Bera (2021)
		Suppression of ethylene biosynthesis by preventing expression of aminocyclopropane-1-carboxylic acid synthase 7 and aminocyclopropane-1-carboxylic acid oxidase 2	Karuppanapandian et al. (2011); Syu et al. (2014)
7	Silver and Aluminium	Upregulation PAB2 and BKR1 and downregulation of PDC; activation of antioxidant defense by downregulation of glyoxalase II 3 in soybean	Mustafa et al. (2015)
8	Silver and silicon	Upregulation of ABA upregulation of NCED3, TAS14 andSnRK2; alleviation of oxidative stress and osmoregulation by overexpression of P5CS1; upregulation of proline biosynthesis; activation of SOS, ROS and MAPK-signalling in tomato	Almutairi (2016)

(continued)

Table 9.2 (continued)

Sr. No	Metal nanoparticle used	Key findings	Reference
9	Titanium	Increase in photosynthesis by upregulation of RUBISCO and PEPC, and by reducing the electrolyte leakage index in *Cicer arietinum*	Amini et al. (2017)
10	Titanium and silver	Translational repression of target genes through non-coding RNA in *Arabidopsis thalliana*	García-Sánchez et al. (2015)
11	Zinc oxide	Alternative splicing of genes in maize	Xun et al. (2017)

and Komatsu 2021). Thus, a thorough knowledge regarding the mechanism of influence of the metal nanoparticles on gene expression and molecular pathways is the need of the hour to tap the potential of this technology in sustainable agriculture.

While some NPs offer potential benefits, others may pose risks, including toxicity and interference with essential physiological processes. Therefore, further study is essential to fully understand these interactions and to harness the potential of NPs in agricultural applications safely.

9.2 Effects of Nanoparticles on Gene Expression

The application of metal nanoparticles can modulate gene expression at different levels, either by interacting with DNA and altering the chromatin assembly or by binding with the promoter region of genes. Further, they can interact with the RNA-binding proteins as well as splicing factors (Ma et al. 2013; Tumburu et al. 2017). Further, it can also modify the activity of the ribosomes to modulate protein synthesis, which may lead to abundance or scarcity of specific proteins. At the post-translational level, it can also alter the localization, stability, and activity through phosphorylation, acetylation, and ubiquitination.

9.3 Omics Approaches to Examine the Effect of Nanoparticles on Plants

Omics is an extensive domain in biology that encompasses the large-scale examination of molecules inside a biological system. It encompasses several subfields, including genomics (where the gene sequences are studied), transcriptomics (RNA molecules are studied), proteomics (the study of proteins), metabolomics (the study of metabolites and metabolic processes), and ionomics, among others. Omics technologies facilitate the comprehension of intricate biological activities, disease mechanisms, and organisms' responses to any external factors. These omics-based methodologies jointly furnish a thorough evaluation of nanomaterial impacts on plant systems, yielding significant insights into toxicity processes, stress adaptability, and prospective agricultural applications (Fan et al. 2025).

9.3.1 Genomic Approaches

Genomics is the examination of the organisation of genes and genetic material inside the genome and how this organisation influences their function. The majority of the produced nanoparticles contain heavy metals. Due to their highly fine dimensions and high surface-to-volume ratio, they can exhibit significant activity within cells and potentially inflict considerable damage to the genetic material of plants (DNA and RNA) (Mascarenhas et al. 2023). Various genomic methodologies exist to trace mutations and discover potential risk factors connected with NPs, including DNA sequencing and methylation/epigenetic analysis.

Various approaches can be roughly categorized according to sequencing technologies, including template preparation (library formation), sequencing (by synthesis or ligation approach), imaging, and subsequent data analysis. However, each technology possesses distinct characteristics that influence the quality of the data generated from them. The template is derived from the complete transcript, or specifically from mRNA, and can be randomly fragmented through nebulization, sonication, or enzymatic digestion to achieve appropriate sizes (if the transcript is extensive), and then adaptor ligation (where they are attached to 3′ and 5′ end of each fragment to enable purification, amplification, and sequencing techniques). The fragments then affixed to a singular capture bead in conjunction with PCR components including enzyme. These are encapsulated in a water-in-oil emulsion, creating a distinct isolated micro-reactor for the PCR process.

(a) **Roche (454) FLX platform**: "454 Life Sciences" (Roche Diagnostics, USA) was initially launched in 2005, and the "Genome Sequencer (GS) FLX System", along with the "GS FLX Titanium series platforms", are presently available (Liu et al. 2012). Subsequent to the preparation of the aforementioned template, the beads, together with the associated DNA fragments, are extracted from the emulsion and transferred into the wells of the "Picotiter Plate". Each well holds a single bead. The principle of pyrosequencing, which involves luciferase-mediated light detection upon pyrophosphate production during the sequencing process, is employed for sequencing. The FLX system, with a read length of 450–500 bases, requires 10 hours per run for template preparation and data processing, producing 400 Mb of sequence data and achieving a consensus accuracy over 99.99%. The newly developed "GS FLX Titanium XL+" platform can produce 1 GB of sequencing data with a read length of 1000 bp simultaneously (Gilles et al. 2011).
(b) **Illumina/Solexa platform (Bridge PCR)**: Illumina was introduced in 2000 and is the most prevalent next-generation sequencing technology to date. Recent platforms such as the "Illumina Genome Analyzer 1 G"b and "HiSeq 600 Gb"

have gained significant popularity (Zhu et al. 2018). This approach involves ligating two distinct adapters to the 5′ and 3′ ends of the cDNA fragments, which are subsequently affixed to a substrate on a flow cell that had a dense array of complementary primers of adapters. The solid-phase rapid PCR is carried out within a flow cell. This is also known as fold-back PCR or bridge PCR, that utilizes reversible terminator technology. The 4 dNTPs with fluorophores are employed for PCR amplification, where the fluorophores act as chain terminators. The Charge-coupled device (CCD) camera records the dNTPs inserted at the 3′ end of the elongating chain primers each time. Illumina read lengths typically range from 35–150 nucleotides, with an accuracy above 98.5%. After 7 to 8 days of single run, the "Illumina HiSeq 2000" platform produces around 400 Gb of sequence data. The "HiSeq X Ten" is a prevalent population-scale whole-genome sequencing device, introduced in 2014, capable of producing 1.8 terabases of sequence data within three days. The technique is applicable for de novo transcriptome sequencing without a reference sequence, resequencing for SNPs/InDels genotyping, detection of copy number variation (CNV), and analysis of structural variation in gene isoforms and transcript expression profiles, among other uses (Wang et al. 2021).

(c) **Ion Torrent—a Semiconductor-based sequencing Technology:** This tool was created by Ion Torrent Systems Inc. (now Thermo Fisher Scientific, USA), and was brought to market in 2010 (Cheng et al. 2013). It employs a semiconductor-based device, known as an ion chip, which detects the H+ ions produced during DNA extension by DNA polymerase, measuring the resultant pH changes due to the release of H+ ions. The ion chip, including wells with a diameter of 3.5 micrometres, is positioned right above the electronic sensing platform. The voltage signal correlates with the quantity of bases integrated into the newly synthesized strand by DNA polymerase, and the successive incorporation of each nucleotide facilitates base judgement via non-optical scanning, thereby significantly accelerating the sequencing process and reducing costs. In 2012, a new high-throughput method named 'Ion Proton' was introduced, enhancing output by 10X, while the read length significantly decreased to 200 bp from 400 bp compared to Ion Torrent (Yuan et al. 2016).

(d) **ABI SOLiD-based platform**: The "SOLiD (sequencing by oligonucleotide ligation detection)" platform employs oligonucleotide probes, individually of 8 base pairs in length, featuring two distinct nucleotides at the 3′ end and labelled with a fluorescence at the 5′ end, for the identification of transcript bases during sequencing. Applied Biosystems commercialized it in 2005 as the "SOLiD 3.0" platform. In this approach, beads are fabricated as previously described and are immobilized in a monolayer within an acrylamide matrix on a glass slide, accompanied by attached DNA molecules. A collection of 16 oligonucleotides (for 4 nucleic acid bases) is necessary for hybridization with template cDNA during each sequencing reaction. During base encoding in sequencing, each distinct base pair at the 3′ end of the probe is designated one of four possible colors for enhanced identification and analysis; for instance, "TT" is designated as blue, "TC" as green, and this pattern continues for all 16 unique oligonucleotides. SOLiD technology is considered to be 100% accurate sequencing, as each nucleotide in the sequencing template is read twice. The "SOLiD 3.0" platform has an ability to give a read length of 50 bases and can generates around 20 Gb of sequencing data in single run of 6 to 7 days. The "SOLiD 5500 and SOLiD 5500 XL" systems were launched to enhance sequencing capacity to 300 Gb data in each run with 99.9% accuracy (Shendure et al. 2011).

(e) **Pacific Biosciences platform/SMRT sequencing:** Single-molecule real-time (SMRT) sequencing was innovated by Nanofluidics, Inc. and subsequently commercialized by Pacific Biosciences, USA. This method involves the production of a template by ligating a single-stranded hairpin-structured adaptor to the ends of cDNA, resulting in a bell-shaped structure known as a SMRT-bell. This technology involves the immobilization of individual DNA polymerase molecules at the base of zeptoliter-sized wells, known as zero-mode waveguides (ZMWs), utilizing biotin–streptavidin interactions. High concentrations of four distinct fluorophore-labelled dNTPs facilitate rapid DNA synthesis through strand-displacing polymerase. The fluorophore is inevitably eliminated, and the signal is identified upon the addition of a dNTP to the 3′ end of the primer in the elongating chain. Each cDNA strand can be sequenced multiple times to enhance the correctness of the sequencing results. Furthermore, direct sequencing, as opposed to clonal amplification, facilitates real-time sequence reading. Every SMRT cell can produce approximately 50,000 reads and yield up to 1 Gb of data in four hours. It exhibits reduced sensitivity to GC content in sequences; yet, its error rate is rather high, around 11%, because of the single-pass sequencing methodology (Hebert et al. 2018).

(f) **Oxford Nanopore platform:** This advanced sequencing technology utilizes a flow cell with several micro-wells, featuring a synthetic bilayer and perforated by biological nanopores (Nanopore 2012). Sequencing is accomplished by accurately detecting the variations in current generated by the incorporation of bases via the nanopore, facilitated by a molecular motor protein. The template library is constructed by attaching adapters to cDNA termini, allowing the first adapter to associate with a motor enzyme and the second adapter (a hairpin oligonucleotide) to bind with an alternative hairpin motor protein. Consequently, complementary strands can be sequenced concurrently from a single molecule, enhancing precision relative to SMRT technology. This method boasts great throughput, and "MinION"—a USB-powered portable sequencer is able of giving over 90 Mb of sequencing data in a single 18-hour run, with maximal read lengths exceeding 60 kb (Brown and Clarke 2016).

(g) **Bisulfite Sequencing for the Detection of Methylated/Epigenetic Alterations in DNA**

Bisulfite sequencing (BS-seq) is an advanced method employed to examine DNA methylation with single-base precision (Piao et al. 2021). It utilizes bisulfite treatment, which changes unmethylated cytosines into uracil, but methylated cytosines stay unaltered. Following sequencing, these alterations facilitate the differentiation between methylated and unmethylated cytosines, yielding a comprehensive methylation map of the genome. BS-seq variants encompass whole-genome bisulfite sequencing (WGBS) for extensive coverage and reduced representation bisulfite sequencing (RRBS) for economical targeted analysis. This technique is extensively employed in epigenetics studies, gene regulation, and plant growth (Abdulraheem et al. 2024).

9.3.2 *Transcriptomics Approaches for Assessing the Impact of NPs on Plants*

All transcripts in an organism studied, at a particular stage or condition, are termed the transcriptome. So, the transcriptome of any organism can change according to the cells/ tissue under consideration and may not be similar in two different conditions. This is unlike the genome, that content the total genetic material of an organism. The genome is fixed for a given organism and is not controlled by prevailing internal or external conditions (McGettigan 2013) (Fig. 9.1).

(a) **Reverse Transcription Polymerase Chain Reaction (RT-PCR)**

RT-PCR is the fundamental procedure of transcriptomics, universally applicable to any method necessitating the change of mRNA into cDNA via the reverse transcriptase enzyme. This is an RNA-dependent DNA polymerase and oligo T (in case of mRNA with a poly-A tail), random hexamer, or tRNA as primers, before employing additional techniques for the storage, quantification, detection, cloning, and analysis of RNA. To enhance the accessibility of genomic biological information, including the locations of open reading frames (ORF), 5′ and 3′ untranslated regions (UTR), and splicing sites, high-quality full-length cDNA sequences are required. Understanding the biological processes that lead to alternative gene structures, such as

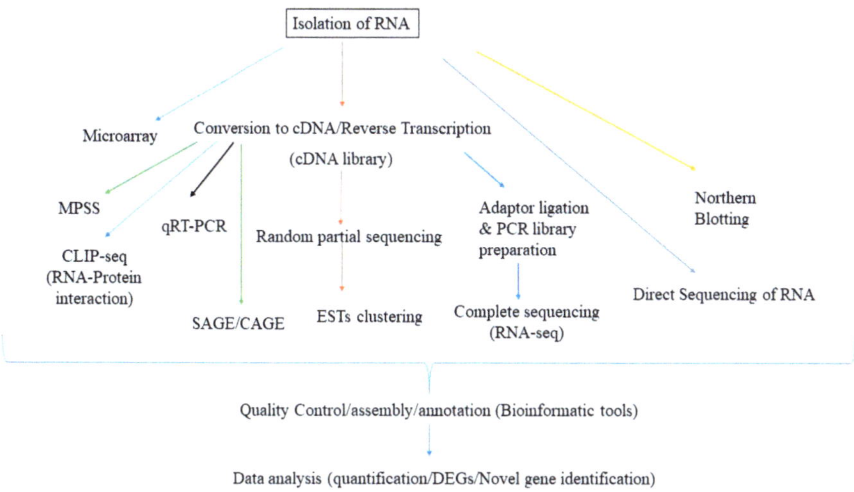

Fig. 9.1 A flowchart of different transcriptomics approaches

RNA editing, trans-splicing, and alternative splicing, is contingent upon the availability of comprehensive cDNA sequence information (Rio 2014).

(b) **Rapid Amplification of cDNA Ends (RACE):**

For the complete sequencing of the full transcript RACE is a potent technique. It facilitates the amplification of an mRNA template between a known internal sequence and unknown sequences at either the 3′ or 5′ end. This process involves the preparation of cDNA from mRNA by RT-PCR with an oligo-dT primer. After that, an internal gene-specific primer (GSP) and the primer that has complementarity with the anchored Oligo dT adapter are used to amplify the region between an unidentified 3'-mRNA end sequence. This is referred to as 3' RACE. Likewise, 5' RACE, also known as anchored PCR, aids in the isolation and characterisation of 5′ end sequences. This is highly beneficial for full-length cloning and characterisation of transcripts when elevated GC content at the 5′ end inhibits standard PCR reactions from achieving complete amplification (Zhang 2003).

(c) **Suppression Subtractive Hybridization (SSH):**

SSH approach was initially conceived by Siebert and Larrick (1992) and termed competitive PCR. Subsequently, modifications were implemented to enhance efficiency, leading to the development of SSH. This method is particularly effective for constructing subtracted cDNA libraries; integrating normalization and subtraction is done in a single process. It normalizes the cDNA abundance within the target population and eliminates common cDNAs from both the target/tester and driver library. The SSH approach relies on the reassociation kinetics of DNA molecules and significantly amplifies the low-abundant cDNA molecules that are derived from differentially expressed genes. This subtracted cDNA library can be used as a probe for obtaining homologous sequences from any genomic or cDNA library and may also be advantageous in positional cloning. According to the SSH protocol, control transcript samples are designated as 'drivers' and treatment samples as 'testers'. The mRNA extracted from both samples is used for cDNA synthesis via RT-PCR. The tester cDNA samples are then digested using a restriction enzyme to produce blunt-end fragments, after which they are separated into two equal volumes and ligated with a distinct adaptor. Tester cDNA is multiplied with adaptor-specific primers, then denaturation, after which a high quantity of driver is introduced to the process to produce driver-tester hybrids of overexpressed or similarly expressed transcripts (normalization). In the subsequent run of PCR, both testers, which possess distinct adaptors, are combined and amplified with adaptor-specific primers. It results in the amplification of differentially expressed genes. The rationale for utilizing an excess of driver cDNA is to inhibit the amplification of shared cDNA found in both driver and tester, hence permitting the amplification of only novel or unique transcripts present in the tester.

(d) **Northern Blotting:**

This accurate transcriptome technique was developed by Stanford University researchers—James Alwine, David Kemp, and George Stark in 1977. It quantifies the abundance of a specific mRNA transcript expressed in tissue with the help of radio-labelled or fluorescent probes (Kumari et al. 2022). This blotting approach uses a denaturing gel to segregate RNA molecules based on size, then transfers them to a nylon membrane utilizing Saline Sodium Citrate (SSC) buffer and capillary action. RNA is subsequently affixed to the membrane, and hybridization is performed with a labelled probe. The probe is designed from the complementary sequence of the target gene. Non-specifically bound probes are subsequently eliminated by washing with diluted buffer, and the dried membrane is subjected to analysis. This method can effectively identify various transcript variants of a gene without the need for PCR or sequencing procedures. However, the total RNA quantity and quality are very critical for this experiment (Josefsen and Nielsen 2010).

(e) **Quantitative Real-Time (qRT) Polymerase Chain Reaction:**

For gene quantification, transcript profiling, gene copy number evaluation, allele discrimination, pathogen detection, and SNP genotyping, real-time PCR is one of the most reliable, sensitive, and popular method. It is very sequence-specific and flexible enough to increase throughput when needed. The moniker "real-time" refers to the fact that, unlike standard PCR procedures, this technique involves measuring amplicons (PCR products) only during PCR cycles. There are two ways to describe the quantification theory of mRNA transcripts in real-time PCR. A one-step reaction, in which cDNA synthesis and PCR amplification take place in the same tube, and a 2-step reaction, in which cDNA synthesis and PCR amplification are carried out in different processes. Although one-step qRT-PCR is considered to be more accurate since it reduces experimental variation, the rapid mRNA decay is a major drawback. Nevertheless, the two-step method is incredibly reproducible and sensitive. A two-step procedure to modify the melting temperature (Tm) reduces the probability of primer dimer detection. Quantification and detection in real-time PCR depend on two important variables: the kinds of probes used, and the quantification technique used (Zhang et al. 2013).

Following each amplification cycle, a popular SYBR Green dye emits light after excitation and binding to double-stranded, giving quantitative information on its amount in samples. When compared to other fluorescent probes, SYBR Green is relatively more affordable. However, the signals that the SYBR dye emits in primer dimers and unwanted PCR products can cause inaccurate and inflated findings. Examining the dissociation curve or melting curve—a graph of fluorescence vs melting temperature—can help to somewhat alleviate this problem (Dragan et al. 2012).

TaqMan probes have a reporter on its 5′ end and a quencher on their 3′ end and are fluorescently labelled probes that are extremely sensitive and efficient. The probe binds to the target template during the denaturation phase of the heat cycler. In the strand elongation step, DNA polymerase breaks the probe using its 5′ exonuclease activity, separating the probes's quencher and reporter dye. Fluorescence is then detected using fluorescence resonance energy transfer (FRET), which measures the interaction between two fluorophores, which is distance-dependent (Liu et al. 2006).

Molecular beacon probes, in contrast to TaqMan probes, do not cleave during the amplification reaction; rather, they hybridize with the target molecule during the denaturation phase of PCR and release fluorescence. This kind of probe has incredibly quick response kinetics. With a reporter at the 5′ end and an internal quencher that is directly linked to the 5′ end of the PCR primer via a blocker, a hairpin-loop structure is employed. The blocker's job is to prevent strands from getting longer. The DNA polymerase lengthens the complementary strand in the first PCR cycle, but in the second cycle, the hairpin loop breaks down during denaturation, enabling the fluorescent probe's loop area to attach to the freshly generated target molecule at complementary bases. Fluorescence emission from the sample is directly proportional to the quantity of DNA target and is recorded after the reporter and quencher have separated (Choijookhuu et al. 2022).

Also, the real-time quantification is categorised into absolute quantification and relative quantification. A standard curve is created by using serial dilution of known values, which is known as absolute quantification. This standard curve shows a linear relationship between the starting quantity of the target cDNA and the Ct value. Ct is the cycle threshold value, which is the PCR cycle at which fluorescence exceeds the background signal. This allows for the calculation of the starting concentration of the unknown sample using the graph shown in relation to the Ct value observed during real-time PCR. In the relative quantitation approach, gene expression is assessed using comparison with a reference sample, sometimes referred to as a calibrator, which can be any housekeeping gene. Results may be articulated as a target/reference ratio. Numerous analytical models exist for calculating mean normalized gene expression; however, the Ct ($2^{-\Delta\Delta Ct}$) model is the most prevalent and accepted one. In this method, reference gene expression and the target gene expression are compared with the Ct values difference to ascertain the ΔΔCt value (Zhang et al. 2013).

(f) **Microarray technology:**

Microarray-based transcriptome analysis is a crucial instrument for characterizing and comprehending gene expression, as well as for analyzing the transcriptional profiles of organisms at the genomic level. It is a hybridization-based instrument that integrates the concurrent analysis of thousands of genes' expression through probe hybridization and fluorescent dye. The efficacy of this technology is based on its high throughput, automation, and rapid capture of quantitative data from many samples at the same time. Data from the DNA microarrays can be assessed in two ways: (1) using a genomic library, cDNA library, or ESTs, in which each gene is indicated by a dot on a solid surface that has been fixed before hybridization, called a spotted DNA microarray; and (2) in-situ synthesis oligonucleotides, in which priorly known the identity and sequences of each transcript using a photolithographic process, called an oligonucleotide microarray, DNA chip, or biochip assay. Every probe has a unique location and gene ID on the chip. The DNA template (also known as the probe) on the chip's surface is currently hybridizing with cDNA/mRNA molecules from the experimental sample. These genes' expression levels are clarified by the fluorescent signals captured during hybridization, which show how much cDNA or mRNA is linked to each spot on the array. Each expressed gene is then given a profile by the detector. The initial computational image-based analysis was developed by P. Brown in 1981. DNA microarray technology was initially developed in 1988 by Edward, and later in 1995, Mark Schena became the first to apply it in gene expression studies in humans (Lenoir and Giannella 2006). It is also an often-utilised method for differential gene expression analysis. The comparative impacts on gene expression under various settings, such as treatments, diseases, the effect of stresses, and different developmental stages, can be readily observed and analysed using multiplexing. Also, this microarray technology can be used for the detection of SNPs (Varun et al. 2025).

(g) **RNA-Seq and data analysis:**

A very efficient high-throughput transcriptomics technique, RNA-seq provides accurate quantification of transcript expression levels in addition to sequencing information (Hrdlickova et al. 2017). Next-generation sequencing (NGS) platforms, including "Illumina IG, Applied Biosystems SOLiD22, Roche 454 Life Science, and Helicos Biosciences tSMS system", involved converting transcripts to cDNA, adaptor ligation, and sequencing. Depending on the depth requirements, time limits, and cost concerns, the sequencing method may employ either single-read or paired-end techniques. Through the use of either de novo or reference sequence algorithms, millions of short reads are then assembled using alignment tools (e.g., Bowtie, TopHat, STAR, BWA, Novoalign, HMMSplicer, Olego, BLAT, etc.) and quality control tools (like "FastQC, Picard, RNA-SeQC, or Qualimap") to evaluate PCR bias, contamination, GC bias, length optimization, etc., and remediation of substandard quality read ends (like "Trimmomatic, fastx-toolkit"). However, a lot of effort and time is required to analyze NGS data generated by RNA-seq. Analytical tools such as Cufflinks, CLC Genomics, Sailfish, RSEM, and BitSeq make it easier to identify transcripts and measure their levels of expression after alignment, while other tools like MISO are useful for identifying alternative splice sites or carrying out more complex analyses. Because bigger and vastly expressed transcripts may be disproportionately counted among RNA-seq reads, even when expression levels are equal for both. The fundamental difficulty with NGS data analysis is scattered read coverage and prejudiced read mapping. To normalize the read counts, software is applied to convert the sequencing data into Reads Per Kilobase per Million Mapped Reads (RPKM), or Fragments Per Kilobase per Million Mapped Reads (FPKM). After comparing RNA-seq data from two or more samples, the critical information gathered from transcriptome profiling is the differentially expressed genes (DEGs). DESeq, EdgeR, and Cuffdiff are useful software for differential expression analysis. Data "outliers" have a substantial impact on all packages, which depend on statistical analysis using "Poisson or negative-binomial models". If sample library sizes are uneven, skewed findings could follow. Therefore, it has also been shown that instruments that employ non-parametric statistical techniques encounter the same problems. Results are more repeatable when gene expression patterns are examined using fold change. When it comes to differential gene expression analysis, the NOISeq, Samseq, and LFCseq algorithms are more reliable. With the highest level of analytical accuracy, RNA-seq is still popular, and essential tool available today. The potential of RNA-seq lies in eQTL mapping, the finding of fusion genes that may not have been previously reported, and the detection and measurement of a large number of distinct or unknown transcripts. In contrast to microarray, which assesses expression levels as continuous intensity values with a log-normal distribution, RNA-seq has a selection advantage since its read counts are non-negative integers that naturally conform to a discrete distribution. In contrast to microarrays, which only use known sequences for probe design, RNA-seq excels at reducing non-intentional interference and is not restrained to known genomic regions. As a result, RNA-Seq provides a comprehensive, numerical, and unbiased understanding of the transcript sequences in the study sample (Haas and Zody 2010).

(h) **Crosslinking and Immunoprecipitation Sequencing (CLIP-Seq):**

The post-transcriptional gene regulation (PTGR) of genes involved in capping, polyadenylation, splicing, editing, destruction, transfer to other organelles, translation, and stability occurs in varied degrees in almost all RNA molecules. Novel approaches to defining the relations between RNA-binding proteins (RBPs) and RNA molecules were made possible by high-throughput sequencing technology. Therefore, after RNA-seq, CLIP procedures are extensively applied in transcriptomics to identify target RNA-binding sites on a large scale to study the protein or RNA of interest (Uhl et al. 2017). In this procedure, after crosslinking of RNA and Protein complexes with the help of UV, breakup occurs via RNase1 digestion and subsequent immobilization on antibody-attached beads. The initial step involves dephosphorylation using phosphatase, followed by the attachment of barcoded adaptors to the 3′ end of RNA and transfer for electrophoresis on a gel. The sample is subsequently transferred to a nitrocellulose membrane and treated with Proteinase K to dissociate RNA from protein, followed by reverse transcription into cDNA. RNA is then eliminated from the cDNA-RNA hybrid, and an adaptor is ligated to the 3′ end of the cDNA. The resultant cDNA is multiplied with PCR for library creation, and the sequence is further analysed for size estimation and characterize protein-binding sites. In this technique, numerous improvements have been implemented to minimize PCR errors and background noise, facilitating the detection of precise single-base interactions. This is a highly effective method for elucidating various transcriptional events involving protein-RNA interactions, including studies on sub-cellular localization. A novel technology, termed as CRISPR-Assisted RNA–Protein Interaction Detection (CARPID), has been developed to identify RNA-binding proteins (RBPs) in living cells (Yi et al. 2020). This technique is efficient, reliable, and rapid. This technique uses CRISPR-dCasRx to precisely target proximity labelling and lncRNA-Protein identification. Monitoring transcript expression in living cells is made easier by the combination of dCasRx (for targeted lncRNA), enhanced green fluorescent protein (eGFP), a self-cleaving T2A peptide, and the biotin ligase BASU (for streptavidin bead capture). Although this tool has been reported for human transcriptomic studies, it needs to be further optimized for use in plants and other organisms.

(i) **Direct RNA Sequencing at the Single-Molecule Level:**

"Helicos® Genetic Analysis System" has developed and commercialized by Helicos Biosciences Corporation (Raz et al. 2011). This method entails entirely impartial sequencing of individual RNA using a hugely parallel sequencing approach, yielding precise quantification and real-time sequence data. This method eliminates the need for RNA to cDNA conversion and offers extensive sequence coverage. This technique is widely used to map polyadenylation sites or profile the expression of transcribed mRNA (genes that code for proteins). This process allows transcripts to be sequenced by synthesis in parallel by hybridizing mRNA molecules with polyA tails with oligo(dT) primers attached to the flow cell. Up to 50 channels are accessible across the two flow cells that can function simultaneously, and the read length average is 35 bp. A single full run has the capacity to produce 800,000 to 8,000,000 readings per channel. Error rates can be up to 5%, mostly as a result of insertions and deletions (Ozsolak and Milos 2011).

9.3.3 Approaches of Proteomics for Assessing the Impact of Nanomaterials on Plants

Proteomics is a further phase of functional genomics, wherein the genome is described at the protein level. Proteins are the fundamental functional molecules that translate genomic information into physiological consequences. Proteomics aims to elucidate the biological state by analyzing both qualitative and quantitative alterations in protein content within cells and extracellular biological materials under varying settings, thereby enhancing the comprehension of biological processes. It entails an extensive characterization and functional investigation of the proteins encoded by a genome. The proteome represents the most dynamic aspect of the genome. Proteomics is often categorized as global proteomics, which identifies and catalogues all potential proteins within an organism, and Targeted proteomics, which focuses on finding and cataloguing a specific subset of proteins associated with a particular pathway, disease, environment, or other specialized area of interest for study. Various methodologies for studying proteomics, including proteome identification, quantification, differential analysis, protein-protein interactions, examination of post-translational changes, and structural proteomics (Patterson and Aebersold 2003). Proteomic methodologies will facilitate the identification of stress-responsive proteins in response to nanoparticle exposure, elucidate mechanisms of oxidative stress and antioxidant defense systems, examine nanoparticle-induced toxicity and alterations in metabolic pathways, and evaluate the effects of nanoparticles on photosynthesis, nutrient absorption, and plant growth. Proteomics examines the entire array of proteins expressed in plants following nanomaterial exposure, facilitating the evaluation of molecular-level functional alterations. Methods include 2- dimensional gel electrophoresis (2D-GE), difference gel electrophoresis (DIGE), and mass spectrometry (MS)-based proteomics. These are extensively employed to discover proteins associated with stress signalling, enzymatic detoxification, and metabolic adaptability. Proteomic investigations have demonstrated that nanoparticles can provoke oxidative stress by modifying the expression of antioxidant enzymes, including SOD and catalase (Munir et al. 2023). Moreover, alterations in photosynthetic and transport proteins indicate that nanoparticles may interfere with energy metabolism and nutritional equilibrium. Proteomics offers critical insights into plant responses and the potential hazards of nanomaterials in agricultural and environmental systems by finding specific protein markers linked to nanoparticle toxicity (Matysiak et al. 2016; Raliya et al. 2015).

(a) **2D-GE**

2D-GE is an effective method employed in proteomics to segregate and examine intricate protein mixtures according to two distinct characteristics:

(a) Isoelectric Focusing (IEF): In this step, proteins are sorted according to their isoelectric point (pI) with a pH gradient, allowing each protein to migrate to the position where its net charge is neutral.
(b) SDS-Polyacrylamide Gel Electrophoresis (SDS-PAGE): Proteins are subsequently segregated according to their molecular weight, facilitating enhanced resolution of individual proteins (Rabilloud and Lelong 2011).

(b) **Mass Spectrometry (MS)**

Mass Spectrometry (MS)-based proteomics is a robust analytical tool employed for the identification and quantification of proteins in biological specimens. The technique entails ionizing protein or peptide molecules and quantifying their mass-to-charge ratio (m/z) to ascertain their composition and structure. Proteins are generally digested with enzymes (e.g., utilizing trypsin) into smaller peptides, which are subsequently examined via mass spectrometry. The resultant spectra are checked against protein databases for identification purposes. Prevalent mass spectrometry techniques encompass Matrix-Assisted Laser Desorption/Ionization (MALDI-MS) and Electrospray Ionization (ESI-MS), frequently integrated with tandem mass spectrometry (MS/MS) for enhanced structural elucidation. MS-Based Proteomics is advantageous in various domains as outlined below: (a) Protein Identification: Mass spectrometry facilitates the identification of proteins inside intricate biological materials, such as plant tissues subjected to nanoparticles. (b) Quantitative Proteomics: Label-based methods (e.g., SILAC, TMT) and label-free techniques assess variations in protein abundance across diverse situations. (c) Post-Translational changes (PTMs): Mass spectrometry identifies changes such as phosphorylation, glycosylation, and ubiquitination, which are essential for protein functionality. (d) Biomarker Discovery: Mass spectrometry is employed to discover stress-related proteins in plants impacted by nanoparticles. (e) Protein-Protein Interactions: Facilitates the examination of interaction networks crucial for biological activities. Mass spectrometry-based proteomics is crucial for elucidating plant responses to environmental stresses, such as nanoparticle exposure, by revealing changes in the proteome and underlying molecular processes (Shuken 2023).

9.3.4 *Metabolomic Methodologies to Evaluate the Effects of Nanomaterials on Plants*

Metabolomics assesses biochemical changes in plant metabolism resulting from nanomaterial exposure, employing techniques like Liquid Chromatography-Mass Spectrometry (LC-MS), Gas Chromatography-Mass Spectrometry (GC-MS), and Nuclear Magnetic Resonance (NMR) spectroscopy. These methodologies facilitate the identification of stress-associated metabolites, including reactive oxygen species (ROS), amino acids, organic acids, and secondary metabolites like flavonoids and alkaloids. Metabolomics research indicates that nanoparticle exposure might induce oxidative stress, resulting in heightened levels of antioxidants such as glutathione and ascorbic acid. Moreover, exposure to nanoparticles may interfere with primary metabolism, impacting essential processes such as glycolysis, the tricarboxylic acid (TCA) cycle, and nitrogen absorption. Comprehending these metabolic alterations elucidates how plants acclimatize to stress generated by nanomaterials and presents possible biomarkers for evaluating nanoparticle toxicity (Silva et al. 2023).

(a) **Gas Chromatography-Mass Spectrometry (GC-MS):**

GC-MS is a prevalent tool for the analysis of volatile and primary metabolites in plants subjected to nanomaterials. It is especially proficient in identifying organic acids, sugars, amino acids, and fatty acids, which are essential in plant stress responses. Due to the non-volatile nature of numerous plant metabolites, derivatization is frequently necessary to enhance their volatility and promote separation. GC-MS offers precise detection and measurement of metabolic abnormalities induced by nanoparticle exposure, including oxidative stress, modifications in primary metabolism, and nutritional assimilation. Studies employing GC-MS have demonstrated alterations in glucose and amino acid profiles in plants exposed to metal-based nanoparticles, signifying metabolic adaptations to stress. Furthermore, it assists in the surveillance of lipid peroxidation and damage generated by ROS. GC-MS is essential in elucidating plant resilience mechanisms and enhancing nanomaterial uses in agriculture through the assessment of toxicity thresholds and physiological responses (Li et al. 2021).

(b) **Assessment of Nanomaterial Impact using Liquid Chromatography-Mass Spectrometry (LC-MS)**

LC-MS is a crucial instrument for the analysis of non-volatile and intricate metabolites in plants influenced by nanomaterials. This method is very effective for identifying secondary metabolites, including flavonoids, alkaloids, polyphenols, and lipids, which are crucial in plant defense mechanisms. In contrast to GC-MS, LC-MS does not necessitate derivatization, rendering it appropriate for a wider array of chemicals. This methodology is extensively employed in both targeted and untargeted metabolomics research to evaluate metabolic alterations produced by nanoparticles. Exposure to metal oxide nanoparticles has been demonstrated to modify flavonoid production, which may be assessed by LC-MS (Liu et al. 2021;

Li and Keller 2024). Furthermore, LC-MS facilitates the detection of biomolecules linked to stress tolerance, including antioxidant chemicals that alleviate nanoparticle toxicity. LC-MS facilitates the elucidation of biochemical pathways affected by nanomaterials through a comprehensive metabolite profile, assisting researchers in formulating strategies to mitigate any detrimental impacts on plant development and productivity.

(c) **Nuclear Magnetic Resonance (NMR) Spectroscopy**

This is a non-invasive analytical method employed to examine the plants' metabolic response to nanomaterial exposure. It offers a thorough examination of plant metabolites, encompassing amino acids, sugars, organic acids, and secondary metabolites, without requiring substantial sample processing (Quintela et al. 2024). NMR is especially beneficial for examining metabolic flux, allowing researchers to monitor the impact of nanoparticles on biochemical pathways over time. NMR-based metabolomics has identified stress markers in plants exposed to silver or carbon-based nanoparticles, indicating alterations in energy metabolism, amino acid synthesis, and osmolyte accumulation. Furthermore, NMR can distinguish between nanoparticle-induced stress and other abiotic stressors by analyzing metabolic signatures (Marbella and Millstone 2015; Acharya et al. 2020). NMR's exceptional reproducibility and quantitative precision render it an invaluable instrument for investigating systemic metabolic changes, thereby facilitating the evaluation of nanomaterial toxicity, the optimization of nanoparticle formulations for agricultural applications, and the comprehension of plant adaptability under nano-enhanced treatments.

(d) **Fourier Transform Infrared Spectroscopy (FTIR)**

FTIR is an expedient and economical method for identifying metabolic alterations in plants subjected to nanomaterials. It operates by detecting functional groups in metabolites, enabling researchers to monitor changes in proteins, lipids, carbohydrates, and cell wall constituents. This technique is very effective for identifying oxidative stress-related alterations, including lipid peroxidation, variations in protein secondary structures, and polysaccharide degradation. FTIR has been used to measure the influence of metal nanoparticles on lignin and cellulose composition in plant cell walls, elucidating structural alterations resulting from nanoparticle interactions (Rico et al. 2015). Furthermore, FTIR can assess nanoparticle-induced metabolic stress by identifying alterations in essential biomolecules such as phenolic compounds, which are fundamental to plant defense mechanisms (Faghihzadeh et al. 2016). The non-destructive characteristics of FTIR render it an invaluable instrument for real-time and in situ investigation of plant responses to nanoparticles, facilitating swift toxicity evaluations and sustainable nanotechnology applications in agriculture (Dendisová et al. 2018).

(e) **Ionomics approaches**

Ionomics focuses on studying the elemental composition of plants to assess how nanoparticles influence nutrient uptake, transport, and homeostasis. Techniques such as Inductively Coupled Plasma-Mass Spectrometry (ICP-MS) and X-ray Fluorescence Spectroscopy (XRF) are commonly used to measure the concentration of essential and toxic elements in plants (Singh et al. 2022; Wang et al. 2024). Nanoparticles can alter ion transport mechanisms by interacting with membrane transporters, leading to imbalances in macro- and micronutrients like N, P, K, and Fe. For instance, exposure to metal-based nanoparticles like silver (Ag) and zinc oxide (ZnO) can result in metal accumulation in plant roots and shoots, potentially disrupting essential nutrient pathways. Ionomics helps determine whether nanoparticles improve or inhibit nutrient absorption and provides crucial data for understanding their impact on plant growth, development, and overall health (Petruccelli et al. 2025).

9.4 Conclusion

Given the wide range of uses for materials in nanoscale, they are relevant to support plants and the agro-industry. Nanoparticles have enormous uses in both industry and the home but also have harmful effects on the environment and human health. Nanotechnology has been used extensively to support plants' ability to withstand abiotic stress. The use of nanoparticles in abiotic stress management has several positive effects. These include improved growth and biomass, chlorophyll content, sugar content, osmolyte accumulation, and antioxidant production. Additionally, genes related to stress pathways are highly expressed. NPs have different interactions with plants depending on their sensitivity, size, surface area, and chemical composition. Significant anatomical, morphological, and physiological modifications are the outcome of these interactions, which are essential for crop plant improvement. To understand the genetic mechanism of NP assimilation and the plant stress response, this chapter discusses the function of transcriptome and proteomic contributions. In this context, a small proteomics study was carried out, which demonstrated the intricate way in which plants react to stress from nanomaterials and the occurrence of differential expression of certain genes that enhance biomolecular processes, such as oxidation-reduction, detoxification, signal transduction, and hormonal signaling. Recent proteomic investigations of silver nanoparticles have shown that particle size plays an important role in determining the degree of cell kinetics in plants, including induced phytotoxicity. This chapter includes a discussion of the mechanisms of NP accumulation in plants as well as how NPs can protect crop plants against a variety of abiotic stresses. Any form of nanoparticle, including nanopesticides, nanoherbicides, and nanofertilizers, can be readily absorbed by plants due to their high reactivity and tiny size. Applications of nanoparticles in plants appear to have a bright future, and studies should focus on creating self-degradable, nontoxic, economical, and environmentally friendly NPs. This will take nanotechnology from the lab to the fields.

References

Abdulraheem MI, Xiong Y, Moshood AY, Cadenas-Pliego G, Zhang H, Hu J (2024) Mechanisms of plant epigenetic regulation in response to plant stress: recent discoveries and implications. Plants 13:163

Acharya P, Jayaprakasha GK, Semper J, Patil BS (2020) 1H nuclear magnetic resonance and liquid chromatography coupled with mass spectrometry-based metabolomics reveal enhancement of growth-promoting metabolites in onion seedlings treated with green-synthesized nanomaterials. J Agric Food Chem 68:13206–13220

Ahmed T, Noman M, Manzoor N, Shahid M, Abdullah M, Ali L, Wang G, Hashem A, Al-Arjani AB, Alqarawi AA, Abd-Allah EF (2021) Nanoparticle-based amelioration of drought stress and cadmium toxicity in rice via triggering the stress responsive genetic mechanisms and nutrient acquisition. Ecotoxicol Environ Saf 209:111829

Akhtar N, Ilyas N, Meraj TA, Pour-Aboughadareh A, Sayyed R, Mashwani ZUR et al (2022) Improvement of plant responses by nanobiofertilizer: a step towards sustainable agriculture. Nano 12:965

Alabdallah NM, Hasan MM (2021) Plant-based green synthesis of silver nanoparticles and its effective role in abiotic stress tolerance in crop plants. Saudi J Biol Sci 28:5631–5639

Ali S, Mehmood A, Khan N (2021) Uptake, translocation, and consequences of nanomaterials on plant growth and stress adaptation. J Nanomat 2021:6677616

Ali S, Mir RA, Tyagi A, Manzar N, Kashyap AS, Mushtaq M, Raina A, Park S, Sharma S, Mir ZA, Lone SA, Bhat AA, Baba U, Mahmoudi H, Bae H (2023) Chromium toxicity in plants: signaling, mitigation, and future perspectives. Plants 12:1502. https://doi.org/10.3390/plants12071502

Al-Khayri JM, Rashmi R, Surya Ulhas R, Sudheer WN, Banadka A, Nagella P, Aldaej MI, Rezk AA, Shehata WF, Almaghasla MI (2023) The role of nanoparticles in response of plants to abiotic stress at physiological, biochemical, and molecular levels. Plants (Basel). 12(2):292. https://doi.org/10.3390/plants12020292. PMID: 36679005; PMCID: PMC9865530

Almutairi ZM (2016) Influence of silver nano-particles on the salt resistance of tomato (*Solanum lycopersicum*) during germination. Int J Agric Biol 18:449–457

Amini S, Maali-Amiri R, Mohammadi R, Shahandashti S-SK (2017) cDNA-AFLP analysis of transcripts induced in ChickpeaPlants by TiO_2 nanoparticles during cold stress. Plant Physiol Biochem 111:39–49

Brown CG, Clarke J (2016) Nanopore development at Oxford nanopore. Nat Biotechnol 34:810–811

Burklew CE, Ashlock J, Winfrey WB, Zhang B (2012) Effects of aluminum oxide nanoparticles on the growth, development, and microrna expression of tobacco (*Nicotiana tabacum*). PLoS One 7:e34783

Cheng CS, Rai K, Garber M, Hollinger A, Robbins D, Anderson S, Macbeth A, Tzou A, Carneiro MO, Raychowdhury R (2013) Semiconductor-based DNA sequencing of histone modification states. Nat Commun 4:2672

Choijookhuu N, Shibata Y, Ishizuka T, Xu Y, Koji T, Hishikawa Y (2022) An advanced detection system for in situ hybridization using a fluorescence resonance energy transfer-based molecular beacon probe. Acta Histochem Cytochem 55:119–128

Dendisová M, Jeništová A, Parchaňská-Kokaislová A, Matějka P, Prokopec V, Švecová M (2018) The use of infrared spectroscopic techniques to characterize nanomaterials and nanostructures: a review. Anal Chim Acta 1031:1–14

Djanaguiraman M, Anbazhagan V, Dhankher OP, Prasad PVV (2024) Uptake, translocation, toxicity, and impact of nanoparticles on plant physiological processes. Plants 13:3137. https://doi.org/10.3390/plants13223137

Dragan A, Pavlovic R, McGivney J, Casas-Finet J, Bishop E, Strouse R, Schenerman M, Geddes C (2012) SYBR green I: fluorescence properties and interaction with DNA. J Fluoresc 22:1189–1199

El-Moneim DA, Dawood MFA, Moursi YS, Farghaly AA, Afifi M, Sallam A (2021) Positive and negative effects of nanoparticles on agricultural crops. Nanotechnol Environ Eng 6:21. https://doi.org/10.1007/s41204-021-00117-0

Faghihzadeh F, Anaya NM, Schifman LA, Oyanedel-Craver V (2016) Fourier transform infrared spectroscopy to assess molecular-level changes in microorganisms exposed to nanoparticles. Nanotechnol Environ Eng 1:1–16

Fan B-L, Chen L-H, Chen L-L, Guo H (2025) Integrative multi-omics approaches for identifying and characterizing biological elements in crop traits: current progress and future prospects. Int J Mol Sci 26:1466

Ferrari E, Barbero F, Busquets-Fit'e M, Franz-Wachtel M, K¨ohler H-R, Puntes V, Kemmerling B (2021) Growth-promoting gold nanoparticles decrease stress responses in *Arabidopsis* seedlings. Nanomaterials 11:3161. https://doi.org/10.3390/nano11123161

Francis DV, Abdalla AK, Mahakham W, Sarmah AK, Ahmed ZF (2024) Interaction of plants and metal nanoparticles: exploring its molecular mechanisms for sustainable agriculture and crop improvement. Environ Int 190:108859

García-Sánchez S, Bernales I, Cristobal S (2015) Early response to nanoparticles in the Arabidopsis transcriptome compromises plant defence and root-hair development through salicylic acid signalling. BMC Genomics 16:341

Ge Y, Priester JH, VanDeWerfhorst LC, Walker SL, Nisbet RM, An Y-J et al (2014) Soybean plants modify metal oxide nanoparticle effects on soil bacterial communities. Environ Sci Tec 48:13489–13496

Ghosh SK, Bera T (2021) Molecular mechanism of nano-fertilizer in plant growth and development: a recent account. In: Advances in nano-fertilizers and nano-pesticides in agriculture. Elsevier, pp 535–560. https://doi.org/10.1016/B978-0-12-820092-6.00022-7

Gilles A, Meglécz E, Pech N, Ferreira S, Malausa T, Martin J-F (2011) Accuracy and quality assessment of 454 GS-FLX titanium pyrosequencing. BMC Genomics 12:245

Haas BJ, Zody MC (2010) Advancing RNA-seq analysis. Nat Biotechnol 28:421–423

Hasan MM, Skalicky M, Jahan MS, Hossain MN, Anwar Z, Nie Z-F et al (2021) Spermine: its emerging role in regulating drought stress responses in plants. Cells 10:261

Hebert PD, Braukmann TW, Prosser SW, Ratnasingham S, deWaard JR, Ivanova NV, Janzen DH, Hallwachs W, Naik S, Sones JE (2018) A sequel to sanger: amplicon sequencing that scales. BMC Genomics 19:219

Hrdlickova R, Toloue M, Tian B (2017) RNA-Seq methods for transcriptome analysis. Wiley Interdiscip Rev RNA 8:e1364

Jha S, Pudake RN (2016) Molecular mechanism of plant–nanoparticle interactions. In: Plant nanotechnology: principles and practices, pp 155–181

Josefsen K, Nielsen H (2010) Northern blotting analysis. In: RNA: Methods and protocols. Springer, pp 87–105

Karuppanapandian T, Wang HW, Prabakaran N, Jeyalakshmi K, Kwon M, Manoharan K, Kim W (2011) 2,4-dichlorophenoxyacetic acid-induced leaf senescence in mung bean (*Vigna radiata* L. Wilczek) and senescence inhibition by co-treatment with silver nanoparticles. Plant Physiol Biochem 49:168–177. https://doi.org/10.1016/j.plaphy.2010.11.007

Khan MN, Li Y, Fu C, Hu J, Chen L, Yan J, Khan Z, Wu H, Li Z (2022) CeO_2 nanoparticles seed priming increases salicylic acid level and ros scavenging ability to improve rapeseed salt tolerance. Glob Chall 6:2200025. https://doi.org/10.1002/gch2.202200025

Khan I, Awan SA, Rizwan M, Akram MA, Zia-ur-Rehman M, Wang X, Zhang X, Huang L (2023) Physiological and transcriptome analyses demonstrate the silver nanoparticles mediated alleviation of salt stress in pearl millet (*Pennisetum glaucum* L). Environ Pollut 318:120863. https://doi.org/10.1016/j.envpol.2022.120863

Kumari M, Khan SS, Pakrashi S, Mukherjee A, Chandrasekaran N (2011) Cytogenetic and genotoxic effects of zinc oxide nanoparticles on root cells of allium cepa. J Haz Mat 190:613–621

Kumari A, Kaur H, Kumar R, Kumar V (2022) Analysis of gene expression at mRNA level: technological evolution. In: Genomic, proteomics, and biotechnology. CRC Press, pp 65–78

Lenoir T, Giannella E (2006) The emergence and diffusion of DNA microarray technology. J Biomed Discov Collab 1:11

Li W, Keller AA (2024) Integrating targeted metabolomics and targeted proteomics to study the responses of wheat plants to engineered nanomaterials. ACS Agric Sci Technol 4:507–520

Li C, Gao Y, He S, Chi H-Y, Li Z-C, Zhou X-X, Yan B (2021) Quantification of nanoplastic uptake in cucumber plants by pyrolysis gas chromatography/mass spectrometry. Environ Sci Technol Lett 8:633–638

Liu H, Wang H, Shi Z, Wang H, Yang C, Silke S, Tan W, Lu Z (2006) TaqMan probe array for quantitative detection of DNA targets. Nucleic Acids Res 34:e4-e4

Liu L, Li Y, Li S, Hu N, He Y, Pong R, Lin D, Lu L, Law M (2012) Comparison of next-generation sequencing systems. Biomed Res Int 2012:251364

Liu J, Ji C, Liu X, Li X, Wu H, Zeng D (2021) Fe3O4 nanoparticles as matrix solid-phase dispersion extraction adsorbents for the analysis of thirty pesticides in vegetables by ultrahigh-performance liquid chromatography-tandem mass spectrometry. J Chromatogr B 1165:122532

Ma C, Chhikara S, Xing B, Musante C, White JC, Dhankher OP (2013) Physiological and molecular response of *Arabidopsis thaliana* (L.) to nanoparticle cerium and indium oxide exposure. ACS Sustain Chem Eng 1:768–778. https://doi.org/10.1021/sc400098h

Mahdy AM, Sherif FK, Elkhatib EA, Fathi NO, Ahmed MH (2020) Seed priming in nanoparticles of water treatment residual can increase the germination and growth of cucumber seedling under salinity stress. J Plant Nutrit 43:1862–1874

Maity A, Natarajan N, Vijay D, Srinivasan R, Pastor M, Malaviya DR (2018) Influence of metal nanoparticles (NPs) on germination and yield of oat (Avena sativa) and berseem (Trifolium alexandrinum). Pro Nat Acad Sci India Sect B Biol Sci 88:595–607

Marbella LE, Millstone JE (2015) NMR techniques for noble metal nanoparticles. Chem Mater 27(8):2721–2739

Mascarenhas R, Mathur T, Maheshwari J, Nagella P (2023) Interaction of nanomaterials with plant macromolecules: nucleic acid, proteins and hormones. In: Nanomaterial interactions with plant cellular mechanisms and macromolecules and agricultural implications. Springer, pp 231–271

Matysiak M, Kapka-Skrzypczak L, Brzóska K, Gutleb AC, Kruszewski M (2016) Proteomic approach to nanotoxicity. J Proteome 137:35–44

McGettigan PA (2013) Transcriptomics in the RNA-seq era. Curr Opin Chem Biol 17:4–11

Mittal D, Kaur G, Singh P, Yadav K, Ali SA (2020) Nanoparticle based sustainable agriculture and food science: recent advances and future outlook. Front Nanotech 2:10

Munir N, Gulzar W, Abideen Z, Hasanuzzaman M, El-Keblawy A, Zhao F (2023) Plant–nanoparticle interactions: transcriptomic and proteomic insights. Agronomy 13:2112

Mustafa G, Komatsu S (2021) Plant proteomic research for improvement of food crops under stresses: a review. Mol Omics 17:860–880. https://doi.org/10.1039/D1MO00151E

Mustafa G, Sakata K, Komatsu S (2015) Proteomic analysis of flooded soybean root exposed to aluminum oxide nanoparticles. J Proteom 128:280–297

Nanopore O (2012) Oxford Nanopore announcement sets sequencing sector abuzz. Nat Biotechnol 30:295–296

Ozsolak F, Milos PM (2011) Single-molecule direct RNA sequencing without cDNA synthesis. Wiley Interdiscip Rev RNA 2:565–570

Patterson SD, Aebersold RH (2003) Proteomics: the first decade and beyond. Nat Genet 33:311–323

PeharecStefani'c P, Jarnevi'c M, Cvjetko P, Biba R, ˇSiki'c S, Tkalec M, Cindri'c M, Letofsky-Papst I, Balen B (2019) Comparative proteomic study of phytotoxic effects of silver nanoparticles and silver ions on tobacco plants. Environ Sci Pollut Res 26:22529–22550. https://doi.org/10.1007/s11356-019-05552-w

Pérez-Labrada F, Hernández-Hernández H, López-Pérez MC, González-Morales S, Benavides-Mendoza A, Juárez-Maldonado A (2020) Nanoparticles in plants: morphophysiological, biochemical, and molecular responses. In: Tripathi DK, Singh VP, Chauhan DK, Sharma S, Prasad SM, Dubey NK, Ramawat N (eds) Plant life under changing environment. AcademicPress, pp 289–322

Petruccelli V, Vaccarella E, Cocco E, Mura F, Rosa D, Talone A, Iazzetta M, Dal Bosco C, Gentili A, Di Palma L (2025) Physiological, metabolic and ionomic responses of Solanum lycopersicum plants to Fe3O4 and FePO4 nanoparticles. Plant Nano Biology 13:100182

Piao Y, Xu W, Park KH, Ryu KH, Xiang R (2021) Comprehensive evaluation of differential methylation analysis methods for bisulfite sequencing data. Int J Environ Res Public Health 18: 7975

Preetha JSY, Sriram D, Premasudha P, Pudake RN, Arun M (2023) Cerium oxide as a nanozyme for plant abiotic stress tolerance: An overview of the mechanisms. Plant Nano Biol 6:100049

Pudake RN (2024) Novel application of bio-based nanomaterials for the alleviation of abiotic stress in crop plants. In: Nanotechnology for abiotic stress tolerance and management in crop plants, pp 181–201

Pudake RN, Mittal J, Tripathi RM, Tyagi J, Mohanta TK (2019) Biochemical responses of maize seedlings exposed to SnNPs. Micro Nano Lett 14:645–649

Quintela AL, Santos MF, de Lima RF, Mayer JL, Marcheafave GG, Arruda MA, Tormena CF (2024) Influence of silver nanoparticles on the metabolites of two transgenic soybean varieties: An NMR-based metabolomics approach. J Agric Food Chem 72:12281–12294

Rabilloud T, Lelong C (2011) Two-dimensional gel electrophoresis in proteomics: a tutorial. J Proteome 74:1829–1841

Rai M, Gupta I (2024) Nanotoxicology for agricultural and environmental applications. Elsevier

Raliya R, Nair R, Chavalmane S, Wang W-N, Biswas P (2015) Mechanistic evaluation of translocation and physiological impact of titanium dioxide and zinc oxide nanoparticles on the tomato (*Solanum lycopersicum* L.) plant. Metallomics 7(12):1584–1594

Rasheed A, Li H, Tahir MM, Mahmood A, Nawaz M, Shah AN, Aslam MT, Negm S, Moustafa M, Hassan MU, Wu Z (2022) The role of nanoparticles in plant biochemical, physiological, and molecular responses under drought stress: a review. Front Plant Sci 13:976179. https://doi.org/10.3389/fpls.2022.976179

Raz T, Causey M, Jones DR, Kieu A, Letovsky S, Lipson D, Thayer E, Thompson JF, Milos PM (2011) RNA sequencing and quantitation using the Helicos genetic analysis system. In: High-throughput next generation sequencing: methods and applications. Springer, pp 37–49

Rico CM, Peralta-Videa JR, Gardea-Torresdey JL (2015) Differential effects of cerium oxide nanoparticles on rice, wheat, and barley roots: a Fourier transform infrared (FT-IR) microspectroscopy study. Appl Spectrosc 69:287–295

Rio DC (2014) Reverse transcription–polymerase chain reaction. Cold Spring Harb Protoc 2014(11):pdb. prot080887

Shah AA, Yasin NA, Mudassir M, Ramzan M, Hussain I, Siddiqui MH, Ali HM, Shabbir Z, Ali A, Ahmed S, Kumar R (2022) Iron oxide nanoparticles and selenium supplementation improve growth and photosynthesis by modulating antioxidant system and gene expression of chlorophyll synthase (CHLG) and protochlorophyllide oxidoreductase (POR) in arsenic-stressed *Cucumis melo*. Environ Pollut 307:119413. https://doi.org/10.1016/j.envpol.2022.119413

Shang H, Ma C, Li C, White JC, Polubesova T, Chefetz B, Xing B (2020) Copper sulfide nanoparticles suppress *Gibberella fujikuroi* infection in rice (*Oryza sativa L.*) by multiple mechanisms: contact-mortality, nutritional modulation and phytohormone regulation. Environ Sci Nano 7:2632–2643. https://doi.org/10.1039/D0EN00535E

Sharif F, Westerhoff P, Herckes P (2013) Sorption of trace organics and engineered nanomaterials onto wetland plant material. Environ Sci Proc Impact 15:267–274

Shendure JA, Porreca GJ, Church GM, Gardner AF, Hendrickson CL, Kieleczawa J, Slatko BE (2011) Overview of DNA sequencing strategies. Curr Protoc Mol Biol 96:7.1. 1–7.1. 23

Shuken SR (2023) An introduction to mass spectrometry-based proteomics. J Proteome Res 22: 2151–2171

Siebert P, Larrick JW (1992) Competitive PCR. Nature 359:557–558

Silva S, Dias MC, Pinto DC, Silva AM (2023) Metabolomics as a tool to understand nano-plant interactions: the case study of metal-based nanoparticles. Plants 12:491

Singh VK, Sharma N, Singh VK (2022) Application of X-ray fluorescence spectrometry in plant science: solutions, threats, and opportunities. X-Ray Spectrom 51(3):304–327

Syu Y, Hung JH, Chen JC, Chuang H (2014) Impacts of size and shape of silver nanoparticles on Arabidopsis plant growth and gene expression. Plant Physiol Biochem 83:57–64. https://doi.org/10.1016/j.plaphy.2014.07.010

Thabet SG, Alqudah AM (2024) Unraveling the role of nanoparticles in improving plant resilience under environmental stress condition. Plant Soil 503:313–330. https://doi.org/10.1007/s11104-024-06581-2

Tiwari M, Krishnamurthy S, Shukla D, Kiiskila J, Jain A, Datta R, Sharma N, Sahi SV (2016) Comparative transcriptome and proteome analysis to reveal the biosynthesis of gold nanoparticles in Arabidopsis. Sci Rep 6:21733. https://doi.org/10.1038/srep21733

Tripathi DK, Tripathi A, Shweta SS, Singh Y, Vishwakarma K, Yadav G, Sharma S, Singh VK, Mishra RK, Upadhyay RG, Dubey NK, Lee Y, Chauhan DK (2017a) Uptake, accumulation and toxicity of silver nanoparticle in autotrophic plants, and heterotrophic microbes: a concentric review. Front Microbiol 08. https://doi.org/10.3389/fmicb.2017.00007

Tripathi DK, Singh S, Singh S, Pandey R, Singh VP, Sharma NC et al (2017b) An overview on manufactured nanoparticles in plants: uptake, translocation, accumulation and phytotoxicity. Plant Physiol Biochem 110:2–12

Tumburu L, Andersen CP, Rygiewicz PT, Reichman JR (2017) Molecular and physiological responses to titanium dioxide and cerium oxide nanoparticles in *Arabidopsis*. Environ Toxicol Chem 36:71–82. https://doi.org/10.1002/etc.3500

Uhl M, Houwaart T, Corrado G, Wright PR, Backofen R (2017) Computational analysis of CLIP-seq data. Methods 118:60–72

Varun P, Shukla V, Jambhulkar PP (2025) Microarray technology: overview and current status. In: Advances in omics technologies: exploring genomics, proteomics, and metabolomics, pp 193–217

Wang X, Li Q, Pei Z, Wang S (2018) Effects of zinc oxide nanoparticles on the growth, photosynthetic traits, and antioxidative enzymes in tomato plants. Biol Plant 62:801–808

Wang C, Wallerman O, Arendt M-L, Sundström E, Karlsson Å, Nordin J, Mäkeläinen S, Pielberg GR, Hanson J, Ohlsson Å (2021) A novel canine reference genome resolves genomic architecture and uncovers transcript complexity. Commun Biol 4:185

Wang C, Hua Y, Liang T, Guo Y, Wang L, Zheng X, Liu P, Zheng Q, Kang Z, Xu Y (2024) Integrated analyses of ionomics, phytohormone profiles, transcriptomics, and metabolomics reveal a pivotal role of carbon-nano sol in promoting the growth of tobacco plants. BMC Plant Biol 24:473

Xun H, Ma X, Chen J, Yang Z, Liu B, Gao X, Li G, Yu J, Wang L, Pang J (2017) Zinc oxide nanoparticle exposure triggers different gene expression patterns in maize shoots and roots. Environ Pollut 229:479–488. https://doi.org/10.1016/j.envpol.2017.05.066

Yi W, Li J, Zhu X, Wang X, Fan L, Sun W, Liao L, Zhang J, Li X, Ye J (2020) CRISPR-assisted detection of RNA–protein interactions in living cells. Nat Methods 17:685–688

Yuan Y, Xu H, Leung RK-K (2016) An optimized protocol for generation and analysis of ion proton sequencing reads for RNA-Seq. BMC Genomics 17:403

Zhang Y. 2003. Rapid amplification of cDNA ends. 13–24. Generation of cDNA libraries: methods and protocols: Springer

Zhang LW, Monteiro-Riviere NA (2009) Mechanisms of quantum dot nanoparticle cellular uptake. Toxicol Sci 110:138–155

Zhang JD, Ruschhaupt M, Biczok R (2013) ddCt method for qRT–PCR data analysis. Citeseer 48: 346–356

Zhu F-Y, Chen M-X, Ye N-H, Qiao W-M, Gao B, Law W-K, Tian Y, Zhang D, Zhang D, Liu T-Y (2018) Comparative performance of the BGISEQ-500 and Illumina HiSeq4000 sequencing platforms for transcriptome analysis in plants. Plant Methods 14:69

Chapter 10
Emerging Systems Biology Approaches in Plant Nanotoxicology: A Mechanism-Based Understanding of Nanomaterial Hazard

Jaya Kumari, Smita Lata, Jutishna Bora, Ridhhi Mishra, Archna Dashmana, and Sumira Malik

Abstract The fast adoption of nanotechnology in agriculture leads to beneficial applications but also generates new problems regarding nanotoxicity in plants. A detailed analysis of engineered nanomaterial (ENM) and plant system interaction is discussed using systems biology as an advanced discovery method throughout this chapter. Plants exhibit both toxic and beneficial consequences from interactions with nanomaterials, which depend on the exposure methods, material composition, and concentration levels. Research using multi-omics methods that combine genomics with transcriptomics, together with proteomics and metabolomics, allows scientists to detect molecular pathways altered by nanoparticle exposures. The combined application of these approaches helps scientists understand how nanoparticles cause oxidative stress while affecting the cells genetically and disrupting signalling and metabolism in plants. Computational modelling and network analysis tools help scientists understand how nanoparticles enter living systems and spread throughout the body and establish predictive methods to evaluate the dangers nanoparticles pose to living beings. The chapter presents protective approaches for nanoparticle risks through designed risk-free nanomaterials alongside bioremediation solutions and discusses why regulatory structures serve as crucial guides for agricultural

---

Jaya Kumari, Smita Lata and Jutishna Bora contributed equally with all other contributors.

J. Kumari · S. Lata · J. Bora · R. Mishra
Amity Institute of Biotechnology, Amity University Jharkhand, Ranchi, Jharkhand, India

A. Dashmana
Himalayan School of Biosciences, Swami Rama University, Dehradun, Uttarakhand, India

S. Malik (✉)
Amity Institute of Biotechnology, Amity University Jharkhand, Ranchi, Jharkhand, India

University Center for Research & Development (UCRD), Chandigarh University, Mohali, Punjab, India
e-mail: smalik@rnc.amity.edu

R. N. Pudake et al. (eds.), *Phyto Nanotoxicology*, Nanotechnology in Plant Sciences 7, https://doi.org/10.1007/978-3-032-24518-2_10

nanotechnology use. The research demonstrates the requirement for holistic strategies to guarantee sustainable and responsible nanomaterial use in plant science.

Keywords Plant nanotoxicology · Nanomaterial hazard · Nanoparticle plant interaction · System biology · Environmental biotechnology · Nanoparticle stress signalling

10.1 Introduction

A conventional transformation in agriculture and plant sciences has occurred through nanotechnology, which provides solutions for plant disease prevention and enhanced crop production systems (Pudake 2024; Pudake et al. 2019; Rohatgi et al. 2021). The unanticipated effects of plant exposure to nanoparticles, known as plant nanotoxicology, create environmental sustainability and food safety risks. The unique physicochemical characteristics of nanoparticles induce multiple biological reactions across molecular through physiological levels inside plants that cause outcomes between positive benefits and dangerous toxicities (Dietz and Herth 2011; Biswas and Pal 2024). A range of these interactions between NPs depends on the NP composition, together with their size, as well as surface charge and functionalization properties (Geisler-Lee et al. 2013; Zafar et al. 2023). Plants absorb nanoparticles through their roots and leaves, which results in their entry into chloroplasts, mitochondria, and nuclear organelles (Zhang et al. 2025). Various research works establish that silver nanoparticles (AgNPs) and zinc oxide (ZnO NPs) generate reactive oxygen species (ROS), which disrupt cellular balance {Abdal Dayem, 2017 #130}. The plant growth and development process gets negatively impacted by toxic effects, which also alter photosynthetic efficiency and cause problems with nutrient intake and enzymatic process functions (Baltasar et al. 2013; Sharma et al. 2020).

Establishing traditional toxicological investigations has brought essential plant stress response information about NPs, yet it lacks a complete systems perspective analysis. The field of systems biology enhances understanding of NP impacts on plants by integrating four multi-omics technologies, namely genomics, transcriptomics, proteomics, and metabolomics (Khosrovyan et al. 2025). Research success depends on this method because it enables scientists to find nanotoxicity molecular fundamentals, as well as NP stress biomarkers and ecological effects prediction. The use of transcriptomic and proteomic analysis techniques shows NP exposure as a factor that changes the gene expression, which controls stress signalling and oxidative defence together with secondary metabolic processes (White et al. 2012). Research using metabolomics demonstrated how NP stress causes changes to both primary and secondary metabolic levels as a biological adaptation. The information derived from these studies enables the essential development of nanomaterials having low ecological risks.

Computational modelling together with network-based analyses serves to forecast plant reactions toward multiple NP types, thus helping establish mechanism-based risk assessment systems (Nasir et al. 2024). Bioinformatics tools that merge experimental data help scientists discover both principal regulatory pathways and specific molecular targets that lead to NP-induced toxicity. The method enhances plant-nanoparticle interaction knowledge and creates new opportunities for nano-enabled sustainable agriculture development.

10.2 Application of Nanoparticles in Plants

Nanoparticles (NPs) have emerged as a transformative tool in plant science, offering various applications in agriculture, plant physiology, and stress management. Their small size and high surface area enable them to improve nutrient uptake, enhance plant growth, and mitigate abiotic stress.

One of the significant applications of nanoparticles in plants is their role in nutrient uptake and bioavailability. Nano-fertilizers have been shown to increase nutrient-use efficiency (NUE), reducing fertilizer waste while improving crop productivity (Rai et al. 2026). These nano-based fertilizers facilitate the slow release of essential nutrients, ensuring prolonged availability for plant roots (Singh et al. 2024; Rana and Pudake 2023). Additionally, nanoparticles such as zinc oxide (ZnO) and titanium dioxide (TiO_2) have been reported to enhance chlorophyll synthesis, leading to better photosynthetic efficiency and plant growth (Abdelhameed et al. 2025).

Nanoparticles also play a crucial role in mitigating abiotic stress in plants (Pudake et al. 2024). For instance, silver nanoparticles (AgNPs) have been found to improve drought and salt stress tolerance by regulating antioxidant enzyme activity (Patel et al. 2023). Similarly, silicon-based nanoparticles (SiNPs) enhance plant resistance to heavy metal toxicity by modulating ion uptake and reducing oxidative stress (Gupta and Sharma 2024). These nanoparticles help maintain cellular homeostasis, allowing plants to survive in harsh environmental conditions (Sharma et al. 2021).

In plant disease management, nanoparticles act as effective antimicrobial agents, reducing pathogen-induced infections (Preetha et al. 2024). Copper and silver nanoparticles have demonstrated strong antifungal and antibacterial properties, minimizing crop loss due to microbial infections (Verma and Bharadvaja 2022). Moreover, nano-pesticides provide a targeted approach to pest control by delivering active compounds directly to plant tissues, reducing chemical wastage and environmental pollution (Mehta et al. 2024).

Another significant advancement is the use of nanoparticles in genetic engineering. Nanoparticle-mediated gene delivery systems enhance the efficiency of plant transformation by allowing DNA and RNA molecules to penetrate plant cells without the need for conventional methods like Agrobacterium-mediated transformation (Das et al. 2025). This approach improves genetic modification techniques, making crop improvement more precise and efficient (Shivashakarappa et al. 2025).

Furthermore, nanoparticles contribute to seed priming, enhancing germination rates and seedling vigour. Studies have indicated that nano-priming with metal oxide nanoparticles improves seed metabolism, leading to faster growth and better stress tolerance during early plant development (Balusamy et al. 2023). This method offers an eco-friendly alternative to traditional seed treatments (Wei et al. 2024). In conclusion, nanoparticles hold immense potential in improving agricultural productivity and sustainability. Their application in nutrient management, stress mitigation, plant protection, genetic transformation, and seed priming highlights their versatility in modern plant science. As research advances, the integration of nanotechnology into agriculture is expected to revolutionize crop production and environmental management (Jha et al. 2023).

10.3 Potential Benefits and Risks

Nanotechnology has attracted attention as one of the potential fields in agriculture with many solutions used to boost plant growth, promote nutrient uptake, and protect plants against biotic and abiotic stressors. Alongside the buoyancy of this approach, the onset of some concern about possible toxicity and environmental risks posed by nanomaterials has begun to develop (Faseela et al. 2024). There is, hence, a need to appraise the positive and negative aspects of NMs in plants to ensure their safe and beneficial application in agriculture.

10.3.1 Benefits of Nanomaterials in Plants

One of the prime benefits of NMs in plant systems is NUE. This may be achieved through enhanced growth and productivity of plants by ameliorating nutrient deficiency through nano-fertilizers (Pudhuvai et al. 2024). Nano-fertilizers help reduce nutrient loss through leaching and volatilization by allowing controlled release. In addition, the correction of deficiencies in microelements in plants is aided by nanomaterials such as nano-iron and nano-zinc, which leads to improved crop quality and yield.

Another important aspect to mention is the ability of NMs to develop insecticidal nanoparticles that allow targeted delivery of active agents with less pesticide loads in the soil to prevent environmental contamination. Compared with traditional certifications, nanopesticides provide added stability and efficacy, ensuring better pest control with reduced toxicity towards non-target organisms. This aspect intrinsically matters for sustainable agricultural practices and helps avoid ecological disasters (Chaud et al. 2023). Moreover, NMs have proved their potential in managing plant stress. Current studies suggest that nano-enabled agrochemicals can increase the resistance of plants toward drought, salinity, and pathogens by modulating plant hormonal pathways and activating defense mechanisms (Faseela et al. 2024). Such achievements will be a nightmare come true in the quest to defeat climate change and ensure food security.

10.3.2 Possible Threats from Nanomaterials for Use in Plants

Despite the rich, beneficial promises they carry, nanomaterials in plants are associated with potential risks. One of the chief concerns is their toxicity, which might be attributed to excessive accumulation in plant tissues. At high concentrations, NMs can interfere with physiological and metabolic processes, producing oxidative stress and inducing damage to the cell (de la Rosa et al. 2017). Studies indicate that some nanoparticles may induce harmful effects on plants, thus interfering with seed germination, root elongation, and photosynthesis (Zhang et al. 2025). There is also an additional massive concern about the environmental effects of nanomaterials. They pose a threat to microbial communities, altering soil quality and affecting plant-microbe signalling if nano-agrochemicals are released uncontrollably into the soil and water (Mishra and Khare 2021). Their entrance into the food chain is another point of concern, raising questions about bioaccumulation and possible health risks for humans and animals. To mitigate these risks, nanoparticle capping approaches have been proposed by many researchers (Pirsaheb et al. 2024) to limit toxicity while retaining the desired benefits. Other enhancements might include the development of biocompatible and biodegradable NMs to offer better safety and environmental sustainability.

Nanomaterials provide some very exciting possibilities for agricultural advancement, and yet, their safe and responsible application is crucial. There is a need for more research on their optimal formulations, their long-term environmental consequence, and the establishment of guidelines for their applications. Any sustainable place for nanotechnology in plant science is conditioned on the balance of maximizing its benefits while minimizing the associated risks.

10.4 Mechanism of Nanomaterial Toxicity in Plants

10.4.1 Oxidative Stress and Reactive Oxygen Species (ROS)

When plants encounter nanoparticles or heavy metals, they develop oxidative stress through increased production of reactive oxygen species (ROS) (Mansoor et al. 2023). These ROS are highly reactive and can damage cellular proteins, lipids and DNA. Normally, plants maintain a balance between ROS production and detoxification through antioxidant defense mechanisms, but environmental stress factors like heat, droughts, and pollutants can disrupt this balance, leading to cellular damage. Oxidative stress in plants is commonly assessed using indicators such as hydrogen peroxide levels, lipid peroxidation assays, and electrolyte leakage tests, which help quantify membrane damage and cell viability (Begum et al. 2011; Rico et al. 2011). Oxidative stress affects plants differently based on the chemical properties of nanoparticles and their size, concentration, and their way of interacting with plant cells. The catalytic function of specific NPs creates ROS, while other NPs possess antioxidant properties that mitigate ROS damage to plant cells. Much effort has been made in research to identify the precise ways through which NPs control ROS levels, but additional investigation is still necessary (Rico et al. 2011).

10.4.1.1 Effects of Metal Oxide Nanoparticles on Oxidative Stress

Cerium Oxide Nanoparticles ($nCeO_2$)

The antioxidant, as well as pro-oxidant features of cerium oxide nanoparticles ($nCeO_2$), stem from their vacant oxygen sites that enable the conversion between Ce^{4+} (oxidized form) and Ce^{3+} (reduced form). Studies show that low $nCeO_2$ dosages at 62.5 mg/L function as antioxidant radicals that remove free radicals to reduce ROS levels. The SOD mimetic properties of these agents at 125–500 mg/L concentration levels induce superoxide dismutase at such strength that results in cellular oxidative stress (Boghossian 2015; Rico et al. 2011).

Titanium Dioxide Nanoparticles ($nTiO_2$)

The free radicals O_2^-, HO, CO_2^- develop from Titanium dioxide nanoparticles ($nTiO_2$) under illumination by UV light. The reduction of ROS accumulation results from redox reactions between Ti^{4+}/Ti^{3+} that convert dangerous ROS species into safer oxygen and hydrogen peroxide compounds. When exposed to UV-B radiation spinach chloroplasts received $nTiO_2$ exposure which led to lower ROS production along with reduced lipid peroxidation damage (Fenoglio et al. 2009; Lei et al. 2008). Plants of chickpea treated with $nTiO_2$ displayed better cold stress tolerance at 4 °C, which was reflected through decreased electrolyte loss measurement (Song et al. 2012; Mohammadi et al. 2013).

Zinc Oxide Nanoparticles (ZnO)

Under ultraviolet light absorption zinc oxide nanoparticles function as photocatalysts to produce highly reactive free radicals. ROs production capability of nanoparticles depends on their morphological structure because nanoparticles presenting higher oxygen vacancy levels on their surface generate greater ROS output (Xia et al. 2008; Kumari et al. 2011). ROS production in NPs arises from the combination effects of NPs and released Zn ions according to research although researchers continue to argue about quantitative ratios (Faisal et al. 2013; Shi et al. 2011; Wang et al. 2012a).

Nickel Oxide Nanoparticles (NiO)

Laboratory evidence demonstrates that Nickel oxide nanoparticles (nNiO) generate oxidative stress in high-concentration solutions. The reaction of tomato root cells to nNiO concentrations from 1.0 to 2.0 mg/mL triggered both excessive ROS formation and a corresponding rise in lipid peroxidation up to 49.5%. The causal factor between oxidative damage and Ni ions remains unconfirmed in this particular research. Research must establish the specific role of nNiO in ROS formation because nickel ions produce hydroxyl radicals (HO) through the Haber-Weiss cycle (Shahid et al. 2014).

Copper Oxide Nanoparticles (CuO)

The role of copper oxide nanoparticles (CuO) in oxidative stress remains controversial. Some studies suggest that CuO itself generates ROS, while others propose that toxicity results from Cu ions released from the nanoparticles. For example, in *Elodea densa*, exposure to nCuO led to increased lipid peroxidation due to ROS generation via the Fenton reaction. Some experiments indicate that nCuO remains stable in plant growth media, implying that the nanoparticles themselves contribute to oxidative damage. However, in other cases, dissolved Cu ions inside plant cells undergo redox reactions (Cu^{2+}-Cu^{+}), leading to ROS accumulation and oxidative stress (Shahid et al. 2014).

10.4.1.2 Membrane Damage and Water Stress Due to ROS Accumulation

Other Nanoparticles and Their Effects on ROS

Oxidative stress may greatly damage the cell membranes of plants by leaching out important cations such as potassium (K^{+}). A specific case can be in maize, where exposure to nCuO (10–100 mg/L) causes severe damage to cell membranes, particularly at the root portions that were directly in contact with the nanoparticle solution. Moreover, high concentrations of nCuO (100 mg/L) were also linked to insufficient water in plants, and some researchers believe water uptake inhibition could lead to an increase in ROS levels via aquaporin closure. These effects were also observed in grasses and pumpkins, which suggested that upon exposure to nFe_3O_4, the treatment may have led to increased lipid peroxidation, most probably due to nFe_3O_4 operating on respiration and water transport.

Silver Nanoparticles (nAg)

Notably, some conditions are seen wherein silver nanoparticles (nAg) act as agents to reduce oxidative stress. In low concentrations (25–50 ppm), nAg reduced ROS levels and lipid peroxidation in Brassica juncea (mustard plants). Scientists think that nAg might be functioning as an electron relay center hence improving the efficiency of redox reactions and hence low oxidative stress (Sharma et al. 2012).

Graphene and Carbon Nanotubes (CNTs)

The conclusions drawn from the use of graphene-based nanomaterials relative to oxidative stress were inconsistent. High concentrations of graphene (500–2000 mg/L) induced a dose-dependent increase in the levels of ROS, cell death, and electrolyte leakage in cabbage, tomato, and red spinach. This damage may ultimately be due to the aggregation of graphene on the surface of roots, disrupting normal root functioning. Similarly, in rice suspension cells, MWCNT also increased the levels of

ROS by three-and-a-half times, resulting in a reduction in cellular viability. Although some modified carbon nanomaterials, such as Fullerene (FMAD), are noted to have antioxidant activity, FMAD was inhibited in the production of ethanedial in *A. thaliana* root tips. These overall results suggest some carbon-based NPs might act as ROS scavengers (Begum et al. 2011).

The nanoparticles influence the generation of ROS and oxidative stress in plants diversely, determined by their specific chemical nature, concentration, and interaction with plant cells. Some nanoparticles, like cerium oxide ($nCeO_2$) and titanium dioxide ($nTiO_2$), have been known to act as antioxidants and alleviate oxidative stress under certain conditions. Others, like zinc oxide (nZnO), nickel oxide (nNiO), and copper oxide (nCuO), induce the production of ROS, hence causing oxidative damage. In many cases, it is still uncertain whether the oxidative stress is induced by the nanoparticles themselves rather than the metal ions released from the nanoparticles. Also, some nanoparticles act in such a way as to maintain disruption of water transport and respiration, thus playing only an indirect role in ROS accumulation. Although considerable progress has been achieved in exploring the effects, further studies are warranted to shed light on the detailed mechanisms involved in the generation and scavenging of ROS in plants treated with nanoparticles (Tan et al. 2009; Liu et al. 2013).

10.4.2 Genotoxicity and DNA Damage

Genotoxicity can be described as having adverse effects of substances, including nanomaterials (NMs), on the genetic material of an organism. In plants, exposure to nanomaterials is associated with DNA damage, chromosomal aberrations, and mutations, which cause several alterations in growth and development. The genotoxicity of NMs has been attributed to the interactions directly with DNA or indirectly through mechanisms that involve oxidative stress and consequent cellular dysfunction. The mechanisms described above should, therefore, be understood while considering the environmental and agricultural risk assessment of exposure to nanomaterials (Chichiriccò and Poma 2015). Nanomaterials have been demonstrated to kill plants via direct mechanisms of nuclear penetration, modification to the base, and disruption of replication or mitosis or indirectly via oxidative stresses caused by the production of ROS (Chichiriccò and Poma 2015). This causes breaks in strands, chromosome aberration, epigenetic change, and formation of micronuclei, killing the growth and even causing the programmed death of cells. (Nel et al. 2006).

Nanomaterials have been reported to influence nutrient uptake and transport in plants. Many studies suggest that certain nanoparticle exposures have shown alteration in the expression of genes involved in nutrient transport, leading to deficiency or toxicity. For instance, micronutrient cerium oxide nanoparticles reportedly affect nodulation by inhibiting nitrogen fixation processes in soybean (Jiao et al. 2025).

10.4.3 Physiological Toxicity

Nanoparticles (NPs) have effects on plant photosynthesis and metabolism that occur via numerous pathways, and the effects depend on the NPs type, size, and concentration. On optimal conditions, certain NPs such as TiO 2 could stimulate light absorption and energy conversion, whereas high concentrations usually would result in negative outcomes in terms of oxidative stress and chloroplast damage (Djanaguiraman et al. 2024). Abundant ROS generation, caused by NPs, including resending CeO 2 and CuO, disrupts chloroplast membranes and inhibits pigment synthesis, and PD II efficacy as shown in rice and duckweed, respectively. Those ROS disrupt thylakoid membranes, preventing electron transport and eliciting a lower production of ATP/NADPH, which ultimately halted the possibility of carbon fixation (Djanaguiraman et al. 2024).

Also, NPs such as ZnO and CuO disrupt the regulation of stomata, which influences the ABA signal by decreasing the CO_2 uptake and photosynthesis. Stomatal conductance and photosynthetic performance of maize were inhibited at 800 mg/kg ZnO NPs, with minimum effects at lower concentrations. The content of chlorophyll is also reduced in sunflower and microalgae by NPs, including Fe3O4, CoFe2O4, and graphene oxide (GO), where the destruction of chloroplasts can be seen (Djanaguiraman et al. 2024). Chloroplast binding of CuO and ZnO NPs to chloroplast carrier disrupts electron transport chains, which is observed in *Chlamydomonas reinhardtii*. The PSII activity and chlorophyll fluorescence are also decreased in Medicago arborea and duckweed in the presence of NPs of CeO_2 and CuO (Djanaguiraman et al. 2024).

Some NPs, such as TiO_2 and $CeO_{2,}$ despite such toxicities, have the potential of increasing photosynthesis via elevation of chlorophyll concentration, water photolysis activity, and ROS, combining with the goal to maintain chloroplast intensity (Djanaguiraman et al. 2024). NPs are also known to influence the metabolism of plants by changing the activities of enzymes. As an example, nano-TiO_2 promoted nitrate reductase and glutamine synthetase in spinach, which elevated the biomass, whereas CdO suppressed the amino acids and sugar content in barley (Yang Fan et al. 2006). Ag NPs enhanced the protein and carbohydrate content in *Bacopa monnieri*, but oxidative stress affected it in higher doses; GO altered the metabolism of the amino acids by damaging mitochondria (Begum and Fugetsu 2013).

When it comes to secondary metabolism, flavonoids, and phenolics in response to stress are regulated by NPs. As an example, nano-CeO_2 and In_2O_3 augmented the glutathione and antioxidant enzymes in Arabidopsis (Ma et al. 2016). TiO_2 NPs produced effects of increasing resveratrol in sage, Ag NPs increased artemisinin in Artemisia annua, both correlating ROS levels with metabolite synthesis. Contrary to resting conditions, some of the NPs generate ROS, and some of them (like CeO_2) scavenge them, based on surface heterogeneities and redox state (Navarro et al. 2008; Thao et al. 2023; Ge et al. 2022; Nele et al. 2024). In general, though NPs such as CuO, ZnO and Fe_3O_4 usually cause oxidative and metabolic stress, other NPs such as TiO_2 and CeO_2 have the potential to encourage photosynthesis and redox if within an appropriate range. The use of agriculture is also to be considered in its dual purposes (Djanaguiraman et al. 2024).

10.5 Systems Biology Approaches in Plant Nanotoxicology

Nanotoxicology in plants is an emerging field focused on understanding how nanoparticles (NPs) interact with cellular and molecular processes. Systems biology combines several high-throughput technologies, including genomics, transcriptomics, proteomics, metabolomics, and computational modelling, to provide a comprehensive view of these interactions. These approaches aid in uncovering the molecular mechanisms behind NP-induced toxicity, stress responses, and the potential benefits in plant systems (Djanaguiraman et al. 2024).

10.5.1 Genomics and Transcriptomics in Plant Nanotoxicology

Nanoparticles are capable of drastically altering the gene expression in plants, leading to either beneficial physiological responses or toxic effects. Owing to their high surface area and reactivity, the NPs accumulate in various sites (especially the plasma membrane, chloroplasts, vacuoles, and endoplasmic reticulum), which may elicit some responses at the transcriptome level. Although the complete molecular mechanisms of these interactions remain underexplored, the studies indicated that both metallic and non-metallic NPs can induce genetic modifications during prolonged exposure (Gowtham et al. 2024).

As an example, the single-walled carbon nanotubes (SWCNT) influenced root development in maize by promoting the expression of seminal root-associated genes (SLR1 and RTCS) and down-regulating root hair-related genes (RTH1 and RTH3), hence suggesting the role of the NPs in modifying root morphology (Yan et al. 2013). Selenium nanoparticles enhanced stress tolerance in *Momordica charantia* by upregulating WRKY1, a principal transcription factor contributing to stress signalling, as well as PAL and 4CL, which regulate secondary metabolite biosynthesis (Rajaee Behbahani et al. 2020). In *Arabidopsis thaliana*, treatment of silver NPs at a concentration of 5 mg/L induced the upregulation of 286 and downregulation of 81 stress-related genes linked to pathogen defence and hormone signalling (Kaveh et al. 2013).

10.5.2 Proteomics and Metabolomics in Plant Nanotoxicology

While various studies have recorded changes in protein abundance in plants due to the action of NPs, it should be noted that they have also provided information concerning NP action through the lens of proteomics in plants on the level where one could see their effect on protein synthesis, degradation, and interaction in plant cells (Rogić et al. 2015). In the proteomic status of soybean roots exposed to silver nanoparticles, a decrease was found in proteins linked to fermentation, including pyruvate decarboxylase (Mustafa et al. 2015). On the contrary, rice studies indicated an increase in proteins associated with cell wall biosynthesis and cell division, possibly arguing in favour of the idea that some NPs can enhance growth by upregulating structural protein expression in plants (Mirzajani et al. 2014). In addition, ZnO nanoparticles altered protein abundance in Brassica napus; this work involved proteins involved in photosynthesis, carotenoid biosynthesis, and light-harvesting complexes, paralleling increased photosynthetic rates and chlorophyll content (Sohail et al. 2022).

Nanoparticles also exert their effects on the expression of stress proteins. Proteomic analysis of *Phaseolus vulgaris* after treatment with CeO_2 nanoparticles shows differential expression of stress proteins, including antioxidant enzymes like ascorbate peroxidase (APX) and glutathione peroxidase (GPX) (Salehi et al. 2018). Much the same, TiO_2 nanoparticles enhanced the expression of superoxide dismutase (SOD) and glutathione transferase, thereby helping to lessen the oxidative stress brought about by stripe rust infection in wheat. These proteomic changes indicate that nanoparticles can promote either stress tolerance or stress responses, depending on their interactions with cellular components.

Metabolomic Adjustments Under Nanoparticle Exposure

At present, the integration of metabolomic analysis has proved beneficial in understanding the effect of nanoparticles on primary and secondary metabolism. Exploiting statistical models, these studies show that nanoparticles induced metabolic reprogramming through alteration of metabolite levels associated with stress resistance, growth, and development (Abideen et al. 2022).

For example, TiO_2-treated hydroponically grown rice showed increased levels of hexose phosphates, fatty acids, and amino acids that attested to the promotion of primary metabolism. Application of TiO_2 nanoparticles into soil raised proline and amino acid content while decreasing fat and organic acid levels, which stood as key indicators of stress adaptation. Seed treatment with CeO_2 nanoparticles led to increased levels of lauric and valeric acid while decreasing lignin content in rice (Djanaguiraman et al. 2024).

Other changes that can be associated with nanoparticles are, of course, the effects on secondary metabolite production. In the case of Salvia officinalis, foliar application of nanoparticles of Cerium Oxide resulted in a change in the levels of their terpenes and alkaloids, thereby altering the metabolism pathway associated with hormone (Imani et al. 2023). Silver nanoparticles modified carnosic acid concentrations in rosemary (*Salvia rosmarinus*), accentuating the antioxidant properties. Along the same lines, within the wheat system, Ag NPs interfered with the soluble sugar and organic acid compositions widely found in germinating seeds, which slowed down coleoptile growth by interfering with sugar metabolism (Hadi Soltanabad et al. 2020).

Nanoparticles are also known to induce the production of reactive oxygen species (ROS), playing significant roles in metabolic regulation. While excess production of ROS can induce oxidative injuries, some other nanoparticles, such as cerium oxide, might also possess mimicking activities to alleviate ROS by mimicking natural antioxidant enzymes. The resolution between generation and scavenging of ROS is affected by NP features, including size, surface charge, solubility, and chemical composition (Djanaguiraman et al. 2024).

10.5.3 Computational Modelling and Network Analysis in Nanotoxicology

Computational methods have empowered researchers to simulate the uptake, distribution, and biochemical interactions of nanoparticles (NPs) within plant tissues. The development of predictive models can assist scientists in determining whether NPs will yield beneficial effects, such as enhanced stress tolerance, or adverse effects, including oxidative stress and metabolic imbalances. Network analysis has been employed to provide a comprehensive view of how nanoparticles influence plant stress responses. By integrating gene expression data with metabolic profiling, it has become evident that metallic nanoparticles induce a coordinated defense response, involving stress-related transcription factors, antioxidant pathways, and hormonal signaling cascades. These models enable researchers to foresee potential risks associated with long-term nanoparticle exposure and facilitate the creation of safer nanoparticle formulations for agricultural applications. Systems biology approaches—including genomics, transcriptomics, proteomics, metabolomics, and computational modeling—offer valuable insights into the interactions between nanoparticles and plant systems. While certain nanoparticles can enhance growth, boost metabolism, and improve stress resistance, they may also disrupt cellular functions, leading to oxidative stress and metabolic imbalances. High-throughput techniques allow for a thorough assessment of nanoparticle toxicity alongside adaptation mechanisms, paving the way for the safe and effective application of nanotechnology in agriculture (Djanaguiraman et al. 2024).

10.6 Risk Assessment and Hazard Evaluation

10.6.1 Exposure Pathways and Bioaccumulation

The growing use of nanomaterials (NMs) in agriculture has raised concerns regarding their potential accumulation in plants and subsequent entry into the food chain. Understanding how NMs are introduced into plant systems, their uptake mechanisms, movement within plant tissues, and accumulation patterns is essential for evaluating potential ecological and health risks (Wang et al. 2022). This section explores the primary exposure routes, NM absorption, translocation, and bioaccumulation in plants. Plants are exposed to NMs through environmental interactions, primarily via soil, water, and air. These pathways influence their bioavailability, movement, and potential toxicity (Atanda et al. 2025). NMs are introduced into soil through direct application in nano-fertilizers, pesticides, and biosensors, as well as indirectly through industrial and agricultural runoff. Soil characteristics, including pH, organic matter, and ionic composition, affects NM dissolution, aggregation, and bioavailability. Certain metal-based nanoparticles, such as silver, zinc oxide, and titanium dioxide, may alter soil microbial communities, influencing nutrient cycling and plant-microbe interactions. Irrigation water containing NMs facilitates their uptake by plant roots. In aquatic environments, NMs undergo various transformations, such as oxidation, aggregation, and dissolution, which impact their toxicity and interactions with plants. Hydroponic studies confirm that metal-based NMs like cerium oxide (CeO_2) and silver nanoparticles (AgNPs) can be absorbed and transported within plants, potentially causing phytotoxic effects. Atmospheric sources, including industrial emissions, agricultural sprays, and vehicular exhaust, contribute to NM exposure through deposition on plant surfaces (Pietrzak et al. 2024). Research suggests that nanoparticles can penetrate plant tissues via cuticles and stomatal openings, enabling direct foliar uptake. Although airborne NM transport remains underexplored, it is an important factor in understanding their environmental movement and effects. After entering plants, NMs can be transported and stored in different tissues, raising concerns about their long-term ecological impact and implications for food safety (White et al. 2012). The primary route for NM entry into plants is through root absorption, occurring via passive (apoplastic) or active (symplastic) transport mechanisms. Root exudates, including organic acids and proteins, influence NM dissolution and uptake dynamics (Djanaguiraman et al. 2024). Once inside the plant, NMs move through the xylem, distributing them to aerial parts such as stems and leaves. Some nanoparticles, including carbon-based and metal oxide NMs, have been detected in the phloem, indicating their potential for systemic circulation within plants (Dietz and Herth 2011). NM accumulation varies depending on their size, chemical properties, and the plant species involved. Studies show that metal-based NMs tend to accumulate in edible plant parts, such as leaves (lettuce), fruits (tomatoes), and seeds (wheat), raising concerns about human and animal exposure through food consumption. Several factors influence the extent of nanomaterial (NM) absorption and storage in plant tissues (Baltasar et al. 2013). The size and surface charge of NMs play a crucial role in their mobility and uptake, with smaller nanoparticles (<50 nm) exhibiting higher absorption efficiency compared to larger ones. Additionally, surface coatings and functionalization can either enhance or restrict NM absorption, as modifications with organic ligands and polymers affect their solubility and interaction with plant cell walls. The ability of plants to absorb and accumulate NMs also varies among species, largely due to differences in root architecture, exudate composition, and transpiration activity, all of which influence NM uptake and distribution. Furthermore, environmental conditions such as soil composition, pH, and temperature significantly impact NM stability, aggregation, and bioavailability, ultimately affecting their absorption efficiency in plants. Understanding these factors is essential for assessing NM behaviour in agricultural systems and mitigating potential risks associated with their accumulation (Ijaz et al. 2020). The presence of nanomaterials (NMs) in edible plant tissues raises significant concerns regarding their potential impact on agricultural sustainability and human health. Research suggests that metal-based NMs, such as silver, zinc oxide, and titanium dioxide, can induce oxidative stress, disrupt DNA integrity, and alter plant metabolism, potentially leading to reduced crop quality and yield (Zhao et al. 2022). Additionally, the consumption of NM-contaminated plant-based foods poses potential risks to human health, as these materials may accumulate in human tissues over time, necessitating further toxicological studies to assess their long-term effects (Khan et al. 2017). Despite these concerns, current agricultural safety regulations lack standardized protocols for evaluating NM residues in crops. There is an urgent need for comprehensive research to establish permissible exposure limits and develop regulatory frameworks that ensure the safe use of nanotechnology in agriculture while mitigating risks to both ecosystems and consumers (Schwab et al. 2016). Assessing the exposure pathways and bioaccumulation patterns of NMs in plants is critical for understanding their potential risks to agriculture, ecosystems, and food safety. While nanotechnology offers promising advancements in plant productivity, addressing concerns related to NM persistence, environmental fate, and health risks is essential for responsible use. Future research should focus on refining risk assessment methodologies, developing safer nanomaterials, and establishing regulatory measures to ensure sustainable and secure agricultural practices.

10.6.1.1 Dose-Response Relationships

Understanding dose-response relationships is essential for assessing how nanomaterials (NMs) influence plant growth, metabolism, and overall health. These relationships illustrate how different NM concentrations trigger physiological, biochemical, and molecular changes, with beneficial effects at lower doses and potential toxicity at higher levels. Investigating these interactions is crucial for optimizing NM applications in agriculture while mitigating associated risks. The study of dose-response relationships in plants exposed to NMs is crucial for:

- **Determining Safe and Toxic Concentrations:** Establishing the threshold between beneficial and harmful NM levels helps guide agricultural applications and regulatory frameworks (Ma et al. 2018).
- **Elucidating Toxicity Mechanisms:** Understanding how plants react at different NM concentrations aids in assessing oxidative stress, DNA damage, and metabolic disruptions (Zhao et al. 2022).
- **Developing Risk Assessment Models:** Data from dose-response studies contribute to risk evaluation frameworks that ensure environmental safety and sustainable NM use (Khan et al. 2017).

At low doses, NMs may stimulate plant growth, improve germination rates, and enhance stress resistance, while higher doses often lead to toxicity. For example, low concentrations of cerium oxide nanoparticles (CeO_2 NPs) have been found to promote root elongation in soybean plants, whereas excessive amounts result in oxidative stress and growth inhibition (Rico et al. 2011). Some NMs exhibit a direct relationship between concentration and toxicity, with increasing levels progressively reducing chlorophyll content, photosynthetic efficiency, and biomass accumulation. Certain NMs have a threshold concentration below which no toxic effects occur, helping to establish safe exposure limits for agricultural use. Several factors determine the extent of NM absorption, accumulation, and toxicity in plants. The physicochemical properties of NMs, including size, shape, surface charge, and surface coatings, significantly influence their uptake, mobility, and potential toxicity within plant systems. Additionally, plant species and developmental stages play a crucial role in NM interactions, as different species and growth phases exhibit varying sensitivities, affecting absorption efficiency and defense mechanisms (Gu et al. 2022). The duration of exposure is another critical factor; short-term (acute) exposure may trigger immediate physiological responses, while long-term (chronic) exposure can lead to gradual NM accumulation and delayed toxic effects. Environmental conditions such as soil composition, pH levels, temperature variations, and microbial activity also impact NM bioavailability, influencing their transport and toxicity in plant systems. Understanding these factors is essential for assessing NM risks and optimizing their safe use in agricultural applications. A comprehensive understanding of dose-response relationships is essential for the safe and effective implementation of NMs in agriculture. Defining safe exposure limits is crucial for determining appropriate NM concentrations that enhance crop productivity while minimizing toxicity risks. Additionally, the development of eco-friendly NMs, such as biodegradable and controlled-release formulations, can help reduce environmental hazards and unintended toxic effects, ensuring sustainable agricultural practices (Singh et al. 2024). Furthermore, integrating dose-response findings into regulatory frameworks is necessary for advancing safety standards, enabling the establishment of guidelines for NM applications in food production and environmental conservation (Patel et al. 2023). These efforts collectively contribute to the responsible use of nanotechnology in agriculture, balancing innovation with ecological and human health considerations. Dose-response studies provide valuable insights into how plants interact with NMs, distinguishing beneficial effects from toxic thresholds. While low NM concentrations may enhance plant growth and resilience, excessive exposure can induce oxidative stress, biomass reduction, and toxicity. Future research should focus on refining toxicity thresholds, unraveling NM-plant interactions at the molecular level, and incorporating these findings into regulatory frameworks to ensure the responsible use of nanotechnology in agriculture (Khan and Hossain 2022).

10.6.2 Regulatory Perspectives and Guidelines

The widespread application of nanomaterials (NMs) in agriculture has necessitated the development of regulatory frameworks to assess potential risks to plants, ecosystems, and human health. These guidelines aim to establish exposure limits, standardize safety assessments, and minimize risks associated with NM accumulation and toxicity. However, due to the intricate interactions of NMs with biological systems, existing regulations encounter difficulties related to standardization, long-term impact evaluation, and efficient detection methodologies. This section reviews current regulatory perspectives, associated challenges, and future directions for NM governance in agriculture. Several global regulatory bodies have established guidelines to monitor the safety of nanomaterials (NMs), with a focus on their environmental behaviour, bioavailability, and associated risks. In the United States, the Environmental Protection Agency (EPA) regulates nanotechnology through various legislative acts. The Toxic Substances Control Act (TSCA) requires NM manufacturers to provide data on physicochemical properties, environmental persistence, and toxicity before commercialization (Keller et al. 2013). Additionally, the Federal Insecticide, Fungicide, and Rodenticide Act (FIFRA) govern nano-based pesticides, ensuring that safety evaluations assess their potential environmental effects. Furthermore, the Nanomaterial Research Strategy (NRS) emphasizes research on NM transformation, plant uptake, and long-term ecological impacts, helping to refine regulatory frameworks and improve risk assessments. Robust regulatory frameworks are essential for the safe and sustainable incorporation of nanotechnology into agriculture. Although existing regulations offer a foundation for NM risk assessments, challenges persist in standardizing toxicity tests, addressing long-term exposure concerns, and harmonizing international policies. Future advancements should emphasize refining NM detection methodologies, adopting systems biology approaches for mechanistic toxicity evaluations, and promoting global regulatory cooperation. Strengthening these regulatory perspectives will facilitate responsible NM application while mitigating potential environmental and human health risks (Joudeh and Linke 2022).

10.7 Mitigation Strategies for Nanomaterial Toxicity

10.7.1 Engineered Safe-by-Design Nanomaterials

The **Safe-by-Design (SbD) approach** for nanomaterials (NMs) aims to reduce their potential risks while maintaining their effectiveness in agricultural applications. This method incorporates risk assessment and material optimization at the design stage to ensure that engineered nanomaterials (ENMs) exhibit lower toxicity, improved biocompatibility, and reduced environmental persistence. By modifying key physicochemical properties such as size, surface charge, shape, and dissolution behavior, SbD strategies minimize hazardous interactions with plant cells, soil microbiota, and surrounding ecosystems. Several innovative strategies have been developed to enhance the safety of nanomaterials used in agricultural applications. One key approach **is surface functionalization and coating,** where nanomaterials are modified with biocompatible coatings such as polymers, biomolecules, or organic acids to reduce their reactivity and minimize oxidative stress and toxicity in plants. For instance, silver nanoparticles (AgNPs) coated with citrate or polyvinylpyrrolidone (PVP) exhibit lower toxicity than uncoated AgNPs due to reduced silver ion release, which is known to be harmful (Khan et al. 2017). Another strategy involves **doping and alloying** nanomaterials with secondary elements to alter their electronic properties and decrease their hazardous effects. Materials such as silicon-based nanoparticles (SiNPs) and chitosan nanoparticles (ChNPs) are naturally biodegradable and exhibit minimal environmental toxicity. Similarly, biopolymer-based nanocarriers, such as starch- or alginate-based materials, provide controlled-release properties while minimizing environmental accumulation (Zhao et al. 2022). A sustainable alternative is **green nanotechnology,** which utilizes plant extracts, microbes, or biomolecules to synthesize nanomaterials with reduced toxicity. These naturally derived nanomaterials contain bioactive capping agents that improve stability and biocompatibility. For instance, gold nanoparticles (AuNPs) synthesized using tea extract have demonstrated lower toxicity in plants compared to chemically synthesized AuNPs (Sharma et al. 2021; Tripathi et al. 2017). Lastly, **controlled release and smart delivery mechanisms** enhance the safety of nanomaterials by ensuring agrochemicals are only released under specific environmental conditions, reducing unintended exposure. Mesoporous silica nanoparticles (MSNs) and polymer-based carriers have been designed to release fertilizers and pesticides in response to triggers such as pH changes or enzymatic activity, effectively minimizing nanoparticle accumulation in plant tissues and soil (Fraceto et al. 2016). These approaches collectively contribute to the development of safer, more sustainable nanomaterials for agricultural applications. Safe-by-Design (SbD) nanomaterials offer several advantages in agricultural applications by addressing environmental and safety concerns while maintaining functionality. One significant benefit is their **reduced environmental impact**, as biodegradable SbD nanomaterials decompose into harmless byproducts, preventing long-term accumulation in soil and water systems. Additionally, **improved biocompatibility** is achieved through surface modifications and functional coatings, which reduce nanomaterial toxicity to plants, beneficial microbes, and soil organisms, ensuring minimal ecological disruption. The integration of SbD nanotechnology also supports **sustainable agricultural practices** by promoting the use of eco-friendly materials and green synthesis methods, thereby decreasing reliance on hazardous chemical inputs and fostering environmentally responsible farming. Furthermore, the adoption of SbD nanomaterials enhances **regulatory compliance,** as these materials are designed to align with evolving nanotoxicology guidelines, facilitating safer commercialization and broader acceptance in agricultural sectors. Safe-by-Design nanomaterials provide a promising solution to mitigate nanotoxicity risks in agricultural applications. By employing surface modifications, green synthesis methods, controlled-release technologies, and predictive toxicity assessments, researchers can develop nanomaterials with improved safety and sustainability. However, challenges related to cost, large-scale production, and long-term safety evaluations need to be addressed. Advancing SbD strategies will support the safe integration of nanotechnology into agriculture while aligning with global environmental and regulatory objectives (Khan and Hossain 2022).

10.7.2 Bioremediation and Phytoremediation Approaches

The widespread use of engineered nanomaterials (ENMs) in agriculture and industry has raised concerns regarding their environmental toxicity and long-term ecological effects. **Bioremediation and phytoremediation** have emerged as effective methods to mitigate nanomaterial contamination, leveraging biological processes to detoxify and eliminate hazardous NMs from soil and water systems. These approaches exploit the natural ability of microorganisms and plants to absorb, transform, and degrade nanomaterials, reducing their environmental persistence and potential toxicity (Khan and Hossain 2022). Bioremediation and phytoremediation are promising strategies for mitigating the environmental toxicity of engineered nanomaterials (ENMs) by leveraging the natural capabilities of microbes, fungi, enzymes, and plants to degrade, immobilize, or transform harmful nanomaterials. Microbial interactions with ENMs occur through mechanisms such as biosorption, bioaccumulation, enzymatic transformation, and redox reactions, which significantly reduce their toxicity. Certain bacteria, such as *Pseudomonas putida* and *Bacillus subtilis*, have demonstrated the ability to convert silver nanoparticles (AgNPs) into silver sulfide (Ag_2S), lowering their environmental risks. Fungal species, particularly white rot fungi like *Phanerochaete chrysosporium*, secrete ligninolytic enzymes that degrade metal oxide nanoparticles such as TiO_2 and ZnO, rendering them less harmful (Dolez 2015). Additionally, enzyme-assisted detoxification using oxidoreductases and peroxidases, including manganese peroxidase and laccase, has been effective in breaking down multi-walled carbon nanotubes (MWCNTs), thereby minimizing their ecological impact. Phytoremediation, a sustainable plant-based remediation strategy, involves mechanisms such as phytoextraction, phytostabilization, phytovolatilization, and rhizofiltration to mitigate nanomaterial toxicity. Some plant species, including *Brassica juncea* (Indian mustard) and *Helianthus annuus* (sunflower), effectively absorb and accumulate metal-based nanoparticles like ZnO and CuO in their tissues, reducing soil contamination. Others, such as *Chrysopogon zizanioides* (vetiver grass), stabilize toxic nanomaterials in the root zone, preventing groundwater contamination. Certain species, including *Medicago sativa* (alfalfa), convert silver nanoparticles into volatile silver compounds through phytovolatilization, minimizing their persistence in soil, while aquatic plants like *Eichhornia crassipes* (water hyacinth) facilitate rhizofiltration by adsorbing and precipitating cadmium-based quantum dots in the rhizosphere (Das et al. 2025).

10.8 Future Perspectives and Challenges

10.8.1 Emerging Research Trends

Advancements in systems biology are transforming plant nanotoxicology by providing a more detailed, mechanism-driven insight into nanomaterial hazards. As this interdisciplinary field continues to progress, it encounters various future opportunities and challenges, particularly in the areas of emerging research directions, technological advancements, and ethical and environmental concerns. The use of multi-omics data, including genomics, transcriptomics, proteomics, and metabolomics, is gaining prominence in plant nanotoxicology. This comprehensive approach provides deeper insights into the impact of nanomaterials (NMs) on plant systems at different biological levels. For example, transcriptomic studies have been employed to examine gene expression changes in plants exposed to NMs, revealing stress response mechanisms and identifying potential biomarkers for NM exposure. Advancements in systems biology are playing a crucial role in plant nanotoxicology by offering a mechanistic insight into the hazards associated with nanomaterials. The integration of genomics, transcriptomics, proteomics, and metabolomics—collectively referred to as multi-omics—is increasingly shaping the field of plant nanotoxicology. This holistic approach allows researchers to evaluate the overall impact of nanomaterials (NMs) on plant systems, supporting the development of predictive and quantitative models for NM behavior in biological environments (Balusamy et al. 2023). Advancements in imaging and analytical techniques have significantly improved the ability to visualize and measure NM presence within plant tissues. Cutting-edge methods such as hyperspectral imaging, Raman spectroscopy, and X-ray computed tomography (X-ray CT) enable real-time, non-invasive tracking of NM uptake, movement, and accumulation in plants. These technologies provide detailed, high-resolution insights into the spatial and temporal aspects of NM-plant interactions. The increasing use of computational models, including machine learning and artificial intelligence, is transforming the prediction of NM toxicity. By processing extensive datasets derived from multi-omics studies, these models can anticipate plant responses to different NMs, facilitating the development of safer nanomaterials and contributing to regulatory decision-making. There is also a rising focus on designing environmentally friendly and biodegradable NMs for agricultural use. Researchers are working on developing nanomaterials that support plant growth and resilience while reducing potential environmental harm. For example, nano-enhanced biostimulants and controlled-release nanofertilizers are being investigated to improve nutrient efficiency and lower the ecological footprint of agricultural practices (González-Melendi et al. 2008; Larue et al. 2012; Wang et al. 2012a, b; Sun et al. 2014). As NM applications in agriculture expand, ethical and environmental concerns are becoming increasingly important. Scientists are examining the long-term effects of NMs on plant genetics, potential hereditary consequences, and broader ecological implications associated with NM accumulation in the environment. Establishing thorough risk assessments and regulatory frameworks is essential to ensure the safe and sustainable incorporation of NMs into agricultural systems. In conclusion, systems biology approaches are driving significant progress in plant nanotoxicology by providing a more detailed, mechanism-based understanding of NM hazards. By adopting these emerging research trends, the field is on track to develop innovative nanotechnology solutions that maximize benefits while protecting plant health and environmental sustainability (Sattelmacher 2001).

10.8.2 Technological Innovations

Emerging technological advancements are revolutionizing systems biology approaches in plant nanotoxicology, enabling a deeper, mechanism-based understanding of nanomaterial (NM) hazards. Nanoparticle-mediated gene delivery has emerged as an efficient and versatile alternative to conventional plant genetic engineering techniques (Shivashakarappa et al. 2025). This method enhances genome manipulation across diverse plant species, allowing for precise genetic modifications and in-depth investigations into NM interactions at the molecular level. The advent of nanosensors capable of detecting specific biomolecules within plant systems has facilitated real-time, in vivo tracking of physiological responses to NM exposure. These advanced sensors generate high-resolution data on NM uptake, distribution, and impact, providing crucial insights into the spatial and temporal dynamics of NM-plant interactions. Nanotechnology is also driving progress in plant genome editing by overcoming existing technical limitations. Nanoparticles enable the targeted delivery of genetic materials such as DNA, miRNA, and siRNA into plant cells, significantly improving the efficiency of genetic modifications. This innovation is instrumental in decoding gene function and regulatory mechanisms involved in NM toxicity responses. Furthermore, the incorporation of microfluidic technology in plant nanotoxicology supports high-throughput analysis of plant responses to NMs under precisely controlled conditions. Microfluidic devices facilitate the manipulation of minute fluid volumes, enabling detailed studies on NM transport, accumulation, and toxicity at both cellular and tissue levels. This technology enhances data collection speed and ensures greater reproducibility of experimental results. These cutting-edge innovations are accelerating advancements in plant nanotoxicology, equipping researchers with sophisticated tools to investigate the intricate interactions between NMs and plant systems. By harnessing these developments, scientists can construct predictive models of NM behavior, evaluate potential risks, and design environmentally safer nanomaterials for sustainable agricultural applications (Pérez-de-Luque 2017). There is an increasing focus on developing environmentally friendly and biodegradable nanomaterials (NMs) for agricultural use to mitigate potential ecological risks. Advancements in green nanotechnology aim to design NMs that promote plant growth while minimizing toxicity-related concerns. Notable examples include nano-enabled biostimulants, controlled-release nanofertilizers, and biocompatible nano-based pesticides, all of which contribute to improved nutrient utilization and enhanced plant resilience. Additionally, artificial intelligence (AI) and machine learning are becoming essential tools for predicting NM toxicity and assessing plant responses to NM exposure. By leveraging extensive datasets from multi-omics studies—encompassing genomics, transcriptomics, proteomics, and metabolomics, AI models can generate predictive insights into NM behaviour, toxicity mechanisms, and potential environmental impacts. These computational approaches are instrumental in guiding the development of safer and more sustainable NMs for agricultural applications (Mirbakhsh 2023; Cui et al. 2016). Advancements in technology are revolutionizing plant nanotoxicology, providing researchers with cutting-edge tools to explore the intricate interactions between nanomaterials (NMs) and plant systems. The integration of nanosensors, microfluidics, AI-powered modeling, and sustainable nanotechnology is driving the development of predictive models for NM behavior, facilitating risk assessments, and enabling the design of safer NMs for agricultural use. Looking ahead, fostering interdisciplinary collaboration and establishing robust regulatory frameworks will be essential to ensure the ethical and responsible application of nanotechnology in plant sciences (Amutha et al. 2024). Technological innovations are revolutionizing plant nanotoxicology by providing advanced tools to study nanomaterial (NM) interactions at molecular, cellular, and whole-plant levels. Key advancements include nanoparticle-mediated gene delivery for precise genome editing, nanosensors for real-time NM tracking, and nano-enabled CRISPR systems for improved genetic modifications. Additionally, microfluidics enables high-throughput screening of plant responses, while artificial intelligence (AI) and machine learning facilitate predictive modeling of NM toxicity. There is also a strong focus on developing sustainable and biodegradable NMs, such as nano-enabled biostimulants and controlled-release nanofertilizers, to minimize environmental risks. Moving forward, interdisciplinary collaboration and regulatory frameworks will be essential for ensuring the safe and responsible application of nanotechnology in plant sciences (Begum et al. 2011; Rico et al. 2011).

10.9 Conclusion

The increasing integration of nanomaterials (NMs) in agriculture and plant sciences necessitates a comprehensive understanding of their potential hazards and regulatory implications. This study explored the field **of** plant nanotoxicology through the lens of systems biology, providing a mechanism-based approach to evaluating NM-induced toxicity. By leveraging multi-omics technologies and computational modeling, this framework offers a holistic perspective on NM interactions with plant systems, bridging gaps in traditional toxicological assessments. Nanomaterials exhibit diverse physicochemical properties that enhance their applicability in agriculture, including their role in nano-enabled fertilizers, pesticides, and plant growth regulators. However, their introduction into plant systems raises concerns about potential bioaccumulation and toxicity. Mechanism of nanomaterial routes, such as foliar absorption, root uptake, and stomatal infiltration, significantly influence NM behavior within plant tissues. While NMs offer agronomic benefits such as enhanced nutrient uptake and stress tolerance, they also pose risks, including cellular toxicity, altered gene expression, and unintended ecosystem impacts. Understanding NM toxicity requires an exploration of molecular and physiological disruptions triggered by their interactions with plant cells. One of the most significant toxicological mechanisms is oxidative stress, wherein NM exposure leads to excessive production of reactive oxygen species (ROS), causing lipid peroxidation, protein oxidation, and cellular dysfunction. Genotoxicity and DNA damage further exacerbate toxicity, as NMs can induce chromosomal aberrations, hinder DNA repair mechanisms, and interfere with normal cellular replication. Additionally, NMs can disrupt cellular and molecular functions, affecting membrane integrity, enzymatic activities, and metabolic pathways. The cumulative effects of these interactions result in physiological toxicity, manifesting as impaired seed germination, stunted growth, and reduced crop productivity. To unravel the complexity of NM-induced toxicity, systems biology approaches offer a high-throughput and integrative assessment of molecular responses. Genomics and transcriptomics reveal NM-induced alterations in gene expression, highlighting stress-response pathways, antioxidant defense mechanisms, and regulatory networks affected by NM exposure. Proteomics and metabolomics provide insights into protein modifications, enzymatic activities, and metabolic shifts, uncovering how plants adapt or succumb to NM-induced stress. Furthermore, computational modeling and network analysis facilitate the prediction of NM toxicity by mapping complex biological interactions, aiding in risk assessment and mitigation strategies. These approaches collectively enhance our mechanistic understanding of plant-NM interactions, leading to more precise toxicity evaluations. The risk assessment of NMs in plant systems requires a thorough examination of exposure pathways, bioaccumulation trends, and dose-dependent toxicological effects. The accumulation of NMs in plant tissues may have downstream implications for food safety and ecosystem stability, necessitating stringent monitoring frameworks. Understanding dose-response relationships is critical for determining NM toxicity thresholds, as both low and high concentrations can trigger distinct physiological effects. Regulatory perspectives and guidelines remain a crucial aspect of NM governance, yet current frameworks face challenges such as the lack of standardized testing protocols, inconsistencies across international regulations, and the complexity of NM fate and transformations in biological systems. Strengthening global regulatory harmonization and enhancing detection methodologies will be essential for ensuring NM safety in agricultural applications. To minimize NM-related risks, engineered safe-by-design nanomaterials present a promising approach. Modifying NM surface properties, functionalizing coatings, and optimizing biodegradability can enhance safety while maintaining functionality. Additionally, bioremediation and phytoremediation approaches leverage plant-microbe interactions to degrade or sequester NMs, reducing their ecological impact. These nature-based solutions offer sustainable remediation options for NM-contaminated soil and water systems. On a broader scale, policy interventions and sustainable practices are imperative to guide NM usage in agriculture, fostering responsible innovation while prioritizing environmental and food safety. As research in plant nanotoxicology advances, several emerging trends are shaping the future of the field. Innovations in nanobiosensors, high-throughput screening techniques, and artificial intelligence-driven toxicity predictions will refine NM risk assessments. Technological advancements in precision nanotechnology and bioengineered nanoparticles have potential for developing safer and more efficient agricultural applications. However, addressing ethical and environmental considerations remains a critical challenge. The long-term ecological impacts of NM accumulation in soils, food chains, and water bodies require continuous monitoring and assessment. Additionally, fostering multidisciplinary collaborations between policymakers, scientists, and industry stakeholders will be essential for developing responsible NM regulations and sustainable agricultural practices. Nanotechnology has the potential to revolutionize agriculture, offering innovative solutions for enhanced crop productivity, stress resilience, and nutrient delivery. However, the potential risks associated with NM exposure necessitate a mechanism-based understanding of toxicity, informed by systems biology approaches and comprehensive risk assessments. Bridging knowledge gaps in NM behavior, toxicity mechanisms, and regulatory frameworks will be crucial for ensuring their safe and sustainable integration into agricultural systems. By prioritizing safer nanomaterial designs, regulatory harmonization, and ethical considerations, future advancements in plant nanotoxicology can balance technological innovation with environmental and human health protection.

References

Abdelhameed RE, Hegazy HS, Abdalla H, Adarosy MH (2025) Efficacy of green synthesized titanium dioxide nanoparticles in attenuation salt stress in Glycine max plants: modulations in metabolic constituents and cell ultrastructure. BMC Plant Biol 25:221

Abideen Z, Hanif M, Munir N, Nielsen BL (2022) Impact of nanomaterials on the regulation of gene expression and metabolomics of plants under salt stress. Plants 11(5):691

Amutha C, Gopan A, Pushbalatatha I, Ragavi M, Reneese JA (2024) Nanotechnology and governance: regulatory framework for responsible innovation. In: Nanotechnology in societal development. Springer, pp 481–503
Atanda SA, Shaibu RO, Agunbiade FO (2025) Nanoparticles in agriculture: balancing food security and environmental sustainability. Discov Agric 3(1):26
Baltasar OSC, Dasgupta-Schubert N, Schubert C (2013) A systems model of the eco-physiological response of plants to environmental heavy metal concentrations. arXiv preprint https://arxiv.org/abs/1304.7496
Balusamy SR, Joshi AS, Perumalsamy H, Mijakovic I, Singh P (2023) Advancing sustainable agriculture: a critical review of smart and eco-friendly nanomaterial applications. J Nanobiotechnol 21(1):372
Begum P, Fugetsu B (2013) Induction of cell death by graphene in Arabidopsis thaliana (Columbia ecotype) T87 cell suspensions. J Hazard Mater 260:1032–1041
Begum P, Ikhtiari R, Fugetsu B (2011) Graphene phytotoxicity in the seedling stage of cabbage, tomato, red spinach, and lettuce. Carbon 49(12):3907–3919
Biswas A, Pal S (2024) Plant-nano interactions: a new insight of nano-phytotoxicity. Plant Physiol Biochem 210:108646
Boghossian AA (2015) Applications of nanoparticles for reactive oxygen species (ROS) scavenging in photosynthetic systems. ECS Trans 66(12):1
Chaud M, Souto EB et al (2023) Nanopesticides in agriculture: benefits and challenges in agricultural productivity, toxicological risks to human health and environment. Toxics 9(6):131
Chichiriccò G, Poma A (2015) Penetration and toxicity of nanomaterials in higher plants. Nano 5(2):851–873
Cui Q, Hernandez R, Mason SE, Frauenheim T, Pedersen JA, Geiger F (2016) Sustainable nanotechnology: opportunities and challenges for theoretical/computational studies. J Phys Chem B 120:7297–7306
Das M, Goswami A, Karmakar M, Dutta S, Naskar S, Das A, Banerjee S (2025) CRISPR in Genetic Engineering: Enhancing Crop Resilience and Food Security Through Genome Editing. In 2025 International Conference on Engineering Innovations and Technologies (ICoEIT) (pp. 1–8). IEEE
de la Rosa G, García-Castañeda C, Vázquez-Núñez E, Alonso-Castro ÁJ, Basurto-Islas G, Mendoza Á, Cruz-Jiménez G, Molina C (2017) Physiological and biochemical response of plants to engineered NMs: implications on future design. Plant Physiol Biochem 110:226–235
Dietz KJ, Herth S (2011) Plant nanotoxicology. Trends Plant Sci 16(11):582–589
Djanaguiraman M, Anbazhagan V, Dhankher OP, Prasad PV (2024) Uptake, translocation, toxicity, and impact of nanoparticles on plant physiological processes. Plants 13(22):3137
Dolez PI (ed) (2015) Nanoengineering: global approaches to health and safety issues. Elsevier
Faisal M, Saquib Q, Alatar AA, Al-Khedhairy AA, Hegazy AK, Musarrat J (2013) Phytotoxic hazards of NiO-nanoparticles in tomato: a study on mechanism of cell death. J Hazard Mater 250:318–332
Faseela P, Joel JM, Johnson R, Janeeshma E, Sameena PP, Sen A, Puthur JT (2024) Paradoxical effects of nanomaterials on plants: phytohormonal perspective exposes hidden risks amidst potential benefits. Plant Physiol Biochem 210:108603
Fenoglio I, Greco G, Livraghi S, Fubini B (2009) Non-UV-induced radical reactions at the surface of TiO2 nanoparticles may trigger toxic responses. Chem A Europ J 15(18):4614–4621
Fraceto LF, Grillo R, de Medeiros GA, Scognamiglio V, Rea G, Bartolucci C (2016) Nanotechnology in agriculture: which innovation potential does it have?. Front Environ Sci 4:20
Ge X, Cao Z, Chu L (2022) The antioxidant effect of the metal and metal-oxide nanoparticles. Antioxidants 11(4):791
Geisler-Lee J, Wang Q, Yao Y, Zhang W, Geisler M, Li K et al (2013) Phytotoxicity, accumulation, and transport of silver nanoparticles by Arabidopsis thaliana. Nanotoxicology 7(3):323–337

González-Melendi P, Fernández-Pacheco R, Coronado MJ, Corredor E, Testillano PS, Risueño MC et al (2008) Nanoparticles as smart treatment-delivery systems in plants: assessment of different techniques of microscopy for their visualization in plant tissues. Ann Bot 101(1):187–195

Gowtham HG, Shilpa N, Singh SB, Aiyaz M, Abhilash MR, Nataraj K et al (2024) Toxicological effects of nanoparticles in plants: mechanisms involved at morphological, physiological, biochemical and molecular levels. Plant Physiol Biochem 210:108604

Gu Q, Lv J, Mo X, Jiang X (2022) High aspect ratio metamaterials and their applications. Sensors Actuators A Phys 335:113376

Gupta R, Sharma P (2024) Sustainable nanomaterials for agriculture: current trends and future directions. Adv Agric Nanotechnol 12(3):87–101

Hadi Soltanabad M, Bagherieh-Najjar MB, Mianabadi M (2020) Carnosic acid content increased by silver nanoparticle treatment in rosemary (Rosmarinus officinalis L.). Appl Biochem Biotechnol 191(2):482–495

Ijaz I, Gilani E, Nazir A, Bukhari A (2020) Detail review on chemical, physical and green synthesis, classification, characterizations and applications of nanoparticles. Green Chem Lett Rev 13(3): 223–245

Imani AF, Gomarian M, Ghorbanpour M, Ramak P, Chavoshi S (2023) Foliar-applied nano-cerium dioxide differentially affect morpho-physiological traits and essential oil profile of Salvia mirzayanii Rech. f. & Esfand under drought stress and post-stress recovery conditions. Plant Physiol Biochem 203:108046

Jha A, Pathania D, Damathia B, Raizada P, Rustagi S, Singh P, ... Chaudhary V (2023) Panorama of biogenic nano-fertilizers: A road to sustainable agriculture. Environ Res 235:116456

Jiao L, Cao X, Wang C, Li X, Xiao Z, Yue L, Wang Z (2025) Deciphering the dynamic interplay between Rhizobacteria and root exudates via cerium oxide nanomaterials modulation for promoting soybean yield and quality. J Agric Food Chem 73:3413–3426

Joudeh N, Linke D (2022) Nanoparticle classification, physicochemical properties, characterization, and applications: a comprehensive review for biologists. J Nanobiotechnol 20(1):262

Kaveh R, Li YS, Ranjbar S, Tehrani R, Brueck CL, Van Aken B (2013) Changes in Arabidopsis thaliana gene expression in response to silver nanoparticles and silver ions. Environ Sci Technol 47(18):10637–10644

Keller AA, McFerran S, Lazareva A, Suh S (2013) Global life cycle releases of engineered nanomaterials. J Nanopart Res 15:1–17

Khan S, Hossain MK (2022) Classification and properties of nanoparticles. In: Nanoparticle-based polymer composites. Woodhead Publishing, pp 15–54

Khan I, Saeed K, Khan I (2017) Nanoparticles: properties, applications, and toxicities. Arab J Chem 12(7):908–931

Khosrovyan A, Vodovnik M, Mortimer M (2025) Omics approaches in environmental effect assessment of engineered nanomaterials and nanoplastics. Environ Sci Nano 12:2551–2579

Kumari M, Khan SS, Pakrashi S, Mukherjee A, Chandrasekaran N (2011) Cytogenetic and genotoxic effects of zinc oxide nanoparticles on root cells of Allium cepa. J Hazard Mater 190(1–3):613–621

Larue C, Veronesi G, Flank AM, Surble S, Herlin-Boime N, Carrière M (2012) Comparative uptake and impact of TiO_2 nanoparticles in wheat and rapeseed. J Toxic Environ Health A 75(13–15): 722–734

Lei Z, Wei X, Zhang L, Bi S (2008) Amphiphilic core–shell particles as carrier systems for metallic nanoparticles. Colloids Surf A Physicochem Eng Asp 317(1–3):705–710

Liu C, Tang J, Chen HM, Liu B, Yang P (2013) A fully integrated nanosystem of semiconductor nanowires for direct solar water splitting. Nano Lett 13(6):2989–2992

Ma C, Liu H, Guo H, Musante C, Coskun SH, Nelson BC, ... Dhankher OP (2016) Defense mechanisms and nutrient displacement in Arabidopsis thaliana upon exposure to CeO_2 and In_2O_3 nanoparticles. Environmental Science: Nano 3(6):1369–1379

Ma X, Geiser-Lee J, Deng Y, Kolmakov A (2018) Interactions between engineered nanoparticles (ENPs) and plants: Phytotoxicity, uptake, translocation, and biotransformation. Environ Sci Technol 46(17):9224–9239

Mansoor S, Ali A, Kour N, Bornhorst J, AlHarbi K, Rinklebe J, Abd El Moneim D, Ahmad P, Chung YS (2023) Heavy metal induced oxidative stress mitigation and ROS scavenging in plants. Plants 12:3003

Mehta A, Yadav A, Kumar A, Kanika K (2024) Role of nanotechnology in horticulture: An overview. Int J Adv Biochem Res 8(1):702–708

Mirbakhsh M (2023) Role of nano-fertilizer in plants plants' nutrient use efficiency (NUE) a mini-review. arXiv

Mirzajani F, Askari H, Hamzelou S, Schober Y, Römpp A, Ghassempour A, Spengler B (2014) Proteomics study of silver nanoparticles toxicity on Oryza sativa L. Ecotoxicol Environ Saf 108: 335–339

Mishra D, Khare P (2021) Emerging nano-agrochemicals for sustainable agriculture: benefits, challenges and risk mitigation. In: Sustainable agriculture reviews 50: emerging contaminants in agriculture. Springer International Publishing, Cham, pp 235–257

Mohammadi R, Maali-Amiri R, Abbasi A (2013) Effect of TiO 2 nanoparticles on chickpea response to cold stress. Biol Trace Elem Res 152:403–410

Mustafa G, Sakata K, Hossain Z, Komatsu S (2015) Proteomic study on the effects of silver nanoparticles on soybean under flooding stress. J Proteome 122:100–118

Nasir S, Berrouk A, Aamir A (2024) Exploring nanoparticle dynamics in binary chemical reactions within magnetized porous media: a computational analysis. Sci Rep 14:25505

Navarro E, Baun A, Behra R, Hartmann NB, Filser J, Miao AJ et al (2008) Environmental behavior and ecotoxicity of engineered nanoparticles to algae, plants, and fungi. Ecotoxicology 17:372–386

Nel A, Xia T, Madler L, Li N (2006) Toxic potential of materials at the nanolevel. Science 311(5761):622–627

Nele V, Melini S, Campani V, Angelillo A, Graziano SF, Pirozzi C et al (2024) Self-assembling nanoparticles with antioxidant activity for ROS scavenging in liver cells. J Drug Delivery Sci Technol 94:105490

Patel N et al (2023) Microfluidics for high-throughput nanotoxicology studies in plants. Environ Sci Technol 57(5):2501–2515

Pérez-de-Luque A (2017) Interaction of nanomaterials with plants: what do we need for real applications in agriculture? Front Environ Sci 5:12

Pietrzak M, Skiba E, Wolf WM (2024) Root-applied cerium oxide nanoparticles and their specific effects on plants: a review. Int J Mol Sci 25(7):4018

Pirsaheb M, Gholami T, Seifi H, Dawi EA, Said EA, Hamoody A-HM, Altimari US, Salavati-Niasari M (2024) Green synthesis of nanomaterials by using plant extracts as reducing and capping agents. Environ Sci Pollut Res 31:24768–24787

Preetha JSY, Sriram D, Premasudha P, Pudake RN, Anand AV, Arun M (2024) Recent advancements and applications of nano-agrochemicals against plant fungal diseases. In: Nanoparticles in plant biotic stress management. Springer, pp 403–418

Pudake RN (2024) Novel application of bio-based nanomaterials for the alleviation of abiotic stress in crop plants. In: Nanotechnology for abiotic stress tolerance and management in crop plants, pp 181–201

Pudake RN, Chauhan N, Kole C (2019) Nanoscience for sustainable agriculture. Springer

Pudake RN, Tripathi RM, Gill SS (2024) Nanotechnology for abiotic stress tolerance and management in crop plants. Elsevier

Pudhuvai B, Koul B, Das R, Shah MP (2024) Nano-fertilizers (NFs) for resurgence in nutrient use efficiency (NUE) a sustainable agricultural strategy. Curr Pollut Rep 11:1

Rai AK, Sharma U, Pathak N, Singh A, Raj S, Panotra N, ... Thakur S (2026) Synergizing nano-urea and organic manure for enhanced physiological efficiency and productivity in wheat. Arch Curr Res Int 26(4):428–440

Rajaee Behbahani S, Iranbakhsh A, Ebadi M, Majd A, Ardebili ZO (2020) Red elemental selenium nanoparticles mediated substantial variations in growth, tissue differentiation, metabolism, gene transcription, epigenetic cytosine DNA methylation, and callogenesis in bittermelon (Momordica charantia); an in vitro experiment. PLoS One 15(7):e0235556

Rana K, Pudake RN (2023) Potential application of nanotechnology in biofertilizer formulation for millets. In: Millet rhizosphere. Springer, pp 331–348

Rico CM, Majumdar S, Duarte-Gardea M, Peralta-Videa JR, Gardea-Torresdey JL (2011) Interaction of nanoparticles with edible plants and their possible implications in the food chain. J Agric Food Chem 59(8):3485–3498

Rogić T, Horvatić A, Tkalec M, Cindrić M, Balen B (2015) Proteomic analysis of Mammillaria gracilis Pfeiff. In vitro-grown cultures exposed to iso-osmotic NaCl and mannitol. Plant Cell Tissue Organ Cult 122:127–146

Rohatgi V, Challagulla NV, Pudake RN (2021) Volatile organic compounds (VOCs) sensors for stress management in crops. In: Biosensors in agriculture: recent trends and future perspectives, pp 81–95

Salehi H, Chehregani A, Lucini L, Majd A, Gholami M (2018) Morphological, proteomic and metabolomic insight into the effect of cerium dioxide nanoparticles to Phaseolus vulgaris L. under soil or foliar application. Sci Total Environ 616:1540–1551

Sattelmacher B (2001) The apoplast and its significance for plant mineral nutrition. New Phytol 149(2):167–192

Schwab F, Zhai G, Kern M, Turner A, Schnoor JL, Wiesner MR (2016) Barriers, pathways and processes for uptake, translocation and accumulation of nanomaterials in plants–Critical review. Nanotoxicol 10(3):257–278

Shahid M, Pourrut B, Dumat C, Nadeem M, Aslam M, Pinelli E (2014) Heavy-metal-induced reactive oxygen species: phytotoxicity and physicochemical changes in plants. Rev Environ Contam Toxicol 232:1–44

Sharma P, Jha AB, Dubey RS, Pessarakli M (2012) Reactive oxygen species, oxidative damage, and antioxidative defense mechanism in plants under stressful conditions. J Bot 2012(1):217037

Sharma A, Kumar V, Shahzad B, Ramakrishnan M, Singh Sidhu GP, Bali AS, Handa N, Kapoor D, Yadav P, Khanna K (2020) Photosynthetic response of plants under different abiotic stresses: a review. J Plant Growth Regul 39:509–531

Sharma P, Tripathi S, Chaturvedi P, Chaurasia D, Chandra R (2021) Newly isolated Bacillus sp. PS-6 assisted phytoremediation of heavy metals using Phragmites communis: potential application in wastewater treatment. Bioresour Technol 320:124353

Shi AM, Li D, Wang LJ, Li BZ, Adhikari B (2011) Preparation of starch-based nanoparticles through high-pressure homogenization and miniemulsion cross-linking: influence of various process parameters on particle size and stability. Carbohydr Polym 83(4):1604–1610

Shivashakarappa K, Marriboina S, Dumenyo K, Taheri A, Yadegari Z (2025) Nanoparticle-mediated gene delivery techniques in plant systems. Front Nanotechnol 7:1516180

Singh M et al (2024) Nano-enabled CRISPR delivery for plant genome editing: implications for agricultural nanotoxicology. Nanotechnol Plant Biol 8(1):55–73

Sohail, Sawati L, Ferrari E, Stierhof YD, Kemmerling B, Mashwani ZUR (2022) Molecular effects of biogenic zinc nanoparticles on the growth and development of Brassica napus L. revealed by proteomics and transcriptomics. Front Plant Sci 13:798751

Song Z, Xu T, Gordin ML, Jiang YB, Bae IT, Xiao Q et al (2012) Polymer–graphene nanocomposites as ultrafast-charge and-discharge cathodes for rechargeable lithium batteries. Nano Lett 12(5):2205–2211

Sun D, Hussain HI, Yi Z, Siegele R, Cresswell T, Kong L, Cahill DM (2014) Uptake and cellular distribution, in four plant species, of fluorescently labeled mesoporous silica nanoparticles. Plant Cell Rep 33:1389–1402

Tan MC, Chow GM, Ren L, Zhang Q (2009) Inorganic nanoparticles for biomedical applications. In: NanoScience in biomedicine, pp 272–289

Thao NTM, Do HDK, Nam NN, Tran NKS, Dan TT, Trinh KTL (2023) Antioxidant nanozymes: mechanisms, activity manipulation, and applications. Micromachines 14(5):1017
Tripathi DK, Tripathi A, Shweta, Singh S, Singh Y, Vishwakarma K, ... Chauhan DK (2017) Uptake, accumulation and toxicity of silver nanoparticle in autotrophic plants, and heterotrophic microbes: a concentric review. Front Microbiol 8:07
Verma A, Bharadvaja N (2022) Plant-mediated synthesis and characterization of silver and copper oxide nanoparticles: antibacterial and heavy metal removal activity. J Clust Sci 33(4): 1697–1712
Wang C, Bobba AD, Attinti R, Shen C, Lazouskaya V, Wang LP, Jin Y (2012a) Retention and transport of silica nanoparticles in saturated porous media: effect of concentration and particle size. Environ Sci Technol 46(13):7151–7158
Wang Z, Xie X, Zhao J, Liu X, Feng W, White JC, Xing B (2012b) Xylem-and phloem-based transport of CuO nanoparticles in maize (Zea mays L.). Environ Sci Technol 46(8):4434–4441
Wang J, Yue L, Zhao J, Cao X, Wang C, Chen F et al (2022) Uptake and bioaccumulation of nanoparticles by five higher plants using single-particle-inductively coupled plasma-mass spectrometry. Environ Sci Nano 9(8):3066–3080
Wei L, Liu J, Jiang G (2024) Nanoparticle-specific transformations dictate nanoparticle effects associated with plants and implications for nanotechnology use in agriculture. Nat Commun 15(1):7389
White JW, Andrade-Sanchez P, Gore MA, Bronson KF, Coffelt TA, Conley MM, ... Wang G (2012) Field-based phenomics for plant genetics research. Field Crops Res 133:101–112
Xia T, Kovochich M, Liong M, Madler L, Gilbert B, Shi H et al (2008) Comparison of the mechanism of toxicity of zinc oxide and cerium oxide nanoparticles based on dissolution and oxidative stress properties. ACS Nano 2(10):2121–2134
Yan S, Zhao L, Li H, Zhang Q, Tan J, Huang M et al (2013) Single-walled carbon nanotubes selectively influence maize root tissue development accompanied by the change in the related gene expression. J Hazard Mater 246:110–118
Yang Fan YF, Hong FaShui HF, You WenJuan YW, Liu Chao LC, Gao FengQing GF, Wu Cheng WC, Yang Ping YP (2006) Influences of nano-anatase TiO2 on the nitrogen metabolism of growing spinach. Biol Trace Elem Res. 110(2):179–190
Zafar H, Javed R, Zia M (2023) Nanotoxicity assessment in plants: an updated overview. Environ Sci Pollut Res 30:93323–93344
Zhang T, Wang Q, Rui Y (2025) The impact of nanomaterials on plant health: a review of exposure, toxicity, and control. Environ Sci Nano 12:2965–2982
Zhao L et al (2022) Advances in nanosensors for in vivo monitoring of nanomaterial interactions in plants. Plant Biotechnol Rep 16(4):287–302

Chapter 11
Metabolomics to Study the Effect of Primary and Secondary Metabolism in Plants Towards Nano Pollution

Sana Awan, Sandeep Sharma, and Ivan Oyege

Abstract The increasing utilization of nanotechnology has resulted in the release of nanoparticles (NPs) into the environment, prompting worries about their effects on plant health and ecosystem stability. Plants are primary producers, and are particularly vulnerable to nano pollution, which can interfere with basic and secondary metabolic pathways that are essential for defense, development, and adaptation. Metabolomics, a comprehensive technique for characterizing small molecules, is a great technique for investigating the metabolic impacts of NPs on plants. Researchers can identify metabolic alterations at high resolution using methods including liquid chromatography-mass spectrometry (LC-MS), gas chromatography-mass spectrometry (GC-MS), and nuclear magnetic resonance (NMR) spectroscopy. This chapter investigates the metabolomic changes caused by NP exposure, emphasizing changes in basic metabolism (photosynthesis, energy production, and nutritional assimilation) and secondary metabolism, which includes defense-related chemicals such as phenolics, flavonoids, and terpenoids. Metabolomics analyzes these metabolite profiles to determine how plants react to oxidative stress, interruption of nutrient absorption, and interference with cellular metabolic pathways brought on by nanopollution. Researchers may build a comprehensive understanding of plant metabolic adaptability to NPs exposure by relating metabolomics data with results of other omics studies. This allows them to make informed decisions about environmental safety and sustainable agriculture practices. This chapter emphasizes the importance of metabolomics in knowing more about plant interactions with NPs and identifies future research options for minimizing the impacts of nano-pollution on plant systems.

S. Awan (✉)
Department of Earth and Environment, Florida International University, Miami, FL, USA
e-mail: sawan005@fiu.edu

S. Sharma
Department of Botany and Plant Sciences, University of California, Riverside, CA, USA

I. Oyege
Department of Earth and Environment, Florida International University, Miami, FL, USA

Department of Chemistry, Busitema University, Tororo, Uganda

R. N. Pudake et al. (eds.), *Phyto Nanotoxicology*, Nanotechnology in Plant Sciences 7, https://doi.org/10.1007/978-3-032-24518-2_11

Keywords Metabolomics · Nano pollution · Plant metabolism · Oxidative stress · Environmental stress response · Antioxidant response

11.1 Introduction

The fast-growing field of nanotechnology has developed extensive application of nanoparticles across multiple industries, such as medicine, electronics, chemicals, and agriculture. As a result, more NPs are being released in surroundings, often entering soil and water systems where they can interact with plants and other organisms. While these particles offer numerous benefits, their environmental persistence and unique physicochemical properties are of concerns when we consider their potential impact on ecological health. This introduction addresses the connection of nanotechnology and plant biology, with a focus on how metabolomics-a high-throughput tool for studying metabolites-enables us to know more how the nano pollution affect plant metabolism, providing insights into both primary and secondary metabolic responses. Nanoparticles are specially designed materials ranging from 1 to 100 nm in size, exhibiting distinct properties compared to their bulk forms due to their large surface area-to-volume ratio and quantum effects. These unique properties make NPs highly reactive and versatile for industrial applications, leading to their proliferation across diverse fields (Khan et al. 2022; Zhu et al. 2020). However, as NPs accumulate in the environment, plants-particularly those in agricultural and natural ecosystems-are exposed to varying concentrations through soil, water, and sometimes air (Khan et al. 2021).

Plants play a central role in ecosystem function and are critical to food security and carbon cycling. Interaction of plants and nanomaterials (NMs) is complex, as these particles can be taken up by the roots, transported to the aerial plant organs, and accumulated in plant tissues (Ma et al. 2010). This entry and deposition in plants had both positive and adverse effects on growth, physiology, and biochemical pathways, subject to the NM type, size, concentration, and plant species involved (Khan et al. 2021). Nano pollution causes significant impact to plant health, as NPs can interfere with critical physiological and biochemical pathways. For example, certain nanomaterials, like zinc oxide (ZnO) and titanium dioxide (TiO_2) were interfering with photosynthesis by affecting chlorophyll synthesis and electron transport chains in chloroplasts, ultimately impacting plant energy production (Hussain et al. 2021; Li et al. 2020). Similarly, research has indicated that NPs can cause oxidative stress in plants by producing reactive oxygen species (ROS), which may result in harm to cellular molecules like proteins and DNA (Ma et al. 2010; Khan et al. 2021).

NMs can also disrupt nutrient absorption and assimilation processes, modifying the accumulation of crucially important nitrogen, phosphorus, and potassium (Tripathi et al. 2017). Additionally, changes in cellular metabolism can lead to the accumulation or depletion of various metabolites, impacting both primary metabolic processes, which are essential for growth and energy production, and secondary metabolic processes, which play roles in defense and adaptation to environmental stresses (Zhu and Thompson 2019). Metabolomics represents a thorough examination of small molecules (metabolites) within a biological system, offering insight into cellular metabolism in response to external influences. By investigating the metabolite profiles of plants exposed to NPs, scientists can gain insights into the chemical and physiological alterations induced by nano pollution (Aliferis and Jabaji 2012). Metabolomics involves cutting-edge analytical methods, including liquid chromatography-mass spectrometry (LC-MS), gas chromatography-mass spectrometry (GC-MS), and nuclear magnetic resonance (NMR) spectroscopy, that enables us to identify and quantify of thousands of metabolites in plant tissues (Fiehn 2002; Arbona et al. 2013; Hao et al. 2025).

The use of metabolomics in nano pollution studies enables the identification of specific metabolic pathways and stress-response mechanisms affected by NPs, shedding light on the adaptive and potentially deleterious effects of NPs on plant systems. For example, metabolomics can reveal changes in carbohydrate metabolism, amino acid profiles, and lipid composition, as well as alterations in secondary metabolites like phenolics, flavonoids, and terpenoids that play critical roles in plant defense (Tripathi et al. 2017). By studying both primary and secondary metabolites, researchers can develop a holistic concern of plant responses to NP exposure, allowing for the identification of key biomarkers and pathways involved in stress adaptation and resilience (Goswami et al. 2010).

11.2 NMs and Their Interactions with Plants

Nanotechnology is reforming agriculture by providing novel methods to improve crop productivity, better input efficiency, and providing solution environmental concerns. NPs described as particles having at least one dimension measuring between 1 and 100 nm are at the forefront of this agricultural transformation (Kumar et al. 2013; Khan et al. 2022; Pudake et al. 2019). These nanoscale materials are being utilized in several applications, like nanofertilizers, nanopesticides, and nanosensors, to optimize farming practices and increase yields (Singh et al. 2021; Pudake 2024). Metal-based NPs are being explored for their antimicrobial use in crop protection. These NMs can be applied to develop more effective and targeted pesticides, reducing the total chemical pollution on agricultural systems (Chen 2018; Luo et al. 2024). Metal oxide NPs, including TiO_2 and ZnO, are finding applications in smart fertilizers and soil amendments, enhancing nutrient delivery and uptake by plants (Tan et al. 2018; Pudake et al. 2024a). These materials can enhance water uptake and nutrient absorption, leading to stronger and more resilient crops (Husen 2020; Aacharya and Chhipa 2020; Pudake et al. 2024b). Polymer-based NMs, made from biodegradable polymers like chitosan, are being used as carriers for pesticides and fertilizers in agriculture, allowing for controlled release and reduced environmental damage (Mondéjar-López et al. 2024). Quantum dots, with their unique optical properties, are being employed in agricultural diagnostics and imaging technologies. These NMs can be used to make advanced sensors for the detection of plant diseases, checking soil conditions, and assessing crop health in real-time

(Gupta et al. 2022; Ma et al. 2018). The application of nanotechnology in agriculture is driving the development of precision farming, enabling customised inputs for individual crop requirements, minimising waste, and enhancing productivity. However, the need to measure the potential environmental and health risks associated with nanoparticles in agricultural practices becomes priority. The novel properties of NPs, like their size, shape, and reactivity, can alter its behaviour in soil and plant tissues (Chen 2018).

11.2.1 Uptake and Translocation of NPs in Plants

NPs interact with plants through complex mechanisms, with their uptake and translocation based on various factors such as NMs' properties, stage and condition of plant, and environmental factors (Vignesh et al. 2024; Wang et al. 2023a). Knowing these processes is important for interpreting the metabolic changes observed in plants exposed to NMs and for developing effective agricultural applications. Uptake of NPs through root is considered the main pathway for its entry into plants. This process begins with the adsorption of NPs to the root surface through electrostatic interactions or specific chemical interactions. Once adhered, NPs must enter the cell wall, which is a natural barrier. Smaller NPs can pass through these pores more easily, while larger ones may face greater resistance (da Cruz et al. 2017; Yusefi-Tanha et al. 2020). The absorption of nanoparticles (NPs) by plant roots is a complex process that begins through adsorption to the root surface with the negative charge, which attracts positively charged NPs due to carboxyl groups in cell wall pectins and organic acids secreted by roots (Navarro et al. 2008). Following adsorption, NPs must overcome several barriers to enter the root, primarily the cell wall with pore sizes typically ranging from 5–20 nm, while some studies indicate that even smaller sizes, ranging from 3.5 to 5.2 nm, can be effective (Carpita et al. 1979; Ma et al. 2010). NPs can enter more easily through areas where protective layers are not fully developed, like in the root tips and newly formed lateral roots (Schwab et al. 2016). Once past the cell wall, NPs may be internalized by root cells through different methods, with endocytosis being a primary pathway. In this process, the cell membrane invaginates to engulf NPs, creating vesicles that transport NPs into the cell (Etxeberria et al. 2006; Shahzad et al. 2024). Small, lipophilic NPs may directly penetrate the plasma membrane, while some metal-based NPs might enter through ion channels or by binding to membrane transport proteins (Pokhrel and Dubey 2013). The entry of NPs can trigger various cellular responses, including the creation of new holes in cell membranes, alteration of gene expression, and changes in cellular metabolism (Judy et al. 2012; Nair et al. 2010; Farooq et al. 2024). Comprehending these complex uptake processes is very important for applications NPs in agriculture and environment, through studying the potential risks attached with NP exposure to plants (Pérez-de-Luque 2017). Once internalized, nanoparticles can be carried to different areas of the plant through the vascular system (Milewska-Hendel et al. 2019). Foliar uptake presents another route for NP entry into plants.

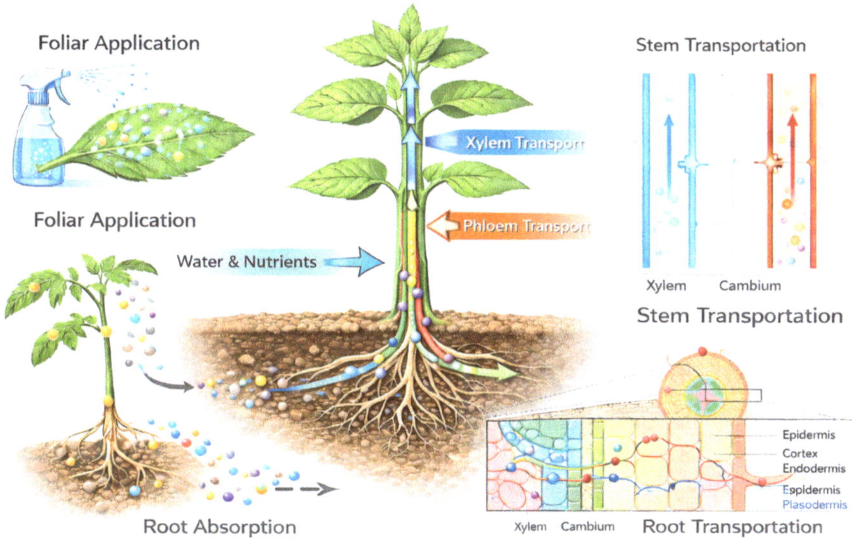

Fig. 11.1 A visual representation of how NPs get absorbed and transported within plants via foliar application or root exposure treatment. (Reproduced with permission from (Wang et al. 2023a, b). Copyright, *Materials*)

This can occur through natural pores like stomatal openings that are used for gas exchange. NPs may enter leaves through these openings, especially when they are open (Fig. 11.1). Numerous studies have shown that NPs are transported to various plant tissues. Researchers have proposed that, in addition to root absorption, NPs can also enter plants through the stomatal pathway, enabling their uptake and translocation via leaf structures. Moreover, some NPs may be able to penetrate the leaf cuticle, particularly if it is damaged or if the NPs are small and lipophilic. In certain plant species, specialized structures called hydathodes can provide another entry point for NPs (Hong et al. 2021). Properties of NP like size, shape, charge, and chemical composition are crucial in deciding their behavior within plant tissues (Rico et al. 2011). Similarly, plant-specific factors like root architecture, leaf surface properties, and vascular structure influence the extent and pattern of NP distribution in plant system (Vignesh et al. 2024).

11.2.2 Factors Affecting the NMs Uptake

The absorption and movement of NMs in plants are determined by a variety of factors, which collectively influence the behavior and effects of these materials within plant systems. The characteristics of the particles are key to this process. For instance, particle size is a major factor, with smaller nanoparticles typically exhibiting a higher uptake rate. Ma et al. (2010) demonstrated that gold NPs less than 50 nm were more easily absorbed by plants compared to larger particles. The shape of NPs also changes their interface with plant surfaces and cellular uptake mechanisms, as evidenced by Judy et al. (2012), who found that rod-shaped gold NPs had lower uptake rates in tobacco plants compared to spherical particles of similar size. Surface charge is another critical factor, with positively charged NPs often demonstrating higher uptake rates from the negatively charged cell walls and membranes of plant. Liu et al. (2019) found that positively charged cerium oxide NPs had greater uptake in cucumber plants compared to negatively charged or neutral particles. However, Wang et al. (2016a), found that while negatively charged gold nanoparticles (Au NPs) exhibited a slower initial accumulation when compared to the positively charged NPs. However, over time, they became increasingly prominent within the xylem, indicating that surface charge has a significant role not only in nanoparticle uptake but also in their movement and distribution within plant tissues. This suggests that negatively charged NPs may experience reduced adhesion to root surfaces or cell walls during uptake but are more efficiently transported through the vascular system. The delayed but steady translocation to the xylem highlights the complex interactions between NPs' surface properties, plant physiology, and internal transport mechanisms, which could have significant implications for nanoparticle-based nutrient delivery or stress mitigation strategies in plants.

The chemical nature of NPs significantly affects their solubility, reactivity, and potential for biotransformation within plants. Selivanov et al. (2017) observed that silver NPs were more toxic to *Arabidopsis thaliana* than gold NPs of similar size, maybe because of the release of ionic silver from NPs. This difference in toxicity emphasizes the importance of judging the specific chemical attributes of NPs when studying their impacts on plant systems. Furthermore, the dissolution rate of NPs can greatly influence their bioavailability and toxicity. As Dimkpa et al. (2012) found, the release of zinc ions from ZnO NPs was a key factor in their toxicity to wheat seedlings, with the dissolution rate being influenced by particle size and soil pH.

Plant species variability is a crucial factor, as different species may have varying capacities for NP uptake and translocation due to variation in root morphology, leaf structure, and physiological characteristics. Roche et al. (2018) studied NP uptake across various crop species and found significant differences, with zucchini showing high uptake of cerium oxide NPs. This variability extends to different cultivars within the same species, as shown by Rico et al. (2015), who observed differences in cerium oxide NP uptake and translocation among various rice cultivars. The physiological state of the plant is also important in stress conditions, and nutrient status affects NP uptake and distribution (Rasheed et al. 2022).

Environmental dynamics, for example temperature, pH, and organic matter can influence NP movement in the soil and ultimately affect the availability for plant uptake. Sun et al. (2023) demonstrated that soil pH significantly impacts the dissolution and plant uptake of ZnO NPs. Additionally, Baalousha et al. (2008) revealed that the organic matter of soil can alter NPs' surface properties and aggregation state, thereby affecting their bioavailability to plants. Soil texture and composition also play a role, with clay content and mineral composition influencing NP retention and mobility in soil (Cornelis et al. 2014). Climate factors such as rainfall and temperature can affect NP behavior in the environment, with increased precipitation potentially leading to greater NP mobility and uptake by plants (Servin and White 2016).

The time of exposure to NPs can affect the level of uptake and potential for accumulated quantity in plant organs. Khan et al. (2021) suggested that continuing exposure to NPs could cause significant accumulation in plant tissues over time, potentially affecting plant metabolism and growth. This chronic exposure scenario is particularly relevant for understanding the long-term impacts of NPs in agriculture and natural environments. The mode of NP application also influences uptake and distribution, with foliar application often resulting in different patterns of accumulation compared to root exposure (Wang et al. 2023a, b).

Furthermore, the contact and reaction of NPs with root exudates, including organic acids, flavonoids, and amino acids, can influence their stability and aggregation, thereby affecting their bioavailability to plants. These exudates can either enhance or inhibit NP uptake by modifying the local chemical environment around the root system. In one study, Dimkpa et al. (2013) found that root exudates from wheat and rape plants could stabilize silver NPs, potentially increasing their bioavailability. The rhizosphere microbiome also plays a significant role in mediating NPs-plant interactions, with soil microorganisms potentially altering NP properties or facilitating their uptake by plants (Jamil et al. 2022). Comprehension of these factors is important for interpreting metabolomic data and elucidating the mechanisms by which NPs induce metabolic changes in plants.

11.3 Metabolomics Approaches in Studying Plant-NMs Interactions

The small molecules that are involved in cellular metabolism, are called metabolites; and their study is termed as metabolomics. It helps identify the chemical signatures of specific cellular processes by capturing all the metabolites in a sample originating from organism. Metabolome can change rapidly in response to various stimuli, making metabolomics a valuable tool for assessing metabolic adaptations and responses under different conditions (Nalbantoglu 2019; Hao et al. 2025).

Metabolites encompass a wide range of chemical substances like sugars, amino acids, lipids, organic acids, and nucleotides. These molecules are essential as building blocks for cellular components and as substrates or products in metabolic processes. By analyzing metabolite profiles, researchers can achieve a deeper knowledge of the physiological state of an organism and how it responds to environmental changes or stressors. This capability allows for a direct functional readout of cellular activity, complementing data from genomics and proteomics that focus on gene expression and protein levels (Nalbantoglu 2019; Judge and Dodd 2020).

11.3.1 Analytical Techniques

11.3.1.1 Nuclear Magnetic Resonance (NMR) Spectroscopy

NMR is an improved analytical method for metabolomics research, offering several key advantages. NMR is highly reproducible and quantitatively accurate, allowing for precise measurement of metabolite concentrations. It can simultaneously detect a wide variety of metabolites that may be challenging to analyze with other methods (Emwas et al. 2019; Nagana Gowda and Raftery 2021). Additionally, NMR requires minimal sample preparation, enabling the analysis of intact biofluid samples or tissues without destructive processing (Nagana Gowda and Raftery 2021). NMR typically detects metabolites in the micromolar range, while mass spectrometry can often detect compounds at nanomolar concentrations. This sensitivity difference means that NMR-based metabolomics studies usually identify and quantify 50–200 metabolites, whereas mass spectrometry approaches can potentially detect over 1000 metabolites (Emwas et al. 2019). Although it has its limitations, NMR is still an important tool in metabolomics because of its non-destructive nature, high reproducibility, and capacity to offer comprehensive structural insights into metabolites (Zia et al. 2019).

11.3.1.2 Gas Chromatography-Mass Spectrometry (GC-MS)

This is a great analytical technique commonly used for plant metabolite profiling. It has high sensitivity and consistency and enables the researchers for the detection and measurement of a broad array of metabolites in plants (Choudhury et al. 2022). It can identify and quantify over 200 compounds in complex biological mixtures and can also detect many additional unknown signals through appropriate data processing workflows (Abadie et al. 2022). One of the major strengths of GC-MS is its excellent separation capabilities, making it ideal for analyzing complex mixtures of plant metabolites. The technique is particularly useful for volatile and semi-volatile compounds (Choudhury et al. 2022). GC-MS also provides access to extensive spectral libraries for metabolite identification, enhancing the ability to annotate and characterize metabolites in plant samples (Abadie et al. 2022). However, GC-MS typically requires derivatization of non-volatile metabolites to make them suitable for gas chromatography analysis (Abadie et al. 2022). This step can introduce additional variability and complexity to the analysis. Despite this limitation, GC-MS remains a cornerstone technique in plant metabolomics due to its ability to provide comprehensive metabolite profiles with high reproducibility and quantitative accuracy (Wang et al. 2015). Recent advancements, such as high-resolution exact mass GC-MS, have further improved the technique's capabilities for plant metabolic profiling, offering enhanced resolution of isotopic patterns and improved identification of sulfur-containing metabolites (Abadie et al. 2022; Mutha et al. 2025).

11.3.1.3 Liquid Chromatography-Mass Spectrometry (LC-MS)

LC-MS is an important analytical method for plant metabolomics, offering several key advantages for studying an array of metabolites in complex plant samples. LC-MS posseses high sensitivity and broad metabolite coverage in a single analysis. According to Fiehn (2016) LC-MS can identify and quantify over 200 compounds per study in plant samples, with the potential to detect more than 300 additional unidentified signals through appropriate data processing workflows. A primary benefit of LC-MS is its capacity to analyze thermally labile compounds, which is particularly important for studying plant hormones and other heat-sensitive metabolites. This capability sets LC-MS apart from techniques like GC-MS, which may require derivatization of certain compounds. Additionally, LC-MS typically requires less sample preparation compared to GC-MS, simplifying the analytical procedure and minimizing the possibility of sample loss or degradation (Kılınc et al. 2023).

11.3.1.4 Fourier-Transform Infrared (FTIR) Spectroscopy

This spectroscopy has emerged as a valuable complementary method in plant metabolomics, offering unique advantages for the analysis of plant metabolites. FTIR provides analysis of functional groups found in metabolites, allowing for rapid and non-destructive analysis of plant samples (Berthomieu and Hienerwadel 2009; Ribeiro da Cunha et al. 2020). One of the key strengths of FTIR in plant metabolomics is its minimal sample preparation requirements. FTIR can often be applied directly to intact plant tissues or crude extracts, reducing the risk of sample alteration during preparation. This feature is particularly advantageous when studying volatile compounds or when preserving the native state of plant metabolites is crucial (Kumar et al. 2016). The development of attenuated total reflection (ATR) FTIR and synchrotron-based FTIR micro spectroscopy has improved spatial resolution and sensitivity, allowing for more detailed analysis of plant tissues (Kumar et al. 2016). Additionally, the combination of FTIR data with multivariate statistical analysis techniques has improved the extraction of meaningful biological information from complex spectral data.

11.4 Metabolomics Workflows

The workflow of metabolomics in studies of nano-pollution involves a systematic approach to understanding how NMs affect biological systems, particularly in plants. This process typically encompasses several key steps, each critical for obtaining reliable and interpretable data. The first step in the metabolomics workflow is to establish an appropriate experimental model. Researchers must choose suitable model organisms that reflect the characteristics of the external stressors-such as specific types of ENMs-being studied. Depending on the research goals, this may involve selecting plants that are either directly exposed to environmental pollutants or those cultivated under controlled laboratory conditions. The concentration and exposure of ENMs are also determined at this stage to ensure relevance to real-world scenarios (Kim and Kang 2021). Once the experimental model is established, the next step is sample collection. This involves harvesting tissues or fluids from the model organisms at predetermined time points during or after exposure to ENMs. It is essential to collect samples in a manner that minimizes degradation or alteration of metabolites, ensuring that the collected material accurately reflects the physiological state of the organism (Silva et al. 2023; Kim and Kang 2021).

Following sample collection, researchers proceed with metabolite extraction. This step is crucial as it aims to isolate metabolites from complex biological matrices. The extraction method must be standardized depending on the polarity of the target metabolites to ensure high recovery rates and minimize losses (Schnackenberg et al. 2012; Barthwal and Mahar 2024; Ghenabzia et al. 2023). After extraction, metabolites are analyzed using complex analytical techniques including LC-MS, GC-MS, and high-resolution NMR spectroscopy. Once data is acquired, it undergoes rigorous processing using bioinformatics tools to identify and quantify significant changes in metabolite levels. This step involves statistical analysis to differentiate between control and treated groups, identifying biomarkers associated with exposure to ENMs. Advanced software can assist in visualizing metabolic pathways impacted by NPs exposure, enabling a more thorough understanding of the responsible mechanisms (Chen et al. 2022; Xia and Wishart 2011). The final step in the workflow involves interpreting the results in conjunction with phenotypic assays and other omics data (e.g., genomics and proteomics). This integrative approach helps to elucidate the biological significance of altered metabolite profiles and their relationship to physiological changes observed in the model organisms. Understanding these interactions can provide insights into adverse outcome pathways (AOPs) related to nanotoxicity (Jendoubi 2021; Sanches et al. 2024). Metabolomics data assisted with additional "omics" data (e.g., transcriptomics and proteomics), can be used by scientists to formulate the comprehensive metabolic networks, identifying key enzymes and regulatory pathways affected by NPs. Software tools like MetaboAnalyst and KEGG Pathway Mapping enable pathway enrichment analysis, which helps pinpoint metabolic pathways most impacted by nano pollution (Steuer et al. 2019; Jiao et al. 2022; Eichner et al. 2014). Network analysis approaches can reveal relationships between metabolites and help identify key nodes or hubs in the metabolic response to NP exposure. These analytical and statistical tools help researchers identify significant metabolic changes and elucidate the underlying biochemical pathways affected by NP exposure.

11.5 Effects of Nano Pollution on Plant Metabolism

Primary metabolism encompasses the essential biochemical processes required for plant growth, development, and survival. Metabolomic studies have revealed that NPs can significantly alter various aspects of primary metabolism in plants. This section explores the effects of nano pollution on key primary metabolic processes.

11.5.1 Photosynthesis and Carbon Metabolism

Photosynthesis is a fundamental process in plants, and its alteration can have far-reaching consequences for plant growth and output. Metabolomic research has found that NPs can affect photosynthetic processes in several ways. The effects of NPs pollution on plant chlorophyll content and photosynthetic efficiency are complex and depend on several factors, including the type of NPs, their concentration, and the plant species involved (Sembada and Lenggoro 2024). Ag NPs have been found to decrease chlorophyll content in various plant species, potentially reducing photosynthetic efficiency. A study by Majumdar et al. (2014) reported a significant reduction of total chlorophyll content when exposed to 1000 μg/mL of Ag NPs after 12 days of treatment. This decrease in chlorophyll content was found to be concentration-dependent, with higher concentrations of Ag NPs causing more severe reductions.

Structural damage to chloroplasts and associated components is one of the primary consequences of NP exposure, disrupting their functionality and reducing photosynthetic efficiency (Rasheed et al. 2022; Djanaguiraman et al. 2024). NPs have been shown to accumulate in chloroplasts, damaging the photosynthetic apparatus and leading to reduced photosynthetic rates. The interaction of NPs with plant cells can trigger ROS production via NADPH oxidases, which may further compromise photosynthetic function (Fig. 11.2) (Wang et al. 2023a, b; Marslin et al. 2017). However, the effects of NPs on chlorophyll content and photosynthesis are not universally negative. Some researchers have found that certain NPs, at lesser dose, may stimulate plant growth and chlorophyll content. For instance, research on common bean (*Phaseolus vulgaris*) and corn (*Zea mays*) demonstrated that Ag NPs at concentrations up to 60 ppm led to enhances in chlorophyll A and B content. In common bean, as compared to control, chlorophyll A and B increased by 49% and 33%, respectively, at 60 ppm Ag NPs concentration (Salama 2012). TiO_2 NPs, for instance, demonstrate a dual effect. At low concentrations, they can enhance photosynthetic activity by improving light absorption and increasing chlorophyll content. However, at higher concentrations, TiO_2 NPs get collected in chloroplasts, and results in chlorophyll degradation and a decline in photosynthetic efficiency (Li et al. 2020). Similarly, ZnO NPs inhibit chlorophyll biosynthesis in *Arabidopsis thaliana*, reducing the plant's capacity for carbon fixation and overall photosynthesis (Wang et al. 2016b).

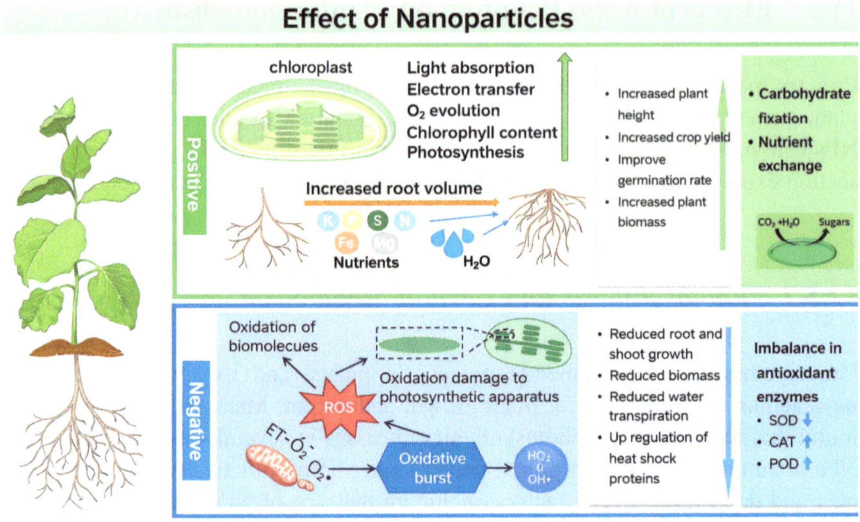

Fig. 11.2 A schematic illustration of the physiological effects of nanomaterials in plants. (Reproduced with permission from Wang et al. (2023a, b). Copyright 2019, *Materials*)

The mechanisms behind these effects include a decrease in chlorophyll content, impairing light-harvesting efficiency, and the inhibition of key enzymes. Oxidative stress caused by NPs further exacerbates structural damage to chloroplasts, disrupting photosynthetic pathways and reducing carbohydrate production. This results in altered levels of sugars, starch, and related metabolites, shedding light on how NPs affect energy metabolism. These findings emphasize the need for cautious evaluation of NPs applications in agriculture to balance their potential benefits against their risks (Djanaguiraman et al. 2024; Vatankhah et al. 2023). The influence of NPs on chlorophyll content and photosynthetic efficiency depends on the type of NP. While Ag NPs often show inhibitory effects at higher concentrations, other NPs like iron-based and titanium-containing nanomaterials have increased chlorophyll and growth of the plant under appropriate conditions (Li et al. 2024). This stresses the intricacy of nanoparticles interactions with plant physiology and the need for careful consideration of NP type and concentration in agricultural applications.

Metabolomic analyses after plants' exposure to NPs have revealed changes in the levels of chlorophyll precursors and degradation products, indicating alterations in chlorophyll biosynthesis and turnover. These changes can have cascading effects on photosynthetic efficiency and overall plant health. For example, increases in malondialdehyde (MDA) content, an indicator of oxidative stress, have been reported in plants treated with increasing concentrations of Fe_2O_3 NPs, coinciding with decreases in chlorophyll content (Feng et al. 2022). In addition, the effects of NPs on chlorophyll content and photosynthetic efficiency are not only depend on NP concentration. Factors such as particle size, plant species, and developmental stage also play crucial roles in determining the outcome of NP exposure. For instance, a study on tomato plants showed that green-synthesized Ag NPs at low concentrations (5 and 10 ppm) significantly increased chlorophyll content and carotenoid production (Ansari et al. 2023).

11.5.2 Carbon Fixation and Calvin Cycle

Nano pollution can significantly impact carbon fixation and the Calvin cycle in plants, with effects that vary based on the type and concentration of NPs involved. CeO_2 NPs have improved photosynthetic carbon assimilation in some plants. A study on wheat (*Triticum aestivum L.*) low concentrations of CeO_2 nanoparticles were observed to positively impacted the net photosynthesis rate, which was linked to an increase in stomatal conductance within the leaves. This improvement in photosynthetic activity may be due to the stimulation of heat shock proteins (HSPs) induced by CeO_2 nanoparticles (Pietrzak et al. 2024). Metabolomic studies have revealed changes in the levels of Calvin cycle intermediates in plants exposed to various NPs. Research using limited proteolysis small molecule mapping (LiP-SMap) identified interactions between metabolites and Calvin cycle enzymes across different bacterial species. Notably, the Calvin cycle intermediate glyceraldehyde-3-phosphate (GAP) was found to interact with many proteins in all four species studied (Sporre et al. 2023). Alterations in the expression of genes encoding key Calvin cycle enzymes have been observed, potentially affecting carbon fixation rates. While the search results do not provide specific examples of gene expression changes, they do indicate that NPs can affect various physiological processes in plants, including photosynthesis (Rasheed et al. 2022). It is worth noting that the effects of NPs on carbon fixation and the Calvin cycle are complex and can differ based on the NPs type, concentration, and the plant. For example, while some studies show positive effects of CeO_2 NPs on photosynthesis at lower concentrations, higher doses can be toxic and decrease plant biomass production (Pietrzak et al. 2024).

11.5.3 Photorespiration

NMs can alter photosynthetic efficiency, which may indirectly affect photorespiration rates. For example, CeO_2 NPs enhanced carbon assimilation by stimulating the Rubisco enzyme. This could potentially reduce photorespiration, as more efficient carbon fixation might decrease the oxygenation reaction of Rubisco (Djanaguiraman et al. 2024). Some NPs can damage chloroplasts and vascular tissues. Since photorespiration occurs in chloroplasts, peroxisomes, and mitochondria, damage to these organelles could disrupt normal photorespiration processes. NPs can affect the electron transport system in photosynthesis. This could potentially influence the NADPH and ATP balance in cells, which might indirectly impact photorespiration rates (Djanaguiraman et al. 2024). Certain NPs can enhance ROS scavenging ability in chloroplasts. This could potentially affect photorespiration, as it is partly a mechanism for protecting against oxidative damage in high-light conditions. Some NPs can enter plant chloroplasts and move to the photosystem-II reaction center and increase electron transmission and light absorption. This enhanced photosynthetic efficiency might indirectly influence photorespiration rates (Rasheed et al. 2022).

11.5.4 Respiration and Energy Metabolism

NMs significantly impact mitochondrial function by stimulating oxidative stress and disturbing the electron transport chain (ETC), leading to detrimental effects on cellular metabolism and energy production. Exposure to NMs like ZnO NPs impairs the activity of mitochondrial complexes, particularly complex I and complex III, disrupting the ETC and reducing ATP synthesis, which is essential for cellular energy metabolism. Additionally, NPs generate oxidative stress within mitochondria, resulting in the elevated level of ROS, causing mitochondrial damage, lipid peroxidation, and compromised mitochondrial integrity, further impairing mitochondrial functionality (Anand et al. 2023; Misra et al. 2023). NMs also interfere with the tricarboxylic acid (TCA) cycle by inhibiting key enzymes, disrupting ATP generation, and altering the levels of intermediates such as citrate, succinate, and malate. These metabolic changes can be detected through metabolomic profiling, providing insight into the extent of NP-induced mitochondrial dysfunction. Moreover, exposure to ZnO NPs has been correlated with structural changes in mitochondria, including increased fission and altered morphology, which reflect a stress response that exacerbates dysfunction and reduces cellular viability. These disruptions have profound implications for energy metabolism, as reduced ATP production hampers energy-dependent cellular activities, impairing growth and function. Furthermore, disruptions in the TCA cycle led to metabolic imbalances, affecting energy production and the synthesis of essential biomolecules, emphasizing the need for careful assessment of NP exposure on cellular health (Misra et al. 2023; Tucci et al. 2013).

11.5.5 Amino Acid Synthesis and Metabolism

NMs significantly influence nitrogen metabolism in plants, a critical process for amino acid and protein synthesis. Exposure to NPs can disrupt key enzymatic activities and alter amino acid profiles, ultimately impairing nitrogen utilization. Studies indicate that certain NPs, such as manganese NPs (Mn NPs), enhance nitrogen assimilation by increasing the enzyme activities that are vital for nitrogen uptake and utilization (Pradhan et al. 2014). This improved efficiency is often superior to that of traditional manganese salts. However, other types of NPs may inhibit these enzymes, reduce nitrogen assimilation and negatively affect plant growth. NPs exposure can substantially change the amino acid profiles in plants. Multi-walled carbon nanotubes (MWCNTs) and nano-SiO_2 have enhanced the activity of essential enzymes responsible for nitrogen metabolism in maize. The enzymes such as glutamine synthetase (GS), glutamate decarboxylase (GAD), glutamate synthase (GOGAT), and glutamate dehydrogenase (GDH) (Hao et al. 2023) have been increased. This enhancement in enzyme activity can lead to increased synthesis of amino acids and improved nitrogen utilization. The impact on amino acid synthesis can vary depending on the dose and kind of nanomaterials. In case of Ag NPs have been found to affect the free amino acids content in tomato plants (Karami Mehrian et al. 2015). The impact of NPs on amino acid synthesis can vary depending on the concentration used. For instance, in maize, the promoting effect of MWCNTs was optimal at 800 mg/L, while for nano-SiO_2, it was 1500 mg/L (Hao et al. 2023).

An overall rise in antioxidant enzyme activity and free amino acid levels in tomatoes has been identified as a key adaptive mechanism to counteract oxidative stress caused by silver nanoparticles (Ag NPs). This response involves the upregulation of enzymes such as superoxide dismutase (SOD), catalase (CAT), and peroxidase (POD), which work collectively to neutralize reactive oxygen species (ROS) and prevent cellular damage. This suggests that changes in amino acid profiles may also be linked to stress response mechanisms in plants exposed to NPs (Guzmán-Báez et al. 2021). Furthermore, NP-induced oxidative stress often leads to reduced nitrogen utilization efficiency, which manifests as stunted growth and decreased photosynthetic efficiency due to impaired synthesis of proteins and other nitrogenous compounds. Metabolomic approaches are invaluable for quantifying these changes and identifying specific disruptions in nitrogen assimilation pathways, providing deeper insights into the physiological impacts of nanotoxicity on plants. These findings emphasize the dual nature of NPs in agriculture, where their effects can vary from beneficial to detrimental based on their type and concentration (Rasheed et al. 2022; Zhang et al. 2024). For instance, CeO_2 NPs at dose of 125 and 500 mg/L significantly increased H_2O_2 concentration in rice seedling roots, leading to lipid peroxidation and electrolyte leakage (Djanaguiraman et al. 2024). Metabolomic studies revealed NPs impact on nitrogen transporters in Arabidopsis; four NRT2 transporters account for 95% of uptake of N under low NO_3 concentrations (Zayed et al., 2023). NPs may alter the expression or function of these transporters, affecting overall nitrogen uptake.

11.5.6 Protein Synthesis and Stress Response

NMs exposure can significantly impact protein synthesis and stress responses in plants through various mechanisms (Rasheed et al. 2022). As mentioned earlier, NPs exposure increases ROS in plants, leading to oxidative stress. This ROS-induced stress response activates various defense mechanisms, including increased production of antioxidant enzymes such as superoxide dismutase (SOD) to scavenge ROS

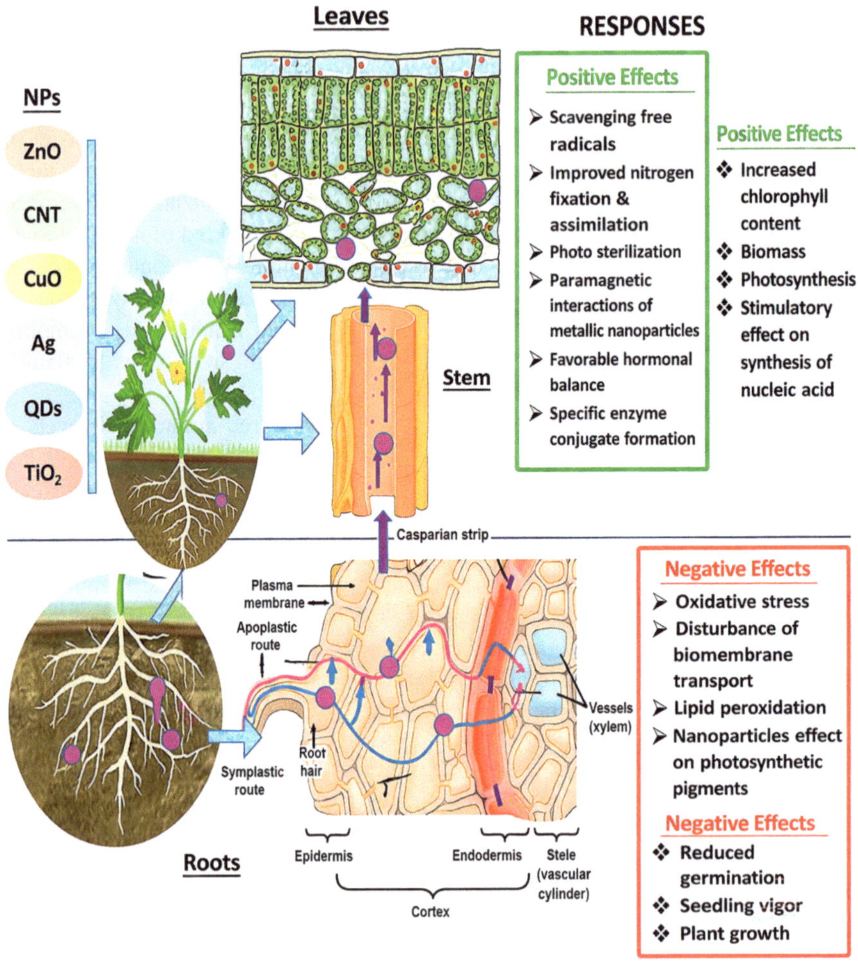

Fig. 11.3 A schematic diagram illustrating plant-NPs interactions and their various responses. Red represents the negative impacts of NPs on plants, while green highlights the beneficial effects of NPs on plant cellular processes. (Reproduced from (Munir et al. 2023). Copyright 2023 Agronomy Journal)

molecules. The distortion in electron flow results in ROS production in chloroplasts and mitochondria, with oxygen taking escaped electrons to generate superoxide radicals (Munir et al. 2023). NP-induced stress can lead to substantial gene expression changes and protein profiles. Studies have shown upregulation of genes involved in hormonal responses, vacuolar proton or cation exchange, pathogen-responsive genes and antioxidant systems (Fig. 11.3). Conversely, genes related to photosynthesis were often downregulated. Proteomic analysis revealed that Ag NPs can alter membrane proteins of vacuoles and endoplasmic reticulum, suggesting these organelles as primary targets for Ag NPs (Munir et al. 2023).

11.6 Effects of Nano Pollution on Secondary Metabolism

Nano pollution has been shown to significantly impact plant secondary metabolism, as revealed by numerous metabolomic studies. These effects are particularly evident in the production of phenolic compounds, flavonoids, terpenoids, and alkaloids, which are an important biomolecule in plant defense and stress responses. Phenolic compounds, including flavonoids, often show increased production in plants exposed to NPs. This increase is likely part of the plant's stress response mechanism. For instance, studies have found that exposure to copper oxide NPs can dramatically increase the content of p-coumaric acid in cucumber plants (Zhang et al. 2022). The biosynthesis of phenolic compounds is closely linked to the phenylpropanoid pathway, where key enzymes such as phenylalanine ammonia-lyase (PAL) and chalcone synthase (CHS) show altered activity in NP-exposed plants (Lala 2021). Flavonoids, a subclass of phenolic compounds, also exhibit changes in response to NP exposure. Total flavonoid content often increases, particularly under conditions of NP-induced oxidative stress (Rasheed et al. 2022). Specific flavonoids, such as quercetin and kaempferol glycosides, have shown altered concentrations in *A. thaliana* treated silver NPs (Lala 2021). These changes in metabolites are regularly associated with modifications in the gene expression that are involved in flavonoid biosynthesis. Terpenoid metabolism is another area affected by NP exposure. Changes in the emission of volatile organic compounds (VOCs), many of which are terpenoids, have been observed in NP-exposed plants. Non-volatile terpenoids, including carotenoids and phytosterols, also show changes in their levels in response to NP exposure (Marslin et al. 2017). Furthermore, NPs can alter the activity of enzymes involved in secondary metabolism. For example, studies have proved that NPs can affect the activity of nitrate reductase, which plays a role in nitrogen metabolism and indirectly influences secondary metabolite production (Vinogradova et al. 2023).

11.7 Mechanisms of NMs-Induced Metabolic Changes

Knowledge of the mechanisms involved in NPs induced metabolic changes in plants is crucial for predicting and mitigating their potential impacts. Several key mechanisms have been identified through metabolomic and complementary studies. NMs can induce a burst of ROS in plant cells, which is linked to oxidative stress. This oxidative stress may result in damage to cellular structures, including membranes, proteins, and, ultimately, impairing metabolic functions and growth (Marslin et al. 2017). The disruption of ROS homeostasis is critical; while moderate ROS levels can stimulate secondary metabolism and growth, excessive ROS can result in phytotoxicity and reduced plant health (Fig. 11.4) (Samynathan et al. 2023). By analyzing the metabolite composition of plants exposed to NPs, researchers can identify specific changes associated with oxidative stress. For example, increased levels of secondary metabolites such as flavonoids and phenolics have been

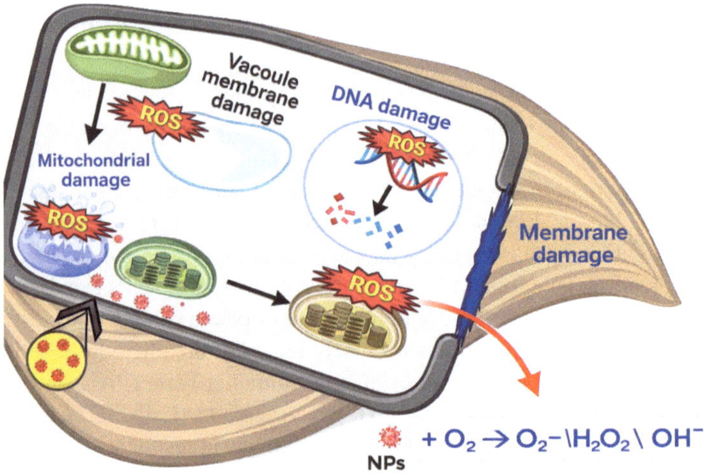

Fig. 11.4 A schematic depicting the oxidative damage caused by NPs. (Reproduced from Djanaguiraman et al. (2024) by permission licensed under CC BY 4.0)

observed in response to ROS production triggered by NPs. Silver NPs (Ag NPs) induce oxidative stress in plants such as tomato. The exposure to Ag NPs leads to increased ROS production, resulting in lipid peroxidation and damage to cellular membranes. This oxidative damage can impair metabolic functions, leading to stunted growth and reduced photosynthetic efficiency (Yan and Chen 2019; Al-Khayri et al. 2023; Komazec et al. 2023). A study found that exposure to CdS-NPs led to significant alterations in the metabolite profiles of broad bean plants. Specifically, there was an increase in flavonoids and other antioxidant metabolites, which are believed to play a crucial role in scavenging of ROS generated by the NPs. Metabolomics offers a promising tool for early detection of Cd contamination in soil (Fig. 11.4). The study suggested that these compounds could serve as markers for stress induced by CdS-NPs in soil, reflecting the plant's defense mechanism (Tian et al. 2020).

Metal and metal oxide NPs can directly generate ROS through surface reactions or by releasing metal ions. For instance, Ag NPs induce ROS production in *A. thaliana* and tomato, leading to increased levels of hydrogen peroxide and superoxide (Marslin et al. 2017). This oxidative damage can impair metabolic functions, leading to reduced growth and photosynthetic efficiency (Yan and Chen 2019; Al-Khayri et al. 2023; Komazec et al. 2023). Plants often respond to NP-induced oxidative stress by upregulating the production of antioxidative metabolites. Metabolomic analyses have revealed increases in phenolic compounds, particularly flavonoids, as well as ascorbic acid, glutathione, and carotenoids in plants exposed to various NPs (Marslin et al. 2017; Kumari et al. 2024). For example, treatment with CuO NPs has been shown to increase the production of phenolic compounds in maize plants (Li et al. 2024). Changes in redox homeostasis are another important aspect of plant responses to NP exposure. Metabolomic studies have revealed alterations in the ratios of reduced to oxidized forms of compounds like glutathione and ascorbate, indicating shifts in cellular redox state (Bao et al. 2024). These changes are often accompanied by alterations in the expression of genes involved in redox homeostasis, such as those encoding glutathione reductase and ascorbate peroxidase (Marslin et al. 2017). Lipid peroxidation, a consequence of oxidative stress, has also been observed in plants exposed to NPs. Metabolomic analyses have shown increases in lipid peroxidation products, such as malondialdehyde, in various plant species treated with NPs. These changes can potentially affect membrane integrity and function, leading to broader impacts on plant physiology (Li et al. 2024).

NMs can significantly impact plant nutrient uptake, translocation, and overall nutrient homeostasis. This disruption can lead to deficiencies of essential nutrients that are necessary for growth and metabolic processes. For example, studies have shown that NPs can alter the absorption efficiency of nutrients like nitrogen and phosphorus (Rasheed et al. 2022). ZnO nanoparticles can block essential ion channels in plant roots, leading to nutrient imbalances, and can reduce the uptake of vital nutrients like potassium and calcium, which are crucial for maintaining cellular functions and overall plant health (Seleiman et al. 2023; Ahmed et al. 2024). Metal-based NPs can release ions, potentially causing toxicity or nutrient imbalances. Metabolomic analyses have revealed changes in metal-binding compounds like phytochelatins and metallothioneins in plants exposed to metal-based NPs (Zhang et al. 2024). For instance, CuO NPs have increased the phenolic compounds in maize plants, which may be part of its defense against metal toxicity (Makarov et al. 2014). Some NPs can physically interfere with nutrient uptake mechanisms in roots. Studies have shown that NPs can alter root morphology and function, impacting nutrient and water uptake (Sembada and Lenggoro 2024). For example, excessive NPs accumulation can obstruct nutrient transport systems, leading to nutrient deficiencies or toxicities (Sembada and Lenggoro 2024). NPs can affect nutrient translocation within plants. A study on barley demonstrated that in both root and leaf tissues, Fe, Zn, and Cu NPs notably increased the content of these elements (Sembada and Lenggoro 2024). This highlights the potential of NPs to alter nutrient distribution and storage within plants.

11.8 Conclusion

Plant nano pollution metabolomics faces numerous complex challenges that researchers must address for better understanding of NPs' interactions with plant metabolism. The effects of NPs on plants are highly variable, depending on factors such as NPs' properties, concentration, plant species, developmental stage, and environmental conditions (Rasheed et al. 2022). This complexity necessitates carefully designed experiments that account for multiple variables simultaneously. The significant metabolic plasticity exhibited by plants in response to NP exposure results in variation of metabolites, making it difficult to identify consistent biomarkers of NPs exposure. Accurately extracting and identifying metabolites from complex plant matrices remains a significant hurdle, particularly for novel or modified metabolites produced in response to NP exposure (Wang et al. 2023a, b). To address this, researchers are evaluating advanced UHPLC-HRMS data analysis tools such as MS-DIAL, XCMS, and MZmine for their performance in analyzing both targeted and untargeted metabolomics in plant samples (Wang et al. 2023a, b). The high-dimensional nature of metabolomics data necessitates sophisticated statistical approaches to identify meaningful patterns and avoid false positives, especially when dealing with the complex interactions between NPs and plant metabolism (Di Minno et al. 2022; Chen et al. 2022). There is also a pressing need for specialized bioinformatics tools and standardized workflows to process and interpret plant metabolomics data in the context of nano pollution studies, including the development of comprehensive databases like Spektraris for both mass spectral and NMR data of plant metabolites (Silva et al. 2023). Additional challenges in this field include increasing analytical coverage of the plant metabolome, integrating metabolomics with other -omics approaches, improving spatial resolution through new imaging tools like MALDI-MSI-PTMA, capturing temporal dynamics of metabolic changes, and translating results from controlled laboratory studies to real-world environmental conditions (Huang et al. 2023). Addressing these multifaceted challenges will be important for advancing our knowing of plant-NP interactions and developing more sustainable approaches to agriculture in the face of increasing nanopollution.

Future directions in plant nano pollution metabolomics are poised to significantly advance our understanding of NPs' interactions with plant systems and their broader ecological impacts. A key focus will be on developing a more mechanistic understanding of how NPs affect specific metabolic pathways and gene expression in plants. This will likely involve integrating metabolomics with other -omics approaches, such as genomics, transcriptomics, and proteomics, to provide a comprehensive view of plant responses to nano pollution. Researchers will continue to refine both targeted and untargeted metabolomics approaches, leveraging advanced analytical techniques like NMR spectroscopy and mass spectrometry to capture the full spectrum of metabolic changes induced by NP exposure. An exciting avenue of research is the exploration of nano-enabled immunomodulation, where engineered nanomaterials could applied to improve plant disease resistance. This goes together with the development of nanoprobe technologies for real-time monitoring of key biomolecular changes in plant immunity and stress responses. Creating broad profiles of genes, proteins, and metabolites that contribute to plant stress resistance, particularly in the context of salinity and drought stress, will be crucial for developing more resilient crops in the face of climate change. As environmental concerns grow, there will be an increased focus on developing environmentally sustainable strategies that use nano-enabled approaches to enhance plant resilience against pathogens while minimizing ecological impact. Researchers will also investigate the impact of NPs on the production of specialized metabolites important for plant adaptation, survival, and potential health-promoting bioactivities. To ensure the comparability and reproducibility of results across studies, there will be efforts to standardize protocols for sample preparation, data acquisition, and data analysis in plant nano pollution metabolomics. Finally, long-term studies will be crucial to understand the chronic effects of nano pollution on plant metabolism and ecosystem functioning, providing valuable insights for sustainable agriculture and environmental management in an increasingly nano-enabled world. The future directions identified for plant nano pollution metabolomics are both exciting and necessary. By integrating metabolomics with other omics approaches, developing more sophisticated analytical and statistical methods, and focusing on real-world applications such as nano-enabled immunomodulation and stress tolerance, researchers can gain deeper insights into the complex interplay between NPs and plant metabolism. This knowledge will be crucial for developing sustainable agricultural practices, assessing environmental risks, and potentially harnessing the benefits of nanotechnology in crop improvement. As we are advancing, it will be essential to keep a balance between basic research and applied studies, ensuring that our growing understanding of plant-NP interactions translates into practical solutions for agriculture and environmental management. Standardization of methods and long-term studies will be key to building a robust and reliable body of knowledge in this field. Ultimately, the advancements in plant nano pollution metabolomics will not only enhance our fundamental knowledge of plant biology but also aid in focusing global food security and environmental sustainability in an increasingly nano-enabled world. As we continue to unravel the complexities of plant responses to nanomaterials, we move closer to harnessing this knowledge for the benefit of agriculture, ecosystems, and human society.

In conclusion, the field of plant nano pollution metabolomics stands at a critical juncture, poised to improve our understanding of plant-NP interactions and their broader ecological implications. The challenges faced by researchers in this field are big, ranging from the complexity of experimental designs to the intricacies of data analysis and interpretation. However, this creates an opportunity for innovations in advancement of analytical techniques, bioinformatics tools, and experimental approaches.

References

Aacharya R, Chhipa H (2020) Nanocarbon fertilizers: implications of carbon nanomaterials in sustainable agriculture production. In: Carbon nanomaterials for agri-food and environmental applications. Elsevier, pp 297–321

Abadie C, Lalande J, Tcherkez G (2022) Exact mass GC-MS analysis: protocol, database, advantages and application to plant metabolic profiling. 45(10):3171–3183

Ahmed M, Marrez DA, Rizk R, Abdul-Hamid D, Tóth Z, Decsi K (2024) Interventional effect of zinc oxide nanoparticles with *Zea mays* L. Plants when compensating irrigation using saline water. Nanomaterials (Basel, Switzerland) 14(16):1341. https://doi.org/10.3390/nano14161341

Aliferis KA, Jabaji S (2012) FT-ICR/MS and GC-EI/MS metabolomics networking unravels global potato sprout's responses to Rhizoctonia solani infection. PLoS One. 7(8):e42576

Al-Khayri JM, Rashmi R, Surya Ulhas R, Sudheer WN, Banadka A, Nagella P, Aldaej MI, Rezk AA-S, Shehata WF, Almaghasla MI (2023) The role of nanoparticles in response of plants to abiotic stress at physiological, biochemical, and molecular levels. Plants 12(2):292. https://doi.org/10.3390/plants12020292

Anand AS, Jain K, Chauhan A, Prasad DN, Kohli E (2023) Zinc oxide nanoparticles trigger dysfunction of mitochondrial respiratory complexes and repair dynamics in human alveolar cells. Toxicol Ind Health 39(3):127–137. https://doi.org/10.1177/07482337231152956

Ansari M, Ahmed S, Abbasi A et al (2023) Plant mediated fabrication of silver nanoparticles, process optimization, and impact on tomato plant. Sci Rep 13:18048. https://doi.org/10.1038/s41598-023-45038-x

Arbona V, Manzi M, de Ollas C, Gómez-Cadenas A (2013) Metabolomics as a tool to investigate abiotic stress tolerance in plants. Int J Mol Sci 14(3):4885–4911

Baalousha M, Manciulea A, Cumberland S, Kendall K, Lead JR (2008) Aggregation and surface properties of iron oxide nanoparticles: influence of pH and natural organic matter. Environ Toxicol Chem Int J 27(9):1875–1882

Bao L, Liu J, Mao T, Zhao L, Wang D, Zhai Y (2024) Nanobiotechnology-mediated regulation of reactive oxygen species homeostasis under heat and drought stress in plants. Front Plant Sci 15: 1418515

Barthwal R, Mahar R (2024) Exploring the significance, extraction, and characterization of plant-derived secondary metabolites in complex mixtures. Meta 14(2):119. https://doi.org/10.3390/metabo14020119

Berthomieu C, Hienerwadel R (2009) Fourier transform infrared (FTIR) spectroscopy. Photosynth Res 101:157–170

Carpita N, Sabularse D, Montezinos D, Delmer DP (1979) Determination of the pore size of cell walls of living plant cells. Science 205(4411):1144–1147. https://doi.org/10.1126/science.205.4411.1144. PMID: 17735052

Chen H (2018) Metal based nanoparticles in agricultural system: behavior, transport, and interaction with plants. Chem Spec Bioavailab 30(1):123–134

Chen Y, Li EM, Xu LY (2022) Guide to metabolomics analysis: a bioinformatics workflow. Metabolites 12(4):357. https://doi.org/10.3390/metabo12040357

Choudhury FK, Pandey P, Meitei R, Cardona D, Gujar AC, Shulaev V (2022) GC-MS/MS profiling of plant metabolites. Methods Mol Biol (Clifton, N.J.) 2396:101–115. https://doi.org/10.1007/978-1-0716-1822-6_9

Cornelis G, Hund-Rinke K, Kuhlbusch T, Van den Brink N, Nickel C (2014) Fate and bioavailability of engineered nanoparticles in soils: a review. Crit Rev Environ Sci Technol 44(24): 2720–2764

da Cruz TN, Savassa SM, Gomes MH, Rodrigues ES, Duran NM, de Almeida E et al (2017) Shedding light on the mechanisms of absorption and transport of ZnO nanoparticles by plants via in vivo X-ray spectroscopy. Environ Sci Nano 4(12):2367–2376

Di Minno A, Gelzo M, Caterino M, Costanzo M, Ruoppolo M, Castaldo G (2022) Challenges in metabolomics-based tests, biomarkers revealed by Metabolomic analysis, and the promise of the application of metabolomics in precision medicine. Int J Mol Sci 23(9):5213. https://doi.org/10.3390/ijms23095213

Dimkpa CO, McLean JE, Latta DE, Manangón E, Britt DW, Johnson WP et al (2012) CuO and ZnO nanoparticles: phytotoxicity, metal speciation, and induction of oxidative stress in sand-grown wheat. J Nanopart Res 14:1–15

Dimkpa CO, McLean JE, Martineau N, Britt DW, Haverkamp R, Anderson AJ (2013) Silver nanoparticles disrupt wheat (*Triticum aestivum* L.) growth in a sand matrix. Environ Sci Technol 47(2):1082–1090

Djanaguiraman M, Anbazhagan V, Dhankher OP, Prasad PVV (2024) Uptake, translocation, toxicity, and impact of nanoparticles on plant physiological processes. Plants 13(22):3137. https://doi.org/10.3390/plants13223137

Eichner J, Rosenbaum L, Wrzodek C, Häring HU, Zell A, Lehmann R (2014) Integrated enrichment analysis and pathway-centered visualization of metabolomics, proteomics, transcriptomics, and genomics data by using the InCroMAP software. J Chromatogr B 966:77–82

Emwas AH, Roy R, McKay RT, Tenori L, Saccenti E, Gowda GAN, Raftery D, Alahmari F, Jaremko L, Jaremko M, Wishart DS (2019) NMR spectroscopy for metabolomics research. Metabolites 9(7):123. https://doi.org/10.3390/metabo9070123

Etxeberria E, Gonzalez P, Baroja-Fernandez E, Romero JP (2006) Fluid phase endocytic uptake of artificial nano-spheres and fluorescent quantum dots by sycamore cultured cells: evidence for the distribution of solutes to different intracellular compartments. Plant Signal Behav 1(4): 196–200

Farooq A, Khan I, Shehzad J, Hasan M, Mustafa G (2024) Proteomic insights to decipher nanoparticle uptake, translocation, and intercellular mechanisms in plants. Environ Sci Pollut Res 31:18313–18339

Feng Y, Kreslavski VD, Shmarev AN, Ivanov AA, Zharmukhamedov SK, Kosobryukhov A, Yu M, Allakhverdiev SI, Shabala S (2022) Effects of iron oxide nanoparticles (Fe_3O_4) on growth, photosynthesis, antioxidant activity and distribution of mineral elements in wheat (*Triticum aestivum*) plants. Plants 11(14):1894. https://doi.org/10.3390/plants11141894

Fiehn O (2002) Metabolomics—the link between genotypes and phenotypes. In: Functional genomics, pp 155–171

Fiehn O (2016) Metabolomics by gas chromatography–mass spectrometry: combined targeted and untargeted profiling. Curr Protoc Mol Biol 114(1):30–34

Ghenabzia I, Hemmami H, Amor IB, Zeghoud S, Seghir BB, Hammoudi R (2023) Different methods of extraction of bioactive compounds and their effect on biological activity: a review. Int J Second Metab 10(4):469–494

Goswami A, Roy I, Sengupta S, Debnath N (2010) Novel applications of solid and liquid formulations of nanoparticles against insect pests and pathogens. Thin Solid Films 519(3): 1252–1257

Gupta A, Mehta SK, Kunal K, Mukhopadhyay K, Singh S (2022) Quantum dots as promising nanomaterials in agriculture. In: Agricultural nanobiotechnology. Woodhead Publishing, pp 243–296

Guzmán-Báez GA, Trejo-Téllez LI, Ramírez-Olvera SM, Salinas-Ruíz J, Bello-Bello JJ, Alcántar-González G, Hidalgo-Contreras JV, Gómez-Merino FC (2021) Silver nanoparticles increase nitrogen, phosphorus, and potassium concentrations in leaves and stimulate root length and number of roots in tomato seedlings in a Hormetic manner. Dose Response 19(4): 15593258211044576. https://doi.org/10.1177/15593258211044576

Hao Y, Yu Y, Sun G, Gong X, Jiang Y, Lv G, Zhang Y, Li L, Zhao Y, Sun D, Gu W, Qian C (2023) Effects of multi-walled carbon nanotubes and Nano-silica on root development, leaf photosynthesis, active oxygen and nitrogen metabolism in maize. Plants 12(8):1604. https://doi.org/10.3390/plants12081604

Hao Y, Zhang Z, Luo E, Yang J, Wang S (2025) Plant metabolomics: applications and challenges in the era of multi-omics big data. aBIOTECH 6:116–132

Hong J, Wang C, Wagner DC, Gardea-Torresdey JL, He F, Rico CM (2021) Foliar application of nanoparticles: mechanisms of absorption, transfer, and multiple impacts. Environ Sci Nano 8(5): 1196–1210

Huang H, Liu H, Ma W, Qin L, Chen L, Guo H et al (2023) High-throughput MALDI-MSI metabolite analysis of plant tissue microarrays. Plant Biotechnol J 21(12):2574–2258

Husen A (2020) Carbon-based nanomaterials and their interactions with agricultural crops. In: Nanomaterials for agriculture and forestry applications. Elsevier, pp 199–218

Hussain F, Hadi F, Rongliang Q (2021) Effects of zinc oxide nanoparticles on antioxidants, chlorophyll contents, and proline in Persicaria hydropiper L. and its potential for Pb phytoremediation. Environ Sci Pollut Res 28(26):34697–34713

Jamil F, Mukhtar H, Fouillaud M, Dufossé L (2022) Rhizosphere signaling: insights into plant–rhizomicrobiome interactions for sustainable agronomy. Microorganisms 10(5):899

Jendoubi T (2021) Approaches to integrating metabolomics and multi-omics data: a primer. Metabolites 11(3):184. https://doi.org/10.3390/metabo11030184

Jiao Y, Zhang J, Pan C (2022) Integrated physiological, proteomic, and metabolomic analyses of pecan cultivar 'Pawnee'adaptation to salt stress. Sci Rep 12(1):1841
Judge A, Dodd MS (2020) Metabolism. Essays Biochem 64(4):607–647. https://doi.org/10.1042/EBC20190041
Judy JD, Unrine JM, Rao W, Wirick S, Bertsch PM (2012) Bioavailability of gold nanomaterials to plants: importance of particle size and surface coating. Environ Sci Technol 46(15):8467–8474
Karami Mehrian S, Heidari R, Rahmani F (2015) Effect of silver nanoparticles on free amino acids content and antioxidant defense system of tomato plants. Indian J Plant Physiol 20:257–263
Khan M, Khan MSA, Borah KK, Goswami Y, Hakeem KR, Chakrabartty I (2021) The potential exposure and hazards of metal-based nanoparticles on plants and environment, with special emphasis on ZnO NPs, TiO2 NPs, and AgNPs: a review. Environ Adv 6:100128
Khan Y, Sadia H, Ali Shah SZ, Khan MN, Shah AA, Ullah N, Ullah MF, Bibi H, Bafakeeh OT, Khedher NB, Eldin SM, Fadhl BM, Khan MI (2022) Classification, synthetic, and characterization approaches to nanoparticles, and their applications in various fields of nanotechnology: a review. Catalysts 12(11):1386. https://doi.org/10.3390/catal12111386
Kılınc H, D'Urso G, Paolillo A, Alankus O, Piacente S, Masullo M (2023) LC-MS and NMR based plant metabolomics: a comprehensive phytochemical investigation of *Symphytum anatolicum*. Metabolites 13(10):1051. https://doi.org/10.3390/metabo13101051
Kim HM, Kang JS (2021) Metabolomic studies for the evaluation of toxicity induced by environmental toxicants on model organisms. Metabolites 11(8):485. https://doi.org/10.3390/metabo11080485
Komazec B, Cvjetko P, Balen B, Letofsky-Papst I, Lyons DM, Peharec Štefanić P (2023) The occurrence of oxidative stress induced by silver nanoparticles in *Chlorella vulgaris* depends on the surface-stabilizing agent. Nanomaterials 13(13):1967. https://doi.org/10.3390/nano13131967
Kumar A, Chen F, Mozhi A, Zhang X, Zhao Y, Xue X et al (2013) Innovative pharmaceutical development based on unique properties of nanoscale delivery formulation. Nanoscale 5(18): 8307–8325
Kumar S, Lahlali R, Liu X, Karunakaran C (2016) Infrared spectroscopy combined with imaging: a new developing analytical tool in health and plant science. Appl Spectrosc Rev 51(6):466–483
Kumari A, Gupta AK, Sharma S, Jadon VS, Sharma V, Chun SC, Sivanesan I (2024) Nanoparticles as a tool for alleviating plant stress: mechanisms, implications, and challenges. Plants 13(11): 1528. https://doi.org/10.3390/plants13111528
Lala S (2021) Nanoparticles as elicitors and harvesters of economically important secondary metabolites in higher plants: a review. IET Nanobiotechnol 15(1):28–57. https://doi.org/10.1049/nbt2.12005
Li Z, Juneau P, Lian Y, Zhang W, Wang S, Wang C, Shu L, Yan Q, He Z, Xu K (2020) Effects of titanium dioxide nanoparticles on photosynthetic and Antioxidative processes of *Scenedesmus obliquus*. Plants 9(12):1748. https://doi.org/10.3390/plants9121748
Li P, Xia Y, Song K, Liu D (2024) The impact of nanomaterials on photosynthesis and antioxidant mechanisms in Gramineae plants: research progress and future prospects. Plants (Basel, Switz) 13(7):984. https://doi.org/10.3390/plants13070984
Liu M, Feng S, Ma Y, Xie C, He X, Ding Y et al (2019) Influence of surface charge on the phytotoxicity, transformation, and translocation of CeO2 nanoparticles in cucumber plants. ACS Appl Mater Interfaces 11(18):16905–16913
Luo L, Huang W, Zhang J, Yu Y, Sun T (2024) Metal-based nanoparticles as antimicrobial agents: a review. ACS Appl Nano Mater 7(3):2529–2545
Ma X, Geiser-Lee J, Deng Y, Kolmakov A (2010) Interactions between engineered nanoparticles (ENPs) and plants: phytotoxicity, uptake and accumulation. Sci Total Environ 408(16): 3053–3061
Ma F, Li CC, Zhang CY (2018) Development of quantum dot-based biosensors: principles and applications. J Mater Chem B 6(39):6173–6190

Majumdar B, Saha AR, Ghorai AK, Sarkar SK, Chowdhury H, Kundu DK, Mahapatra BS (2014) Effect of fertilizer treatments on jute (Chorchorus olitorius), microbial dynamics in its rhizosphere and residual fertility status of soil. Indian J Agric Sci 84(4):503–508

Makarov VV, Love AJ, Sinitsyna OV, Makarova SS, Yaminsky IV, Taliansky ME, Kalinina NO (2014) "Green" nanotechnologies: synthesis of metal nanoparticles using plants. Acta Nat 6(1): 35–44

Marslin G, Sheeba CJ, Franklin G (2017) Nanoparticles alter secondary metabolism in plants via ROS burst. Front Plant Sci 8:832

Milewska-Hendel A, Zubko M, Stróż D, Kurczyńska EU (2019) Effect of nanoparticles surface charge on the *Arabidopsis thaliana* (L.) roots development and their movement into the root cells and protoplasts. Int J Mol Sci *20*(7):1650. https://doi.org/10.3390/ijms20071650

Misra SK, Rosenholm JM, Pathak K (2023) Functionalized and nonfunctionalized nanosystems for mitochondrial drug delivery with metallic nanoparticles. Molecules 28(12):4701. https://doi.org/10.3390/molecules28124701

Mondéjar-López M, García-Simarro MP, Navarro-Simarro P, Gómez-Gómez L, Ahrazem O, Niza E (2024) A review on the encapsulation of "eco-friendly" compounds in natural polymer-based nanoparticles as next generation nano-agrochemicals for sustainable agriculture and crop management. Int J Biol Macromol 280:136030

Munir N, Gulzar W, Abideen Z, Hasanuzzaman M, El-Keblawy A, Zhao F (2023) Plant–nanoparticle interactions: transcriptomic and proteomic insights. Agronomy 13(8):2112. https://doi.org/10.3390/agronomy13082112

Mutha RE, Kalaskar M, Khan ZG (2025) Modern analytical techniques for quality control and chemical identification of phytochemicals. In: Pharmacognosy and phytochemistry: principles, techniques, and clinical applications. Wiley, pp 167–188

Nagana Gowda GA, Raftery D (2021) NMR-based metabolomics. Adv Exp Med Biol 1280:19–37. https://doi.org/10.1007/978-3-030-51652-9_2

Nair R, Varghese SH, Nair BG, Maekawa T, Yoshida Y, Kumar DS (2010) Nanoparticulate material delivery to plants. Plant Sci 179(3):154–163

Nalbantoglu S (2019) Metabolomics: basic principles and strategies. Mol Med 10(1)

Navarro E, Baun A, Behra R, Hartmann NB, Filser J, Miao AJ, Quigg A, Santschi PH, Sigg L (2008) Environmental behavior and ecotoxicity of engineered nanoparticles to algae, plants, and fungi. Ecotoxicology 17(5):372–386. https://doi.org/10.1007/s10646-008-0214-0. Epub 2008 May 7

Pérez-de-Luque A (2017) Interaction of nanomaterials with plants: what do we need for real applications in agriculture? Front Environ Sci 5:12

Pietrzak M, Skiba E, Wolf WM (2024) Root-applied cerium oxide nanoparticles and their specific effects on plants: a review. Int J Mol Sci 25(7):4018. https://doi.org/10.3390/ijms25074018

Pokhrel LR, Dubey B (2013) Evaluation of developmental responses of two crop plants exposed to silver and zinc oxide nanoparticles. Sci Total Environ 452:321–332

Pradhan S, Patra P, Mitra S, Dey KK, Jain S, Sarkar S, Roy S, Palit P, Goswami A (2014) Manganese nanoparticles: impact on non-nodulated plant as a potent enhancer in nitrogen metabolism and toxicity study both in vivo and in vitro. J Agric Food Chem 62(35):8777–8785

Pudake RN (2024) Novel application of bio-based nanomaterials for the alleviation of abiotic stress in crop plants. In: Nanotechnology for abiotic stress tolerance and management in crop plants, pp 181–201

Pudake RN, Chauhan N, Kole C (2019) Nanoscience for sustainable agriculture. Springer

Pudake RN, Mohanta TK, Mahato N (2024a) Opportunities and challenges for nanotechnology in sustainable agri-food production. Front Nanotechnol 6:1420192

Pudake RN, Tripathi RM, Gill SS (2024b) Nanotechnology for abiotic stress tolerance and management in crop plants. Elsevier

Rasheed A, Li H, Tahir MM, Mahmood A, Nawaz M, Shah AN, Aslam MT, Negm S, Moustafa M, Hassan MU, Wu Z (2022) The role of nanoparticles in plant biochemical, physiological, and molecular responses under drought stress: a review. Front Plant Sci 13:976179

Ribeiro da Cunha B, Fonseca LP, Calado CR (2020) Metabolic fingerprinting with Fourier-transform infrared (FTIR) spectroscopy: towards a high-throughput screening assay for antibiotic discovery and mechanism-of-action elucidation. Meta 10:145

Rico CM, Majumdar S, Duarte-Gardea M, Peralta-Videa JR, Gardea-Torresdey JL (2011) Interaction of nanoparticles with edible plants and their possible implications in the food chain. J Agric Food Chem 59(8):3485–3498

Rico CM, Barrios AC, Tan W, Rubenecia R, Lee SC, Varela-Ramirez A et al (2015) Physiological and biochemical response of soil-grown barley (Hordeum vulgare L.) to cerium oxide nanoparticles. Environ Sci Pollut Res 22:10551–10558

Roche RDLT, Pagano L, Majumdar S, Eitzer BD, Zuverza-Mena N, Ma C et al (2018) Co-exposure of imidacloprid and nanoparticle Ag or CeO2 to Cucurbita pepo (zucchini): contaminant bioaccumulation and translocation. NanoImpact 11:136–145

Salama HM (2012) Effects of silver nanoparticles in some crop plants, common bean (Phaseolus vulgaris L.) and corn (Zea mays L.). Int Res J Biotechnol 3(10):190–197

Samynathan R, Venkidasamy B, Ramya K, Muthuramalingam P, Shin H, Kumari PS, Thangavel S, Sivanesan I (2023) A recent update on the impact of nano-selenium on plant growth, metabolism, and stress tolerance. Plants 12(4):853. https://doi.org/10.3390/plants12040853

Sanches PHG, de Melo NC, Porcari AM, de Carvalho LM (2024) Integrating molecular perspectives: strategies for comprehensive multi-omics integrative data analysis and machine learning applications in transcriptomics, proteomics, and metabolomics. Biology 13(11):848. https://doi.org/10.3390/biology13110848

Schnackenberg LK, Sun J, Beger RD (2012) Metabolomics techniques in nanotoxicology studies. In: Nanotoxicity: methods and protocols, pp 141–156

Schwab F, Zhai G, Kern M, Turner A, Schnoor JL, Wiesner MR (2016) Barriers, pathways and processes for uptake, translocation and accumulation of nanomaterials in plants–critical review. Nanotoxicology 10(3):257–278

Seleiman MF, Ahmad A, Alhammad BA, Tola E (2023) Exogenous Applicationf zinc oxide nanoparticles improved antioxidants, photosynthetic, and yield traits in salt-stressed maize. Agronomy 13(10):2645. https://doi.org/10.3390/agronomy13102645

Selivanov NY, Selivanova OG, Sokolov OI, Sokolova MK, Sokolov AO, Bogatyrev VA, Dykman LA (2017) Effect of gold and silver nanoparticles on the growth of the Arabidopsis thaliana cell suspension culture. Nanotechnol Russia 12:116–124

Sembada AA, Lenggoro IW (2024) Transport of nanoparticles into plants and their detection methods. Nanomaterials (Basel, Switzerland) 14(2):131. https://doi.org/10.3390/nano14020131

Servin AD, White JC (2016) Nanotechnology in agriculture: next steps for understanding engineered nanoparticle exposure and risk. NanoImpact 1:9–12

Shahzad R, Bilal S, Khan A, Shehzad A, Al-Suhaimi EA (2024) Emerging concept on cellular uptake mechanism of nanoparticles. In: Molecular impacts of nanoparticles on plants and algae. Elsevier, pp 31–40

Silva S, Dias MC, Pinto DCGA, Silva AMS (2023) Metabolomics as a tool to understand nano-plant interactions: the case study of metal-based nanoparticles. Plants (Basel, Switz) 12(3):491. https://doi.org/10.3390/plants12030491

Singh H, Sharma A, Bhardwaj SK, Arya SK, Bhardwaj N, Khatri M (2021) Recent advances in the applications of nano-agrochemicals for sustainable agricultural development. Environ Sci: Processes Impacts 23(2):213–239

Sporre E, Karlsen J, Schriever K, Asplund-Samuelsson J, Janasch M, Strandberg L et al (2023) Metabolite interactions in the bacterial Calvin cycle and implications for flux regulation. Commun Biol 6(1):947

Steuer AE, Brockbals L, Kraemer T (2019) Metabolomic strategies in biomarker research-new approach for indirect identification of drug consumption and sample manipulation in clinical and forensic toxicology? Front Chem 7:319. https://doi.org/10.3389/fchem.2019.00319

Sun H, Guo W, Zhou Q, Gong Y, Lv Z, Wang Q et al (2023) Uptake, transformation, and environmental impact of zinc oxide nanoparticles in a soil-wheat system. Sci Total Environ 857:159307

Tan W, Peralta-Videa JR, Gardea-Torresdey JL (2018) Interaction of titanium dioxide nanoparticles with soil components and plants: current knowledge and future research needs–a critical review. Environ Sci Nano 5(2):257–278

Tian L, Zhang H, Zhao X, Gu X, White JC, Zhao L, Ji R (2020) CdS nanoparticles in soil induce metabolic reprogramming in broad bean (Vicia faba L.) roots and leaves. Environ Sci Nano 7(1):93–104

Tripathi DK, Mishra RK, Singh S, Singh S, Vishwakarma K, Sharma S et al (2017) Nitric oxide ameliorates zinc oxide nanoparticles phytotoxicity in wheat seedlings: implication of the ascorbate–glutathione cycle. Front Plant Sci 8:1

Tucci P, Porta G, Agostini M et al (2013) Metabolic effects of TiO_2 nanoparticles, a common component of sunscreens and cosmetics, on human keratinocytes. Cell Death Dis 4:e549. https://doi.org/10.1038/cddis.2013.76

Vatankhah A, Aliniaeifard S, Moosavi-Nezhad M et al (2023) Plants exposed to titanium dioxide nanoparticles acquired contrasting photosynthetic and morphological strategies depending on the growing light intensity: a case study in radish. Sci Rep 13:5873

Vignesh A, Amal TC, Sivalingama R, Selvakumar S, Vasanth K (2024) Unraveling the impact of nanopollution on plant metabolism and ecosystem dynamics. Plant Physiol Biochem:108598

Vinogradova N, Vinogradova E, Chaplygin V, Mandzhieva S, Kumar P, Rajput VD, Minkina T, Seth CS, Burachevskaya M, Lysenko D, Singh RK (2023) Phenolic compounds of the medicinal plants in an Anthropogenically transformed environment. Molecules (Basel, Switzerland) 28(17):6322. https://doi.org/10.3390/molecules28176322

Wang Y, Xu L, Shen H, Wang J, Liu W, Zhu X et al (2015) Metabolomic analysis with GC-MS to reveal potential metabolites and biological pathways involved in Pb & Cd stress response of radish roots. Sci Rep 5(1):18296

Wang W, Ding X, Xu Q, Wang J, Wang L, Lou X (2016a) Zeta-potential data reliability of gold nanoparticle biomolecular conjugates and its application in sensitive quantification of surface absorbed protein. Colloids Surf B: Biointerfaces 148:541–548

Wang X, Yang X, Chen S, Li Q, Wang W, Hou C et al (2016b) Zinc oxide nanoparticles affect biomass accumulation and photosynthesis in Arabidopsis. Front Plant Sci 6:1243

Wang XC, Ma XL, Liu JN, Zhang Y, Zhang JN, Ma MH, Ma FL, Yu YJ, She Y (2023a) A comparison of feature extraction capabilities of advanced UHPLC-HRMS data analysis tools in plant metabolomics. Anal Chim Acta 1254:341127. https://doi.org/10.1016/j.aca.2023.341127

Wang X, Xie H, Wang P, Yin H (2023b) Nanoparticles in plants: uptake, transport and physiological activity in leaf and root. Materials (Basel, Switzerland) 16(8):3097. https://doi.org/10.3390/ma16083097

Xia J, Wishart DS (2011) Metabolomic data processing, analysis, and interpretation using MetaboAnalyst. Curr Protoc Bioinformatics 34(1):14–10

Yan A, Chen Z (2019) Impacts of silver nanoparticles on plants: a focus on the phytotoxicity and underlying mechanism. Int J Mol Sci 20(5):1003. https://doi.org/10.3390/ijms20051003

Yusefi-Tanha E, Fallah S, Rostamnejadi A, Pokhrel LR (2020) Root system architecture, copper uptake and tissue distribution in soybean (Glycine max (L.) Merr.) grown in copper oxide nanoparticle (CuONP)-amended soil and implications for human nutrition. Plants 9(10):1326

Zayed O, Hewedy OA, Abdelmoteleb A, Ali M, Youssef MS, Roumia AF, ... Yuan ZC (2023) Nitrogen journey in plants: From uptake to metabolism, stress response, and microbe interaction. Biomolecules 13(10):1443

Zhang Y, Qi G, Yao L, Huang L, Wang J, Gao W (2022) Effects of metal nanoparticles and other preparative materials in the environment on plants: from the perspective of improving secondary metabolites. J Agric Food Chem 70(4):916–933

Zhang H, Zheng T, Wang Y, Li T, Chi Q (2024) Multifaceted impacts of nanoparticles on plant nutrient absorption and soil microbial communities. Front Plant Sci 15:1497006

Zhu J, Thompson CB (2019) Metabolic regulation of cell growth and proliferation. Nat Rev Mol Cell Biol 20(7):436–450

Zhu J, Li J, Shen Y, Liu S, Zeng N, Zhan X et al (2020) Mechanism of zinc oxide nanoparticle entry into wheat seedling leaves. Environ Sci Nano 7(12):3901–3913

Zia K, Siddiqui T, Ali S, Farooq I, Zafar MS, Khurshid Z (2019) Nuclear magnetic resonance spectroscopy for medical and dental applications: a comprehensive review. Eur J Dent 13(1): 124–128. https://doi.org/10.1055/s-0039-688654

Chapter 12
Exploring the Use of Flow Cytometry for Studying Nanomaterials' Phytotoxicity Assessment

Tanushree Maharana, Caleb Fernandes, Prabhakar Mishra, and Yuvashree Muralidaran

Abstract As the integration of engineered nanomaterials into agriculture and industry expands, understanding their phytotoxic effects on plants and other economically important crops becomes increasingly critical. Flow cytometry has hence emerged as a vital tool in assessing the phytotoxic effects of these particles. The technique accounts for rapid, high-throughput in-depth analysis of various cellular processes impacted by nanoparticle exposure. It has been widely applied to measure cell viability, oxidative stress, apoptosis, autophagy, DNA damage, and changes in the cell cycle, amongst other applications. These insights are crucial for understanding the physiological and biochemical disruptions caused by nanomaterials. Through ongoing and continuous advances in instrumentation and analytical tools, flow cytometry continues to gain importance as a reliable method in nanotoxicology and plant-environment interaction research.

Keywords Flow cytometry · Nanomaterial · Phytotoxicity · Reactive oxygen species · Apoptosis · Genotoxicity

12.1 Introduction

Nanomaterials are gaining increasing prominence across diverse sectors, notably in agriculture (Bhatia 2017; Pudake et al. 2019a, 2024a). Their unique properties, arising from their size-dependent characteristics at the nanoscale (typically 1–100 nm), present opportunities for innovations such as enhanced nutrient delivery (Dutta et al. 2022; Pudake et al. 2024b), more precise pesticide application (Jalil and Ansari 2020; Rana et al. 2019; Ray et al. 2023), and overall improvements in crop yields (Khan et al. 2021). For example, NMs can enhance the absorption of bioactive compounds in plants, hence elevating their nutritional value and resilience

T. Maharana · C. Fernandes · P. Mishra · Y. Muralidaran (✉)
Department of Biotechnology, School of Applied Sciences, REVA University, Bengaluru, Karnataka, India

R. N. Pudake et al. (eds.), *Phyto Nanotoxicology*, Nanotechnology in Plant Sciences 7, https://doi.org/10.1007/978-3-032-24518-2_12

(Wunderlich et al. 2017). However, the escalating utillization of NMs also raises concerns about potential risks to both environmental and human health (Hong et al. 2013; Rodríguez-Delgado et al. 2019; Pudake et al. 2019b). These concerns emphasize the necessity for thorough and accurate assessments of their potential impact, particularly concerning phytotoxicity.

Phytotoxicity, which refers to the negative effects of a substance on plant health, manifests in a spectrum of impacts ranging from inhibited growth and reduced photosynthetic activity to visible symptoms like chlorosis and necrosis, and in severe cases, plant death. Ensuring sustainable agricultural practices and maintaining healthy ecosystems hinges on understanding and accurately assessing phytotoxicity (Vuković et al. 2014; Kodaparthi et al. 2025). Traditional methods for phytotoxicity assessment, including germination tests, seedling growth assays, and measurements of chlorophyll content, often fall short due to limitations in throughput, sensitivity, and the ability to elucidate comprehensive insights into cellular understanding. (Joshua et al. 2020; Hu et al. 2022).

Flow cytometry presents a promising alternative as a powerful technique for analyzing the physical and chemical characteristics of cells or particles in a fluid stream (Arze et al. 2020; Robinson et al. 2023). With widespread applications in biological research, including immunology, cancer biology, and environmental toxicology, flow cytometry offers several advantages for phytotoxicity assessment (Lara et al. 2018; Robinson et al. 2023). These advantages include high-throughput analysis, rapid data acquisition, the ability to simultaneously measure multiple parameters at the single-cell level, and the capacity to analyze large populations of cells, yielding statistically robust data. This chapter will explore the application of flow cytometry in studying the phytotoxicity of nanomaterials, highlighting its strengths and limitations, and discussing future research directions.

12.2 Nanomaterial Exposure in Plants

The increasing use of nanotechnology across various sectors is leading to a greater release of engineered nanomaterials into the environment (Chichiriccò and Poma 2015; Huang et al. 2024). Plants, being primary producers, are directly exposed to ENMs through various pathways, encompassing contaminated soil, water, and air (Kumar et al. 2019; Kodaparthi et al. 2025). Hence, awareness of the mechanisms of ENM uptake and their subsequent phytotoxic effects is crucial for assessing the ecological risks associated with nanotechnology.

12.2.1 Mechanisms of Nanomaterial Uptake in Plants

The underlying mechanisms by which nanomaterials enter flora depend on the nanomaterials' properties (size, shape, surface charge, and composition), the plant species, and environmental conditions (Montes et al. 2017). Nanomaterials can be internalized by plants through several pathways. Root uptake is a significant pathway, where nanomaterials in the soil can be assimilated by plant roots via the apoplastic and symplastic pathways (Schwab et al. 2015; Djanaguiraman et al. 2024). The apoplastic pathway facilitates movement through the cell walls and intercellular spaces of the root cortex. However, this is limited by the Casparian strip. To bypass this barrier, nanomaterials can enter the symplast by crossing the cell membrane of endodermal cells through endocytosis or membrane transporters (Predoi et al. 2020). For instance, prior research indicates that gold NP’s s can be taken up by *Arabidopsis thaliana* roots and translocated to other parts of the plant. The size of the nanoparticles affects their uptake, with smaller nanoparticles being more readily taken up via the cell walls (Schwab et al. 2015; Wu et al. 2024).

Foliar uptake is another important entry point, where nanomaterials can deposit on leaf surfaces through atmospheric deposition or direct application, such as in agricultural practices. Nanomaterials may enter through stomatal pores on the leaf epidermis, although the size of the stomatal pores can limit the entry of larger particles (Chichiriccò and Poma 2015; Li et al. 2024). Nanomaterials can also penetrate the cuticle, the waxy layer covering the epidermis, through diffusion or aqueous pores. For instance, copper oxide nanoparticles have been observed to penetrate the and accumulate within the leaves. Once inside the plant, nanomaterials can be translocated throughout the plant via the vascular system (xylem and phloem) (Tripathi et al. 2016). Functionalized nanomaterials, which have chemical groups attached to their surfaces, may exhibit different uptake behaviors compared to non-functionalized nanomaterials (Chichiriccò and Poma 2015).

12.2.2 Phytotoxic Effects of Nanomaterials

The phytotoxic effects of nanomaterials depend on the type and concentration of nanomaterials, plant species, exposure conditions, and the plant’s developmental stage (Miralles et al. 2012). Commonly observed effects include growth inhibition, where nanomaterials can reduce root elongation, shoot development, and overall biomass (Gao et al. 2023). For example, exposure to titanium dioxide nanoparticles (TiO_2 NPs) can inhibit the growth of soybean plants. The specific mechanisms of toxicity vary depending on the type of nanomaterial; some may disrupt cell division (Kumar et al. 2019; Siddiqui 2024).

Oxidative stress is another significant phytotoxic effect. Nanomaterials can induce the production of reactive oxygen species in plant cells, leading to oxidative damage to lipids, proteins, and DNA (Sevim et al. 2020). This oxidative stress can disrupt cellular functions and lead to cell death. For instance, zinc oxide nanoparticles have been shown to induce oxidative stress in wheat plants. Furthermore, nanomaterials can interfere with photosynthesis by damaging chlorophyll, disrupting chloroplast structure, or inhibiting photosynthetic enzymes. Some nanomaterials can also affect the uptake and transport of essential nutrients, potentially leading to nutrient deficiencies (Predoi et al. 2020). Some studies suggest that nanomaterials can disrupt plant iron uptake, leading to chlorosis.

Interestingly, some studies have reported positive effects of nanomaterials on plant growth and development, particularly at low concentrations, sometimes referred to as "nano-fertilization" (Predoi et al. 2020; Pudake 2024). For example, research indicates that tomato plants grown in soil supplemented with 50 μg/L multiwalled CNTs could produce twice as much fruit as a control group (Zheng et al. 2017). Also, nano-functionalized carbon nanotubes enhanced root elongation in onion and cucumber (Mahajan et al. 2011). Water-soluble carbon nanotubes have been found to positively affect the growth of various crops (Tripathi et al. 2009, 2011; Tripathi and Sarkar 2015). However, the potential risks associated with nanomaterial exposure should be carefully considered, and further research is needed to fully understand the long-term effects of nanomaterials on plant health and ecosystems (Barkataki and Singh 2019; Heliopoulos et al. 2020).

12.3 Principles of Flow Cytometry

Flow cytometry is a powerful technique used to analyze cells' or particles' physical and chemical characteristics as they flow in a fluid stream through a laser beam (Robinson and Grégori 2007). It allows for rapid, multiparametric analysis of individual cells, providing valuable data for various applications, including phytotoxicity assessment of nanomaterials. Flow cytometric analyses are well suited for pre-screening for toxic effects of chemicals (Tuschl and Schwab 2004; Mary et al. 2023).

12.3.1 Basic Components of a Flow Cytometer

A flow cytometer consists of three primary systems: fluidics, optics, and electronics (Fig. 12.1). *Fluidics System* is responsible for transporting the sample, usually a cell suspension, to the interrogation point. For phytotoxicity assessment, this system needs to be capable of handling plant cells or protoplasts. The *optics system* comprises light sources (typically lasers), lenses, and filters. The laser beam illuminates the cells, and the lenses collect the scattered and emitted light. Filters are used to select specific wavelengths of light for detection. The side-scattered light of FCM can be used as a guide to measure the uptake potential of nanoparticles (Ibuki and Toyooka 2012; Garcia Romeu et al. 2021). The *electronics system* converts the detected light signals into electronic signals, which are then amplified and processed. This system also includes computer hardware and software for data acquisition, analysis, and display.

Cells in suspension pass single file through a laser beam. Light scatter and fluorescence signals are detected to analyze cell size, complexity, and markers.

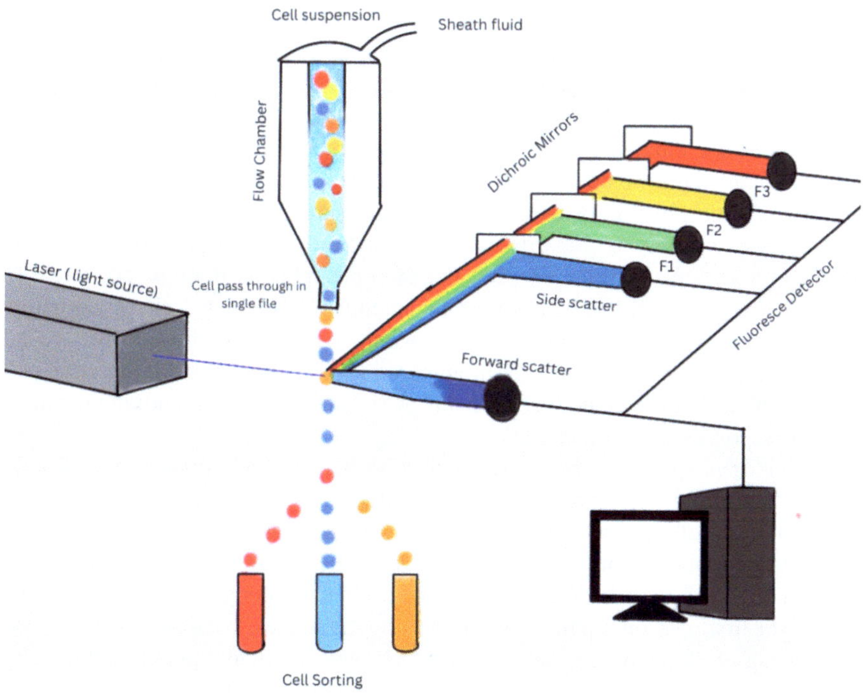

Fig. 12.1 Schematic representation of a flow cytometry unit

Forward scatter (FSC) and side scatter (SSC) are detected to determine cell size and internal complexity. Fluorescent emissions are split by dichroic mirrors and filtered into different detectors (F1, F2, F3) for multi-parameter analysis. Detectors capture these signals, enabling analysis and sorting of cells based on their properties (McKinnon 2018).

12.3.2 Measurement Parameters in Flow Cytometry

Flow cytometry offers a multitude of parameters that can be measured to assess the characteristics of cells, making it a versatile tool for phytotoxicity assessment.

Forward Scatter (FSC) is primarily related to the size of the cell or particle. When a cell passes through the laser beam, it diffracts light in a forward direction. The amount of forward-scattered light is proportional to the cell's size. In phytotoxicity studies, changes in FSC can indicate cell swelling or shrinkage due to exposure to nanomaterials. *Side Scatter* (SSC) provides information about the internal complexity or granularity of the cell. It measures the light scattered at approximately 90 degrees to the laser beam. The amount of side-scattered light is influenced by the presence of intracellular structures such as organelles, vacuoles, and cell wall components. An increase in SSC may indicate increased granularity due to the accumulation of nanomaterials or cellular damage. The side-scattered light of FCM can be used as a guide to measure the uptake potential of nanoparticles (Ibuki and Toyooka 2012).

Fluorescence emitted by fluorescent probes (fluorochromes) bound to specific cellular components can be detected. Fluorochromes can be conjugated to antibodies to label specific proteins or used as dyes to measure various cellular parameters. *Fluorescence intensity* is proportional to the amount of the target molecule or the degree of a particular cellular process. For example, a higher fluorescence intensity of dye that binds to DNA indicates a greater amount of DNA in the cell. Modern flow cytometers can detect multiple fluorescent signals simultaneously, called *multiplexing*, allowing for the measurement of several parameters in the same cell (Nyland 2012; Francis et al. 2023). This is particularly useful for studying complex cellular responses to nanomaterials.

Gating strategies are crucial for analyzing specific cell populations. Gating involves selecting cell subpopulations based on FSC and SSC characteristics, which indicate cell size and internal complexity, respectively, as well as *fluorescence intensity* from labeled antibodies or dyes. Sequential gating is a common approach where a series of gates is created to isolate specific cell populations. For example, one might first gate on single cells to exclude aggregates, then on a specific size range using FSC, and finally, select cells expressing a particular fluorescent marker. To accurately define the boundaries of positive and negative populations, especially when using multiple fluorochromes, *Fluorescence-Minus-One controls* are used. An FMO control contains all the antibodies in the panel except for one, which helps determine the level of background fluorescence and allows for the proper setting of gates for the missing marker.

Other parameters, such as time and the area, height, and width of signals, can also provide valuable information. Time can be useful for identifying changes in cellular responses over time, while the area, height, and width of signals can be used to discriminate between single cells and cell aggregates.

12.3.3 Types of Flow Cytometric Measurements

Flow cytometry enables a variety of measurements that are valuable in assessing the phytotoxic effects of nanomaterials. These measurements can provide insights into cell viability, stress responses, and specific cellular damage. Cell viability assays are crucial for determining the proportion of live and dead cells in a sample (Tuschl and Schwab 2004). Common dyes used include propidium iodide, which enters cells with compromised membranes, and fluorescein diacetate, which is cleaved by esterases in live cells to produce a fluorescent product. Reactive oxygen species detection is also important, as nanomaterials can induce oxidative stress in plant cells, leading to the production of ROS. Dyes like DCFH-DA can be used to detect

ROS levels (Liu et al. 2025). Furthermore, changes in membrane potential, which can indicate cellular stress or damage, can be assessed using dyes like DiOC6 (Araniti et al. 2018). Cell cycle analysis can be performed to analyze the distribution of cells in different phases of the cell cycle (Tuschl and Schwab 2004), and disruption of the cell cycle can be an indicator of toxicity. Additionally, measuring chlorophyll fluorescence can provide information on the photosynthetic activity of plant cells (Franqueira et al. 2000), with changes indicating damage to the photosynthetic apparatus. Finally, alterations in esterase activity can be a sensitive indicator of the effects of toxic compounds (Eigemann et al. 2012).

12.3.4 Troubleshooting of the Flow Cytometer

Flow cytometry is a valuable asset for assessing phytotoxicity, but several challenges can arise during experimentation. Plant cells often exhibit high levels of autofluorescence, which can interfere with the detection of specific fluorescence signals; using fluorophores with emission spectra well-separated from the autofluorescence spectrum or pre-treating cells with autofluorescence-reducing dyes can help mitigate this (Kay et al. 2013). Debris and cell aggregates can clog the flow cytometer and interfere with data analysis, but proper filtration and gating strategies can exclude these from analysis (Leary 2023). Incomplete digestion of cell walls can lead to clumping and inaccurate cell counts, which can be resolved by optimizing the enzymatic digestion protocol. Non-selective antibody binding can lead to false-positive signals, but using appropriate blocking agents and properly titrating antibodies can minimize this. Variations in experimental conditions, such as temperature, light exposure, and media composition, can affect the results, thus careful control and standardization are essential. Finally, incorrect gating strategies or inappropriate statistical analyses can lead to inaccurate conclusions, so using appropriate gating strategies based on the cells' FSC and SSC characteristics and consulting with a flow cytometry expert are recommended (Robinson and Grégori 2007).

12.4 Applications of Flow Cytometry in Nanomaterial Phytotoxicity Research

12.4.1 Impact of Nanomaterials on Plant Cell Viability

In a study conducted by researchers (Thuesombat et al. 2014), calcein-AM and propidium iodide (PI) staining were used to evaluate the cytotoxic effects of silver nanoparticles on *Oryza sativa* (rice) seedlings. Calcein-AM is a cell-permeant dye that is hydrolyzed by intracellular esterases to produce a bright green fluorescence,

Table 12.1 Flow cytometry analysis of cell viability and apoptosis

Quadrant	Cell type
Annexin V−/ PI−	Viable cells
Annexin V+/ PI−	Early apoptotic cells
Annexin V+/ PI+	Late apoptotic/secondary necrotic cells
Annexin V−/ PI+	Necrotic cells

thereby indicating metabolic activity and membrane integrity. In contrast, Propidium Iodide is a red-fluorescent nucleic acid stain that cannot cross intact membranes and therefore selectively stains dead or membrane-compromised cells by binding to their DNA. The rice cells, after nanoparticle exposure and laser excitation, were able to be distinguished into live (calcein-positive, PI-negative) and dead (PI-positive) cells. This high-throughput approach provided statistically robust data on cell viability, making it possible to assess the dose-dependent toxicity of silver nanoparticles with greater precision than manual microscopy-based methods.

In a foundational study by O'Brien et al. (1997), researchers explored the cytotoxic effects of gold (Au) and silver (Ag) nanostructures on plant cells, focusing not only on general cell death but specifically on the mechanisms of apoptosis and necrosis. To achieve this, they utilized a dual staining protocol involving annexin V tagged with fluorescein isothiocyanate (FITC) and counterstained with PI, followed by flow cytometric analysis. Annexin V-FITC exhibits high-affinity binding to phosphatidylserine (PS), a phospholipid typically localized to the inner leaflet of the plasma membrane. but externalized to the outer surface during early apoptosis. The presence of green fluorescence thus indicates early apoptotic cells. Flow cytometry allowed for the simultaneous detection of annexin V-FITC and PI fluorescence at the single-cell level. The researchers could effectively classify cells into four distinct populations by analyzing these fluorescence signals in two separate channels, as shown in Table 12.1.

Annexin V-/PI- cells are healthy and show no signs of apoptosis or membrane damage. Annexin V+/PI− cells have initiated apoptosis but still maintain intact membranes. Annexin V+/PI+ cells are in late-stage apoptosis or have progressed to secondary necrosis, while Annexin V−/ PI+ cells have lost membrane integrity due to primary necrosis without signs of early apoptosis.

12.4.2 Cytometric Analysis of ROS Generation in Plants Exposed to Nanomaterials

Flow cytometry, combined with fluorescent probes like 2′,7′-dichlorofluorescin diacetate (DCFDA), offers a sensitive and rapid method for quantifying ROS at the single-cell level. The methodology employed flow cytometry to assess cell-cycle progression in root cells of *Allium cepa* after exposure to ZnO nanoparticles. The stained nuclei were then analyzed by a BD FACS Aria III flow cytometer, recording at least 10,000 events per sample to determine the cell population distribution across different cell-cycle phases. This approach enabled quantification of cell-cycle arrest, particularly the accumulation of cells in the G2/M phase and an increase in the sub-G0 population indicative of apoptosis or necrosis (Ghosh et al. 2016). ZnO nanoparticle exposure in plant root cells significantly increases reactive oxygen species (ROS) production, driven by both Zn^{2+} ion release and the nanoparticles' surface reactivity, disrupting electron transport chains. This oxidative stress causes lipid peroxidation, depletion of antioxidants like glutathione and protein thiols, and DNA damage, as confirmed by increased malondialdehyde levels and comet assays. ROS thus play a central role in ZnO nanoparticle-induced phytotoxicity. Figure 12.2 shows ROS scavenging pathways in plants.

In another study by Melegari et al. (2013), advanced cytometric analysis was essential in exploring the cellular toxicity of nano-sized copper oxide (nCuO) and ionic copper (Cu^{2+}) on *Chlamydomonas reinhardtii*. The researchers employed a BD Accuri™ C6 flow cytometer by BD Biosciences to measure key cellular parameters such as reactive oxygen species (ROS) production, membrane disruption, and changes in metabolic function. Specific fluorescent markers, including H_2DCFDA to track oxidative stress, propidium iodide (PI) for membrane integrity, and fluorescein diacetate (FDA) for assessing esterase activity, were utilized to distinguish between healthy and stressed cells. Engineered nickel oxide nanoparticles (NiO-NPs) disrupt ROS–NO balance in onion roots, inducing oxidative stress, antioxidant gene upregulation (50–250%), and cellular damage in a dose-dependent manner. Higher concentrations mimic bulk nickel toxicity, overwhelming defense mechanisms and threatening agricultural systems through food chain contamination (Manna et al. 2021).

12.4.3 Cell Cycle Arrest and Proliferation Inhibition

The cell cycle, comprising G0/G1, S, and G2/M phases, is is indispensable for proper plant growth and development, and its disruption can have severe consequences. Engineered nanoparticles (NPs) have been shown to interfere with these phases, causing cell cycle arrest, reduced mitotic activity, and impaired plant growth (Siddiqui 2024). Flow cytometry with DNA-specific stains like propidium iodide (PI) enables single-cell analysis of DNA content in plant tissues, allowing researchers to assess cell cycle distribution and detect abnormalities such as phase-specific accumulation or sub-G1 peaks, indicative of DNA damage and apoptosis (Fomicheva and Domblides 2023) Flow cytometric analysis revealed that TiO_2 nanoparticles induced a significantincreased proportion of cells at the G2/M checkpoint, indicating a mitotic delay likely caused by oxidative stress or DNA damage. DNA content histograms showed distinct peaks for each cell cycle phase, enabling precise quantification of cell distribution. This high-resolution, reproducible technique allowed for population-wide profiling, linking subtle cell cycle disruptions to phenotypic effects such as inhibited root elongation and reduced germination rates (Marmiroli et al. 2022).

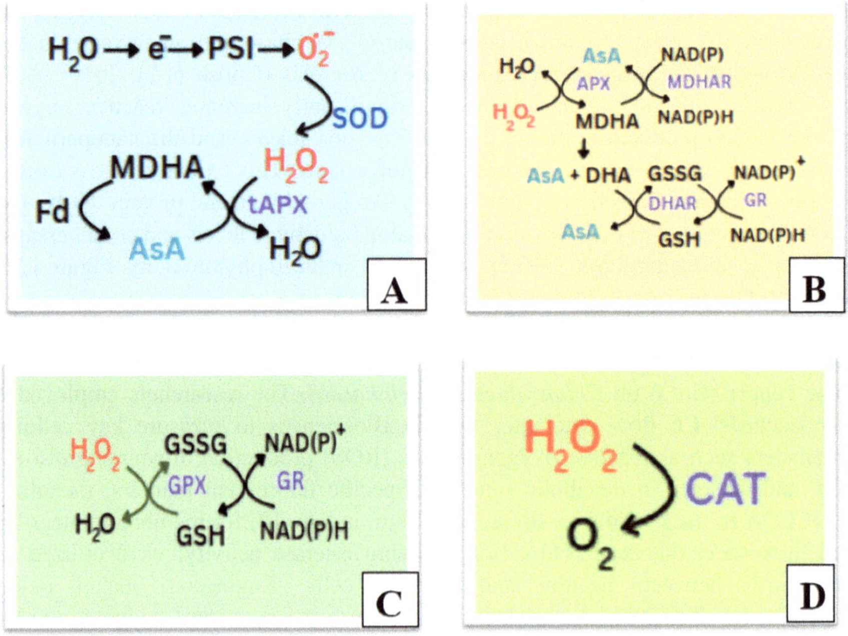

Fig. 12.2 Different pathways for reactive oxygen species (ROS) scavenging in plants. (**a**) During photosynthesis, the Water-Water Cycle (Mehler Reaction) involves water (H_2O) donating electrons (e^-) to Photosystem I (PSI), leading to the formation of superoxide anion (O_2^-). Superoxide Dismutase (SOD) converts O_2^- into hydrogen peroxide (H_2O_2), which is then reduced to water (H_2O) by Thylakoid-bound Ascorbate Peroxidase (tAPX) using Ascorbate (AsA), generating Monodehydroascorbate (MDHA); Ferredoxin (Fd) helps regenerate AsA from MDHA to maintain antioxidant balance (Foyer and Shigeoka 2011). (**b**) In the Ascorbate-Glutathione Cycle, Ascorbate Peroxidase (APX) detoxifies H_2O_2 using AsA to produce MDHA, which is then reduced back to AsA by Monodehydroascorbate Reductase (MDHAR) using NAD(P)H; if MDHA disproportionates into Dehydroascorbate (DHA), it is reduced by Dehydroascorbate Reductase (DHAR) using Glutathione (GSH), with oxidized Glutathione Disulfide (GSSG) recycled to GSH by Glutathione Reductase (GR) consuming NADPH (Nicontinamide Adenine Dinucletide Phospahte) in the process (Foyer and Kunert 2024). (**c**) In the Glutathione Peroxidase Pathway, Glutathione Peroxidase (GPX) detoxifies H_2O_2 into H_2O by utilizing GSH, and the resulting GSSG is regenerated to GSH by GR with the help of NADPH to maintain redox balance (Lubos et al. 2011). (**d**) In the Catalase Pathway, Catalase (CAT) directly converts H_2O_2 into H_2O and oxygen (O_2) without requiring additional reducing agents, providing a rapid and efficient detoxification route (Heck et al. 2010)

Another study (Yan and Chen 2019) compared the negative effects of biosynthesized AgNPs (using *Aloe vera*) and $AgNO_3$ on mustard seedlings. Usage of flow cytometry in tandem with propidium iodide staining enabledresearchers to assess cell viability and cell cycle dynamics. $AgNO_3$ caused more severe cytotoxicity, with greater cell death and reduced S and G2 phase populations, compared to AgNPs. Both treatments led to increased sub-G1 populations, indicating DNA fragmentation and apoptosis. Higher concentrations and prolonged exposure intensified these effects, reducing proliferating cells and disrupting mitosis. Flow cytometry enabled precise, large-scale analysis linking these cellular changes to physiological outcomes like stunted growth and altered root morphology (Yan and Chen 2019).

12.4.4 Membrane Integrity and Permeability Changes

The plasma membrane factors as an essential role in ion transport, nutrient uptake, and signaling, and its disruption by nanoparticles can lead todysfunction of metabolic activity, impaired signaling, and apoptotic death. Nanoparticles can compromise membrane integrity through oxidative stress, physical disruption, and ROS generation (Mustafa et al. 2024). Recent advances in nanotoxicology, particularly the use of fluorescent dyes combined with flow cytometry, have enabled precise, high-throughput analysis of membrane damage in plant cells. These techniques allow researchers to assess changes in membrane integrity, potential, and permeability across large cell populations, offering detailed insights into nanoparticle-induced toxicity and their environmental impact.

Zinc oxide nanoparticles (ZnO NPs) are widely used in industrial and consumer products, but their toxicity to marine organisms, such as the microalga *Chlorella vulgaris*, remains a critical area of investigation. The study by Suman et al. (2015) evaluated the cytotoxic effects of ZnO NPs on *C. vulgaris* using a multidisciplinary approach, including flow cytometry, oxidative stress analysis, and microscopic techniques. To evaluate membrane integrity, researchers once again utilized propidium iodide (PI) that intercalates with DNA only in cells where the membrane is compromised, producing a fluorescent signal. By applying various concentrations of ZnO NPs to plant root cells, the team observed a dose-dependent increase in PI-positive cells, signaling higher levels of membrane damage with increasing nanoparticle exposure. These findings were supported by oxidative stress markers (e.g., elevated SOD activity, GSH depletion, and lipid peroxidation). Through this technique, the researchers confirmed that exposure to higher concentrations of ZnO NPs resulted in significant membrane rupture, directly correlating with cell death.

Mitochondrial dysfunction, particularly the loss of mitochondrial membrane potential ($\Delta\Psi m$), was recognized as a critical early biomarker of nanoparticle-induced cellular stress and apoptosis. A study by Hussain (2018) employed MitoTracker® Red CMXRos (Thermo Fisher Scientific, USA), essentially a fluorescent dye that accumulated in mitochondria with intact membrane potential, emitting red fluorescence. Upon membrane depolarization, indicative of mitochondrial damage and early apoptosis, the dye shifted to green fluorescence. By measuring changes in the red-to-green fluorescence ratio through flow cytometry, researchers were able to accurately assess mitochondrial health. In the referenced study, exposure to increasing concentrations of CB-NPs (Carbon Black Nanoparticles) led to a significant, dose-dependent drop in red fluorescence, with 20 $\mu g/cm^2$ of CB-NPs causing mitochondrial depolarization in approximately

30–40% of the cell population. The inclusion of controls, such as FCCP (a known mitochondrial uncoupler) and catalase pretreatment (to mitigate oxidative stress), confirmed that oxidative damage played a central role in the observed dysfunction.

12.4.5 Evaluation of Nanomaterial Uptake by Plants

The assessment of nanoparticle uptake and internal transport in plants is critical to understanding the ecological risks associated with increasing nanomaterial exposure (Kang et al. 2024). Engineered nanomaterials, under their size and surface chemistry, can traverse plant barriers and accumulate in tissues, influencing both physiology and development (Schwab et al. 2016). Detecting and quantifying this uptake requires precise and sensitive techniques capable of resolving intracellular localization and movement across tissues. By combining fluorescent labeling strategies with flow cytometric analysis, researchers can detect nanomaterials within plant cells based on fluorescence signal intensity, spectral properties, and cell population gating, and those that are internalized, providing insights into uptake mechanisms and tissue specificity (Proença et al. 2022).

In a foundational study by Fratoddi et al. (2021), fluorescent labeling with FITC (fluorescein isothiocyanate) was employed to investigate the internalization of silver nanoparticles (AgNPs) in plant cells. By conjugating FITC to AgNPs, researchers achieved strong and stable green fluorescence, enabling precise detection of nanoparticles within root epidermal and cortical cells. Flow cytometry was then used to analyze fluorescence in thousands of individual cells, distinguishing between treated and control samples. Advanced gating techniques helped identify specific cell populations most prone to nanoparticle entry and differentiated between surface-bound and internalized particles, especially after rigorous washing or enzymatic treatments. The combined use of FITC labeling and flow cytometry offered high sensitivity, specificity, and scalability, proving especially effective in complex plant tissues for tracking nanoparticle behavior. FACS enabled sorting based on chlorophyll autofluorescence, showing that mesophyll protoplasts with higher chloroplast content internalized more nanoparticles than epidermal cells. The study also confirmed that nanoparticle charge influenced localization—highly charged particles targeted chloroplasts, while neutral ones were excluded. Another key study by Grauda et al. (2024) utilized flow cytometry to measure fluorescence in pollen cells of Tilia sp. and assess their response to SiO_2 nanoparticles (SiNPs) and UV stress in urban and greenhouse environments. It found greenhouse trees were more sensitive to stress, while urban trees showed muted responses, suggesting adaptation to urban pollutants. SiNPs had minimal impact on urban trees, likely due to pre-existing tolerance mechanisms (Grauda et al. 2024).

12.4.6 Detection of Plant Immune Responses

Nanoparticles, when introduced into plant systems, may act as abiotic elicitors that activate innate immune mechanisms. This process is analogous to the way pathogens trigger immune responses via recognition of pathogen-associated molecular patterns (PAMPs) (Aseel et al. 2024). In plants, such immune activation involves a cascade of defense responses, including the synthesis of pathogenesis-related (PR) proteins, bursts of reactive oxygen species (ROS), and hormonal signaling shifts. In foundational studies like that of Oberdörster et al. (2005), the immune-stimulating potential of nanomaterials was discussed in the context of biological systems, including plant models. This technique allows for rapid, single-cell-level detection of immune markers by measuring fluorescence from tagged molecules within individual plant cells. These cells are then passed through a flow cytometer, where laser-based interrogation detects and quantifies fluorescence intensity. This output typically results in a histogram or scatter plot showing shifts in marker expression, with peaks corresponding to upregulated immune activity in exposed cells. Flow cytometry enables the quantification of DNA methylation (e.g., 5-methylcytosine) and histone modifications (e.g., H3K4me3, H3K9ac), both of which are involved in activating defense genes during pathogen attacks (Somasundaram et al. 2021). Flow cytometry can assess chromatin openness, a key step in enabling defense gene expression (Ding et al. 2021). When paired with methods like ATAC-seq (Assay for Transposable Chromatin) allows profiling of accessible chromatin regions linked to immune gene loci (Wang et al. 2022).

12.4.7 Apoptosis and Autophagy Detection

Programmed cell death mechanisms, such as apoptosis and autophagy, are vital to maintaining cellular homeostasis and enabling plant adaptation to stress {Ebeed, 2021 #82}. Exposure to nanomaterials often triggers these pathways as part of the plant's defensive response to oxidative stress, membrane disruption, and cytotoxicity. In recent years, flow cytometry has emerged as a pivotal tool for distinguishing and quantifying apoptosis and autophagy in plant cells with high precision and throughput.

One key study (Awashra and Młynarz 2023) highlighted that those cells, when treated with zinc-phosphate nanoparticles (ZnPNPs), demonstrated a significant shift from viable to apoptotic states, with apoptotic cells increasing from 0.19% in the control group to 44.8% after NP treatment, and necrotic cells also showed a modest increase. This methodological approach was crucial because conventional colorimetric and fluorometric cytotoxicity assays often suffer from nanoparticle interference due to the optical properties or surface reactivity of the NPs, which can lead to false positives or negatives. Flow cytometry, on the other hand, is largely unaffected by such interference, making it an indispensable tool for applications such as this.

Positively charged NPs showed enhanced cellular uptake and a higher tendency to induce mitochondrial stress, leading to apoptosis, whereas neutral particles were more associated with necrotic outcomes. These findings not only underscore the sensitivity and accuracy of flow cytometry in evaluating NP-induced cytotoxicity but also emphasize the need to consider nanoparticle properties when designing safer nanomaterials for biomedical or industrial use.

In their study, Kulaphisit et al. (2023) introduced a refined methodology for quantifying autophagy by integrating spot counting with imaging flow cytometry. Conventional techniques such as immunoblotting or fluorescence microscopy often fall short in distinguishing between autophagic stages or in effectively analyzing heterogeneous cell populations. To address these limitations, the authors utilized imaging flow cytometry to simultaneously evaluate two autophagy-related markers: LC3, indicative of autophagosomes, and LAMP1, a lysosomal marker. By combining LC3 spot enumeration with bright detail similarity (BDS)—a colocalization metric that quantifies the overlap between LC3 and LAMP1—they successfully discriminated between autophagosomes and autolysosomes. This dual-parameter strategy enabled the classification of cells into three distinct categories: those with minimal autophagic activity, those with autophagosome accumulation (characterized by high LC3 spot counts and low BDS), and those with autolysosome accumulation (exhibiting both high LC3 spot counts and high BDS) (Kulaphisit et al. 2023, 2015).

12.4.8 DNA Damage and Genotoxicity Studies

Nanomaterials have raised concerns about their potential genotoxic effects on plants. Due to their small size and high surface area, nanoparticles can easily penetrate plant cells and interact with cellular components, leading to DNA damage (Gowtham et al. 2024). The primary mechanism for this damage is oxidative stress, where nanoparticles generate reactive oxygen species (ROS) that can attack DNA, causing strand breaks, base modifications, and chromosomal fragmentation. Such genetic instability can impair plant growth, reproduction, and survival. To assess DNA damage, techniques like the comet assay and micronuclei formation assays, often combined with flow cytometry, are used to quantify the extent of genotoxic effects in plant cells exposed to nanoparticles (Kumari et al. 2023).

The comet assay, also known as single-cell gel electrophoresis, is a well-established technique widely used to assess DNA damage at the individual cell level. This assay allows for the quantification of DNA strand breaks, a primary indicator of genetic damage (Tutty et al. 2022). The extent of migration, measured as the tail length, is directly proportional to the level of DNA damage. A study by Santos et al. (2015) comprehensively discusses the methodology, applications, and key findings surrounding the comet assay as a sensitive and versatile tool to detect DNA damage in plant cells exposed to various environmental stressors, appearing as a "comet tail" under fluorescence microscopy. Key findings in the article highlight the growing relevance of the comet assay in detecting genotoxic impacts from a wide range of stressors, including heavy metals, nanoparticles, pesticides, radiation, and complex environmental matrices. The assay has shown particular value in revealing oxidative DNA damage and strand breaks due to pollutants such as cadmium, lead, and chromium, and has been employed to assess DNA repair kinetics, especially using enzymatic variants like formamidopyrimidine DNA glycosylase (FPG). Notably, the assay's high sensitivity allows for the detection of sub-lethal DNA damage levels, which might not be revealed by other genotoxicity assays.

A key study by Tomaszewska-Sowa et al. (2023) employed flow cytometry to assess nuclear DNA content and the degree of endoreduplication in roots, hypocotyls, and leaves of five crop species subjected to silver nanoparticle (AgNP's) exposure. It allowed precise quantification of nuclei with varying ploidy levels (e.g., 2C, 4C, 8C, etc.), revealing how AgNP treatment altered cell cycle dynamics without changing genome size. It was found that AgNPs at 50–100 ppm significantly increased endoreduplication in most crops—especially rapeseed, white mustard, and sugar beet—while promoting root elongation without affecting germination (Tomaszewska-Sowa et al. 2023). Micronuclei (MN) are key indicators of chromosomal damage and genomic instability, arising from chromosome fragments or entire chromosomes that fail to incorporate during cell division (Adams et al. 2024). Their presence signals genotoxic stress, often tied to mutagenic or carcinogenic agents (MacDonald et al. 2024). As reviewed by Kwasniewska and Bara (2022), the MN assay has evolved from classical cytogenetic microscopy methods to more advanced approaches incorporating fluorescence-based DNA staining, FISH, and molecular markers, which improve specificity and mechanistic understanding of DNA damage in plants. Although traditional flow cytometry struggled with distinguishing micronuclei from debris in plant cells, methods like the Flow Cytometry Micronucleus (FCMN) assay and ImageStream technology (Harte et al. 2024), now enable accurate, high-throughput detection. These platforms combine statistical robustness with image-based validation, enhancing genotoxicity screening across stressors such as heavy metals, radiation, and nanoparticles (Kwasniewska and Bara 2022).

12.4.9 Vascular and Root Development Alterations

The development of plant vascular systems and root architecture is vital for water and nutrient transport and structural integrity. Engineered nanomaterials can disrupt these processes, affecting root growth, cell division, and vascular differentiation. Flow cytometry has emerged as a key tool for studying these impacts, providing detailed insights into cell cycle changes and tissue development. Flow cytometry can be used to understand nanoparticle-caused modulations in vascular behavior (Kamel et al. 2024).

Maurer-Jones et al. (2013) highlighted flow cytometry's value in evaluating nanoparticle toxicity across diverse biological systems. Though not the primary methodology in all cases, it was used to assess functional endpoints such as respiration and protein synthesis in bacterioplankton exposed to silver nanoparticles, using redox dye uptake and radiolabeled leucine incorporation. Their findings showed that metal-based nanoparticles like Ag, ZnO, and TiO_2 negatively affect plants, bacteria, and aquatic organisms through mechanisms such as oxidative stress, ion dissolution, and membrane disruption. In plants, root elongation and DNA integrity were compromised, while bacterial communities exhibited altered respiration and enzyme activity. Flow cytometry provided high-resolution, real-time insights into these effects, outperforming traditional assays in detecting functional shifts in complex ecosystems.

The study (Nedosekin et al. 2011) introduced *in vivo* plant flow cytometry as a noninvasive method to monitor nanoparticle transport in plant vascular tissues (xylem and phloem). A key technique was Photoacoustic Flow Cytometry (PAFC), which utilized laser pulses to generate acoustic signals from nanoparticles like quantum dot-carbon nanotube conjugates (QD-CNTs). These signals, detected by ultrasound transducers, enabled real-time tracking of nanoparticles, even in deep tissues, with high sensitivity and without requiring fluorescent labels. PT measured heat-induced optical changes, enabling noninvasive, real-time tracking in vascular tissues. Despite limitations from scattering and background noise, fluorescence imaging aided in distinguishing QD-CNTs from autofluorescence. PT/PA methods, with high signal-to-noise ratios (up to 30), outperformed fluorescence in deep tissues. Innovations included label-free, multiplexed monitoring in chlorophyll-rich tissues, a portable fiber-based PAFC system, and performance-enhancing techniques like nonlinear PT amplification and spectral tuning.

12.4.10 Influence on Secondary Metabolism

Secondary metabolism involves the production of compounds like phenolics, alkaloids, flavonoids, terpenoids, and phytoalexins, crucial for plant defense and stress adaptation. These pathways are highly responsive to external stimuli, including engineered nanoparticles, which can enhance or inhibit metabolite biosynthesis (Zhang et al. 2022). Flow cytometry has proved to be a valuable tool for studying these interactions, offering rapid, sensitive analysis of individual. The interaction between nanomaterials and plant metabolism is complex and can lead to either enhancement or inhibition of these biochemical pathways. It provides insights not just into the presence of metabolites, but also their relative abundance and cellular distribution under different treatment conditions.

Cui et al. (2019) employed LC-MS-based metabolomics to assess the cytotoxic effects of metal oxide nanoparticles (ZnO, SiO_2, TiO_2, CeO_2) on human bronchial epithelial cells and mouse lung tissues. Although flow cytometry was not the primary technique, the study's insights complement cytometry-based approaches by revealing molecular-level disruptions. Functional assays (MTS, ATP, EdU, ROS, GSH, ELISA) measured viability, oxidative stress, and inflammation, paralleling emerging uses of photoacoustic flow cytometry in plants. ZnO NPs caused pronounced metabolic collapse, while other MOx NPs induced subtler, dose-dependent effects, notably impacting folate signaling. These findings, validated *in vivo,* support the integration of metabolomic profiling with cytometric tools to evaluate nanoparticle-induced stress and transport in plant systems.

A recent study by Selvakesavan et al. (2023) discussed several flow cytometry-relevant methodologies, particularly those based on high-throughput omics (metabolomics, proteomics, transcriptomics) and functional assays that are analogous to cytometric analyses. These approaches capture dynamic cellular responses like ROS generation, signal transduction, and the modulation of secondary metabolite biosynthesis upon NM exposure. The integration of nanoparticle-induced signaling cascades, such as ROS, calcium, nitric oxide (NO), and phytohormone signaling, provides mechanistic insight into how nanoparticles elicit changes in cell metabolism (Marslin et al. 2017), some of which could be profiled in future using advanced cytometric tools like photoacoustic flow cytometry in plant systems. Metal NPs (Ag, CuO, ZnO, CeO_2) influenced the production of flavonoids, alkaloids, terpenoids, and glucosinolates, with effects dependent on concentration and oxidative stress. Notably, NMs enhanced pharmacologically relevant metabolites like artemisinin and quercetin. These findings underscore the potential for coupling metabolomics with advanced cytometry to monitor nanomaterial impacts on plant metabolism.

12.5 Challenges in Nanomaterial Phytotoxicity Assessment Using Flow Cytometry

Nanomaterials exhibit size-dependent reactivity, surface charge changes, and agglomeration tendencies, which complicate exposure standardization (Richard et al. 2008). These properties, along with nanoparticle interactions with cell surfaces or fluorescent dyes, can skew results. Nanoparticles may also transform cells, such as generating reactive oxygen species (ROS), adding further complexity to the analysis and interpretation of cellular responses through flow cytometry (Fallah et al. 2024). The lack of standardized protocols for preparing plant samples for flow cytometry is a significant challenge. Plant cells' rigid cell walls and secondary metabolites require tailored methods for consistent sample preparation (Sliwinska et al. 2022). The presence of nanomaterials can interfere with enzymatic digestion and staining protocols, leading to variability in data and complicating cross-laboratory comparisons (Kyriakides et al. 2021).

Flow cytometry generates complex, multi-parametric data, where distinguishing real biological signals from artifacts caused by nanoparticles is critical. Nanoparticles can emit fluorescence or scatter light, mimicking cellular signals, leading to false positives (Bohmer et al. 2018). Additionally, plant tissues' intrinsic autofluorescence, combined with nanoparticle interference, makes gating specific cell populations difficult and requires careful interpretation (Donaldson 2020). Current flow cytometers may lack the sensitivity required to detect subtle changes caused by nanoparticle interactions. Advanced cytometers with better spectral resolution and sensitivity can address these limitations but are often expensive and require specialized training. Integration with techniques like imaging flow cytometry or microscopy may also be necessary for validating nanoparticle internalization and toxicity (Botha et al. 2021).

Most studies focus on short-term effects of nanoparticle exposure, but plants often face long-term, chronic exposure in natural environments. Long-term studies are essential to uncover cumulative effects like epigenetic changes or developmental interference. Flow cytometry alone is insufficient for assessing such effects, and integrating other methods is needed for a holistic understanding of long-term phytotoxicity (Pogribna and Hammons 2021).

12.6 Future Directions in Nanomaterial Phytotoxicity Assessment

To enhance specificity, future assays should target nanomaterial-specific interactions, using plant biomarkers like stress-responsive transcription factors. These assays, combined with reporter plant lines and flow cytometry, could enable real-time monitoring of nanotoxicity, improving screening precision for diverse nanomaterials (McCarrick et al. 2020). Long-term studies are essential to understand chronic exposure effects, such as epigenetic changes or developmental disruptions (Valente et al. 2023). Implementing multi-generational studies or seasonal cycle analyses, coupled with flow cytometry, could reveal delayed or cumulative toxicity, providing insights into the persistence of nanoparticle-induced stress over extended periods (Gao et al. 2023).

Combining flow cytometry with molecular, biochemical, and imaging techniques will provide a more comprehensive view of nanomaterial-induced plant changes. This multidimensional approach could link cellular responses to gene expression patterns and clarify nanoparticle internalization, leading to more accurate toxicological models (Browning et al. 2022). Artificial Intelligence and machine learning could enhance the interpretation of complex, multi-parametric flow cytometry data, identifying correlations and predicting plant responses to nanomaterials (Dhoble and Wu 2024). Integrating flow cytometry with multi-omic approaches will allow systems-level insights, creating predictive models of plant-nanoparticle interactions. (Aragoneses-Cazorla et al. 2022).

Real-world exposures often involve nanoparticle mixtures, which may have synergistic or antagonistic effects. Flow cytometry is well-suited for studying these mixtures by simultaneously monitoring multiple cellular parameters. This approach will help identify emergent properties of mixed exposures and improve risk assessments for complex nanomaterial mixtures (Li et al. 2020).

12.7 Conclusion

Flow cytometry has become an essential tool for evaluating how nanomaterials affect plant health and development. To overcome limitations in their present use, flow cytometry is often coupled with other techniques like fluorescent-based assays, comet assays and photoacoustic to provide a highly efficient yet thorough analysis. Thus, the technique is often utilized to assess plant cell viability, apoptosis, necrosis, generation of reactive oxygen species (ROS) and degree of mitochondrial damage following exposure to engineered nanoparticles. The document also examines how nanoparticles influence plant vascular structure, root growth, and the production of secondary metabolites. While flow cytometry offers powerful analytical capabilities, challenges such as nanoparticle behaviour variability, complex data interpretation, and a lack of long-term exposure data remain. Future advancements are needed to tackle these research gaps include the development of tailored assays, integration with omics technologies, and studies on mixed nanoparticle exposures to better assess real-world risks.

References

Adams DJ, Barlas B, McIntyre RE, Salguero I, van der Weyden L, Barros A et al (2024) Genetic determinants of micronucleus formation in vivo. Nature 627(8002):130–136

Aragoneses-Cazorla G, Buendia-Nacarino MP, Mena ML, Luque-Garcia JL (2022) A multi-omics approach to evaluate the toxicity mechanisms associated with silver nanoparticles exposure. Nano 12(10):1762

Araniti F, de la Peña TC, Sánchez-Moreiras AM (2018) Flow cytometric measurement of different physiological parameters. In: Springer eBooks. Springer Nature, p 195. https://doi.org/10.1007/978-3-319-93233-0_11

Arze AR, Manier N, Châtel A, Mouneyrac C (2020) Characterization of the nano–bio interaction between metallic oxide nanomaterials and freshwater microalgae using flow cytometry. Nanotoxicology 14(8). https://doi.org/10.1080/17435390.2020.1808106

Aseel DG, Ibrahim OM, Abdelkhalek A (2024) Biosynthesized silver nanoparticles mediated by Ammi visnaga extract enhanced systemic resistance and triggered multiple defense-related genes, including SbWRKY transcription factors, against tobacco mosaic virus infection. BMC Plant Biol 24(1):756

Awashra M, Młynarz P (2023) The toxicity of nanoparticles and their interaction with cells: an in vitro metabolomic perspective. Nanoscale Adv 5(10):2674–2723

Barkataki MP, Singh T (2019) Plant-nanoparticle interactions: mechanisms, effects, and approaches. In: Comprehensive analytical chemistry. https://doi.org/10.1016/bs.coac.2019.09.007

Bhatia SC (2017) Nanotechnology in agriculture and food industry. In: Food biotechnology. https://doi.org/10.1201/9781315156491-23

Bohmer N, Rippl A, May S, Walter A, Heo MB, Kwak M et al (2018) Interference of engineered nanomaterials in flow cytometry: a case study. Colloids Surf B: Biointerfaces 172:635–645

Botha J, Pugsley HR, Handberg A (2021) Conventional, high-resolution and imaging flow cytometry: benchmarking performance in characterisation of extracellular vesicles. Biomedicine 9(2):124

Browning AP, Ansari N, Drovandi C, Johnston AP, Simpson MJ, Jenner AL (2022) Identifying cell-to-cell variability in internalization using flow cytometry. J R Soc Interface 19(190): 20220019
Chichiriccò G, Poma A (2015) Penetration and toxicity of nanomaterials in higher plants [review of penetration and toxicity of nanomaterials in higher plants]. Nanomaterials 5(2). https://doi.org/10.3390/nano5020851
Cui L, Wang X, Sun B, Xia T, Hu S (2019) Predictive metabolomic signatures for safety assessment of metal oxide nanoparticles. ACS Nano 13(11):13065–13082
Dhoble S, Wu TH, Kenry (2024) Decoding nanomaterial-biosystem interactions through machine learning. Angew Chem Int Ed 63(16):e202318380
Ding P, Sakai T, Krishna Shrestha R, Manosalva Perez N, Guo W, Ngou BPM et al (2021) Chromatin accessibility landscapes activated by cell-surface and intracellular immune receptors. J Exp Bot 72(22):7927–7941
Djanaguiraman M, Anbazhagan V, Dhankher OP, Prasad PV (2024) Uptake, translocation, toxicity, and impact of nanoparticles on plant physiological processes. Plants 13:3137
Donaldson L (2020) Autofluorescence in plants. Molecules 25(10):2393
Dutta S, Pal S, Panwar P, Sharma RK, Bhutia PL (2022) Biopolymeric nanocarriers for nutrient delivery and crop biofortification [review of biopolymeric nanocarriers for nutrient delivery and crop biofortification]. ACS Omega 7(30). https://doi.org/10.1021/acsomega.2c02494
Eigemann F, Hilt S, Schmitt-Jansen M (2012) Flow cytometry as a diagnostic tool for the effects of polyphenolic allelochemicals on phytoplankton. Aquat Bot 104. https://doi.org/10.1016/j.aquabot.2012.10.005
Fallah S, Yusefi-Tanha E, Peralta-Videa JR (2024) Interaction of nanoparticles and reactive oxygen species and their impact on macromolecules and plant production. Plant Nano Biol:100105
Fomicheva M, Domblides E (2023) Mastering DNA content estimation by flow cytometry as an efficient tool for plant breeding and biodiversity research. Methods Protoc 6(1):18
Foyer CH, Kunert K (2024) The ascorbate-glutathione cycle coming of age. J Exp Bot 75(9): 2682–2699
Foyer CH, Shigeoka S (2011) Understanding oxidative stress and antioxidant functions to enhance photosynthesis. Plant Physiol 155(1):93–100
Francis DV, Jayakumar MN, Ahmad H, Gokhale T (2023) Antimicrobial activity of biogenic metal oxide nanoparticles and their synergistic effect on clinical pathogens. Int J Mol Sci 24:9998
Franqueira D, García MO, Torres E, Herrero C, Cid Á (2000) Potential use of flow cytometry in toxicity studies with microalgae. Sci Total Environ 247. https://doi.org/10.1016/s0048-9697(99)00483-0
Fratoddi I, Battocchio C, Iucci G, Catone D, Cartoni A, Paladini A et al (2021) Silver nanoparticles functionalized by fluorescein isothiocyanate or rhodamine B isothiocyanate: fluorescent and plasmonic materials. Appl Sci 11(6):2472
Gao M, Chang J, Wang Z, Zhang H, Wang T (2023) Advances in transport and toxicity of nanoparticles in plants. J Nanobiotechnol 21(1):75
Garcia Romeu H, Deville S, Salvati A (2021) Time-and space-resolved flow-cytometry of cell organelles to quantify nanoparticle uptake and intracellular trafficking by cells. Small 17: 2100887
Ghosh K, De S, Das S, Mukherjee S, Sengupta Bandyopadhyay S (2016) Withaferin A induces ROS-mediated paraptosis in human breast cancer cell-lines MCF-7 and MDA-MB-231. PloS one 11(12):e0168488
Gowtham HG, Shilpa N, Singh SB, Aiyaz M, Abhilash MR, Nataraj K et al (2024) Toxicological effects of nanoparticles in plants: mechanisms involved at morphological, physiological, biochemical and molecular levels. Plant Physiol Biochem 210:108604
Grauda D, Kataševs A, Belogrudova I, Čekstere-Muižniece G, Butkauskas D, Krasņevska N et al (2024, September) Influence of SiO2 nanoparticles on gametic cells of lime trees from urban area detected by flow cytometry. Proc Latvian Acad Sci Sect B Nat Exact Appl Sci 78(5–6)

Harte DS, Lynch AM, Verma J, Rees P, Filby A, Wills JW, Johnson GE (2024) A multi-biomarker micronucleus assay using imaging flow cytometry. Arch Toxicol 98(9):3137–3153

Heck DE, Shakarjian M, Kim HD, Laskin JD, Vetrano AM (2010) Mechanisms of oxidant generation by catalase. Ann N Y Acad Sci 1203:120–125

Heliopoulos NS, Kythreoti G, Lyra KM, Panagiotaki KN, Papavasiliou A, Sakellis E, Papageorgiou S, Kouloumpis A, Gournis D, Katsaros FK (2020) Cytotoxicity effects of water-soluble multi-walled carbon nanotubes decorated with quaternized hyperbranched poly (ethyleneimine) derivatives on autotrophic and heterotrophic gram-negative bacteria. Pharmaceuticals 13:293

Hong J, Peralta-Videa JR, Gardea-Torresdey JL (2013) Nanomaterials in agricultural production: benefits and possible threats? In: ACS symposium series. https://doi.org/10.1021/bk-2013-1124.ch005

Hu D, Bai Y, Li L, Ai M, Jin J, Song K, Liu D (2022) Phytotoxicity assessment study of gold nanocluster on broad bean (Vicia Faba L.) seedling. Environ Pollut Bioavailab 34:284–296

Huang X, Auffan M, Eckelman MJ, Elimelech M, Kim J-H, Rose J, Zuo K, Li Q, Alvarez PJ (2024) Trends, risks and opportunities in environmental nanotechnology. Nat Rev Earth Environ 5: 572–587

Hussain S (2018) Measurement of nanoparticle-induced mitochondrial membrane potential alterations. In: Nanotoxicity: methods and protocols. Springer New York, New York, pp 123–131

Ibuki Y, Toyooka T (2012) Nanoparticle uptake measured by flow cytometry. Methods Mol Biol. https://doi.org/10.1007/978-1-62703-002-1_11

Jalil SU, Ansari MI (2020) Role of nanomaterials in weed control and plant diseases management. In: Elsevier eBooks. https://doi.org/10.1016/b978-0-12-817852-2.00017-2

Joshua PE, Chukwuka RS, Arazu AV, Godfrey N, Ogwuegbu MC, Cosmas S (2020) Phytochemical and toxicity analysis of Ricinus communis. Asian J Biol 34. https://doi.org/10.9734/ajob/2020/v10i330109

Kamel MM, Badr A, Alkhalifah DHM, Mahmoud R, GadelHak Y, Hozzein WN (2024) Unveiling the impact of eco-friendly synthesized nanoparticles on vegetative growth and gene expression in Pelargonium graveolens and Sinapis alba L. Molecules 29(14):3394

Kang M, Liu Y, Weng Y, Wang H, Bai X (2024) A critical review on the toxicity regulation and ecological risks of zinc oxide nanoparticles to plants. Environ Sci Nano 11(1):14–35

Kay P, Choudhury R, Nel M, Orellana MV, Durand P (2013) Multicolour flow cytometry analyses and autofluorescence in chlorophytes: lessons from programmed cell death studies in Chlamydomonas reinhardtii. J Appl Phycol 25(5). https://doi.org/10.1007/s10811-013-9988-4

Khan MR, Khan AU, Bogdanchikova N, Garibo D (2021) Antibacterial and antifungal studies of biosynthesized silver nanoparticles against plant parasitic nematode Meloidogyne incognita, plant pathogens Ralstonia solanacearum and fusarium oxysporum. Molecules 26(9). https://doi.org/10.3390/molecules26092462

Kodaparthi A, Shivani T, Akhila K, Susmitha B, Chiranjeevi P (2025) Nanoparticle phytotoxicity. In: Plant-based nanoparticle synthesis for sustainable agriculture. CRC Press, pp 159–180

Kulaphisit M, Pomlok K, Saenjum C, Mungkornasawakul P, Trisuwan K, Wipasa J et al (2023) The anti-leukemic activity of a luteolin-apigenin enriched fraction from an edible and ethnomedicinal plant, Elsholtzia stachyodes, is exerted through an ER stress/autophagy/cell cycle arrest/apoptotic cell death signaling axis. Biomed Pharmacother 160:114375

Kumar V, Lakkaboyana SK, Sharma N, Abdelaal AS, Maitra SS, Pant D (2019) Engineered nanomaterials uptake, bioaccumulation and toxicity mechanisms in plants. Comprehens Analyt Chem. https://doi.org/10.1016/bs.coac.2019.09.005

Kumari A, Chokheli VA, Lysenko VS, Mandzhieva SS, Minkina TM, Mazarji M et al (2023) Genotoxic and morpho-physiological responses of ZnO macro-and nano-forms in plants. Environ Geochem Health 45(12):9345–9357

Kwasniewska J, Bara AW (2022) Plant cytogenetics in the micronuclei investigation—the past, current status, and perspectives. Int J Mol Sci 23(3):1306

Kyriakides TR, Raj A, Tseng TH, Xiao H, Nguyen R, Mohammed FS et al (2021) Biocompatibility of nanomaterials and their immunological properties. Biomed Mater 16(4):042005

Lara VM, Roballo KCS, Gotardo AT, Ambrósio CE (2018) Flow cytometry to evaluate potential developmental toxicants in the embryonic stem cell. Methods Mol Biol 165. https://doi.org/10.1007/978-1-4939-7883-0_7

Leary JF (2023) Flow cytometry-based detection. In: Single biomolecule detection and analysis. CRC Press, pp 123–164

Li M, Liu W, Slaveykova VI (2020) Effects of mixtures of engineered nanoparticles and metallic pollutants on aquatic organisms. Environments 7(4):27

Li R, Zhang R, Li Y, Liu C, Wang P, Sun H, Wang L (2024) Foliar uptake and distribution of metallic oxide nanoparticles in maize (Zea mays L.) leaf. Environ Sci Technol 58:16994–17003

Liu B, Peng B, Jin Y, Tao Y, Xu W, Zhang Y, Li Z (2025) Developmental toxicity and cardiotoxicity of N, N-dimethylaniline in zebrafish embryos. Toxics 13(2). https://doi.org/10.3390/toxics13020125

Lubos E, Loscalzo J, Handy DE (2011) Glutathione peroxidase-1 in health and disease: from molecular mechanisms to therapeutic opportunities. Antioxid Redox Signal 15(7):1957–1997

MacDonald KM, Khan S, Lin B, Hurren R, Schimmer AD, Kislinger T, Harding SM (2024) The proteomic landscape of genotoxic stress-induced micronuclei. Mol Cell 84(7):1377–1391

Mahajan PV, Dhoke SK, Khanna AS (2011) Effect of Nano-ZnO particle suspension on growth of mung (Vigna radiata) and gram (Cicer arietinum) seedlings using plant agar method. J Nanotechnol 2011. https://doi.org/10.1155/2011/696535

Manna I, Sahoo S, Bandyopadhyay M (2021) Effect of engineered nickel oxide nanoparticle on reactive oxygen species–nitric oxide interplay in the roots of Allium cepa L. Front Plant Sci 12: 586509

Marmiroli M, Marmiroli N, Pagano L (2022) Nanomaterials induced genotoxicity in plant: methods and strategies. Nanomaterials (Basel, Switzerland) 12(10):1658

Marslin G, Sheeba CJ, Franklin G (2017) Nanoparticles alter secondary metabolism in plants via ROS burst. Front Plant Sci 8:832

Mary LS, Truong L, Bieberich AA, Fatig RO III, Rajwa B, Tanguay RL (2023) Comparative analysis between zebrafish and an automated live-cell assay to classify developmental neurotoxicant chemicals. Toxicol Appl Pharmacol 476:116659

Maurer-Jones MA, Gunsolus IL, Murphy CJ, Haynes CL (2013) Toxicity of engineered nanoparticles in the environment. Anal Chem 85(6):3036–3049

McCarrick S, Cappellini F, Kessler A, Moelijker N, Derr R, Hedberg J et al (2020) ToxTracker reporter cell lines as a tool for mechanism-based (geno) toxicity screening of nanoparticles—metals, oxides and quantum dots. Nanomaterials 10(1):110

McKinnon KM (2018) Flow cytometry: an overview. Curr Protoc Immunol 120:5.1.1–5.1.11

Melegari SP et al (2013) Evaluation of toxicity and oxidative stress induced by copper oxide nanoparticles in the green alga Chlamydomonas reinhardtii. Aquat Toxicol 142:431–440

Miralles P, Church TL, Harris AT (2012) Toxicity, uptake, and translocation of engineered nanomaterials in vascular plants [review of toxicity, uptake, and translocation of engineered nanomaterials in vascular plants]. Environ Sci Technol 46(17). https://doi.org/10.1021/es202995d

Montes A, Bisson MA, Gardella JA, Aga DS (2017) Uptake and transformations of engineered nanomaterials: critical responses observed in terrestrial plants and the model plant Arabidopsis thaliana [review of uptake and transformations of engineered nanomaterials: critical responses observed in terrestrial plants and the model plant Arabidopsis thaliana]. Sci Total Environ. https://doi.org/10.1016/j.scitotenv.2017.06.190

Mustafa G, Chaudhari SK, Manzoor M, Batool S, Hatami M, Hasan M (2024) Zinc oxide nanoparticles mediated salinity stress mitigation in Pisum sativum: a physio-biochemical perspective. BMC Plant Biol 24(1):835

Nedosekin DA, Khodakovskaya MV, Biris AS, Wang D, Xu Y, Villagarcia H et al (2011) In vivo plant flow cytometry: a first proof-of-concept. Cytometry A 79(10):855–865
Nyland JF (2012) Determining biological activity of nanoparticles as measured by flow cytometry. Methods Mol Biol:167. https://doi.org/10.1007/978-1-62703-002-1_12
Oberdörster G, Oberdörster E, Oberdörster J (2005) Nanotoxicology: an emerging discipline evolving from studies of ultrafine particles. Environ Health Perspect 113(7):823–839
O'Brien IE, Reutelingsperger CP, Holdaway KM (1997) Annexin-V and TUNEL use in monitoring the progression of apoptosis in plants. Cytometry 29(1):28–33
Pogribna M, Hammons G (2021) Epigenetic effects of nanomaterials and nanoparticles. J Nanobiotechnol 19:1–18
Predoi D, Ghiță RV, Iconaru SL, Cîmpeanu C, Raită Ş (2020) Application of nanotechnology solutions in plants fertilization. In: IntechOpen eBooks. https://doi.org/10.5772/intechopen.91240
Proença PLF, Carvalho LB, Campos EVR, Fraceto LF (2022) Fluorescent labeling as a strategy to evaluate uptake and transport of polymeric nanoparticles in plants. Adv Colloid Interf Sci 305:102695
Pudake RN (2024) Novel application of bio-based nanomaterials for the alleviation of abiotic stress in crop plants. In: Nanotechnology for abiotic stress tolerance and management in crop plants, pp 181–201
Pudake RN, Chauhan N, Kole C (2019a) Nanoscience for sustainable agriculture. Springer
Pudake RN, Mittal J, Tripathi RM, Tyagi J, Mohanta TK (2019b) Biochemical responses of maize seedlings exposed to SnNPs. Micro Nano Lett 14:645–649
Pudake RN, Mohanta TK, Mahato N (2024a) Opportunities and challenges for nanotechnology in sustainable agri-food production. Front Nanotechnol 6:1420192
Pudake RN, Tripathi RM, Gill SS (2024b) Nanotechnology for abiotic stress tolerance and management in crop plants. Elsevier
Rana K, Luksiene Z, Pudake RN (2019) Recent advances in plant pathogen control by nanocides. In: Nanoscience for sustainable agriculture, pp 101–137
Ray MK, Mishra AK, Mohanta YK, Mahanta S, Chakrabartty I, Kungwani NA, Avula SK, Panda J, Pudake RN (2023) Nanotechnology as a promising tool against phytopathogens: a futuristic approach to agriculture. Agriculture 13:1856
Richard C, Doan BT, Beloeil JC, Bessodes M, Tóth É, Scherman D (2008) Noncovalent functionalization of carbon nanotubes with amphiphilic Gd3+ chelates: toward powerful T1 and T2 MRI contrast agents. Nano Letters 8(1):232–236
Robinson JP, Grégori G (2007) Principles of flow cytometry, p 19. https://doi.org/10.1002/9783527610921.ch2
Robinson JP, Ostafe R, Iyengar SN, Rajwa B, Fischer R (2023) Flow cytometry: the next revolution. Cells 12:1875
Rodríguez-Delgado MM, Martinez-Ledezma C, Villarreal-Chiu JF (2019) Advances in bio-coaters for nanoparticles and biodegradable delivery systems in agriculture and food industry: toward a safer and eco-friendly nanotechnology. In: Nanotechnology in the life sciences. https://doi.org/10.1007/978-3-030-17061-5_14
Santos CL, Pourrut B, Ferreira de Oliveira JM (2015) The use of comet assay in plant toxicology: recent advances. Front Genet 6:216
Schwab F, Zhai G, Kern M, Turner A, Schnoor JL, Wiesner MR (2015) Barriers, pathways and processes for uptake, translocation and accumulation of nanomaterials in plants—critical review [review of barriers, pathways and processes for uptake, translocation and accumulation of nanomaterials in plants—critical review]. Nanotoxicology 10(3). https://doi.org/10.3109/17435390.2015.1048326
Schwab F, Zhai G, Kern M, Turner A, Schnoor JL, Wiesner MR (2016) Barriers, pathways and processes for uptake, translocation and accumulation of nanomaterials in plants–critical review. Nanotoxicology 10(3):257–278

Selvakesavan RK, Kruszka D, Shakya P, Mondal D, Franklin G (2023) Impact of nanomaterials on plant secondary metabolism. In: Nanomaterial interactions with plant cellular mechanisms and macromolecules and agricultural implications, pp 133–170

Sevim Ç, Doğan E, Çomaklı S (2020) Cardiovascular disease and toxic metals. Curr Opin Toxicol 19. https://doi.org/10.1016/j.cotox.2020.01.004

Siddiqui S (2024) DNA damage, cell death, and alteration of cell proliferation insights caused by copper oxide nanoparticles using a plant-based model. Biology 13:805

Sliwinska E, Loureiro J, Leitch IJ, Šmarda P, Bainard J, Bureš P et al (2022) Application-based guidelines for best practices in plant flow cytometry. Cytometry A 101(9):749–781

Somasundaram S, Satheesh V, Singh M, Anandhan S (2021) A simple flow cytometry-based assay to study global methylation levels in onion, a non-model species. Physiol Mol Biol Plants 27(8): 1859–1865

Suman TY, Rajasree SR, Kirubagaran R (2015) Evaluation of zinc oxide nanoparticles toxicity on marine algae Chlorella vulgaris through flow cytometric, cytotoxicity and oxidative stress analysis. Ecotoxicol Environ Saf 113:23–30

Thuesombat P, Hannongbua S, Akasit S, Chadchawan S (2014) Effect of silver nanoparticles on rice (Oryza sativa L. cv. KDML 105) seed germination and seedling growth. Ecotoxicol Environ Saf 104:302–309

Tomaszewska-Sowa M, Rewers M, Bocian K, Sliwinska E (2023) Endoreduplication changes in five in-vitro-grown crops upon treatment with silver nanoparticles. Plant Cell Tissue Organ Cult 155(2):455–466

Tripathi S, Sarkar S (2015) Influence of water soluble carbon dots on the growth of wheat plant. Appl Nanosci 5:609–616

Tripathi S, Sonkar SK, Kumar A, Khan MY, Sarkar S (2009) Water soluble carbon nanotubes affect growth of the common gram (Cicer arietinum). Nat Prec. https://doi.org/10.1038/npre.2009.4056.1

Tripathi S, Sonkar SK, Sarkar S (2011) Growth stimulation of gram (Cicer arietinum) plant by water soluble carbon nanotubes. Nanoscale 3:1176–1181

Tripathi DK, Shweta S, Singh S, Singh S, Pandey R, Singh VP, Sharma N, Prasad SM, Dubey NK, Chauhan DK (2016) An overview on manufactured nanoparticles in plants: uptake, translocation, accumulation and phytotoxicity [review of an overview on manufactured nanoparticles in plants: uptake, translocation, accumulation and phytotoxicity]. Plant Physiol Biochem 110. https://doi.org/10.1016/j.plaphy.2016.07.030

Tuschl H, Schwab C (2004) Flow cytometric methods used as screening tests for basal toxicity of chemicals. Toxicol In Vitro 18(4):483. https://doi.org/10.1016/j.tiv.2003.12.004

Tutty MA, Vella G, Vennemann A, Wiemann M, Prina-Mello A (2022) Evaluating nanobiomaterial-induced DNA strand breaks using the alkaline comet assay. Drug Deliv Transl Res 12(9):2243–2258

Valente A, Vieira L, Silva MJ, Ventura C (2023) The effect of nanomaterials on DNA methylation: a review. Nanomaterials (Basel, Switzerland) 13(12):1880

Vuković S, Indjic D, Gvozdenac S (2014) Phytotoxic effects of fungicides, insecticides and nonpesticidal components on pepper depending on water quality. Pesticidi i Fitomedicina 29(2):145. https://doi.org/10.2298/pif1402145v

Wang X, Chen C, He C, Chen D, Yan W (2022) Mapping open chromatin by ATAC-seq in bread wheat. Front Plant Sci 13:1074873

Wu Y, Wang Y, Liu X, Zhang C (2024) Unveiling key mechanisms: transcriptomic meta-analysis of diverse nanomaterial applications addressing biotic and abiotic stresses in Arabidopsis Thaliana. Sci Total Environ 928:172476

Wunderlich AC, Zica É d OP, Ayres VF d S, Guimarães AC, Takeara R (2017) Plant-derived compounds as an alternative treatment against parasites in fish farming: a review [review of plant-derived compounds as an alternative treatment against parasites in Fish farming: a review]. In: InTech eBooks. https://doi.org/10.5772/67668

Yan A, Chen Z (2019) Impacts of silver nanoparticles on plants: a focus on the Phytotoxicity and underlying mechanism. Int J Mol Sci 20(5):1003

Zhang Y, Qi G, Yao L, Huang L, Wang J, Gao W (2022) Effects of metal nanoparticles and other preparative materials in the environment on plants: from the perspective of improving secondary metabolites. J Agric Food Chem 70(4):916–933

Zheng Y, Xie G, Zhang X, Chen Z, Cai Y, Yu W, Liu H, Shan J, Li R, Liu Y, Lei B (2017) Bioimaging application and growth-promoting behavior of carbon dots from pollen on hydroponically cultivated Rome lettuce. ACS Omega 2(7). https://doi.org/10.1021/acsomega.7b00657

Chapter 13
Long-Term Evaluation of Nanoparticle Contamination with the Help of Tree Ring Chemistry

Ivan Oyege, Nelson Mwebesa, Sana Awan, Sandeep Sharma, and Dennis Ssekimpi

Abstract The increasing prevalence of nanoparticles in the environment, resulting from both natural and anthropogenic activities, has raised important questions about their long-term ecological impacts. While conventional pollution monitoring techniques often focus on short-term or localized contamination, tree-ring chemistry offers a powerful means to investigate historical nanoparticle deposition across time and space. This chapter explores how trees can serve as long-term environmental recorders, capturing nanoparticle accumulation in annual growth layers, Tree-Rings. Emphasis is placed on the potential of tree ring chemistry, also known as dendrochemical analysis, to trace the movement, retention, and stability of metal-based nanoparticles introduced through atmospheric deposition and soil uptake. This chapter outlines the methodological advances and challenges of using tree rings as a natural archive for nanoparticle contamination. This interdisciplinary approach is significantly beneficial in environmental monitoring, regulation, and the development of bioindicators for the future in a nanoscale pollution era.

Keywords Nanotechnology · Dendrochemistry · Nanomaterials · Bioindicators · Phytoremediation · Nanotoxicology

---

I. Oyege (✉)
Department of Earth and Environment, Florida International University, Miami, FL, USA

Department of Chemistry, Busitema University, Tororo, Uganda

N. Mwebesa
Department of Chemistry, Makerere University, Kampala, Uganda

S. Awan
Department of Earth and Environment, Florida International University, Miami, FL, USA

S. Sharma
Department of Botany and Plant Sciences, University of California, Riverside, CA, USA

D. Ssekimpi
Department of Civil and Environmental Engineering, Florida State University, Tallahassee, FL, USA

R. N. Pudake et al. (eds.), *Phyto Nanotoxicology*, Nanotechnology in Plant Sciences 7, https://doi.org/10.1007/978-3-032-24518-2_13

13.1 Introduction

Environmental pollution has left lasting impacts on ecosystems worldwide, prompting scientists to develop innovative methods for reconstructing contamination histories. Among these, dendrochemistry, the study of chemical signatures preserved in tree rings (Balouet et al. 2023), has emerged as a powerful approach. Tree rings, the annual growth layers of woody plants, are naturally time-stamped records of environmental conditions (Wimmer 2002). As trees grow, they absorb and store elements and compounds from the surrounding air, water, and soil, leaving behind a chemical archive that can be read decades or even centuries later (Petkewich 2008). By analyzing the elemental and isotopic composition of these growth layers, researchers can reconstruct not only when a tree was exposed to pollutants but also the intensity and persistence of those exposures (Canning et al. 2023; Petkewich 2008). Tree-ring sampling is relatively inexpensive and minimally invasive, typically involving the extraction of a narrow core from the trunk without causing permanent harm to the tree (Canning et al. 2023; Petkewich 2008). Advances in analytical chemistry, including techniques capable of detecting trace elements at extremely fine spatial scales, now enable the measurement of subtle chemical signals in individual rings with high precision (Canning et al. 2023). This combination of practicality and sensitivity has established dendrochemistry as a valuable tool for long-term environmental monitoring, especially for contaminants that lack historical documentation (Binda et al. 2021).

One area of growing interest is the application of dendrochemistry to the study of nanoparticle (NP) contamination. Nanoparticles, defined as particles with at least one dimension smaller than 100 nm, represent a rapidly expanding class of pollutants with unique physicochemical properties and complex ecological interactions (Ballikaya et al. 2022b). Their sources encompass both natural events and a diverse range of anthropogenic activities, including combustion, vehicle emissions, industrial manufacturing, and engineered nanomaterial production (Aslam and Roeffaers 2022). Engineered nanomaterials like titanium dioxide (TiO_2), silver (Ag-NPs), gold (AuNPs), and zinc oxide (ZnO) are widely used in products ranging from sunscreens to pesticides, while incidental nanoparticles are produced in large quantities through processes like fuel combustion and metallurgy (Ballikaya et al. 2022b; Huber 2022). With industrialization and urbanization continuing to expand worldwide, atmospheric nanoparticle pollution has become a pressing environmental concern.

Over the past two decades, toxicological research has established that nanoparticles can pose risks to human health and ecological systems (Xuan et al. 2023). However, our understanding of their long-term interactions with plants, particularly whether they leave detectable and retrievable traces in tree biomass, remains limited (Ballikaya et al. 2022b; Huber 2022). This knowledge gap is critical because nanoparticles, as persistent pollutants, can integrate into biological systems and remain within air, water, and soil matrices for extended periods (Tran et al. 2024). Tracking their environmental fate, therefore, requires tools and analytical techniques capable of covering long timescales.

The longevity of trees and their natural capacity to record environmental conditions in annual growth layers offer a unique opportunity for such retrospective analysis (Hevia et al. 2018; Housset et al. 2018). Tree-ring chemistry has already proven helpful in reconstructing historical contamination by heavy metals, organic pollutants, and isotopes (Ballikaya et al. 2022b; Cocozza et al. 2019). Recent studies suggest it may also be effective for monitoring nanoparticle uptake and accumulation, especially of metal-based nanoparticles, showing that trees can act as passive bioindicators of both airborne and soil-based contamination (Cocozza et al. 2019). Compared to conventional phytoscreening methods, which detect recent or ongoing pollution, dendrochemistry offers the advantage of high-resolution, time-stamped records of past exposures (Balraju et al. 2025). Tree rings thus serve as chronological archives that not only reveal long-term trends in contamination but also allow for correlations with industrial activity, urban expansion, or regulatory interventions (Hevia et al. 2018).

Therefore, this chapter examines the potential of tree-ring chemistry as a tool for evaluating the long-term environmental impact of NPs. It introduces the fundamentals of tree-ring formation and contaminant uptake, reviews primary sources and types of nanoparticle pollutants, and examines current evidence for their deposition in woody tissues. Special attention is given to the pathways through which nanoparticles enter trees (via leaves, roots, and bark), their stability in woody matrices, and the influence of environmental factors such as humidity, soil chemistry, and species-specific traits. The chapter also discusses analytical challenges, future directions, and the promise of using dendrochemistry as a long-term monitoring system for nanopollution, thereby serving as "silent witnesses" to nanoparticle contamination in their environments.

13.2 Tree-Ring Chemistry as an Environmental Archive

Dendrochemistry, a sub-discipline of dendrochronology, deals with analyzing the chemical composition of dated tree rings (Fig. 13.1) to reconstruct past environmental conditions (Canning et al. 2023). Trees in temperate and many tropical regions produce visible annual growth rings, each ring typically comprising earlywood formed in the spring and latewood formed later in the season. The vascular cambium creates these rings as it adds new xylem tissue each year, building a sequential record of growth. Because the rings are formed in chronological order and remain largely undisturbed, the trunk of a tree functions as a natural archive of environmental history. When water, nutrients, or airborne substances are available during a given year, trees may incorporate these into the newly formed ring (Huber 2022; Petkewich 2008). Pollutants present in soil, water or air, such as heavy metals, organic compounds, or nanoparticles, can be taken up through roots or foliage and once deposited in woody tissues, may bind to cellulose, lignin, or cell walls (Petkewich 2008). In this way, each ring effectively locks in the chemical signals

Fig. 13.1 Cross-section of a tree trunk with distinct annual rings. Each ring marks 1 year, and ring chemistry can reconstruct pollution histories, including spikes linked to industrial emissions. (The figure was created in BioRender (https://BioRender.com))

of its growth year. Over time, successive rings accumulate outward, preserving the chronology of exposure (Petkewich 2008).

A cross-section or increment core from a tree reveals a layered chemical record, which can then be linked to historical events. Dendrochemical studies have successfully identified peaks of contaminants, such as lead, mercury, or cadmium, that correspond to industrial emissions, traffic density, or known pollution episodes (Ballikaya et al. 2022b; Huber 2022). In areas where instrumental records are sparse or entirely absent, tree-ring chemistry provides important information on how pollution has changed over decades or centuries (Canning et al. 2023). Beyond heavy metals, tree-ring archives have been utilized to reconstruct the impacts of acid rain, forest nutrient dynamics, wildfire histories, volcanic activity, and even forensic investigations of toxic releases (Canning et al. 2023; Petkewich 2008).

The reliability of tree-ring chemistry, however, depends on the complex interactions between tree physiology, species-specific traits, and environmental conditions. Elements differ in their mobility within wood, as some remain fixed in the ring of deposition, while others may migrate radially between rings, a process known as translocation, which can blur chronological signals (Canning et al. 2023). Different species also vary in their ability to absorb and sequester elements, and factors such as soil chemistry, drought stress, and growth rate, among others, further influence how contaminants are incorporated each year (Canning et al. 2023). As early as 1993, Cutter and Guyette emphasized that dendrochemical interpretations must account for variables such as soil type, xylem transport, and tree physiology, which may complicate the signal (Cutter and Guyette 1993). Despite these challenges, advances in analytical methods have greatly improved the reliability of dendrochemical reconstructions. High-resolution techniques, such as laser ablation ICP-MS and X-ray fluorescence mapping, now enable precise measurements of trace elements in individual rings (Canning et al. 2023). These developments have reinforced dendrochemistry as a credible environmental archive, comparable to other long-term records, such as lake sediments or ice cores, for reconstructing pollution histories from the peri-industrial to the post-industrial era (Canning et al. 2023).

In recent years, attention has turned to the potential of dendrochemistry to record NP contamination. Trees act as natural "archival recorders" of nanoscale pollution, capturing both natural and engineered NPs in their tissues (Binda et al. 2021). While early studies concentrated on trace metals such as Pb, Hg, Cd, and Zn from industrial emissions (Arnold et al. 2018; Bellis et al. 2004; Gačnik and Gustin 2023), recent findings suggest that trees can also preserve signatures of NP contamination (Ballikaya et al. 2022a, b; Cocozza et al. 2019). Because tree rings provide a high-resolution, time-stamped archive, they are especially promising for tracking nanoparticle exposures from both atmospheric deposition and soil uptake (Padilla and Anderson 2002). Once nanoparticles, particularly metal oxides, are taken up, they may remain stable in woody tissues for decades (Handy et al. 2008; Martin-Ortigosa et al. 2013). This stability increases the potential for long-term nanoparticle archiving within the layered porous structure of wood, which naturally supports retention and stabilization (Cui et al. 2025). Recent work indicates that natural nanoparticles are five times more abundant than engineered ones (Ballikaya et al. 2023a). While natural nanoscale particles have existed in the environment for millions of years, industrial activities have intensified nanoscale processes and increased contamination risks (Handy et al. 2008). Engineered nanoparticles, such as Ag, ZnO, and TiO_2, are of particular concern due to their persistence and reactivity, which may influence how they integrate into tree tissues (Cocozza et al. 2019).

Supporting evidence continues to grow; for example, Pulatoglu et al. (2025) demonstrated that tree rings can record airborne Cr pollution over time. Although Cr itself is not a nanoparticle, its deposition behavior, particularly its concentration in outer bark and restricted mobility into wood, resembles that of metal-based NPs. Similarly, Koç (2025) reported that heavy metals such as As, Pd, V, and Sr remain confined to the annual rings in which they were first deposited, with minimal migration, thereby reinforcing the reliability of tree rings as chronological archives. These findings indicate that dendrochemistry provides a robust and underutilized tool for monitoring nanoparticle contamination. By combining the long-lived structure of trees with advanced analytical techniques, researchers can reconstruct historical nanoparticle exposures and assess their persistence in ecosystems. In this way, trees extend their role from biological witnesses of conventional pollution to silent recorders of nanoscale contaminants.

13.3 Importance of Tree-Ring Chemistry for Environmental Monitoring of Nanoparticles

The importance of dendrochemistry in nanoparticle research lies in its capacity to reconstruct historical contamination with high resolution. Tree rings serve as chronological records, with their chemical signals often correlating with independent data sources such as emissions inventories, groundwater records, or soil contamination maps (Burken et al. 2011; Lv et al. 2019; Schwab et al. 2016; Su et al. 2019). Engineered nanoparticles such as CeO_2 and Fe_2O_3 exhibit low solubility and are stably retained in woody tissues, making them excellent markers of long-term exposure (Wojcieszek et al. 2020). Tree species selection plays a crucial role in dendrochemical monitoring because different tree species vary in their ability to absorb, retain, and redistribute contaminants (Cutter and Guyette 1993). *Tilia tomentosa*, for example, has been identified as a reliable biomonitor of airborne Cr due to its limited translocation, enabling precise temporal mapping (Pulatoglu et al. 2025).

In contrast, *Robinia pseudoacacia* and *Cedrus atlantica* accumulate higher levels of Cr, which makes them valuable for phytoremediation but less reliable for monitoring (Pulatoglu et al. 2025). Similarly, *Pinus nigra* has demonstrated a strong capacity to retain Pd and V compared to other conifers, such as *Picea orientalis*, and can therefore, be used as a biomonitor and a phytoremediation species (Koç 2025). *Fraxinus excelsior*, *Tilia tomentosa*, and *Pseudotsuga menziesii* demonstrated substantial Sr accumulation, despite internal redistribution limiting the precision of their monitoring; nevertheless, *Fraxinus excelsior* recorded the highest wood Sr concentrations, making it a suitable candidate for remediation strategies in polluted urban areas (Kulac et al. 2025).

Long-term studies strengthen these conclusions. For instance, a 27-year dendrochemical study in Mizoram, India, using *Magnolia champaca*, documented steady increases in Zn, Pb, Fe, Cu, Mn, and Ni in roadside plantations, linking them to traffic and industrial activity (Balraju et al. 2025). These findings confirm the value of dendrochemistry for reconstructing histories of both conventional metals and nanoscale contaminants. In forest ecosystems, evidence suggests that trees absorb airborne NPs through deposition on foliage and subsequent translocation into woody tissues, resulting in their incorporation into annual rings (Alizadeh et al. 2025). These examples demonstrate why tree-ring chemistry should be considered a crucial component of NP monitoring. Trees provide a long-term, spatially distributed, and cost-effective record of contamination, complementing instrumental monitoring systems and filling temporal gaps. By treating trees as biological archives, researchers can not only track historical NP exposures but also evaluate the effectiveness of environmental regulations and anticipate future ecological risks.

13.4 Sources of Nanoparticles

The small size of NPs makes them exhibit unique chemical and physical behaviors, such as high reactivity, catalytic activity, and unusual optical properties (Altammar 2023; Khan et al. 2019). These novel characteristics have made engineered NPs increasingly valuable in modern society. Over the past few decades, their production and use have grown exponentially. Metal oxides like TiO_2 and ZnO are incorporated into sunscreens and paints (Banerjee and Samanta 2022), Ag NPs are used in antimicrobial coatings (Dube and Okuthe 2025), carbon-based nanomaterials such as nanotubes and fullerenes are used in electronics, and composite nanomaterials are integrated into industrial processes (Park et al. 2020; Yu and Feng 2024). Alongside these engineered nanomaterials, incidental NPs are continuously generated as by-products of human activity (Hochella et al. 2019). Combustion engines and power plants produce ultrafine soot and metal-rich particles, while metallurgical and welding operations emit nanoscale fumes. Additionally, the abrasion of tires, brake pads, and construction materials releases nanoscale debris into the environment (Brink et al. 2022; Fussell et al. 2022; Gonet and Maher 2019). These particles disperse widely and now represent a pervasive form of pollution across air, water, and soil (Ballikaya et al. 2022b).

Environmental monitoring efforts have led to the detection of NPs not only in urban atmospheres but also in remote locations, surface and groundwater, soils, and even in Arctic ice (Materić et al. 2022a, b; Weber et al. 2023). Their small size allows long-range transport through air currents and water systems, while their high surface-area-to-volume ratio increases the likelihood of biological reactivity and toxicity (Weber et al. 2023). Concerns about the impacts of NP on human health, particularly respiratory and cardiovascular risks from inhaled ultrafine particles, and their potential to disrupt plant and animal physiology, have intensified in recent years (Portugal et al. 2024; Xuan et al. 2023).

Traditional monitoring frameworks were designed to track dissolved contaminants or larger particulate matter (PM_{10} and $PM_{2.5}$). NPs, being smaller than 0.1 μm, often pass undetected through these systems, and their behavior differs fundamentally from that of larger particulates. Moreover, NP contamination is usually chronic and diffuse, originating from routine processes such as traffic, wastewater effluent, or industrial emissions, rather than from singular catastrophic events (Rönkkö and Timonen 2019; Smita et al. 2012). This makes long-term monitoring essential, necessitating the need to establish how NP concentrations have evolved across decades and how they respond to technological changes or regulatory interventions.

Tree rings, with their annual resolution and capacity to record environmental inputs over the lifespan of the tree, offer a promising retrospective archive. If NPs are indeed incorporated into tree tissues, dendrochemistry can provide historical baselines, capturing conditions before the rise of nanotechnology and contrasting them with contemporary pollution levels (Ballikaya et al. 2022b; Kumah et al. 2023). For example, a tree located near a busy roadway might register elevated platinum-group

NPs in the years following the adoption of catalytic converters. In contrast, a tree near a factory could record TiO_2 NPs corresponding to industrial expansion.

Importantly, NPs relevant to tree-ring studies extend beyond engineered nanomaterials. Metallic and metal oxide particles (Pb, Cd, Al-rich dust from smelters or coal burning (Ballikaya et al. 2022b)), combustion-derived soot and black carbon, nanoplastics derived from polymer degradation, and naturally occurring ultrafine particles from volcanic eruptions or desert dust all represent nanoscale contaminants that trees could absorb (Yu et al. 2024). Heavy metal studies provide indirect confirmation, as dendrochemistry has consistently demonstrated that trees archive pollutants transported on fine particles. Historical analyses have shown Pb declines in tree rings after the phase-out of leaded gasoline and Hg peaks coinciding with industrial expansion (Binda et al. 2021; Huber 2022). In São Paulo, Brazil, tree rings revealed a marked reduction in Cd, Cu, Ni, and Pb between the 1980s and 2000s, indicating the success of emission controls (Binda et al. 2021). Hence, trees are already recording nanoscale pollutants, and the next step is to focus on engineered and incidental NPs as distinct targets explicitly.

Environmental gradients influence the accumulation of NPs in tree rings, and comparative research between roadside plantations and natural forests has revealed significantly higher concentrations of metals in urban-exposed trees, which show local vehicular and industrial activity (Balraju et al. 2025). NPs reach forest ecosystems through atmospheric deposition, rainfall, surface runoff, and soil interactions (Weber et al. 2023). Once inside ecosystems, they can disrupt microbial communities, alter nutrient cycles, and affect physiological processes in trees, including transpiration and carbon sequestration (Alizadeh et al. 2025; Kara et al. 2019). Dendrochemistry captures these cumulative interactions and preserves them chronologically, allowing for the reconstruction of NP impacts on ecosystem functions, such as fertility and carbon balance.

13.5 Nanoparticle Contamination and Their Environmental Fate

Nanoparticles persist in the environment due to their physicochemical properties, including small size, large surface area, and high reactivity. These features increase their ability to interact with biological surfaces and enhance their likelihood of being taken up by plants, including trees (Wang et al. 2013). Metal-based NPs such as TiO_2, CeO_2, ZnO, and carbon black exhibit remarkable stability in soils and waters, which makes them suitable candidates for long-term monitoring through dendrochemistry (Bundschuh et al. 2018). In natural environments, NPs transform such as aggregation, dissolution, oxidation, and surface modification. Their interactions with organic matter and colloids determine their mobility in soils and uptake by plants (Ju-Nam and Lead 2008). Some transformations, such as those involving Ag and Cu NPs, can enhance toxicity. For instance, these nanoparticles can damage chloroplasts in pine needles, altering plastoglobule formation even when root structures remain largely unaffected (Aleksandrowicz-Trzcinska et al. 2018).

Retention is also influenced by matrix interactions, such as embedding NPs in layered clay structures, which can reduce leachability and enhance long-term stability (Xie et al. 2025). Similar processes may occur in trees, where NPs bind to lignin and cellulose, remaining immobilized in xylem cell walls. Studies on Cr uptake show a parallel behavior where Cr adheres to particulates, is deposited on bark, and remains immobile in wood tissues (Pulatoglu et al. 2025). Furthermore, although not yet observed in NPs, Tree-ring records often show multi-element clustering, with Pd, V, and As frequently co-occurring with Tl, Sb, Se, and Pb, which suggests that trees can reflect complex contamination zones rather than single pollutants (Koç 2025). This capacity adds value for risk assessment, as tree rings provide an integrated record of environmental exposure.

13.6 Dendrochemical Evidence of Past Nanoparticle Pollution

Although direct studies of NPs in tree rings are still emerging, decades of dendrochemical research on heavy metals and other pollutants provide strong precedent. Many pollutants of the past were effectively NPs or submicron particles, even if they were not described in those terms at the time. For example, tree-ring studies in the 1980s linked elevated lead concentrations to emissions from leaded gasoline, which were primarily particulate in form (Doucet et al. 2012; Marty et al. 2024). Similarly, trees near smelters and coal-fired power plants show spikes of As, S, and Hg during peak industrial years (Ballikaya et al. 2022b; Huber 2022; Scanlon et al. 2020). Hg is especially illustrative since it can deposit in oxidized form on leaves or soil and enter trees either as a gas or bound to ultrafine aerosols. Tree rings from industrial regions have been used to reconstruct historical Hg contamination with striking accuracy (Baroni et al. 2023; Chen et al. 2021).

A global survey of dendrochemistry (Binda et al. 2021) confirmed that environmental contamination studies remain one of its most powerful applications, alongside climate and ecological reconstructions. Heavy metals such as Pb, Hg, Cd, Zn, and Cu are most frequently examined. Still, rings have also recorded constituents of acid rain, nitrogen deposition, hydrocarbons, and even petroleum residues from oil spills (Canning et al. 2023). In some cases, tree-ring data have been admitted as legal evidence, such as in lawsuits over gasoline leaks, where spikes of petroleum-derived chemicals in rings identified the timing of contamination (Petkewich 2008), thereby showing that tree rings function as impartial witnesses of pollution events.

Recent reviews on NPs suggest growing optimism, for instance Ballikaya et al. (2022b) asked whether tree rings could track atmospheric NP contamination over time. They concluded that trees can likely do so, but further validation is needed. They emphasized that dendrochemistry has already documented particulate emissions from traffic, coal burning, and metal refineries, all of which involve NPs. The review also summarized plant uptake experiments with engineered NPs, including Au and Ag, showing that ultrafine particles can indeed enter and persist in woody tissues. With the rapid advance of analytical techniques such as Laser Ablation Inductively Coupled Plasma Mass Spectrometry (LA-ICP-MS) and synchrotron probes, researchers now have the tools to test these questions directly. Unlike modern monitoring stations, which are limited in both time and space, trees have been sampling their environment passively for decades to centuries, providing the only way to reconstruct NP pollution in the pre-monitoring era.

The "Trees as Witnesses to Environmental Pollution" report published by WSL (Switzerland) in 2022 echoed these points for a broader audience (Huber 2022). It highlighted cases where tree rings recorded pollutants from car exhaust, metal smelting, and coal combustion, clearly showing that dendrochemistry captures contamination histories. The report called for integrating tree-ring analysis into environmental monitoring networks to address NP pollution more systematically. However, concrete case studies directly linking rings to engineered NP emissions are still scarce; the conceptual and methodological groundwork is, nonetheless, firmly in place. If trees have recorded heavy metals, hydrocarbons, and even petroleum residues, there is every reason to expect they can archive NPs. For example, sampling trees near a TiO_2 NP factory could reveal low baseline Ti concentrations before operations began, followed by elevated levels in subsequent decades. Similarly, urban trees may one day reveal the trajectory of NP exposures linked to catalytic converters, low-emission zones, or the transition to electric vehicles.

13.7 Factors Influencing Tree-Ring Chemistry for Nanoparticle Monitoring

The effectiveness of dendrochemistry for NP monitoring depends on a combination of environmental, biological, and methodological factors. Environmental conditions such as humidity, soil pH, ionic strength, and organic matter influence NP stability and uptake (Bundschuh et al. 2018; Handy et al. 2008). Climate variability, including shifts in rainfall and temperature, alters elemental mobility in wood and must be considered when interpreting tree-ring data (Hevia et al. 2018). NP properties also matter, such as particle coatings, size, and reactivity, which affect detectability and retention in woody tissues (Su et al. 2020), and standardizing sampling protocols, analytical methods, and data corrections is critical to ensure reproducibility across studies (Ballikaya et al. 2023b; Balouet et al. 2023). Without consistency, inter-study comparisons become unreliable.

Furthermore, species-specific traits add another layer of complexity. For instance, *Pinus nigra* shows strong retention of Pd, V, and Sr, making it a valuable biomonitor (Koç 2025). By contrast, *Fraxinus excelsior* accumulates high Sr concentrations but shows internal redistribution, which undermines chronological fidelity (Kulac et al.

2025). Monitoring requires species with stable retention and low translocation, while phytoremediation favors species with high uptake regardless of temporal stability. Moreover, engineered wood treatments provide functional analogues; for instance, Cu-MMT–tung oil composites demonstrate slow release and strong retention of NPs, mimicking possible behavior in natural tree tissues (Xie et al. 2025). Similarly, $NiFe_2O_4/Ni_3Fe$ composites embedded in biomass exhibit complex electromagnetic behaviors, including interfacial polarization and dielectric loss (Cui et al. 2025). While trees are not engineered for such functions, these findings show new possibilities for NP detection in dendrochemical studies.

Additionally, tree physiology and climate interact to influence NP uptake, as factors such as vapor pressure deficit, stomatal conductance, and water availability regulate how contaminants enter leaves and roots (Monaci and Baroni 2025). As climate change shifts forest composition and physiology, NP assimilation patterns may also change, complicating long-term comparisons. Therefore, integrating eco-physiological data with dendrochemical records will be essential for interpreting NP chronologies in future studies.

13.8 Mechanisms of Nanoparticle Uptake and Transport in Trees

For tree rings to function as reliable archives of NP contamination, the particles or their elemental constituents must first enter the tree and then be sequestered within the woody tissues that form the annual rings. Research over the past two decades has demonstrated that trees, like other plants, can indeed absorb NPs from their environment. The two main entry pathways are through the foliage and through the roots, although bark and stem surfaces also play an important role in direct deposition (Wang et al. 2023).

13.8.1 Foliar Uptake Pathways

Leaves are often the first interface between trees and airborne NPs (Fig. 13.2). Particles deposited on leaf surfaces may enter through stomata, typically 10–100 μm in diameter, large enough to admit nano- and even some micro-sized particles (Wang et al. 2023). In addition, NPs can penetrate the cuticle either through hydrophilic pores (for very small hydrophilic NPs, typically <5 nm) or through lipid-associated pathways (for hydrophobic particles) (Wang et al. 2023). Once inside leaf tissues, NPs can move with the transpiration stream in the xylem or, in some cases, transfer into the phloem, allowing bidirectional redistribution of particles within the plant (Wang et al. 2023; Wu and Li 2022).

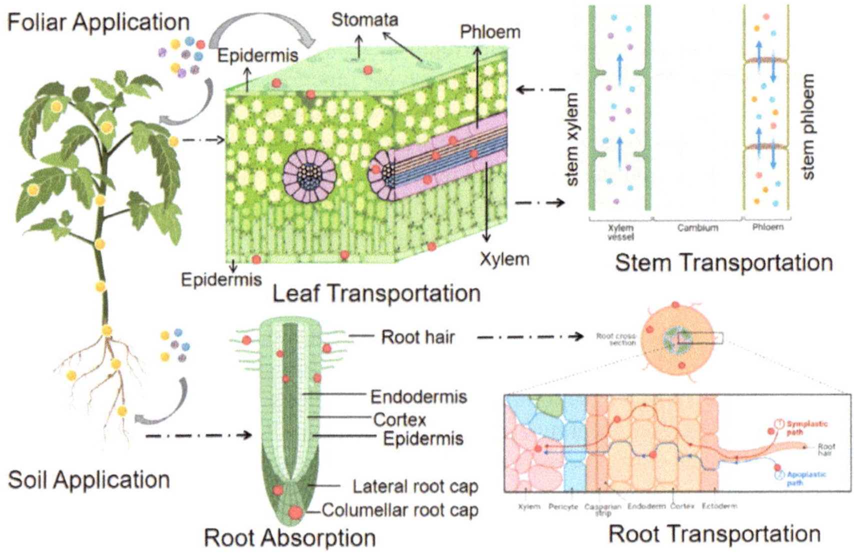

Fig. 13.2 Pathways of nanoparticle uptake in plants. Nanoparticles can enter through leaf stomata or cuticles and move via xylem or phloem or be absorbed by roots through apoplastic and symplastic routes before translocation to stems and leaves. (Adapted from Wang et al. (2023))

A landmark experiment by Ballikaya et al. (2022a) provided the first direct evidence of NP uptake and internal translocation in mature trees. In this study, Au NPs were sprayed onto leaves of European beech (*Fagus sylvatica*) and Scots pine (*Pinus sylvestris*) saplings. Au was selected as a tracer due to its chemical inertness and ease of detection. Within 20 days, Au was detected not only on leaf surfaces but also in internal leaf tissues, stems, and even roots. The results indicated stomatal entry followed by systemic transport through the vascular system, ultimately depositing in woody tissues where they could later be archived in tree rings (Huber 2022). Foliar application produced higher concentrations in stems than root exposure, showing the efficiency of leaf pathways for NP transport.

Other studies confirm the foliar uptake of Ag, ZnO, and CeO_2 NPs in tree species, as well as in herbaceous plants such as *Arabidopsis*, rice, and wheat (Wang et al. 2023). Birch trees, for example, have been shown to internalize Ag NPs in their leaves, with imaging techniques confirming their presence inside leaf cells (Ballikaya et al. 2023a; Weber 2022). Micro- and nanoplastics also penetrate through foliage, and experiments have demonstrated uptake of plastic particles smaller than 10 μm, with birch trees retaining fragments in both root and stem tissues (Austen et al. 2022; Weber et al. 2023); therefore, trees can capture a broad spectrum of ultrafine pollutants deposited from the atmosphere.

13.8.2 Root Uptake Pathways

Roots provide the second central entry point for NPs (Fig. 13.2), particularly those present in contaminated soil or groundwater. Uptake occurs mainly at young root tips and hairs, where NPs cross the epidermis and cortex either apoplastically (between cells) or symplastically (through cells), eventually entering the xylem stream (Wang et al. 2023). Studies with herbaceous plants have documented root absorption of CeO_2 and ZnO NPs, which are then translocated to shoots and leaves (Wang et al. 2023). In trees, Ballikaya et al. (2022a) experiments showed that roots can indeed absorb Au NPs, although less efficiently than foliar routes.

Soil conditions strongly influence NP root uptake, and factors such as pH, organic matter content, and redox state determine whether NPs remain in particulate form or dissolve into ions, which can then bind to xylem tissues (Wu and Li 2022). Some particles undergo chemical transformations within roots, such as dissolution, sulfidation, or binding to cell walls, which affect their transport and long-term stability (Yan and Chen 2019). There is also evidence that NPs absorbed by roots can be re-released into the rhizosphere through exudation, especially in fibrous root systems, creating a dynamic two-way exchange between trees and their soils (Alizadeh et al. 2025).

13.8.3 Bark and Direct Stem Uptake

In addition to leaves and roots, bark surfaces act as a critical interface for atmospheric deposition (Fig. 13.2). Rough-textured bark traps airborne particles, and outer bark tissues often show contaminant concentrations up to ten times higher than wood (Kulac et al. 2025). Some NPs, especially soluble or very fine particles, may migrate into the inner bark or even into the cambial region. Cr uptake studies illustrate this process, where airborne Cr adhered to bark surfaces, with limited but measurable movement inward, a pattern that likely applies to other metal-based NPs as well (Pulatoglu et al. 2025).

13.8.4 Internal Transport and Retention

Once NPs enter the vascular system, they are carried in the transpiration stream of the xylem or distributed via the phloem (Sembada and Lenggoro 2024). Their transport and final deposition depend on particle size, chemistry, and environmental conditions. Tiny particles (<5 nm) can move through phloem transport proteins, while larger aggregates often accumulate on cell walls or in intercellular spaces (Etxeberria et al. 2016). Experimental studies have shown that CeO_2 NPs remain largely intact during transport, while TiO_2 penetration into leaves is influenced by environmental conditions such as wind speed (Hong et al. 2014; Räsänen et al. 2017). Humidity fluctuations and hydrogen bonding also play a role in deposition. For example, soot NPs can be retained in leaf surfaces within hours, with larger clusters migrating slowly into tissues (Burkhardt et al. 2012; Tao et al. 2023). Once inside woody tissues, NPs may either remain in a solid form or partially dissolve, releasing ions that then bind to lignin, cellulose, or other polymers (Yan and Chen 2019). Such binding contributes to the long-term stabilization of NPs in tree rings. Conifers, with their long-lived xylem tissues, tend to retain NPs more effectively than deciduous species, where internal cycling may be more dynamic (Cocozza et al. 2016). Trunk injection studies have shown uneven distribution, with the highest concentrations often in stems, followed by roots and branches. Internal filters, such as pit membranes, restrict movement between conduits, while seasonal cycles, like freeze-thaw events, have a minimal effect on NP retention (Guan et al. 2022; Su et al. 2020).

Besides the NP uptake pathways described, evidence from wood composites and other engineered systems helps explain how NPs behave once inside tree tissues. Experiments with $NiFe_2O_4/Ni_3Fe$-loaded carbon composites showed that NPs form strong interfacial bonds with carbon substrates, stabilizing them structurally (Cui et al. 2025). Similar interactions may occur in natural wood, where NPs bind to lignocellulosic polymers and remain chemically locked in place. Studies with Cu-MMT–tung oil composites also suggest that NPs can fill voids and form surface films, processes that likely parallel retention in xylem cell walls (Xie et al. 2025). Furthermore, although Hg is not an NP, its uptake through stomata, intracellular oxidation, and binding to thiol-rich compounds mirrors many of the pathways described for NPs (Monaci and Baroni 2025).

Furthermore, the dynamics of NP uptake and retention vary widely across species and conditions. Traits such as stomatal conductance, phloem loading efficiency, and xylem architecture all affect NP mobility (Monaci and Baroni 2025). For example, *Tilia tomentosa* effectively archives Cr due to low internal redistribution, while *Robinia pseudoacacia* accumulates large quantities but redistributes them too extensively for accurate monitoring (Pulatoglu et al. 2025). Similarly, *Pinus nigra* shows strong retention of Pd and V, making it a promising biomonitor (Koç 2025). Recent work has also highlighted the bidirectional movement of NPs. Au-NPs applied to the foliage of beech and pine trees were later found in root tissues and even in the surrounding soil, suggesting phloem-mediated downward transport and exudation (Alizadeh et al. 2025). This complexity emphasizes the need for careful interpretation of dendrochemical records, as tree rings preserve contamination signals that reflect a combination of uptake, internal redistribution, and species-specific filtering.

13.9 Analytical Techniques for Determining Nanoparticles in Tree Rings

Detecting NPs in tree rings poses a formidable challenge due to their small size, low environmental concentrations, and potential transformations within woody tissues (Zhang et al. 2024). Over the past half-century, however, dendrochemistry has developed from relatively simple bulk methods into a sophisticated suite of high-resolution analytical techniques capable of tracing NPs and their chemical fingerprints.

13.9.1 Historical Approaches and Their Limitations

Early dendrochemical studies, conducted from the 1970s to the 1990s, relied on bulk analyses, where wood samples were often dissolved or ashed, and the extracts were analyzed using atomic absorption spectroscopy or inductively coupled plasma mass spectrometry (ICP-MS) (Tafur-Marinos et al. 2016). While these approaches provided total metal concentrations, they lacked spatial resolution and often blurred the chronological signal by mixing adjacent rings (Canning et al. 2023). Despite these drawbacks, they laid the foundation by showing that trees could archive heavy metals and other pollutants in their growth rings.

13.9.2 Advances in Solid Sampling and High-Resolution Methods

The advent of direct solid sampling transformed dendrochemistry. LA-ICP-MS enables researchers to ablate microscopic spots or lines on polished cores, then feed the vapor into an ICP-MS for elemental detection. This method can achieve sub-millimeter resolution, enabling the imaging of element distributions, such as copper, zinc, and lead, across annual boundaries (Kim et al. 2020; USGS 2004). Its sensitivity, in the parts-per-billion range, makes it particularly suited to NP-derived elements. For example, a spike in Ti from TiO_2 NPs in a given year could be detected and localized within specific ring sectors, hinting at a sudden deposition event (Amais et al. 2021).

X-ray fluorescence (XRF) analysis has also become a valuable tool since core scanners equipped with XRF can rapidly screen entire wood cores for heavy metals, providing quick, non-destructive surveys (Petkewich 2008; Rocha et al. 2020). Energy-dispersive XRF has even been used in "forest forensics," identifying chlorine and Pb as markers of solvent spills and gasoline contamination (Petkewich 2008). While XRF generally has higher detection limits than LA-ICP-MS, synchrotron-based micro-XRF can achieve the sensitivity required for NPs. Additionally, although not directly applicable to inorganic NPs, Mass spectrometry of evolved vapors has been applied to volatile pollutants. For instance, heating cores in sealed vials and analyzing vapors with gas chromatography have revealed chlorinated solvent contamination in tree rings (Petkewich 2008). With these creative and analytical approaches, tree tissues retain even volatile compounds, and novel adaptations could one day extend to nanoplastics or engineered organics.

13.9.3 Microscopy and Direct Visualization

To confirm the physical presence of NPs, electron microscopy remains indispensable. Scanning electron microscopy (SEM) with energy-dispersive spectroscopy (EDS) enables the visualization of particle aggregates and confirmation of their elemental composition. In contrast, transmission electron microscopy (TEM) resolves particles as small as 10 nm within cell walls or organelles. In NP uptake studies, TEM images have confirmed Ag NPs lodged in birch tissues and Au NPs in beech leaves (Ballikaya et al. 2023a; Huber 2022; Weber 2022). Although Microscopy is low throughput, it provides qualitative confirmation that complements chemical analyses.

13.9.4 Isotope Ratio and Fingerprinting

In addition to concentration data, isotopic fingerprints help distinguish between natural and anthropogenic sources. For example, ratios such as $^{206}Pb/^{207}Pb$ have been used to trace Pb from gasoline additives (Binda et al. 2021; Bindler et al. 2004). Similarly, engineered NPs could carry unique isotopic or molecular signatures, for instance, enriched isotopes in manufactured ZnO or additives in nanoplastics, that provide a powerful source of attribution.

13.9.5 Multi-Method Workflows

In practice, dendrochemistry increasingly employs multi-method workflows in which a core may first be dated, mounted, and screened with XRF to identify anomalies, followed by LA-ICP-MS to resolve high-resolution elemental records. Microscopy may then confirm the presence of NPs, and isotopic analyses may pinpoint their source. These stepwise approaches not only enhance accuracy but also allow cross-validation with independent environmental data such as emissions records or soil contamination surveys (Binda et al. 2021). Recent adaptations include single-particle ICP-MS, which can confirm whether detected elements are present as NPs or dissolved ions (Ballikaya et al. 2022b). Other techniques such as X-ray photoelectron spectroscopy (XPS), Fourier-transform infrared spectroscopy (FTIR), and synchrotron-based microspectroscopy have been applied to NP-treated woods, revealing bonding states, microstructural changes, and transformation processes (Kulac et al. 2025; Xie et al. 2025). It will therefore be possible in the future to repurpose these tools for dendrochemistry to distinguish engineered from natural NPs and to track chemical transformations over time.

13.9.6 Emerging Approaches and Challenges

Detecting organic NPs, such as nanoplastics, presents additional challenges because their elemental composition (C, H, O) is similar to that of wood. Potential solutions include pyrolysis-GC/MS, which breaks down embedded polymers into identifiable fragments, and infrared microspectroscopy, which can detect characteristic polymer signals (Binda et al. 2021). Although still experimental, these methods represent the next frontier for dendrochemistry. Recent studies have also showed how NP–wood interactions influence detection. Experiments with $NiFe_2O_4/Ni_3Fe$ composites in wood matrices have shown electromagnetic behaviors, such as Debye relaxation and interfacial polarization, which may complicate spectroscopic signals but also provide diagnostic markers (Cui et al. 2025). Similarly, Cu-MMT composites demonstrate that layered structures slow NP release, indicating analogous processes in tree tissues (Xie et al. 2025). Furthermore, mathematical modeling of radial translocation further refines chronological resolution by accounting for limited elemental migration between rings. Therefore, with advanced instruments like thermal decomposition amalgamation atomic absorption spectroscopy and stable isotope profiling, these approaches are enhancing the precision of dendrochemical reconstructions (Monaci and Baroni 2025).

13.10 Limitations of Tree-Ring Chemistry for Nanoparticle Pollution Monitoring

While tree-ring chemistry holds great promise for reconstructing NP pollution histories, it is not without its challenges. These limitations, including analytical, biological, and environmental dimensions, each influence how reliably dendrochemical data can be interpreted.

13.10.1 Analytical and Detection Challenges

The first difficulty lies in detection and attribution, where NPs themselves are rarely measured directly; instead, dendrochemists often rely on elemental proxies for their detection. This works well for metal-based NPs such as Ag or Au, where elevated concentrations of Ag or Au in rings can be interpreted as NP signatures. However, the task becomes far more complex for particles composed of organic polymers, composite materials, or engineered nanostructures that lack unique elemental markers. Detecting nanoplastics, for example, is complicated by the fact that wood and polymers are both primarily composed of carbon, hydrogen, and oxygen. Even when elements are detectable, concentrations in tree tissues may be extremely low, often at parts-per-billion or less, requiring ultra-sensitive analytical methods and strict contamination control. Modern dust contamination, for example, can easily confound historical signals if cores are not handled with care. This makes methodological rigour a prerequisite for dendrochemical research in NP.

13.10.2 Biological Redistribution and Stability

A second major limitation is biological redistribution, as trees are dynamic organisms in which water and solutes move not only vertically through the xylem and phloem but also sometimes radially across rings. Mobile elements, such as ZN or Cu, can migrate from older rings into newer tissues, especially during processes like sapwood aging or heartwood formation (Canning et al. 2023). When NPs dissolve into ionic form, redistribution may blur short-term pollution signals into a diffuse background. The physical size of NPs can both aid and hinder interpretation, as particles that lodge in cell walls may remain fixed. At the same time, those that dissolve into ions are more likely to move and reprecipitate elsewhere. Stability depends on the particle type, where inert Au may persist in a solid form indefinitely, while ZnO or Ag NPs may partially dissolve and redistribute over time. Understanding the chemical transformations of NPs once inside tree tissues remains a critical research area.

13.10.3 Differentiating Natural from Anthropogenic Inputs

Another challenge is distinguishing between natural nutrient uptake and pollution signals. Trees naturally require and accumulate elements such as Zn, Cu, and Mn, which are also common in engineered or incidental NPs. Elevated levels in rings may therefore reflect natural physiology rather than contamination. To separate the two, researchers often compare trees from polluted sites with those from background locations, looking for excess concentrations in the impacted trees. Isotopic analyses provide another powerful tool because anthropogenic pollutants often carry distinctive isotopic signatures compared with natural sources. The use of $^{206}Pb/^{207}Pb$ ratios to trace lead from gasoline additives is a classic example (Binda et al. 2021; Bindler et al. 2004), and similar strategies may be applied to NPs, particularly if they contain isotopically enriched materials or unusual element ratios.

13.10.4 Tree-to-Tree and Species-Specific Variability

Tree-to-tree variability adds further complexity because even individuals of the same species growing side by side can show different baseline concentrations due to microenvironmental factors or genetic variation (Canning et al. 2023). For this reason, dendrochemical studies rely on replication, and pollution events are only considered credible if they appear in multiple trees from the same site and time period (Canning et al. 2023). Species traits also matter, as some trees exhibit stable contaminant retention, while others redistribute elements internally, thereby weakening chronological resolution. Diffuse-porous species such as sycamore may provide more stable records compared to ring-porous species like oak or Scots pine (Gačnik and Gustin 2023; Scharnweber et al. 2016). Conifers such as *Larix decidua* and *Picea abies* often produce clearer archives due to rapid heartwood formation, whereas some *Pinus* species show significant radial migration of contaminants (Monaci and Baroni 2025).

13.10.5 Temporal Resolution and Dating Issues

Dendrochemistry theoretically provides annual resolution, but in practice, this can be compromised as some tropical trees lack distinct yearly rings. In contrast, others may produce missing or false rings during years of climatic stress. Cross-dating methods, which compare ring-width patterns across trees, are crucial for avoiding misaligned chronologies. Even when rings are correctly dated, contamination events may affect only part of the growing season, influencing only earlywood or latewood. Intra-ring analysis has been used to capture this seasonal timing (Petkewich 2008), but it requires very fine-scale resolution.

13.10.6 Environmental and Physiological Constraints

Environmental conditions such as temperature, rainfall, soil pH, and organic matter content also affect NP uptake and mobility (Alterio et al. 2020; Dinis et al. 2021; Locosselli et al. 2018). These background influences can mask or exaggerate pollution signals unless they are carefully accounted for through statistical or isotopic corrections. Climate variability alters not just uptake rates but also tree physiology, changing stomatal conductance, transpiration, and xylem transport patterns. However, not all NPs are equally bioavailable since large agglomerates or strongly hydrophobic particles may adhere to bark or leaf surfaces but never cross biological barriers. In contrast, smaller particles (<50 nm) or those with surface coatings designed for solubility are more likely to enter trees (Wang et al. 2023). Null results in dendrochemical studies may therefore reflect either the absence of

NPs in the environment or their failure to penetrate the tree. Controlled exposure experiments, such as the Au-NP study (Ballikaya et al. 2022a), remain crucial for calibrating which particle types, trees can effectively archive.

Therefore, these limitations indicate the need for cautious and multidisciplinary interpretation of dendrochemical data. Analytical constraints, physiological complexity, species-specific variability, and environmental influences all mean that tree-ring records of NPs are rarely straightforward. Robust conclusions require replication across trees, careful choice of species, ultra-sensitive analyses, and corroboration with other environmental records.

13.11 Conclusion

The potential applications of tree-ring chemistry for NP monitoring are far-reaching despite available challenges. As analytical technologies improve and controlled calibration studies expand, dendrochemistry could become a standard tool for environmental surveillance. One promising vision is the integration of trees into pollution monitoring networks. Just as air filters and weather stations provide routine data, sentinel trees in industrial and urban areas could serve as living archives. Incrementing cores collected at regular intervals would reveal whether NP deposition is increasing or declining. Fast-growing species such as poplars or willows could be planted around factories or landfills for this purpose.

Species selection and genetic approaches also hold promise, for instance, if certain trees demonstrate a superior ability to archive NPs with minimal redistribution, they could be prioritized for monitoring. Conversely, trees that accumulate NPs rapidly but redistribute them internally may be better suited for phytoremediation. Tree-ring NP records could also enhance ecotoxicology studies. If periods of high NP deposition coincide with reduced growth or physiological stress in trees, dendrochemistry could reveal long-term ecological effects of NP exposure. Because trees are keystone species, their records may indirectly indicate risks for entire ecosystems. Beyond present-day monitoring, dendrochemistry offers a means to reconstruct historical pollution. Historical NP releases, undocumented in instrumental records, may still be preserved in old trees. Such information could be vital for environmental forensics, legal cases, or remediation planning.

Future applications may expand into contexts related to climate change and natural disturbances. Wildfires, for example, generate nanoscale soot and ash that may be archived in nearby forests, providing a combined record of both climate-driven events and nanopollution. Furthermore, advances in remote sensing, artificial intelligence, and big data analysis could further enhance dendrochemistry, enabling the analysis of large datasets of multi-element profiles for subtle patterns and pollution source attribution.

In essence, trees act as "silent witnesses" to environmental change, extending dendrochemistry into the NP domain. This not only builds on decades of research into heavy metals and other pollutants but also aligns with pressing concerns about the environmental footprint of nanotechnology. As methodologies improve, tree-ring chemistry may become a cornerstone for global NP monitoring, bridging past and present records and guiding future environmental policy.

Acknowledgments Ivan Oyege is grateful to the Fulbright Foreign Students Program Uganda for providing him with funding.

References

Aleksandrowicz-Trzcinska M, Szaniawski A, Studnicki M, Bederska-Blaszczyk M, Olchowik J, Urban A (2018) The effect of silver and copper nanoparticles on the growth and mycorrhizal colonisation of Scots pine (Pinus sylvestris L.) in a container nursery experiment. iForest 11(5): 690–697. https://doi.org/10.3832/ifor2855-011

Alizadeh M, Musazade E, Qaderi S, Qarachal JF, Siahpoush S, Abbod M, Siahpoush S, Ghasemi H (2025) The environmental and anthropogenic impacts of nanoparticles on forest trees. J Nanopart Res 27(6):143

Altammar KA (2023) A review on nanoparticles: characteristics, synthesis, applications, and challenges. Front Microbiol 14:1155622. https://doi.org/10.3389/fmicb.2023.1155622

Alterio E, Cocozza C, Chirici G, Rizzi A, Sitzia T (2020) Preserving air pollution forest archives accessible through dendrochemistry. J Environ Manag 264:110462

Amais RS, Moreau PS, Francischini DS, Magnusson R, Locosselli GM, Godoy-Veiga M, Ceccantini G, Ortega Rodriguez DR, Tomazello-Filho M, Arruda MAZ (2021) Trace elements distribution in tropical tree rings through high-resolution imaging using LA-ICP-MS analysis. J Trace Elem Med Biol 68:126872. https://doi.org/10.1016/j.jtemb.2021.126872

Arnold J, Gustin MS, Weisberg PJ (2018) Evidence for nonstomatal uptake of Hg by aspen and translocation of Hg from foliage to tree rings in Austrian pine. Environ Sci Technol 52(3): 1174–1182

Aslam I, Roeffaers MBJ (2022) Carbonaceous nanoparticle air pollution: toxicity and detection in biological samples. Nanomaterials 12(22):3948. https://doi.org/10.3390/nano12223948

Austen K, MacLean J, Balanzategui D, Hölker F (2022) Microplastic inclusion in birch tree roots. Sci Total Environ 808:152085. https://doi.org/10.1016/j.scitotenv.2021.152085

Ballikaya P, Brunner I, Cocozza C, Grolimund D, Kaegi R, Murazzi ME, Schaub M, Schönbeck LC, Sinnet B, Cherubini P (2022a) First evidence of nanoparticle uptake through leaves and roots in beech (Fagus sylvatica L.) and pine (Pinus sylvestris L.). Tree Physiol 43(2):262–276. https://doi.org/10.1093/treephys/tpac117

Ballikaya P, Marshall J, Cherubini P (2022b) Can tree-ring chemistry be used to monitor atmospheric nanoparticle contamination over time? Atmos Environ 268:118781

Ballikaya P, Mateos JM, Brunner I, Kaech A, Cherubini P (2023a) Detection of silver nanoparticles inside leaf of European beech (Fagus sylvatica L.) [Brief Research Report]. Front Environ Sci 10:1107005. https://doi.org/10.3389/fenvs.2022.1107005

Ballikaya P, Song W, Bachmann O, Guillong M, Wang X, Cherubini P (2023b) Chemical elements recorded by Quercus mongolica Fisch. ex Ledeb. tree rings reveal trends of pollution history in Harbin, China. Forests 14(2):187. https://www.mdpi.com/1999-4907/14/2/187

Balouet C, Burken J, Martelain J, Lageard J, Karg F, Megson D (2023) Dendrochemical forensics as material evidence in courts: how could trees lie? Environ Forensic 24(1–2):21–27. https://doi.org/10.1080/15275922.2021.1940381

Balraju W, Upadhyay KK, Dhyani R, Joshi R, Sharma A, Tripathi SK (2025) Dendrochemical analysis of heavy metals in Magnolia champaca tree rings in Mizoram, Northeast India. Environ Monit Assess 197(4):421. https://link.springer.com/article/10.1007/s10661-025-13903-z

Banerjee T, Samanta A (2022) Chapter 4 – Metal oxides for cosmetics and sunscreens. In: Mondal K (ed) Metal oxides for biomedical and biosensor applications. Elsevier, pp 119–135. https://doi.org/10.1016/B978-0-12-823033-6.00002-8

Baroni D, Ancora S, Franzaring J, Loppi S, Monaci F (2023) Tree-rings analysis to reconstruct atmospheric mercury contamination at a historical mining site [Original Research]. Front Plant Sci 14:1260431. https://doi.org/10.3389/fpls.2023.1260431

Bellis DJ, Satake K, McLeod CW (2004) A comparison of lead isotope ratios in the bark pockets and annual rings of two beech trees collected in Derbyshire and South Yorkshire, UK. Sci Total Environ 321(1–3):105–113

Binda G, Di Iorio A, Monticelli D (2021) The what, how, why, and when of dendrochemistry: (paleo) environmental information from the chemical analysis of tree rings. Sci Total Environ 758:143672

Bindler R, Renberg I, Klaminder J, Emteryd O (2004) Tree rings as Pb pollution archives? A comparison of 206Pb/207Pb isotope ratios in pine and other environmental media. Sci Total Environ 319(1):173–183. https://doi.org/10.1016/S0048-9697(03)00397-8

Brink T, Milanese E, Molinari J-F (2022) Effect of wear particles and roughness on nanoscale friction. Phys Rev Mater 6(1):013606. https://doi.org/10.1103/PhysRevMaterials.6.013606

Bundschuh M, Filser J, Lüderwald S, McKee MS, Metreveli G, Schaumann GE, Schulz R, Wagner S (2018) Nanoparticles in the environment: where do we come from, where do we go to? Environ Sci Eur 30(1):6. https://doi.org/10.1186/s12302-018-0132-6

Burken JG, Vroblesky DA, Balouet JC (2011) Phytoforensics, dendrochemistry, and phytoscreening: new green tools for delineating contaminants from past and present. Environ Sci Technol 45(15):6218–6226. https://doi.org/10.1021/es2005286

Burkhardt J, Basi S, Pariyar S, Hunsche M (2012) Stomatal penetration by aqueous solutions – an update involving leaf surface particles. New Phytol 196(3):774–787. https://doi.org/10.1111/j.1469-8137.2012.04307.x

Canning CM, Laroque CP, Muir D (2023) Critical analysis of the past, present, and future of dendrochemistry: a systematic literature review. Forests 14(10):1997. https://www.mdpi.com/1999-4907/14/10/1997

Chen S, Yao Q, Chen X, Liu J, Chen D, Ou T, Liu J, Dong Z, Zheng Z, Fang K (2021) Tree-ring recorded variations of 10 heavy metal elements over the past 168 years in southeastern China. Elementa 9(1):00075. https://doi.org/10.1525/elementa.2020.20.00075

Cocozza C, Ravera S, Cherubini P, Lombardi F, Marchetti M, Tognetti R (2016) Integrated biomonitoring of airborne pollutants over space and time using tree rings, bark, leaves and epiphytic lichens. Urban For Urban Green 17:177–191

Cocozza C, Perone A, Giordano C, Salvatici MC, Pignattelli S, Raio A, Schaub M, Sever K, Innes JL, Tognetti R, Cherubini P (2019) Silver nanoparticles enter the tree stem faster through leaves than through roots. Tree Physiol 39(7):1251–1261. https://doi.org/10.1093/treephys/tpz046

Cui A, Miao Y, Wang C, Tian Z, Jia Q, Liu J, Song G, Wang T, Li Z, Zhang M (2025) NiFe2O4/Ni3Fe nanoparticles Decorate wood carbon strategies for efficient electromagnetic absorption. Chem Eng J 507:160354

Cutter BE, Guyette RP (1993) Anatomical, chemical, and ecological factors affecting tree species choice in dendrochemistry studies. J Environ Qual 22:611–619

Dinis L, Bégin C, Savard MM, Parent M (2021) Impacts of smelter atmospheric emissions on forest nutrient cycles: evidence from soils and tree rings. Sci Total Environ 751:141427

Doucet A, Savard MM, Bégin C, Marion J, Smirnoff A, Ouarda TBMJ (2012) Combining tree-ring metal concentrations and lead, carbon and oxygen isotopes to reconstruct peri-urban atmospheric pollution. Tellus B 64(1):19005. https://doi.org/10.3402/tellusb.v64i0.19005

Dube E, Okuthe GE (2025) Silver nanoparticle-based antimicrobial coatings: sustainable strategies for microbial contamination control. Microbiol Res 16(6):110. https://www.mdpi.com/2036-7481/16/6/110

Etxeberria E, Gonzalez P, Bhattacharya P, Sharma P, Ke PC (2016) Determining the size exclusion for nanoparticles in citrus leaves. HortScience 51(6):732–737. https://doi.org/10.21273/hortsci.51.6.732

Fussell JC, Franklin M, Green DC, Gustafsson M, Harrison RM, Hicks W, Kelly FJ, Kishta F, Miller MR, Mudway IS, Oroumiyeh F, Selley L, Wang M, Zhu Y (2022) A review of road traffic-derived non-exhaust particles: emissions, physicochemical characteristics, health risks, and mitigation measures. Environ Sci Technol 56(11):6813–6835. https://doi.org/10.1021/acs.est.2c01072

Gačnik J, Gustin MS (2023) Tree rings as historical archives of atmospheric mercury: a critical review. Sci Total Environ 898:165562

Gonet T, Maher BA (2019) Airborne, vehicle-derived Fe-bearing nanoparticles in the urban environment: a review. Environ Sci Technol 53(17):9970–9991. https://doi.org/10.1021/acs.est.9b01505

Guan X, Schenk HJ, Roth MR, Welti R, Werner J, Kaack L, Trabi CL, Jansen S (2022) Nanoparticles are linked to polar lipids in xylem sap of temperate angiosperm species. Tree Physiol 42(10):2003–2019. https://doi.org/10.1093/treephys/tpac054

Handy RD, Owen R, Valsami-Jones E (2008) The ecotoxicology of nanoparticles and nanomaterials: current status, knowledge gaps, challenges, and future needs. Ecotoxicology 17(5):315–325. https://doi.org/10.1007/s10646-008-0206-0

Hevia A, Sánchez-Salguero R, Camarero JJ, Buras A, Sangüesa-Barreda G, Galván JD, Gutiérrez E (2018) Towards a better understanding of long-term wood-chemistry variations in old-growth forests: a case study on ancient Pinus uncinata trees from the Pyrenees. Sci Total Environ 625: 220–232

Hochella MF, Mogk DW, Ranville J, Allen IC, Luther GW, Marr LC, McGrail BP, Murayama M, Qafoku NP, Rosso KM, Sahai N, Schroeder PA, Vikesland P, Westerhoff P, Yang Y (2019) Natural, incidental, and engineered nanomaterials and their impacts on the Earth system. Science 363(6434):eaau8299. https://doi.org/10.1126/science.aau8299

Hong J, Peralta-Videa JR, Rico C, Sahi S, Viveros MN, Bartonjo J, Zhao L, Gardea-Torresdey JL (2014) Evidence of translocation and physiological impacts of foliar applied CeO2 nanoparticles on cucumber (Cucumis sativus) plants. Environ Sci Technol 48(8):4376–4385. https://doi.org/10.1021/es404931g

Housset JM, Nadeau S, Isabel N, Depardieu C, Duchesne I, Lenz P, Girardin MP (2018) Tree rings provide a new class of phenotypes for genetic associations that foster insights into adaptation of conifers to climate change. New Phytol 218(2):630–645. https://doi.org/10.1111/nph.14968

Huber L (2022) Trees as witnesses to environmental pollution. Available online: https://www.slf.ch/en/news/trees-as-witnesses-to-environmental-pollution/#:~:text=pollution%20caused%20by%2C%20for%20example%2C,or%20modern%20pesticides%2C%20for%20example. Last accessed on 29 Aug 2025

Ju-Nam Y, Lead JR (2008) Manufactured nanoparticles: an overview of their chemistry, interactions and potential environmental implications. Sci Total Environ 400(1–3):396–414

Kara M, Odabasi M, Dumanoglu Y, Falay EO, Tuygun GT, Altiok H, Bayram A, Tolunay D, Elbir T (2019) Investigation of atmospheric pollution by biomonitoring of major and trace elements in an industrial region. J Environ Sci Eng Technol 7:16–25

Khan I, Saeed K, Khan I (2019) Nanoparticles: properties, applications and toxicities. Arab J Chem 12(7):908–931. https://doi.org/10.1016/j.arabjc.2017.05.011

Kim JY, Park J, Choi J, Kim J (2020) Determination of metal concentration in roadside trees from an industrial area using laser ablation inductively coupled plasma mass spectrometry. Minerals 10(2):175. https://www.mdpi.com/2075-163X/10/2/175

Koç İ (2025) Chronological levels of As, Pd, V, and Sr in 356-year-old Pinus nigra annual rings in northern Türkiye. Bioresources 20(1):2215–2233

Kulac S, Pulatoglu AO, Koç İ, Sevik H, Ozel HB (2025) Assessing tree species for monitoring and mitigating strontium pollution in urban environments. Water Air Soil Pollut 236(9):605

Kumah EA, Fopa RD, Harati S, Boadu P, Zohoori FV, Pak T (2023) Human and environmental impacts of nanoparticles: a scoping review of the current literature. BMC Public Health 23(1): 1059. https://doi.org/10.1186/s12889-023-15958-4

Locosselli GM, Chacón-Madrid K, Arruda MAZ, de Camargo EP, Moreira TCL, de André CDS, de André PA, Singer JM, Saldiva PHN, Buckeridge MS (2018) Tree rings reveal the reduction of Cd, Cu, Ni and Pb pollution in the central region of São Paulo, Brazil. Environ Pollut 242:320–328

Lv J, Christie P, Zhang S (2019) Uptake, translocation, and transformation of metal-based nanoparticles in plants: recent advances and methodological challenges. Environ Sci Nano 6(1):41–59. https://doi.org/10.1039/C8EN00645H

Martin-Ortigosa S, Peterson DJ, Valenstein JS, Lin VS-Y, Trewyn BG, Lyznik LA, Wang K (2013) Mesoporous silica nanoparticle-mediated intracellular Cre protein delivery for maize genome editing via loxP site excision. Plant Physiol 164(2):537–547. https://doi.org/10.1104/pp.113.233650

Marty C, Houle D, Bilodeau-Gauthier S, Gagnon C (2024) Using sugar maple tree rings to trace historic lead pollution in eastern Canada temperate forest. Appl Geochem 160:105855. https://doi.org/10.1016/j.apgeochem.2023.105855

Materić D, Kjær HA, Vallelonga P, Tison J-L, Röckmann T, Holzinger R (2022a) Nanoplastics measurements in northern and southern polar ice. Environ Res 208:112741. https://doi.org/10.1016/j.envres.2022.112741

Materić D, Peacock M, Dean J, Futter M, Maximov T, Moldan F, Röckmann T, Holzinger R (2022b) Presence of nanoplastics in rural and remote surface waters. Environ Res Lett 17(5):054036

Monaci F, Baroni D (2025) Leaves and tree rings as biomonitoring archives of atmospheric mercury deposition: an ecophysiological perspective. Plants 14(9):1275. https://pmc.ncbi.nlm.nih.gov/articles/PMC12073167/

Padilla KL, Anderson KA (2002) Trace element concentration in tree-rings biomonitoring centuries of environmental change. Chemosphere 49(6):575–585

Park S-J, Deshmukh MA, Kang B-C, Jeon J-Y, Chen C, Ha T-J (2020) Review—A review of advanced electronic applications based on carbon nanomaterials. ECS J Solid State Sci Technol 9(7):071002. https://doi.org/10.1149/2162-8777/abb035

Petkewich R (2008) Trees testify to pollution. Chem Eng News 86(36):37–38

Portugal J, Bedia C, Amato F, Juárez-Facio AT, Stamatiou R, Lazou A, Campiglio CE, Elihn K, Piña B (2024) Toxicity of airborne nanoparticles: facts and challenges. Environ Int 190:108889. https://doi.org/10.1016/j.envint.2024.108889

Pulatoglu AO, Koç İ, Özel HB, Şevik H, Yıldız Y (2025) Using trees to monitor airborne Cr pollution: effects of compass direction and woody species on Cr uptake during phytoremediation. Bioresources 20(1):121

Räsänen JV, Leskinen JT, Holopainen T, Joutsensaari J, Pasanen P, Kivimäenpää M (2017) Titanium dioxide (TiO2) fine particle capture and BVOC emissions of Betula pendula and Betula pubescens at different wind speeds. Atmos Environ 152:345–353

Rocha E, Gunnarson B, Kylander ME, Augustsson A, Rindby A, Holzkämper S (2020) Testing the applicability of dendrochemistry using X-ray fluorescence to trace environmental contamination at a glassworks site. Sci Total Environ 720:137429. https://doi.org/10.1016/j.scitotenv.2020.137429

Rönkkö T, Timonen H (2019) Overview of sources and characteristics of nanoparticles in urban traffic-influenced areas. J Alzheimers Dis 72(1):15–28. https://doi.org/10.3233/jad-190170

Scanlon TM, Riscassi AL, Demers JD, Camper TD, Lee TR, Druckenbrod DL (2020) Mercury accumulation in tree rings: observed trends in quantity and isotopic composition in Shenandoah National Park, Virginia. J Geophys Res Biogeosci 125(2):e2019JG005445. https://doi.org/10.1029/2019JG005445

Scharnweber T, Hevia A, Buras A, van der Maaten E, Wilmking M (2016) Common trends in elements? Within-and between-tree variations of wood-chemistry measured by X-ray fluorescence—a dendrochemical study. Sci Total Environ 566:1245–1253

Schwab F, Guangshu Z, Meaghan K, Amalia T, Schnoor JL, Wiesner MR (2016) Barriers, pathways and processes for uptake, translocation and accumulation of nanomaterials in plants – critical review. Nanotoxicology 10(3):257–278. https://doi.org/10.3109/17435390.2015.1048326

Sembada AA, Lenggoro IW (2024) Transport of nanoparticles into plants and their detection methods. Nanomaterials 14(2):131. https://doi.org/10.3390/nano14020131

Smita S, Gupta SK, Bartonova A, Dusinska M, Gutleb AC, Rahman Q (2012) Nanoparticles in the environment: assessment using the causal diagram approach. Environ Health 11(Suppl 1):S13. https://doi.org/10.1186/1476-069x-11-s1-s13

Su Y, Ashworth V, Kim C, Adeleye AS, Rolshausen P, Roper C, White J, Jassby D (2019) Delivery, uptake, fate, and transport of engineered nanoparticles in plants: a critical review and data analysis. Environ Sci Nano 6(8):2311–2331. https://doi.org/10.1039/C9EN00461K

Su Y, Ashworth VETM, Geitner NK, Wiesner MR, Ginnan N, Rolshausen P, Roper C, Jassby D (2020) Delivery, fate, and mobility of silver nanoparticles in citrus trees. ACS Nano 14(3): 2966–2981. https://doi.org/10.1021/acsnano.9b07733

Tafur-Marinos JA, Ginepro M, Pastero L, Zelano V (2016) Digestion procedures for the elemental analysis of wood by inductively coupled plasma optical emission spectrometry. Anal Lett 49(11):1722–1733. https://doi.org/10.1080/00032719.2015.1120309

Tao M, Xu Y, Liu Q, Liu Y, Tian S, Schauer JJ (2023) Penetration of submicron amino-functionalized graphene quantum dots in plant stomata, implication for the depollution of atmospheric soot particles. Environ Chem Lett 21(3):1281–1286. https://doi.org/10.1007/s10311-022-01535-5

Tran T-K, Nguyen M-K, Lin C, Hoang T-D, Nguyen T-C, Lone AM, Khedulkar AP, Gaballah MS, Singh J, Chung WJ, Nguyen DD (2024) Review on fate, transport, toxicity and health risk of nanoparticles in natural ecosystems: emerging challenges in the modern age and solutions toward a sustainable environment. Sci Total Environ 912:169331. https://doi.org/10.1016/j.scitotenv.2023.169331

USGS (2004) Laser ablation ICP-mass spectrometry—a new tool for analyzing metals in tree rings. Innovative research methods for environmental studies

Wang W-N, Tarafdar JC, Biswas P (2013) Nanoparticle synthesis and delivery by an aerosol route for watermelon plant foliar uptake. J Nanopart Res 15(1):1417. https://doi.org/10.1007/s11051-013-1417-8

Wang X, Xie H, Wang P, Yin H (2023) Nanoparticles in plants: uptake, transport and physiological activity in leaf and root. Materials 16(8):3097. https://doi.org/10.3390/ma16083097

Weber K (2022) Foliar penetration of silver nanoparticles in 3 tree species: role of macro and micro-structures of leaves in foliar penetration. Master's thesis, University of Zurich

Weber CJ, Rillig MC, Bigalke M (2023) Mind the gap: forest soils as a hidden hub for global micro- and nanoplastic pollution. Microplast Nanoplast 3(1):19. https://doi.org/10.1186/s43591-023-00067-1

Wimmer R (2002) Wood anatomical features in tree-rings as indicators of environmental change. Dendrochronologia 20(1–2):21–36

Wojcieszek J, Jiménez-Lamana J, Ruzik L, Szpunar J, Jarosz M (2020) To-do and not-to-do in model studies of the uptake, fate and metabolism of metal-containing nanoparticles in plants. Nanomaterials 10(8):1480. https://www.mdpi.com/2079-4991/10/8/1480

Wu H, Li Z (2022) Nano-enabled agriculture: how do nanoparticles cross barriers in plants? Plant Commun 3(6):100346. https://doi.org/10.1016/j.xplc.2022.100346

Xie F, Pu Y, Liang Y, Qiu J, Zheng K, Gao J (2025) Preparation of highly stable wood with fast curing, ultraviolet, and mildew resistance through copper-montmorillonite nanoparticles hybridized with tung oil. ACS Appl Mater Interfaces 17(5):8231–8247

Xuan L, Ju Z, Skonieczna M, Zhou PK, Huang R (2023) Nanoparticles-induced potential toxicity on human health: applications, toxicity mechanisms, and evaluation models. MedComm (2020) 4(4):e327. https://doi.org/10.1002/mco2.327

Yan A, Chen Z (2019) Impacts of silver nanoparticles on plants: a focus on the phytotoxicity and underlying mechanism. Int J Mol Sci 20(5):1003. https://pmc.ncbi.nlm.nih.gov/articles/PMC6429054/pdf/ijms-20-01003.pdf

Yu H, Feng W (2024) Carbon-based multifunctional nanomaterials. Nanomaterials 14(19):1600. https://doi.org/10.3390/nano14191600

Yu Z, Xu X, Guo L, Jin R, Lu Y (2024) Uptake and transport of micro/nanoplastics in terrestrial plants: detection, mechanisms, and influencing factors. Sci Total Environ 907:168155. https://doi.org/10.1016/j.scitotenv.2023.168155

Zhang K, Liu Q, Wang Y, Liu X, Zhou X, Yan B (2024) Advances and challenges in tracking interactions between plants and metal-based nanoparticles. Nanomaterials 14(23):1939. https://www.mdpi.com/2079-4991/14/23/1939

Chapter 14
Impact of Nanomaterials on Plant Cell Suspension Models

Anjali Bharti, Utkarsh Jain, and Nidhi Chauhan

Abstract Nanomaterials (NMs) offer several advantages due to their fascinating properties and are being widely used in agriculture. This involves their widespread applications in various fields, including biomedicine, engineering, food, cosmetics, sensing, and energy. Nanotechnology offers unique opportunities for plant scientists and agricultural biotechnologists to develop and implement innovative tools and procedures for enhancing commercially valuable plants and crops. More specifically, the implementation of NMs within plant tissue culture and cell suspension systems has shown substantial potential to stimulate cell growth, increase nutrient absorption, and enhance the secondary metabolites yield. These applications contribute to the efficiency of *in vitro* plant culture and offer prospects for mass production of biologically derived active compounds for pharmaceutical or industrial applications. The current chapter aims to gather the latest advancements in nanotechnology and cell suspension culture, focusing on the favourable and negative impacts of the nanoparticles (NPs). The toxicity and safety issues associated with exposing plants and their surrounding environment to NPs have been outlined. Advancing plant nanobiotechnology necessitates further research to resolve unanswered questions and clarify underlying mechanisms, which is essential for achieving significant scientific progress in the agriculture field.

Keywords Nanomaterials · Cell suspension culture · Toxicity · Plant growth regulators · Secondary metabolites

14.1 Introduction

Plants have remained as an accessible and renewable source of food, flavour agents, biochemical dyes, therapeutic proteins, and even stem cells. However, their natural environments are rapidly disappearing nowadays due to the growing number of environmental pressures and human interventions. This reinforces the urgent need

A. Bharti · U. Jain · N. Chauhan (✉)
School of Health Sciences & Technology (SoHST), UPES, Dehradun, Uttarakhand, India

R. N. Pudake et al. (eds.), *Phyto Nanotoxicology*, Nanotechnology in Plant Sciences 7, https://doi.org/10.1007/978-3-032-24518-2_14

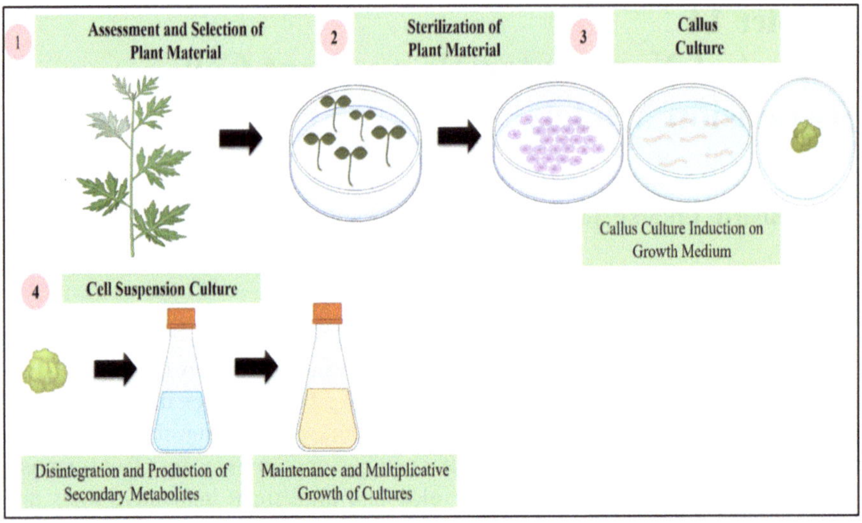

Fig. 14.1 Plant tissue culture involves various stages, including selection and sterilisation of the plant material, followed by the callus proliferation initiation, to the final step of cell suspension culture for the production of secondary metabolites and cell growth

for alternative technologies for the reliable production of plant-derived compounds used in food and healthcare (Bapat et al. 2023). Among the 50,000 plant species on earth, merely 1% are producing bioactive compounds, causing over-reaping, habitat loss, and species deterioration (Motolinía-Alcántara et al. 2021). Nowadays, approximately a quarter of advanced medicines are derived from nature, including algae, fungi, and higher plants (Zafar et al. 2019; Elkordy et al. 2021). Traditional approaches to acquiring bioactive compounds have their limitations and are affected by various internal and external factors. However, plant tissue cultures provide a consistent source for targeted metabolite extraction. This has led to the increasing applications of plant tissue, organ, and cell suspension cultures in the biosynthesis and transformation of numerous phytochemicals, as well as the conversion of vegetatively inactive compounds into pharmacologically active compounds (Shriram et al. 2010; Doma et al. 2012; Ochoa-Villarreal et al. 2016; Espinosa-Leal et al. 2018; Jawahar et al. 2018; Arya et al. 2020). Plant cell suspension culture serves as a practical and inexpensive system for cultivating large quantities of plant-based secondary metabolites, as mentioned in Fig. 14.1. This production method utilises totipotent cells of the same plant that produce chemically identical bioactive compounds and may result in a higher yield of plant secondary metabolites (Yue et al. 2016; Gonçalves and Romano 2018). Currently, both academia and industry are investigating plant cell culture as a means to produce bioactive compounds on a commercial scale, utilising callus, immobilised cells, embryos, roots, shoots, and suspended cells (Ochoa-Villarreal et al. 2016; Süntar 2020). Plant cell suspension cultures also represent an economical and safer alternative to mammalian cells for synthesising complex recombinant proteins. It has the advantages of both scalability and precision while eliminating the need of animal-derived components. However, plant cells have a non-uniform shape and tend to aggregate or clump together, so obtaining an accurate quantification of cell-specific productivity can be difficult, and conventional cell counting methods often do not work, which makes it hard to accurately monitor throughout the process (Havenith et al. 2014).

During the following decades, research activity in this area was outcompeted by advances in recombinant protein production in whole plant systems. As a result, there were few published studies before the early 2000s focused on using cultured plant cells as recombinant protein production platforms (Feizi 2023). Further, the research on *Capsicum baccatum* L. var. pendulum cells was concentrated on optimising the culture medium to enhance cell growth, antioxidant activity, and secondary metabolites that could have crucial uses in the pharmaceutical and food industries (Fidemann et al. 2018).

14.2 Formulation and Media Composition

Murashige and Skoog (1962) and Gamborg's B5 are among the most commonly utilised growth media for plant cell culture. The MS salt medium is readily available and has been used for many years. The MS salts initially were developed to culture tobacco tissues. Since then, they have gained popularity and can be used in a variety of concentrations. The MS salts can consistently support growth across varied plant cell types (Greenway et al. 2012; Monfort et al. 2018). The B5 medium was first created to promote the suspension culture of soybean and has now been accepted as a staple in plant tissue culture (Greenway et al. 2012), along with other common media such as White's, Schenk & Hildebrandt's, and Chu's N6 medium (George et al. 2008). All plant tissue culture media generally have the following essential components, including macronutrients, micronutrients, trace elements, a carbohydrate source, plant growth regulators or phytohormones, and organic substances, including vitamins and organic acids (George et al. 2008). Phytohormones used in growth media are illustrated in Table 14.1. In addition to these, synthetic compounds that mimic their effects or substances can affect plant cell growth through various mechanisms that are widely used. Some common synthetic analogues include 2,4-dichlorophenoxyacetic acid (2,4-D), indole-3-butyric acid (IBA), kinetin, and 2-hydroxybenzyladenosine (George et al. 2008).

Plant cell cultures mainly involve callus, immobilised cells, embryos, roots, shoots, and cell suspensions. They facilitate easier handling as well as further understanding of the intricate processes of plant-derived secondary metabolite (PDSM) synthesis (Siddiqui and Al-Whaibi 2014). Callus buildup results from rapid cell division and the accumulation of undifferentiated cell masses. Undifferentiated cells can eventually resume their ability to regenerate into a whole plant by acquiring characteristics similar to meristematic cells, and can eventually differentiate into new stem cells and grow into complete plants (Kwon et al. 2021). Meanwhile, plant cell suspension culture serves as a simple and

Table 14.1 Applications of the plant growth promoters (PGPs) in *in vitro* culture medium

PGPs	Chemical name	Abbreviation	Function in plant physiology	References
Auxin	Indole-3-acetic acid	IAA	Cell proliferation and elongation	Duca et al. (2014)
	Indole-3-butyric acid	IBA	Root development	Ludwig-Müller (2000)
Cytokinin	6-(benzylamino)-purine	BAP	Shoot development	Jafari et al. (2011)
	6-(furfurylamino)-purine	(KIN) Kinetin	Promotes cell division	Miller et al. (1956)
Gibberellin	Gibberellic acid	GA_3	*In vitro* floral morphogenesis	Chaari-Rkhis et al. (2006)
Abscisic acid	Abscisic acid	ABA	Somatic embryogenesis	Rai et al. (2011)
Ethylene	Ethylene	–	Organogenesis and tissue patterning	Biddington (1992); Li and Guo (2018)

cost-effective method of a biotechnological approach utilised for the large-scale synthesis of PDSMs. This method uses the biosynthetic totipotency of plant cells to produce bioactive compounds that are chemically identical to those found in plant stem cells, offering potential advantages in terms of both quality and quantity (Yue et al. 2016; Gonçalves and Romano 2018).

14.3 Various Nanomaterials Used in Plant Cell Suspension Models

Nanomaterials (NMs) are very small materials, generally between 10 and 100 nm in size, that have unique properties linked to their tiny size. Properties associated with NMs include a high surface-area-to-volume ratio and unusual optical properties (Dutschk et al. 2014; Al-Halafi 2014). NMs include many types of materials, including carbon-based materials (Baughman et al. 2002; Lang et al. 2011), metal oxides, semiconductors (Lang et al. 2011; Rizzello and Pompa 2014), lipids (Diao and Yao 2009; Yang and Ma 2010), zero-valent metals (Diao and Yao 2009), quantum dots, nanopolymers, and dendrimers (Astruc 2012; Ljubimova and Holler 2012). It can be prepared *via* chemical, physical, and biological methods.

NMs have shown great promise in supporting different stages of *in vitro* plant tissue cultures. This includes starting callus, developing organs, forming somatic embryos, and increasing the synthesis of secondary metabolites (Kim et al. 2017). Many studies have shown the helpful impacts of nanotechnology in plant biotechnology. For instance, researchers have successfully applied nanoparticles (NPs) to improve seed germination percentage, improve plant growth and productivity, genetic transformation, increase the yield of valuable bioactive compounds, and provide protection against various stresses (Siddiqui and Al-Whaibi 2014; Wang et al. 2016).

14.4 Role of Nanomaterials on Plant Cell Suspension Models

Nanotechnology, one of the most significant advances of the twenty-first century (Abasi et al. 2022), consists of designing and utilising materials and devices by manipulating matter on the nanometer scale (10^{-9} m). Hence, the creation of innovative and advanced NMs is essential for the effective utilisation in several applications, such as nano-fertilisers, nano-pesticides, water treatment, and detecting hazardous/toxic metals (Bharti et al. 2024). Engineered NMs can be specifically designed with target functionalities, thus offering unique possibilities for bioengineering applications. It can be organic, inorganic, or composite materials developed in various structural formats, however, it is not limited to nanofibers, NPs, and nanomembranes. Protoplasts, which have no cell wall, are often used as model cells because they have a higher capacity for NMs uptake through the process of endocytosis (Osmani et al. 2024). Gene delivery *via* NMs is generally categorized according to their chemical composition, structure and gene transfer mechanisms. The primary categories of NMs involve carbon-based (carbon nanotubes, graphene, carbon nanofibers), inorganic (metallic NPs such as Cu, Ag, Au, or metal oxide NPs), organic (e.g. polymeric NPs, nanocellulose, nanostarch), composite (such as hybrid nanofibers or mixed metal oxides), and bio-based NMs, which encompass nanobacteria and NPs derived from enzymes (Tuominen and Schultz 2010; Jeevanandam et al. 2018). NMs are capable of overcoming many challenges inherent to traditional methods of delivering biomolecules in plant cells. It enables targeted and accurate delivery while protecting and stabilizing DNA and significantly enhances the efficiency of plant cell uptake, with often 1000X less DNA needed than traditional genetic transformation approaches. Additionally, NPs can be coated with other materials such as polyethene glycol (PEG) to further limit immune responses associated with their surface charge. The incorporation of NMs can also minimize labour efforts and cost, therefore, it provide a sustainable and scalable alternative that enable passive entry in plant cells lacking any external tools or enhanced force (Milewska-Hendel et al. 2017; Wang et al. 2019). In another study, NPs directly modified plant embryos or gametes by eliminating the regeneration steps commonly required during plant tissue culture (Zhao et al. 2017). Additionally, it has the capability of targeting specific cellular compartments, thereby enabling the targeted modification of nuclear and non-nuclear genomes found in chloroplasts and mitochondria, respectively (Cunningham et al. 2018).

Further, exposure of 3 mg/L of copper oxide (CuO) NPs for 2 days, *Gymnema sylvestre* cell suspension cultures produced a high yield of gymnemic acid, and the amounts of total phenolics and flavonoids were also increased. However, a higher concentration of Cu would decrease the production of gymnemic acid. Cu in excess amounts, *i.e.*, 5 mg/L, can be toxic to plant cells and interfere with the activity of enzymes, potentially leading to antagonism with other essential nutrients. Therefore, 5 mg/L not only decreased the production of gymnemic acid but also decreased the levels of phenolics and flavonoids (Chung et al. 2019). Further, the addition of zinc oxide (ZnO) NPs to *Linum usitatissimum* cell suspension cultures stimulated lignan and neolignan production, as well as total phenolics and flavonoids (Karimzadeh et al. 2019; Abbasi et al. 2019). Taghizadeh et al. (2019) examined the impact of both static magnetic field (30 mT), iron oxide (Fe_2O_3) magnetic NPs (100 ppm), and both static magnetic fields and Fe_2O_3 magnetic NPs on the suspension culture of *Dracocephalum polychaetum* Bornm. HPLC and UV–visible spectroscopy were used to evaluate variations in phenolic metabolism and in the concentration of some key compounds with medicinal applications. The treatments caused changes in the activities of the enzymes (polyphenol oxidase and phenylalanine ammonia-lyase) and led to increased levels of total phenolics, flavonoids, anthocyanins, lignin, and malondialdehyde in comparison to untreated cells (Taghizadeh et al. 2019). Further, the study evaluated the effect of ZnO and titanium dioxide (TiO_2) NPs on enzyme activity and secondary metabolites synthesis in flax cell cultures. The results showed that 30 mg/L of ZnO significantly increased phenylalanine ammonia lyase enzyme activity after 48 h, and 60 mg/L of ZnO increased cinnamate-4-hydroxylase (CAD) enzyme activity. Further, TiO_2 at 150 mg/L also increased CAD enzyme activity and total phenol levels. Therefore, NPs facilitated increased lignan content with ZnO at 30–60 mg/L, providing the greatest phenol and lignan production. From these findings, it was concluded that in flax cell suspension cultures, the effects of NPs on enzymatic activity and secondary metabolite production were dependent on the types and concentration of NPs used (Karimzadeh et al. 2019).

Further, the incorporation of 20-nm gold (Au) and silver (Ag) NPs into the growth medium improves biomass accumulation in *Arabidopsis thaliana* cell suspension cultures. Notably, silver NPs can cause the culture medium to become more acidic, while gold NPs make it more alkaline. MTT or 3-(4,5-dimethylthiazol-2-yl)-2,5-diphenyl tetrazolium bromide assay outcome showed a mild but constant reduction in specific respiratory activity. Additionally, NPs treatment raises intracellular levels of stress-related free amino acids, including alanine, γ-aminobutyric acid, and valine. Therefore, the significant changes in the extracellular protein profile indicate physiological responses to abiotic stress (Selivanov et al. 2017). Furthermore, many studies have documented organogenesis tracked by various NPs. For instance, when organogenesis was performed at varying concentrations of zinc (Zn), and ZnO NPs with callus derived from banana shoot tips on shoot regeneration media (MS medium supplemented with different concentrations of NPs), the number of shoots per callus was increased with rising NPs concentration, reaching to a maximum range of 100 mg/L. This observation in shoot formation can be linked with the properties of the Zn NPs, which are an essential micronutrient for higher plants that potentially promote plant growth (Helaly et al. 2014). The same study revealed that the greatest rooting response (89%), the highest average of roots per shoot (6.57),

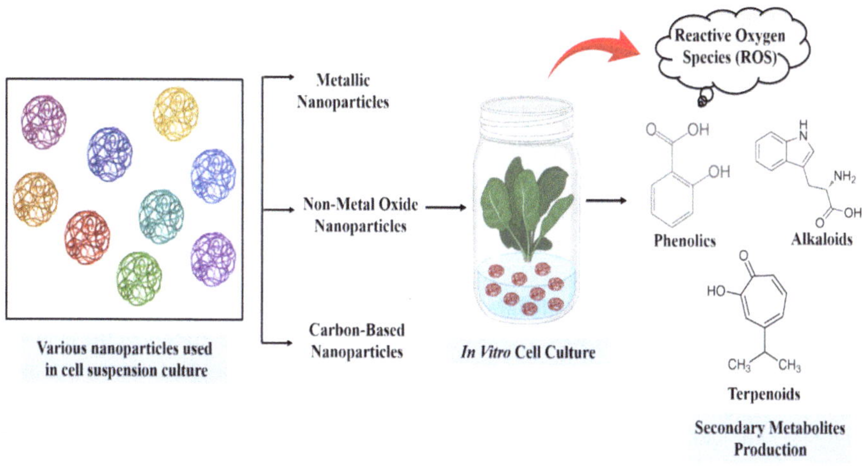

Fig. 14.2 The figure represents various nanoparticles that stimulate the reactive oxygen species (ROS) production in plant cell suspension cultures, which in turn activate the biosynthesis of secondary metabolites

and the maximum root length (2.93 cm) for *in vitro* banana explants were obtained on MS medium containing 100 mg/L of Zn NPs. Biochemical study showed that proline content increased linearly throughout the stages of *in vitro* banana explant regeneration with increasing concentrations of nano Zn and nano ZnO. Figure 14.2 illustrates the implementation of NMs that induce ROS activity in plant cell suspension cultures, consequently triggering the biosynthesis of secondary metabolites (phenolics, terpenoids, and alkaloids), respectively.

14.5 Toxicity and Safety Assessment of Nanomaterials

Nanomaterials in the environment can be released from several sources, including natural processes (e.g., microbial processes, incomplete burning, and chemical precipitation), human processes (e.g., intentional uses like nanofertilizers or nanopesticides, geological events), and accidental/unintentional emissions (e.g., from oxidation processes or disposal of waste material). Nonetheless, widespread utilisation of NPs has raised concerns regarding their potentially harmful effects on the environment, plants, and human health (Szőllősi et al. 2020). In simple terms, leftover NPs typically enter the food chain through absorption by plants (de Melo et al. 2020). Moreover, NPs that are accumulated in plants and microbes may get transferred into the human food chain and may also affect the quality of food and health (Ali et al. 2020). Nanotechnology is being increasingly utilised in agriculture to promote crop productivity, strengthening plant resilience to stress, and reducing the use of agricultural chemicals. However, the uncontrolled NPs pollution in the environment has generated increasing concern regarding their potential toxic effects on plants. Plants may undergo phytotoxicity under elevated NPs levels that can negatively impact seed germination, biomass growth, photosynthesis, and oxidative balance of plants. It may also alter membranes, modify gene regulation, result in DNA damage, and even cause epigenetic changes (Gowtham et al. 2024). NPs can significantly affect the yield and the quality of food crops. Rico et al. (2013, 2014) reported the toxic impact of CeO_2 NPs on rice plants, including changes in profiles of amino acids, fatty acids, phenolic compounds, and non-reducing sugars (Rico et al. 2013, 2014). With a rapid increase in nanotechnology, increasing amounts of NMs, particularly metals and metal oxide NPs, are introduced into the environment and taken up by plants. However, it is critical to understand the effects of these NMs on plants because NPs absorption can adversely affect crop yields, food safety, and cause long-term ecological problems through bioaccumulation. Research studies suggest that certain metals or metal oxides, including TiO_2, can help plants to grow when present in trace amounts; however, they become harmful and toxic at high concentrations (Zhang et al. 2025). Once NPs penetrate inside the plant, they may induce changes in the organisation of cells, molecules, membranes, and even disrupt the plant's defence mechanism. In addition, their properties, solubility, and ability to induce the synthesis of reactive oxygen species (ROS) that can lead to unintended effects (Dharmalingam et al. 2020; Balážová et al. 2018). Further, NPs' toxicity arises from various mechanisms that include ROS induction, cytotoxicity, genotoxicity, and neurotoxicity, which depend greatly on their characteristics, like "type, size, surface area, shape, surface coating, crystallinity, dissolution, and agglomeration" (Egbuna et al. 2021). Engineered NPs can generate elevated levels of ROS because of their high reactivity due to higher surface area, leading to cytotoxicity and genotoxicity (Kim and Ryu 2013; Morsy et al. 2021). ZnO NPs are widely used in industry and medicine, but there is limited information about their effects on the environment. Research on tobacco BY-2 cells showed that ZnO NPs could impact cell growth and survival through ROS production with autophagy and programmed cell death, thus highlighting the need for more research into potential ZnO toxicities (Balážová et al. 2020). Similarly, Ag NPs having a size of 27 nm were synthesized using the Tollens procedure and characterized *via* TEM, DLS, AFM, and spectrophotometry. Their toxic and genetic effects were assessed with mammalian (SVK14 and NIH3T3) and plant (*Solanum* species) cells. NIH3T3 cells were more susceptible to Ag NPs, with NPs accumulation, exhibiting damaged DNA and cell death. Among plants, *S. habrochaites* experienced the highest oxidative stress and membrane damage, indicating that Ag NPs toxicity was dependent on both inherent characteristics and the type of cells exposed (Jiravova et al. 2016). Similarly, the effects of aluminium oxide NPs were also studied on tobacco BY-2 cell cultures, an alternative model plant system similar to HeLa cells used in animal cell studies. The NPs reduced cell growth in both time and dosage-dependent manners and increased reactive oxygen and nitrogen species. They can also induce programmed cell death, indicated by plasma membrane permeability damage, harm to mitochondrial membrane potential, increased caspase-like activity, and DNA breakage. Thus, the study concludes that aluminium oxide NPs induced oxidative stress in BY-2 tobacco cells, ultimately resulting to cell death (Poborilova et al. 2013). Furthermore, the study investigated the impacts of magnetic NPs and their derivatives, specifically modified magnetic nanoparticles (MNPs) effects on tobacco BY-2 cell culture systems. Although their influence on cell proliferation was minimal, prominent alterations occurred in protein levels, phytochelatin levels, GST (glutathione S-transferase) activity, and antioxidant mechanisms in NPs-treated BY-2 cell culture systems. However, MNPs altered cellular metabolism and rapidly activated the protective compounds synthesis in BY-2 cell suspension cultures, suggesting their potential to induce stress-responsive biochemical pathways in plant cells (Krystofova et al. 2013). Similarly, CuO NPs were synthesised using chemical (polymer pyrolysis) and biological (green) synthesis (30 and 44 nm average size) and were tested on *Nicotiana tabacum* (tobacco) cell cultures. Both types led to a reduction in cell viability depending on the time and dose, more antioxidant enzyme activity and stimulated the ROS and MDA production (Malondialdehyde). Chemically synthesized CuO NPs exhibited more toxic effects than biologically synthesized NPs. This is due to the presence of some organic compounds that are derived from the plant source on the surface of the green NPs, which likely function as capping agents that help to reduce their toxicity (Mahjouri et al. 2018).

In another research on NPs with a validated mathematical model of lipid exchange envelope penetration, revealed that its uptake in plant protoplasts is a function of both particle size and zeta potential. This model applies to a wide range of NPs, including quantum dots, gold, silica, nanoceria, and SWNTs (single-walled carbon nanotubes). Furthermore, imaging flow cytometry identified that chloroplast glycerolipids associated more strongly with NPs than the protoplast phospholipids. Based on the DNA-wrapped SWNTs as a model, these findings provide guidance in the design of NPs for selective delivery to plant cell compartments without any chemical or mechanical interventions, as an initial step toward next-generation plant engineering (Lew et al. 2018). Further, *Medicago sativa* cell suspension culture was used to study the impact of mercaptopropanoic acid-coated CdSe/ZnS quantum dots (QDs). The cell growth rate decreased rapidly after exposure to QDs at 100 nM, with less than 50% of the cells remaining viable after 72 h. Confocal imaging confirmed QDs internalisation and accumulation of QDs in the cell cytoplasm and nucleus. Additionally, the cells exhibited a dose- and time-dependent increase in ROS, with fluorescence showing a gradual increase in ROS at 5, 50, and 180 nM QD concentrations (Santos et al. 2010). The potential harmful effects of multiwalled carbon nanotubes (MWCNTs) on rice cells were examined using suspension cultures, and its exposure increased the generation of ROS and reduced cell viability. However, the addition of ascorbic acid, a relevant antioxidant, lowered ROS and increased cell viability. Transmission electron microscopy (TEM) discovered that individual nanotubes were found to be in direct contact with cell walls, indicating that rice cells invoked a hypersensitive, ROS-based defence similar to that triggered by microbial pathogens (Tan et al. 2009).

14.6 Conclusion

Nanotechnology has provided a new way for developing new strategies and scientific interventions to improve the quality of agricultural products or agri-based products. However, certain negative implications of this technology need to be recognised and reduced to utilise it sustainably and successfully. The observed interactions of NPs in plant tissue culture highlights their substantial opportunities in plant biotechnology. A plant tissue culture medium supports cells and organs that are undergoing nutrition to grow, and further it needs to optimize the supply of nutrients, vitamins, plant growth regulators, organic supplements, and elicitors to the dynamic exogenous supply. The nutrition requirements of explants will differ even among plant species. Thus, the plant type and the explants used are important measures for selecting the culture medium best suited for the selected plant species. NPs are shown to have positive influences at all levels, including sterilization, callus formation, genetic transformation, somaclonal variation, and secondary metabolite production. Plant tissue cultures are a foundational principle of plant biology and are an important tool for plant conservation, large-scale propagation, genetic modification, bioactive compounds production, and overall plant growth and development. However, the different applications of plant cell suspension cultures, including recombinant protein production, large-scale cultivation within bioreactors, creating transgenic plants, germplasm conservation *in vitro*, mutation, bioprinting, cosmetics, and biotransformation as depicted in Fig. 14.3. Recently, NPs have been successfully used to eliminate microbial contamination from explants and have also demonstrated significant roles in promoting callus initiation and growth, organogenesis, somatic embryogenesis, creation of somaclonal variation, and genetic transformation. However, the incorporation of NPs caused increased bioactive compounds in

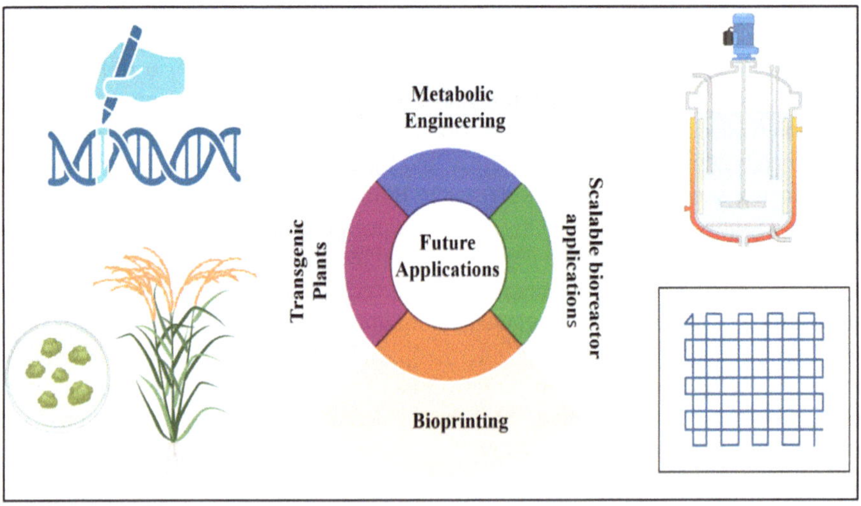

Fig. 14.3 Future applications of plant cell suspension cultures

cell, shoot, and root cultures, due to possible pathways involving ROS generation, antioxidant enzymes activation, regulation of specific genes, or a combination of these mechanisms. However, the exact pathways are still unknown therefore, it is important to pursue further studies to investigate the role of different NPs on production of secondary metabolite in different plant species, as well as to investigate the chemical, physical, and biological properties of NPs-treated culture compounds. There should be an in-depth investigation of the interactions and ultimate fate of NPs in plant systems, since significant progress cannot be achieved without any clarity on their interaction or consequences in real time.

References

Abasi F, Raja NI, Mashwani ZUR et al (2022) Biogenic silver nanoparticles as a stress alleviator in plants: a mechanistic overview. Molecules 27:3378. https://doi.org/10.3390/molecules27113378

Abbasi BH, Zahir A, Ahmad W et al (2019) Biogenic zinc oxide nanoparticles-enhanced biosynthesis of lignans and neolignans in cell suspension cultures of *Linum usitatissimum* L. Artif Cells Nanomed Biotechnol 47:1367–1373. https://doi.org/10.1080/21691401.2019.1596942

Al-Halafi AM (2014) Nanocarriers of nanotechnology in retinal diseases. Saudi J Ophthalmol 28: 304–309. https://doi.org/10.1016/j.sjopt.2014.02.009

Ali MA, Ahmed T, Wu W et al (2020) Advancements in plant and microbe-based synthesis of metallic nanoparticles and their antimicrobial activity against plant pathogens. Nanomaterials 10:1146. https://doi.org/10.3390/nano10061146

Arya SS, Rookes JE, Cahill DM, Lenka SK (2020) Next-generation metabolic engineering approaches towards development of plant cell suspension cultures as specialized metabolite producing biofactories. Biotechnol Adv 45:107635. https://doi.org/10.1016/j.biotechadv.2020.107635

Astruc D (2012) Electron-transfer processes in dendrimers and their implication in biology, catalysis, sensing and nanotechnology. Nat Chem 4:255–267. https://doi.org/10.1038/nchem.1304

Balážová Ľ, Babula P, Baláž M et al (2018) Zinc oxide nanoparticles phytotoxicity on halophyte from genus Salicornia. Plant Physiol Biochem 130:30–42. https://doi.org/10.1016/j.plaphy.2018.06.013

Balážová Ľ, Baláž M, Babula P (2020) Zinc oxide nanoparticles damage tobacco BY-2 cells by oxidative stress followed by processes of autophagy and programmed cell death. Nanomaterials 10(6):1066. https://doi.org/10.3390/nano10061066

Bapat VA, Kavi Kishor PB, Jalaja N et al (2023) Plant cell cultures: biofactories for the production of bioactive compounds. Agronomy 13:858. https://doi.org/10.3390/agronomy13030858

Baughman RH, Zakhidov AA, de Heer WA (2002) Carbon nanotubes--the route toward applications. Science (1979) 297:787–792. https://doi.org/10.1126/science.1060928

Bharti A, Jain U, Chauhan N (2024) Metal–organic-frameworks: as revolutionizing material utilized for distinct applications in precision farming. Bionanoscience 14:1798–1818. https://doi.org/10.1007/s12668-024-01356-6

Biddington NL (1992) The influence of ethylene in plant tissue culture. Plant Growth Regul 11: 173–187. https://doi.org/10.1007/BF00024072

Chaari-Rkhis A, Maalej M, Messaoud SO, Drira N (2006) In vitro vegetative growth and flowering of olive tree in response to GA3 treatment. Afr J Biotechnol 5:2097–2302. https://doi.org/10.4314/ajb.v5i22.55975

Chung I-M, Rajakumar G, Subramanian U et al (2019) Impact of copper oxide nanoparticles on enhancement of bioactive compounds using cell suspension cultures of Gymnema sylvestre (Retz.) R. Br. Appl Sci 9:2165. https://doi.org/10.3390/app9102165

Cunningham FJ, Goh NS, Demirer GS, Matos JL, Landry MP (2018) Nanoparticle-mediated delivery towards advancing plant genetic engineering. Trends Biotechnol 36(9):882–897. https://doi.org/10.1016/j.tibtech.2018.03.009

de Melo APZ, de Oliveira Brisola Maciel MV, Sganzerla WG et al (2020) Antibacterial activity, morphology, and physicochemical stability of biosynthesized silver nanoparticles using thyme (Thymus vulgaris) essential oil. Mater Res Express 7:015087. https://doi.org/10.1088/2053-1591/ab6c63

Dharmalingam K, Bordoloi D, Kunnumakkara AB, Anandalakshmi R (2020) Preparation and characterization of cellulose-based nanocomposite hydrogel films containing $CuO/Cu_2O/Cu$ with antibacterial activity. J Appl Polym Sci 137:49216. https://doi.org/10.1002/app.49216

Diao M, Yao M (2009) Use of zero-valent iron nanoparticles in inactivating microbes. Water Res 43:5243–5251. https://doi.org/10.1016/j.watres.2009.08.051

Doma M, Abhayankar G, Reddy VD, Kavi Kishor PB (2012) Carbohydrate and elicitor enhanced withanolide (withaferin A and withanolide A) accumulation in hairy root cultures of Withania somnifera (L.). Indian J Exp Biol 50:484–490

Duca D, Lorv J, Patten CL et al (2014) Indole-3-acetic acid in plant–microbe interactions. Antonie Van Leeuwenhoek 106:85–125. https://doi.org/10.1007/s10482-013-0095-y

Dutschk V, Karapantsios T, Liggieri L et al (2014) Smart and green interfaces: from single bubbles/drops to industrial environmental and biomedical applications. Adv Colloid Interf Sci 209:109–126. https://doi.org/10.1016/j.cis.2014.02.020

Egbuna C, Parmar VK, Jeevanandam J et al (2021) Toxicity of nanoparticles in biomedical application: nanotoxicology. J Toxicol 2021:1–21. https://doi.org/10.1155/2021/9954443

Elkordy AA, Haj-Ahmad RR, Awaad AS, Zaki RM (2021) An overview on natural product drug formulations from conventional medicines to nanomedicines: past, present and future. J Drug Delivery Sci Technol 63:102459. https://doi.org/10.1016/j.jddst.2021.102459

Espinosa-Leal CA, Puente-Garza CA, García-Lara S (2018) In vitro plant tissue culture: means for production of biological active compounds. Planta 248:1–18. https://doi.org/10.1007/s00425-018-2910-1

Feizi S (2023) Role of nanomaterials in plant cell and tissue culture. In: Nanomaterial interactions with plant cellular mechanisms and macromolecules and agricultural implications. Springer International Publishing, Cham, pp 359–397. https://doi.org/10.1007/978-3-031-20878-2_14

Fidemann T, de Araujo Pereira GA, Heluy TR et al (2018) Handling culture medium composition for optimizing plant cell suspension culture in shake flasks. Plant Cell Tissue Organ Cult 133:137–146. https://doi.org/10.1007/s11240-017-1368-3

George EF, Hall MA, De Klerk G-J (2008) The components of plant tissue culture media II: organic additions, osmotic and pH effects, and support systems. In: Plant propagation by tissue culture. Springer Netherlands, Dordrecht, pp 115–173. https://doi.org/10.1007/978-1-4020-5005-3_4

Gonçalves S, Romano A (2018) Production of plant secondary metabolites by using biotechnological tools. In: Secondary metabolites – sources and applications, vol 5. IntechOpen, pp 81–99. https://doi.org/10.5772/intechopen.76414

Gowtham HG, Shilpa N, Singh SB et al (2024) Toxicological effects of nanoparticles in plants: mechanisms involved at morphological, physiological, biochemical and molecular levels. Plant Physiol Biochem 210:108604. https://doi.org/10.1016/j.plaphy.2024.108604

Greenway MB, Phillips IC, Lloyd MN et al (2012) A nutrient medium for diverse applications and tissue growth of plant species in vitro. In Vitro Cell Dev Biol Plant 48:403–410. https://doi.org/10.1007/s11627-012-9452-1

Havenith H, Raven N, Di Fiore S et al (2014) Image-based analysis of cell-specific productivity for plant cell suspension cultures. Plant Cell Tissue Organ Cult 117:393–399. https://doi.org/10.1007/s11240-014-0448-x

Helaly MN, El-Metwally MA, El-Hoseiny H, Omar SA, El-Sheery NI (2014) Effect of nanoparticles on biological contamination of "in vitro" cultures and organogenic regeneration of banana. Aust J Crop Sci 8:612–624. https://search.informit.org/doi/10.3316/informit

Jafari N, Othman RY, Khalid N (2011) Effect of benzylaminopurine (BAP) pulsing on in vitro shoot multiplication of Musa acuminata (banana) cv. Berangan. Afr J Biotechnol 10:2446–2450. https://doi.org/10.4314/ajb.v10i13

Jawahar G, Punita DL, Rajasheker G et al (2018) Feeding elicitors and precursors enhance colchicine accumulation in morphogenic cultures of Gloriosa superba L. Plant Cell Tissue Organ Cult 135:235–245. https://doi.org/10.1007/s11240-018-1459-9

Jeevanandam J, Barhoum A, Chan YS et al (2018) Review on nanoparticles and nanostructured materials: history, sources, toxicity and regulations. Beilstein J Nanotechnol 9:1050–1074. https://doi.org/10.3762/bjnano.9.98

Jiravova J, Tomankova KB, Harvanova M, Malina L, Malohlava J, Luhova L, ... Kolarova H (2016). The effect of silver nanoparticles and silver ions on mammalian and plant cells in vitro. Food Chem Toxicol 96:50–61. https://doi.org/10.1016/j.fct.2016.07.015

Karimzadeh F, Haddad R, Garoosi G, Khademian R (2019) Effects of nanoparticles on activity of lignan biosynthesis enzymes in cell suspension culture of Linum usitatissimum L. Russ J Plant Physiol 66:756–762. https://doi.org/10.1134/S1021443719050078

Kim S, Ryu D (2013) Silver nanoparticle-induced oxidative stress, genotoxicity and apoptosis in cultured cells and animal tissues. J Appl Toxicol 33:78–89. https://doi.org/10.1002/jat.2792

Kim DH, Gopal J, Sivanesan I (2017) Nanomaterials in plant tissue culture: the disclosed and undisclosed. RSC Adv 7:36492–36505. https://doi.org/10.1039/C7RA07025J

Krystofova O, Sochor J, Zitka O, Babula P, Kudrle V, Adam V, Kizek R (2013) Effect of magnetic nanoparticles on tobacco BY-2 cell suspension culture. Int J Environ Res Public Health 10(1): 47–71. https://doi.org/10.3390/ijerph10010047

Kwon Y-W, Lee S-H, Kim A-R et al (2021) Plant callus-derived shikimic acid regenerates human skin through converting human dermal fibroblasts into multipotent skin-derived precursor cells. Stem Cell Res Ther 12:346. https://doi.org/10.1186/s13287-021-02409-3

Lang X, Hirata A, Fujita T, Chen M (2011) Nanoporous metal/oxide hybrid electrodes for electrochemical supercapacitors. Nat Nanotechnol 6:232–236. https://doi.org/10.1038/nnano.2011.13

Lew TTS, Wong MH, Kwak SY, Sinclair R, Koman VB, Strano MS (2018) Rational design principles for the transport and subcellular distribution of nanomaterials into plant protoplasts. Small 14(44):1802086. https://doi.org/10.1002/smll.201802086

Li Z, Guo H (2018) Ethylene treatment in studying leaf senescence in Arabidopsis. Methods Mol Biol 1744:105–112. https://doi.org/10.1007/978-1-4939-7672-0_8

Ljubimova JY, Holler E (2012) Biocompatible nanopolymers: the next generation of breast cancer treatment? Nanomedicine 7:1467–1470. https://doi.org/10.2217/nnm.12.115

Ludwig-Müller J (2000) Indole-3-butyric acid in plant growth and development. Plant Growth Regul 32:219–230. https://doi.org/10.1023/A:1010746806891

Mahjouri S, Movafeghi A, Divband B, Kosari-Nasab M (2018) Toxicity impacts of chemically and biologically synthesized CuO nanoparticles on cell suspension cultures of Nicotiana tabacum. Plant Cell, Tissue and Organ Culture (PCTOC) 135(2):223–234. https://doi.org/10.1007/s11240-018-1458-x

Milewska-Hendel A, Zubko M, Karcz J, Stróż D, Kurczyńska E (2017) Fate of neutral-charged gold nanoparticles in the roots of the Hordeum vulgare L. cultivar Karat. Scientific Reports 7(1): 3014. https://doi.org/10.1038/s41598-017-02965-w

Miller CO, Skoog F, Okumura FS et al (1956) Isolation, structure and synthesis of kinetin, a substance promoting cell division[1,2]. J Am Chem Soc 78:1375–1380. https://doi.org/10.1021/ja01588a032

Monfort LEF, Bertolucci SKV, Lima AF et al (2018) Effects of plant growth regulators, different culture media and strength MS on production of volatile fraction composition in shoot cultures of Ocimum basilicum. Ind Crop Prod 116:231–239. https://doi.org/10.1016/j.indcrop.2018.02.075

Morsy EA, Hussien AM, Ibrahim MA et al (2021) Cytotoxicity and genotoxicity of copper oxide nanoparticles in chickens. Biol Trace Elem Res 199:4731–4745. https://doi.org/10.1007/s12011-021-02595-4

Motolinía-Alcántara EA, Castillo-Araiza CO, Rodríguez-Monroy M et al (2021) Engineering considerations to produce bioactive compounds from plant cell suspension culture in bioreactors. Plants 10:2762. https://doi.org/10.3390/plants10122762

Murashige T, Skoog F (1962) A revised medium for rapid growth and bio assays with tobacco tissue cultures. Physiol Plant 15(3). https://doi.org/10.1111/j.1399-3054.1962.tb08052.x

Ochoa-Villarreal M, Howat S, Hong S et al (2016) Plant cell culture strategies for the production of natural products. BMB Rep 49:149–158. https://doi.org/10.5483/BMBRep.2016.49.3.264

Osmani Z, Wang L, Xiao W, Kulka M (2024) Nanomaterials as tools in plant transformation: a protoplast-centric perspective. Plant Nano Biol 10:100100. https://doi.org/10.1016/j.plana.2024.100100

Poborilova Z, Opatrilova R, Babula P (2013) Toxicity of aluminium oxide nanoparticles demonstrated using a BY-2 plant cell suspension culture model. Environ Exp Bot 91:1–11. https://doi.org/10.1016/j.envexpbot.2013.03.002

Rai MK, Shekhawat NS, Harish et al (2011) The role of abscisic acid in plant tissue culture: a review of recent progress. Plant Cell Tissue Organ Cult 106:179–190. https://doi.org/10.1007/s11240-011-9923-9

Rico CM, Hong J, Morales MI et al (2013) Effect of cerium oxide nanoparticles on rice: a study involving the antioxidant defense system and in vivo fluorescence imaging. Environ Sci Technol 47:5635–5642. https://doi.org/10.1021/es401032m

Rico CM, Lee SC, Rubenecia R et al (2014) Cerium oxide nanoparticles impact yield and modify nutritional parameters in wheat (*Triticum aestivum* L.). J Agric Food Chem 62:9669–9675. https://doi.org/10.1021/jf503526r

Rizzello L, Pompa PP (2014) Nanosilver-based antibacterial drugs and devices: mechanisms, methodological drawbacks, and guidelines. Chem Soc Rev 43:1501–1518. https://doi.org/10.1039/C3CS60218D

Santos AR, Miguel AS, Tomaz L, Malhó R, Maycock C, Vaz Patto MC, Oliva A (2010) The impact of CdSe/ZnS quantum dots in cells of Medicago sativa in suspension culture. J Nanobiotechnology 8(1):24. https://doi.org/10.1186/1477-3155-8-24

Selivanov NY, Selivanova OG, Sokolov OI et al (2017) Effect of gold and silver nanoparticles on the growth of the Arabidopsis thaliana cell suspension culture. Nanotechnol Russ 12:116–124. https://doi.org/10.1134/S1995078017010104

Shriram V, Kumar V, Kishor PBK et al (2010) Cytotoxic activity of 9,10-dihydro-2,5-dimethoxyphenanthrene-1,7-diol from Eulophia nuda against human cancer cells. J Ethnopharmacol 128:251–253. https://doi.org/10.1016/j.jep.2009.12.031

Siddiqui MH, Al-Whaibi MH (2014) Role of nano-SiO2 in germination of tomato (Lycopersicum esculentum seeds Mill.). Saudi J Biol Sci 21:13–17. https://doi.org/10.1016/j.sjbs.2013.04.005

Süntar I (2020) Importance of ethnopharmacological studies in drug discovery: role of medicinal plants. Phytochem Rev 19:1199–1209. https://doi.org/10.1007/s11101-019-09629-9

Szőllősi R, Molnár Á, Kondak S, Kolbert Z (2020) Dual effect of nanomaterials on germination and seedling growth: stimulation vs. phytotoxicity. Plants 9:1745. https://doi.org/10.3390/plants9121745

Taghizadeh M, Nasibi F, Kalantari KM, Ghanati F (2019) Evaluation of secondary metabolites and antioxidant activity in Dracocephalum polychaetum Bornm. cell suspension culture under magnetite nanoparticles and static magnetic field elicitation. Plant Cell Tissue Organ Cult 136:489–498. https://doi.org/10.1007/s11240-018-01530-1

Tan XM, Lin C, Fugetsu B (2009) Studies on toxicity of multi-walled carbon nanotubes on suspension rice cells. Carbon 47(15):3479–3487. https://doi.org/10.1016/j.carbon.2009.08.018

Tuominen M., & Schultz, E. (2010). Environmental aspects related to nanomaterials. Finnish Environment Institute, Helsinki. http://hdl.handle.net/10138/37969.

Wang P, Lombi E, Zhao F-J, Kopittke PM (2016) Nanotechnology: a new opportunity in plant sciences. Trends Plant Sci 21:699–712. https://doi.org/10.1016/j.tplants.2016.04.005

Wang JW, Grandio EG, Newkirk GM, Demirer GS, Butrus S, Giraldo JP, Landry MP (2019) Nanoparticle-mediated genetic engineering of plants. Mol Plant 12(8):1037–1040. https://doi.org/10.1016/j.molp.2019.06.010

Yang K, Ma Y-Q (2010) Computer simulation of the translocation of nanoparticles with different shapes across a lipid bilayer. Nat Nanotechnol 5:579–583. https://doi.org/10.1038/nnano.2010.141

Yue W, Ming Q, Lin B et al (2016) Medicinal plant cell suspension cultures: pharmaceutical applications and high-yielding strategies for the desired secondary metabolites. Crit Rev Biotechnol 36:215–232. https://doi.org/10.3109/07388551.2014.923986

Zafar T, Shrivastava VK, Shaik B (2019) Pharmaceutical biotechnology in herbal neuroprotection. In: Biotechnology products in everyday life. Springer, Cham, pp 221–228. https://doi.org/10.1007/978-3-319-92399-4_15

Zhang T, Wang Q, Rui Y (2025) The impact of nanomaterials on plant health: a review of exposure, toxicity, and control. Environ Sci Nano 12:2965–2982. https://doi.org/10.1039/D5EN00037H

Zhao X, Meng Z, Wang Y, Chen W, Sun C, Cui B, ... Cui H (2017) Pollen magnetofection for genetic modification with magnetic nanoparticles as gene carriers. Nat Plants 3(12):956–964. https://doi.org/10.1038/s41477-017-0063-z

Chapter 15
IoTs: Internet of Things Systems for Smart Plant Health Monitoring Under Nanoparticle Stresses

Anjali Bharti, Vinita Hooda, Utkarsh Jain, and Nidhi Chauhan

Abstract The crops are subjected to environmental stressors like drought, salinity, extreme temperatures, heavy metals and nanoparticles (NPs), which reduce the quality of crops and interfere with their overall health. These issues are addressed by the farmers seeking assistance with smart technologies, including precise and real-time data analysis, facilitating field management. Cloud computing, biosensors, machine learning (ML) and the Internet of Things (IoT) are all making a major contribution to the implementation of more efficient and sustainable agricultural practices. The latest developments in the field of agriculture are the high-tech greenhouse systems and field sensors, which enable farmers to continuously monitor the health of soil as well as to measure other important parameters like pH, temperature, salinity, heavy metals, and nutrient content. This information is further utilised in controlling the irrigation management, application of the fertiliser and the general consumption of the resources. Along with this, the ML algorithms are also essential in precision farming as they give live data of the field and subsequently predict the future requirement of crops. IoT and AI lead to an immense advancement in the agricultural system, consequently allowing farmers to make informed decisions. Moreover, the use of NPs in the agricultural sector using such intelligent and accurate methods enables farmers to boost production without impacting the environment. The studies in this field require additional research, which certainly guarantees the creation of efficient farming methods that enhance the development of crops by using environmentally friendly methods. Thus, the introduction of smart technologies is defining a new age of precision farming in which three crucial parameters, including efficiency, sustainability, and environmental protection, will significantly achieve the growing global food demand.

Keywords Environmental stress · Internet of Things (IoT) · Machine learning (ML) · Biosensors · Precision farming

A. Bharti · U. Jain · N. Chauhan (✉)
School of Health Sciences & Technology (SoHST), UPES, Dehradun, Uttarakhand, India

V. Hooda
Department of Botany, Maharshi Dayanand University, Rohtak, Haryana, India

R. N. Pudake et al. (eds.), *Phyto Nanotoxicology*, Nanotechnology in Plant Sciences 7, https://doi.org/10.1007/978-3-032-24518-2_15

15.1 Introduction

Nanotechnology encompasses numerous applications in the agricultural sector, ranging from enhanced nutrient use efficiency to advanced disease and pest management, which can help to reshape sustainable farming methods (Gupta et al. 2023). Nanoparticles (NPs) can affect plant processes at the molecular level and thus may impact their growth, development, and stress tolerance (Vijayakumar et al. 2022). Today, the world population is growing at a rapid rate and is expected to reach six billion by 2050. However, the agricultural output is declining as a result of several biotic and abiotic stresses (Wang et al. 2021c). Abiotic (non-living) stresses, including drought, salinity, alkalinity, temperature changes, and metal or mineral toxicity, as well as biotic stresses, are caused by living organisms, including pest infestation, pathogens, nematodes, and parasitic weeds. NPs can replicate the function of natural antioxidant enzymes, thereby acting as nano-enzymes that scavenge and neutralise reactive oxygen species (ROS) (Sharifi et al. 2020). Currently, nanomaterials (NMs) are widely utilised in agricultural fields, mainly in the form of nanofertilizers and nanopesticides. Consequently, farmers benefit from the implementation of NMs in agriculture *via* higher crop yield, improved soil health, and reduced water consumption (Dwivedi et al. 2016). For instance, when strawberry plants were treated with nano-silicon, the epicuticular wax exhibited thicker and better-structured wax compared to the controls that were treated with salt stress (Avestan et al. 2019). The response of plants to abiotic stress has been studied by several researchers, in which the application of SiO_2 NPs improved the transpiration rate, water use efficiency, total chlorophyll content, and carbonic anhydrase activity of the salt-stressed *Cucurbita pepo*, contributing to plant adaptation to salt stress (Siddiqui et al. 2014). In another research on medicinal plants, Karamian et al. reported that application of methyl jasmonate (200 μM), salicylic acid (100 μM), and TiO_2 NPs (20 ppm) mitigated drought stress. From these results, it was affirmed that both enzymatic and non-enzymatic antioxidant defence systems improve tolerance to water stress (Karamian et al. 2020). In agriculture, different parameters, including soil moisture, temperature, and humidity, with modern monitoring systems based on IoT technology, will support farmers to grow healthier and higher-yielding crops. The sensors enable farmers with rapid and useful information regarding the health of their crops by transmitting real-time updates directly *via* a mobile application. The system provides a smart and useful solution for modern farming systems, as it can be integrated with other tools to manage irrigation and monitor greenhouse conditions (Absar et al. 2023). The present agri-food sector has transformed from traditional farming practices to modern and intelligent practices with the incorporation of smart sensors and IoT tools. It can assist farmers to gather data related to soil fertility, soil erosion, crop quality evaluation, and fertiliser needs (Sishodia et al. 2020). The sensors that work with the help of the IoTs gather the data, allow real-time analysis and provide the farmers and the researchers with a valuable insight into the health of the plants, practices of precision farming, the applications of nutrients, and the effective ways of irrigating the plants (FuBing 2012), as depicted in Fig. 15.1. As

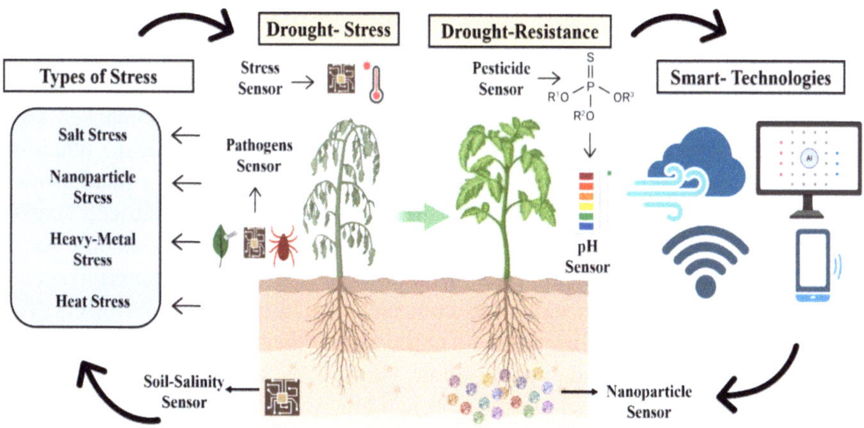

Fig. 15.1 An overview of the smart technologies supporting drought-resistant crops. IoT-based technologies, including wireless communication devices, cloud platforms, soil moisture sensors, and temperature sensors, enable real-time monitoring of stress factors, enhancing crop resilience

discussed earlier, IoT platforms incorporated with smart tools can empower farmers to integrate more enhanced planting strategies (Liu et al. 2016). Farmers can incorporate these technologies to detect pest infestations and plant diseases at a very early stage. Furthermore, deep learning technologies promise great potential for comprehensive data analysis obtained by various sensors. For instance, IoT sensors, together with platforms including ZigBee and Arduino, can analyse soil moisture, crop species, temperature, and water content at a particular location (Rodríguez-Robles et al. 2020).

15.2 Nanoparticles in Agriculture: An Overview

Today, the world's population continues to grow, and agricultural nanotechnology has become an important tool for enhancing crop performance. The study and manipulation of material at the nanoscale range, typically between 1 and 100 nm, is referred to as nanotechnology. Nanotechnology holds the promise to transform food security and safety through improved plant development and increased crop yield (Bharti et al. 2025). These NPs are classified as either organic or inorganic, depending on their basic composition. Metal-based NPs mainly composed of metals including Al, Bi, Co, Cu, Au, Fe, In, Mo, Ni, Si, Ag, Sn, Ti, W, Zn, along with metal oxide-based NPs such as, Al_2O_3, CeO_2, CuO, Cu_2O, In_2O_3, La_2O_3, MgO, NiO, SiO_2, TiO_2, SnO_2, ZnO, and ZrO_2, are further classified under the category of inorganic NPs. The most extensively used NPs in agriculture include copper oxide (CuO), iron oxides (Fe_2O_3, Fe_3O_4), titanium dioxide (TiO_2), zinc oxide (ZnO), silver (Ag), and silver oxides (Ag_2O). Their undesirable biological consequences on the diversity, quantity, and biological activity of both flora and fauna are being carefully investigated (Rajput et al. 2018). However, NPs involving a high surface area-to-volume ratio offer several advantages over conventional farming practices if applied appropriately in the field. It mainly involves enhanced mass transfer and nutrients or pesticides delivery in smaller or more precise doses, resulting in enhanced crop productivity (Hussain et al. 2023). Copper, iron, magnesium, silicon, titanium, manganese, zinc, and silver are some of the key elements required for plant growth and yield. They help to protect plants from diseases, promote nutrient uptake, maintain plant health, and increase crop yields (Rastogi et al. 2019; Nandhini et al. 2019; Singh et al. 2021; Hazarika et al. 2022; Bansal et al. 2022). Nanofertilizers are effective even at lower concentrations compared to traditional bulk fertilisers. The smaller dosage of NPs can improve the nutrient use efficiency by reducing nutrient losses from leaching and volatilisation (Raliya et al. 2018). Additionally, nanofertilizers can be applied to plants through foliar spray on the leaves or irrigated directly into the soil, which increases plant productivity as depicted in Fig. 15.2. It was observed that applying NPs in the form of nanofertilizers can increase crop growth and productivity (Merghany et al. 2019; Babu et al. 2022). However, the biosynthesised NPs can also be additionally applied *via* the seed priming method, which involves treating seeds before planting in the soil (Korishettar 2016; Mittal et al. 2020). In addition, NPs can also be formulated in the form of nanopesticides to defend crops against bacterial and fungal infestation (Kumar et al. 2025).

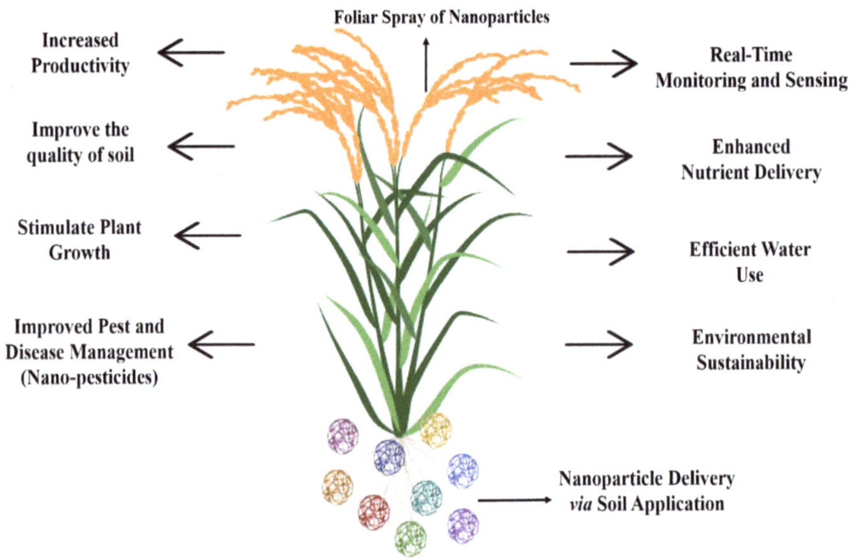

Fig. 15.2 The beneficial effects of nanoparticle applications in the agricultural field

15.3 Plant Responses to Nanoparticle-Induced Stress

Global farming is profoundly affected by several abiotic stresses, including high or low water levels, saline soil, UV radiation, heavy metal pollution, and extreme temperature fluctuations (Seleiman et al. 2020; Badawy et al. 2021). Different horticultural crops, including fruits, vegetables, and spices, can be adversely impacted due to salinity stress. It can also lead to oxidative damage, nutrient imbalances, and impaired water absorption. Excessive levels of salt can also disrupt the plant's internal ionic balance and hamper cell division, which disrupts vital functions related to growth, metabolism, and overall health (Dong et al. 2021; Sinha and Verma 2021). Drought stress affects the cell membranes and water balance in plants that severely hindering plant growth (Umair Hassan et al. 2020). It can significantly increase the formation of ROS, which affects cell membranes, oxidises lipids, and increases malondialdehyde (MDA) levels (Das and Roychoudhury 2014). Interestingly, ZnO NPs were reported to reduce oxidative stress by minimising MDA and hydrogen peroxide (H_2O_2) accumulation, maintaining cell membrane stability, and decreasing the loss of important osmolytes (Mohamed et al. 2017; Sun et al. 2020; El-Zohri et al. 2021). NPs may exhibit phytotoxicity to the plant mainly due to oxidative stress, often expressed as a reduction in root length, damage to the tips of the roots, reduced biomass, degradation of chlorophyll, and other detrimental changes. Nevertheless, some NPs do not possess negative impacts. For instance, the effects of TiO_2 NPs in *Cucumis sativus* and ZnO NPs in *Cyamopsis tetragonoloba* exhibit improved chlorophyll content (Judy et al. 2012; Tarafdar et al. 2014). In response to NPs penetration, plants frequently produce ROS, which may stimulate many toxic responses. These toxic effects can range from ion leakage to programmed cell death, predominantly driven by oxidative stress and ROS-induced lipid peroxidation, which leads to mechanically damaged membranes. For instance, exposure to CeO_2 NPs in *Zea mays* induced lipid peroxidation through ion leakage, whereas no effect was found in *Oryza sativa* under the same NP concentration, i.e., 0–500 mg/L (Rico et al. 2013). Besides, the relationships between plants and NPs may negatively impact the secondary metabolism, hormonal balance, and plant development. Further, the transcriptomic analysis of *Arabidopsis thaliana* has shown that there is a downregulation of gene expression during the time of NPs exposure that controls phosphate regulation, pathogen defence, and stress response genes. Reduction of genes may affect the plant defence and developmental processes, especially at the roots. This points to the potential adverse effects of NPs exposure on the physiology of plants (Sanzari et al. 2019). The plants can absorb NPs through the roots and leaves. Once NPs are introduced into the plant system, they may cause biochemical, molecular, physiological, or even structural changes to occur that can influence total plant growth and productivity. Nonetheless, the NPs' effect on plants will greatly rely on the concentration, size and mode of application (through soil/foliar) that play a significant role in influencing the physiological and biochemical responses aroused by the plant system. Moreover, recent studies on NPs indicate that it is capable of causing drought stress tolerance and enhancing the growth and yield of plants (Khan et al. 2019; Ahmed et al. 2021). The soil and foliar application of ZnO NPs have greatly improved the uptake of nitrogen (N), phosphorus (P), potassium (K), and zinc (Zn), which mitigated the adverse effects of drought stress (Mangena 2018; Dimkpa et al. 2019; Fatollahpour Grangah et al. 2020; Semida et al. 2021; Mustafa et al. 2021; Akhtar et al. 2022). Further, chitosan NPs enhanced the cold stress tolerance of banana plants by reducing detrimental ROS concentrations and promoting the accumulation of protective substances (Wang et al. 2021a). In rice, applying ZnO NPs on vegetative parts can help to alleviate chilling stress by increasing the antioxidant activity and stimulating important transcription factors (Song et al. 2021). Similarly, it was observed that Si NPs enhanced the photosynthetic activity of sugarcane during exposure to chilling stress (Elsheery et al. 2020). NPs' application to the soil may also help to reduce the movement and accumulation of heavy metals. For instance, Fe_3O_4 NPs can stabilise cadmium (Cd) in the soil, thus decreasing its bioavailability (Wang et al. 2020). Similarly, Hydroxyapatite NPs can mitigate the harmful effects of metals in soil and maintain the pH balance *via* the release of phosphate ions (Cui et al. 2018). NPs can promote apoplast barrier formation within roots, ultimately reducing the accumulation of heavy metals. It can also influence gene expression related to metal transporters, ensuring metals are intercepted and complexed (Wang et al. 2021b), hence, reducing their translocation through the plant. For instance, Si NPs stimulate the secretion of organic acids while reducing heavy metal stress (Wu et al. 2017). Furthermore, NPs can also activate the antioxidant defence mechanism in the plant system to minimise the damage caused by ROS (Wu et al. 2017). Moreover, Se NPs used on the sorghum under high-temperature stress decrease adverse effects, such as membrane damage, and reduce pollen germination, while improving its antioxidant defence mechanism (Djanaguiraman et al. 2018). Similarly, AgNPs protected wheat plants from heat stress, thereby enhancing growth and developmental processes (Iqbal et al. 2019). Thus, nanomaterials have two main effects on plant systems. Firstly, they shield the plants against the damaging ROS activity. Secondly, they induce a weak oxidative stress and result in the stimulation of the antioxidant machinery in the plant (González-García et al. 2021).

15.4 Benefits of Internet of Things (IoT) Integrated with Artificial Intelligence (AI) for Real-Time Data Analytics

Agricultural IoT is defined as the application of intelligent sensors and smart devices that exchange information *via* the Internet to enhance the management of farms. Such intelligent technologies have the potential to automate a number of operations in the agriculture industry, besides integrating sensors that can gather real-time information about key variables such as soil quality, temperature, and moisture content (Rudrakar and Rughani 2024). This can facilitate farmers to have greater knowledge of the management of irrigation in their fields. It also helps in the implementation of improved water conservation methods in the agriculture field, ensuring the appropriate quantity of moisture needed for the healthy growth and high-yielding harvests (Devkota et al. 2022; Chamara et al. 2023). For instance, soil sensors may also enable farmers to use fertilisers only when it is required. Moreover, soil sensors can also be used to measure various soil characteristics like pH, salinity, and nitrogen levels (Yin et al. 2021; Sarkar et al. 2022; Elashmawy and Uysal 2023). Using AI and machine learning, farmers can be able to detect the initial signs of stress and the disease in the plants promptly and avoid losses in the harvest (Molina-Rotger et al. 2023; Praveen et al. 2024). Khan et al. (2020) designed a smart greenhouse system based on IOT that can identify plant diseases and real-time environmental conditions. The key applications of the system are climate data recording, irrigation automation, plant imaging, and leaf analysis to forecast diseases. Thus, this smart technology will transform a traditional greenhouse into a hi-tech and more efficient greenhouse, which will be an advantage for farmers and agricultural enterprises (Khan et al. 2020). Shrestha and Wei (2024) considered the application of real-time nitrogen detection and an IoT-based technology, which can improve the use of nitrogen in agriculture (Shrestha and Wei 2024). Agricultural IoT has many smart farming applications, which comprise livestock, greenhouse, fisheries, and weather. Nevertheless, Wi-Fi, Zigbee, LoRa, Bluetooth, and 5G are the primary communication technologies created in the context of IoT systems (Pratama et al. 2020). The creation of the Wireless Sensor Networks (WSNs) has revolutionised agricultural activities, bringing about great advantages to farmers and agronomists. The use of sensors, GPS and data loggers is common in modern farming equipment has the ability to monitor the soil characteristics, plants and the field conditions easily (Keskin et al. 2021). These systems can receive and centralise information from each field to make informed judgments as individual systems. They are also able to check soil moisture, compaction, fertility, biomass, leaf area, temperature, chlorophyll content, irrigation requirement, local climate, pests, and crop yields (Thakur et al. 2019). Moreover, as a system, Geographic Information Systems (GIS) are a significant aspect of the modern agriculture industry since they include computer hardware, computer software, geographic data, and personnel skills. The primary tasks of GIS include capturing, storing, updating, analysing, and visualising the information with geographic references, which have the potential to make superior decisions in farming (Shafi et al. 2019). In addition to core mapping functions, GIS also supports spatial reasoning, data management, and modelling. It uses base maps showing topography, soil characteristics, nutrient loads, soil moisture, pH, fertility levels, weed and pest pressure levels (Mani et al. 2021). GIS combines data from numerous sources and uses decision support measures to make a more precise examination and impose the right dosages of nutrients or pesticides. Moreover, NPs can help plants to deal with stress, reduce the negative impact of environmental factors, and enhance growth and yield. However, it can affect morphology, anatomy, physiology, biochemistry, and gene expression of different plant traits. IoT-based smart farming revolves around soil and water sensors, which measure critical factors that include moisture, pH, temperature, and nutrient levels (Yin et al. 2021; Lloret et al. 2021), giving farmers precise information about the soil's condition. Accuracy is especially important in precision irrigation, which saves water by changing irrigation schedules based on soil conditions and weather reports. The implementation of AI water sensors can monitor contaminants, salinity levels, and wasteful irrigation practices, ultimately supporting farmers in more sustainable water management (Placidi et al. 2021; Kashyap et al. 2021; Bwambale et al. 2022). Precision agriculture using IoT technologies offers enormous benefits in terms of better water management, smarter fertiliser use, and effective pest control. However, the capital cost of setting up smart farming systems is still an obstacle, particularly for smallholder farmers (da Silveira et al. 2023; Mallinger et al. 2024). Currently, there is a widespread adoption of intelligent sensing technology solutions for grassland management and urban agriculture. In urban areas, systems based on IoT devices and sensors used for automatic irrigation and climate management in vertical farms, hydroponic systems, etc., are increasing food production (Wani 2023). AI can assess data to determine the effects of different types of fertilisers, in different amounts, and help to identify the solutions that reduce damage and maximise yields. This leads to better environmental management practices that support sustainability on the farms (Mkrttchian 2021; Chen and Yu 2022).

15.5 Future Perspectives

Farmers are changing the way to monitor the health of their crops, irrigate water, and fight disease due to the utilisation of intelligent technologies and interconnected systems. In order to make sure that these technologies are utilised in a responsible and effective way, one should consider further elaboration of the sophisticated sensing devices, which will contribute to sustainable farming practices. The recent technologies like Edge IoT sensors and drones have enabled farmers to take real-time pictures of their crops, identify pests, and control the irrigation processes without relying on a far-off data server, as illustrated in Fig. 15.3. This is especially useful among farmers who are in areas where internet connection is weak or unavailable. The constant technological advancement and the establishment of coherent moral and functional guidelines have contributed to the successful implementation of these tools in agricultural farming. The ongoing advancement in the implementation of new technologies, such as blockchain, Edge computing, and automation, is projected to be influential when it comes to defining the new generation of smart and efficient farming systems. Nevertheless, there are still several problems associated with their widespread adoption. Farmers should be encouraged and offered practical assistance to embrace such new practices, high policy support and adequate training to utilise such new and smart models. Further, the affordability of advanced farming equipment and the accessibility of digital tools should remain the greatest priorities, and these technologies should be easy and accessible to all

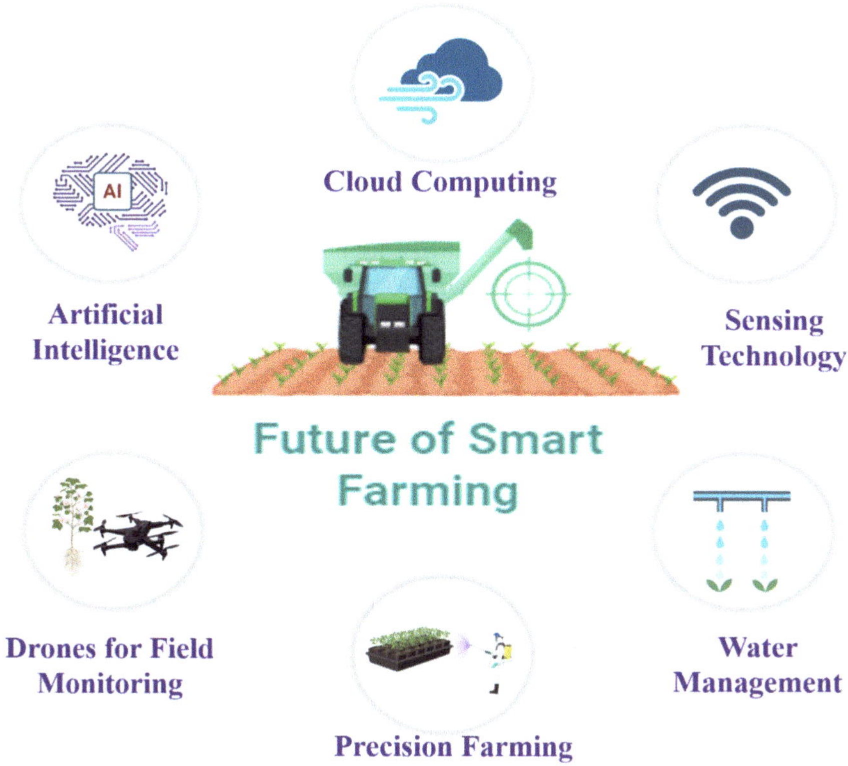

Fig. 15.3 Precision agriculture is enhanced through the use of smart farming technologies, including drones, cloud computing, and artificial intelligence (AI), which optimise crop productivity and resource management by offering real-time data analysis

farmers, irrespective of their size and income. By focusing on these priorities, it will enable agriculture to have a more sustainable, equitable and transparent future.

15.6 Conclusion

Nanotechnological studies, especially in the agricultural and food systems, have been very crucial in enhancing food security and safety. It provides accurate and efficient methods of managing weeds, pests, and harmful microbes. Modern smart farming practices can also transform the industry by making agriculture safer, more reliable and profitable, while increasing crop growth and yield. Connected devices and data-based tools can now enable farmers to develop customised farm management plans to enhance decision-making related to their fields. Real-time field data allows fast and informed decisions, with respect to changing irrigation techniques, fertiliser management, and pest control, to ensure sustainable farming practices.

Nanotechnology is useful in the food industry, where food stability and food safety can be enhanced, along with enhancing food shelf life, and aids in better detection and monitoring mechanisms. As a measure to ensure food security, more attention should be given to the development and implementation of optimal approaches regarding the management of crop diseases. Research across the globe indicates that digital and automated technologies would make farming more efficient and less harmful to the environment. The solutions to these challenges should involve close collaboration of governments, agricultural experts, technology developers, and research institutions. The use of IoT-based sensors in agriculture and food production is projected to be more automated, data-driven, and sustainable, which will bring a robust and precision farming system.

References

Absar MH, Mirza GF, Zakai W, John Y, Mansoor N (2023) Novel IoT-based plant monitoring system. Eng Proc 32(1):12. https://doi.org/10.3390/engproc2023032012

Ahmed T, Noman M, Manzoor N et al (2021) Nanoparticle-based amelioration of drought stress and cadmium toxicity in rice via triggering the stress responsive genetic mechanisms and nutrient acquisition. Ecotoxicol Environ Saf 209:111829. https://doi.org/10.1016/j.ecoenv.2020.111829

Akhtar N, Ilyas N, Meraj TA et al (2022) Improvement of plant responses by nanobiofertilizer: a step towards sustainable agriculture. Nanomaterials 12:965. https://doi.org/10.3390/nano12060965

Avestan S, Ghasemnezhad M, Esfahani M, Byrt CS (2019) Application of nano-silicon dioxide improves salt stress tolerance in strawberry plants. Agronomy 9:246. https://doi.org/10.3390/agronomy9050246

Babu S, Singh R, Yadav D et al (2022) Nanofertilizers for agricultural and environmental sustainability. Chemosphere 292:133451. https://doi.org/10.1016/j.chemosphere.2021.133451

Badawy SA, Zayed BA, Bassiouni SMA et al (2021) Influence of nano silicon and nano selenium on root characters, growth, ion selectivity, yield, and yield components of rice (Oryza sativa L.) under salinity conditions. Plants 10:1657. https://doi.org/10.3390/plants10081657

Bansal K, Hooda V, Verma N et al (2022) Stress alleviation and crop improvement using silicon nanoparticles in agriculture: a review. Silicon 14:10173–10186. https://doi.org/10.1007/s12633-022-01755-y

Bharti A, Hooda V, Singh A et al (2025) Advancing sustainable crop yield enhancement through environmentally friendly nanoparticle synthesis: a comprehensive review. J Soil Sci Plant Nutr 25:2314–2345. https://doi.org/10.1007/s42729-025-02269-3

Bwambale E, Abagale FK, Anornu GK (2022) Smart irrigation monitoring and control strategies for improving water use efficiency in precision agriculture: a review. Agric Water Manag 260:107324. https://doi.org/10.1016/j.agwat.2021.107324

Chamara N, Bai G, Ge Y (2023) AICropCAM: deploying classification, segmentation, detection, and counting deep-learning models for crop monitoring on the edge. Comput Electron Agric 215:108420. https://doi.org/10.1016/j.compag.2023.108420

Chen T-C, Yu S-Y (2022) The review of food safety inspection system based on artificial intelligence, image processing, and robotic. Food Sci Technol 42:e35421. https://doi.org/10.1590/fst.35421

Cui H, Shi Y, Zhou J et al (2018) Effect of different grain sizes of hydroxyapatite on soil heavy metal bioavailability and microbial community composition. Agric Ecosyst Environ 267:165–173. https://doi.org/10.1016/j.agee.2018.08.017

da Silveira F, da Silva SLC, Machado FM et al (2023) Farmers' perception of the barriers that hinder the implementation of agriculture 4.0. Agric Syst 208:103656. https://doi.org/10.1016/j.agsy.2023.103656
Das K, Roychoudhury A (2014) Reactive oxygen species (ROS) and response of antioxidants as ROS-scavengers during environmental stress in plants. Front Environ Sci 2:53. https://doi.org/10.3389/fenvs.2014.00053
Devkota KP, Devkota M, Rezaei M, Oosterbaan R (2022) Managing salinity for sustainable agricultural production in salt-affected soils of irrigated drylands. Agric Syst 198:103390. https://doi.org/10.1016/j.agsy.2022.103390
Dimkpa CO, Singh U, Bindraban PS et al (2019) Zinc oxide nanoparticles alleviate drought-induced alterations in sorghum performance, nutrient acquisition, and grain fortification. Sci Total Environ 688:926–934. https://doi.org/10.1016/j.scitotenv.2019.06.392
Djanaguiraman M, Belliraj N, Bossmann SH, Prasad PVV (2018) High-temperature stress alleviation by selenium nanoparticle treatment in grain sorghum. ACS Omega 3:2479–2491. https://doi.org/10.1021/acsomega.7b01934
Dong F, Yang F, Liu Y et al (2021) Calmodulin-binding transcription activator (CAMTA)/factors in plants. In: Calcium transport elements in plants. Elsevier, pp 249–266. https://doi.org/10.1016/B978-0-12-821792-4.00017-5
Dwivedi S, Saquib Q, Al-Khedhairy AA, Musarrat J (2016) Understanding the role of nanomaterials in agriculture. In: Microbial inoculants in sustainable agricultural productivity. Springer India, New Delhi, pp 271–288. https://doi.org/10.1007/978-81-322-2644-4_17
Elashmawy R, Uysal I (2023) Precision agriculture using soil sensor driven machine learning for smart strawberry production. Sensors 23:2247. https://doi.org/10.3390/s23042247
Elsheery NI, Sunoj VSJ, Wen Y et al (2020) Foliar application of nanoparticles mitigates the chilling effect on photosynthesis and photoprotection in sugarcane. Plant Physiol Biochem 149: 50–60. https://doi.org/10.1016/j.plaphy.2020.01.035
El-Zohri M, Al-Wadaani NA, Bafeel SO (2021) Foliar sprayed green zinc oxide nanoparticles mitigate drought-induced oxidative stress in tomato. Plants 10:2400. https://doi.org/10.3390/plants10112400
Fatollahpour Grangah M, Rashidi V, Mirshekari B et al (2020) Effects of nano-fertilizers on physiological and yield characteristics of pinto bean cultivars under water deficit stress. J Plant Nutr 43:2898–2910. https://doi.org/10.1080/01904167.2020.1799000
FuBing (2012) Research on the agriculture intelligent system based on IOT. In 2012 international conference on image analysis and signal processing (pp. 1–4). IEEE. https://doi.org/10.1109/IASP.2012.6425066
González-García Y, González-Moscoso M, Hernández-Hernández H et al (2021) Induction of stress tolerance in crops by applying nanomaterials. In: Nanotechnology in plant growth promotion and protection. Wiley, pp 129–169. https://doi.org/10.1002/9781119745884.ch8
Gupta A, Rayeen F, Mishra R et al (2023) Nanotechnology applications in sustainable agriculture: an emerging eco-friendly approach. Plant Nano Biol 4:100033. https://doi.org/10.1016/j.plana.2023.100033
Hazarika A, Yadav M, Yadav DK, Yadav HS (2022) An overview of the role of nanoparticles in sustainable agriculture. Biocatal Agric Biotechnol 43:102399. https://doi.org/10.1016/j.bcab.2022.102399
Hussain M, Zahra N, Lang T et al (2023) Integrating nanotechnology with plant microbiome for next-generation crop health. Plant Physiol Biochem 196:703–711. https://doi.org/10.1016/j.plaphy.2023.02.022
Iqbal M, Raja NI, Mashwani Z-U-R et al (2019) Effect of silver nanoparticles on growth of wheat under heat stress. Iran J Sci Technol Trans A: Sci 43:387–395. https://doi.org/10.1007/s40995-017-0417-4
Judy JD, Unrine JM, Rao W et al (2012) Bioavailability of gold nanomaterials to plants: importance of particle size and surface coating. Environ Sci Technol 46:8467–8474. https://doi.org/10.1021/es3019397

Karamian R, Ghasemlou F, Amiri H (2020) Physiological evaluation of drought stress tolerance and recovery in Verbascum sinuatum plants treated with methyl jasmonate, salicylic acid and titanium dioxide nanoparticles. Plant Biosyst-An Int J Deal All Asp Plant Biol 154(3): 277–287. https://doi.org/10.1080/11263504.2019.1591535

Kashyap PK, Kumar S, Jaiswal A et al (2021) Towards precision agriculture: IoT-enabled intelligent irrigation systems using deep learning neural network. IEEE Sensors J 21:17479–17491. https://doi.org/10.1109/JSEN.2021.3069266

Keskin M, Say SM, Sekerli YE, Sehri M (2021) Appraisal of three proximal sensing systems to estimate macronutrient contents of detached soybean leaves. Commun Soil Sci Plant Anal 52: 1943–1953. https://doi.org/10.1080/00103624.2021.1900862

Khan I, Saeed K, Khan I (2019) Nanoparticles: properties, applications and toxicities. Arab J Chem 12:908–931. https://doi.org/10.1016/j.arabjc.2017.05.011

Khan FA, Ibrahim AA, Zeki AM (2020) Environmental monitoring and disease detection of plants in smart greenhouse using Internet of Things. J Phys Commun 4:055008. https://doi.org/10.1088/2399-6528/ab90c1

Korishettar P, Vasudevan SN, Shakuntala NM, Doddagoudar SR, Hiregoudar S, Kisan B (2016) Seed polymer coating with Zn and Fe nanoparticles: an innovative seed quality enhancement technique in pigeonpea. J Appl Nat Sci 8(1):445. https://doi.org/10.31018/jans.v8i1.814

Kumar H, Kumari N, Yadav A (2025) Synthesis and application of nanopesticides. Nanoformulations Agric Appl 91–118. https://doi.org/10.1002/9781119819127.ch4

Liu C, Cutforth H, Chai Q, Gan Y (2016) Farming tactics to reduce the carbon footprint of crop cultivation in semiarid areas. A review. Agron Sustain Dev 36(4):69. https://doi.org/10.1007/s13593-016-0404-8

Lloret J, Sendra S, Garcia L, Jimenez JM (2021) A wireless sensor network deployment for soil moisture monitoring in precision agriculture. Sensors 21:7243. https://doi.org/10.3390/s21217243

Mallinger K, Corpaci L, Neubauer T et al (2024) Breaking the barriers of technology adoption: explainable AI for requirement analysis and technology design in smart farming. Smart Agric Technol 9:100658. https://doi.org/10.1016/j.atech.2024.100658

Mangena P (2018) Water stress: morphological and anatomical changes in soybean (Glycine max L.) plants. In: Plant, abiotic stress and responses to climate change. http://dx.doi.org/10.5772./intechopen.72899

Mani PK, Mandal A, Biswas S et al (2021) Remote sensing and geographic information system: a tool for precision farming. In: Geospatial technologies for crops and soils. Springer Singapore, Singapore, pp 49–111. https://doi.org/10.1007/978-981-15-6864-0_2

Merghany MM, Shahein MM, Sliem MA, Abdelgawad KF, Radwan AF (2019) Effect of nano-fertilizers on cucumber plant growth, fruit yield and it's quality. Plant Arch 19:165–172

Mittal D, Kaur G, Singh P, Yadav K, Ali SA (2020) Nanoparticle-based sustainable agriculture and food science: recent advances and future outlook. Front nanotechnol 2:579954. https://doi.org/10.3389/fnano.2020.579954

Mkrttchian V (2021) Artificial and natural intelligence techniques as IoP- and IoT-based technologies for sustainable farming and smart agriculture. In Artificial Intelligence and IoT-Based Technologies for Sustainable Farming and Smart Agriculture (pp. 40-53). IGI Global Scientific Publishing. https://doi.org/10.4018/978-1-7998-1722-2.ch003

Mohamed AKSH, Qayyum MF, Abdel-Hadi AM et al (2017) Interactive effect of salinity and silver nanoparticles on photosynthetic and biochemical parameters of wheat. Arch Agron Soil Sci 63: 1736–1747. https://doi.org/10.1080/03650340.2017.1300256

Molina-Rotger M, Morán A, Miranda MA, Alorda-Ladaria B (2023) Remote fruit fly detection using computer vision and machine learning-based electronic trap. Front Plant Sci 14:1241576. https://doi.org/10.3389/fpls.2023.1241576

Mustafa H, Ilyas N, Akhtar N et al (2021) Biosynthesis and characterization of titanium dioxide nanoparticles and its effects along with calcium phosphate on physicochemical attributes of wheat under drought stress. Ecotoxicol Environ Saf 223:112519. https://doi.org/10.1016/j.ecoenv.2021.112519

Nandhini M, Rajini SB, Udayashankar AC et al (2019) Biofabricated zinc oxide nanoparticles as an eco-friendly alternative for growth promotion and management of downy mildew of pearl millet. Crop Prot 121:103–112. https://doi.org/10.1016/j.cropro.2019.03.015

Placidi P, Morbidelli R, Fortunati D et al (2021) Monitoring soil and ambient parameters in the IoT precision agriculture scenario: an original modeling approach dedicated to low-cost soil water content sensors. Sensors 21:5110. https://doi.org/10.3390/s21155110

Pratama MT, Kim S, Ozawa S et al (2020) Deep learning-based object detection for crop monitoring in soybean fields. In 2020 International joint conference on neural networks (IJCNN). IEEE, pp 1–7. https://doi.org/10.1109/IJCNN48605.2020.9207400

Praveen R, Hemavathi U, Sathya R et al (2024) AI powered plant identification and plant disease classification system. In 2024 4th International conference on sustainable expert systems (ICSES). IEEE, pp 1610–1616. https://doi.org/10.1109/ICSES63445.2024.10763167

Rajput VD, Minkina T, Sushkova S et al (2018) Effect of nanoparticles on crops and soil microbial communities. J Soils Sediments 18:2179–2187. https://doi.org/10.1007/s11368-017-1793-2

Raliya R, Saharan V, Dimkpa C, Biswas P (2018) Nanofertilizer for precision and sustainable agriculture: current state and future perspectives. J Agric Food Chem 66:6487–6503. https://doi.org/10.1021/acs.jafc.7b02178

Rastogi A, Tripathi DK, Yadav S et al (2019) Application of silicon nanoparticles in agriculture. 3 Biotech 9:90. https://doi.org/10.1007/s13205-019-1626-7

Rico CM, Hong J, Morales MI et al (2013) Effect of cerium oxide nanoparticles on rice: a study involving the antioxidant defense system and in vivo fluorescence imaging. Environ Sci Technol 47:5635–5642. https://doi.org/10.1021/es401032m

Rodríguez-Robles J, Martin Á, Martin S, Ruipérez-Valiente JA, Castro M (2020) Autonomous sensor network for rural agriculture environments, low cost, and energy self-charge. Sustainability 12(15):5913. https://doi.org/10.3390/su12155913

Rudrakar S, Rughani P (2024) IoT based agriculture (Ag-IoT): a detailed study on architecture, security and forensics. Inf Process Agric 11:524–541. https://doi.org/10.1016/j.inpa.2023.09.002

Sanzari I, Leone A, Ambrosone A (2019) Nanotechnology in plant science: to make a long story short. Front Bioeng Biotechnol 7:120. https://doi.org/10.3389/fbioe.2019.00120

Sarkar MR, Masud SR, Hossen MI, Goh M (2022) A comprehensive study on the emerging effect of artificial intelligence in agriculture automation. In 2022 IEEE 18th international colloquium on signal processing & applications (CSPA). IEEE, pp 419–424. https://doi.org/10.1109/CSPA55076.2022.9781883

Seleiman MF, Alotaibi MA, Alhammad BA et al (2020) Effects of ZnO nanoparticles and biochar of rice straw and cow manure on characteristics of contaminated soil and sunflower productivity, oil quality, and heavy metals uptake. Agronomy 10:790. https://doi.org/10.3390/agronomy10060790

Semida WM, Abdelkhalik A, Mohamed Gamal F et al (2021) Foliar application of zinc oxide nanoparticles promotes drought stress tolerance in eggplant (Solanum melongena L.). Plants 10:421. https://doi.org/10.3390/plants10020421

Shafi U, Mumtaz R, García-Nieto J et al (2019) Precision agriculture techniques and practices: from considerations to applications. Sensors 19:3796. https://doi.org/10.3390/s19173796

Sharifi M, Faryabi K, Talaei AJ et al (2020) Antioxidant properties of gold nanozyme: a review. J Mol Liq 297:112004. https://doi.org/10.1016/j.molliq.2019.112004

Shrestha MM, Wei L (2024) Review—Perspectives on the roles of real time nitrogen sensing and IoT integration in smart agriculture. J Electrochem Soc 171:027526. https://doi.org/10.1149/1945-7111/ad22d8

Siddiqui MH, Al-Whaibi MH, Faisal M, Al Sahli AA (2014) Nano-silicon dioxide mitigates the adverse effects of salt stress on *Cucurbita pepo* L. Environ Toxicol Chem 33:2429–2437. https://doi.org/10.1002/etc.2697

Singh RP, Handa R, Manchanda G (2021) Nanoparticles in sustainable agriculture: an emerging opportunity. J Control Release 329:1234–1248. https://doi.org/10.1016/j.jconrel.2020.10.051

Sinha RK, Verma SS (2021) Proteomics approach in horticultural crops for abiotic-stress tolerance. *In Stress tolerance in horticultural crops*. Elsevier, pp 371–385. https://doi.org/10.1016/B978-0-12-822849-4.00003-6

Sishodia RP, Ray RL, Singh SK (2020) Applications of remote sensing in precision agriculture: a review. Remote sens 12(19):3136. https://doi.org/10.3390/rs12193136

Song Y, Jiang M, Zhang H, Li R (2021) Zinc oxide nanoparticles alleviate chilling stress in rice (Oryza sativa L.) by regulating antioxidative system and chilling response transcription factors. Molecules 26:2196. https://doi.org/10.3390/molecules26082196

Sun L, Song F, Guo J et al (2020) Nano-ZnO-induced drought tolerance is associated with melatonin synthesis and metabolism in maize. Int J Mol Sci 21:782. https://doi.org/10.3390/ijms21030782

Tarafdar JC, Raliya R, Mahawar H, Rathore I (2014) Development of zinc nanofertilizer to enhance crop production in pearl millet (Pennisetum americanum). Agric Res 3:257–262. https://doi.org/10.1007/s40003-014-0113-y

Thakur D, Kumar Y, Kumar A, Singh PK (2019) Applicability of wireless sensor networks in precision agriculture: a review. Wirel Pers Commun 107:471–512. https://doi.org/10.1007/s11277-019-06285-2

Umair Hassan M, Aamer M, Umer Chattha M et al (2020) The critical role of zinc in plants facing the drought stress. Agriculture 10:396. https://doi.org/10.3390/agriculture10090396

Vijayakumar MD, Surendhar GJ, Natrayan L et al (2022) Evolution and recent scenario of nanotechnology in agriculture and food industries. J Nanomater 2022:1280411. https://doi.org/10.1155/2022/1280411

Wang Y, Liu Y, Zhan W et al (2020) Long-term stabilization of Cd in agricultural soil using mercapto-functionalized nano-silica (MPTS/nano-silica): a three-year field study. Ecotoxicol Environ Saf 197:110600. https://doi.org/10.1016/j.ecoenv.2020.110600

Wang A, Li J, AL-Huqail AA et al (2021a) Mechanisms of chitosan nanoparticles in the regulation of cold stress resistance in banana plants. Nanomaterials 11:2670. https://doi.org/10.3390/nano11102670

Wang K, Wang Y, Wan Y et al (2021b) The fate of arsenic in rice plants (Oryza sativa L.): influence of different forms of selenium. Chemosphere 264:128417. https://doi.org/10.1016/j.chemosphere.2020.128417

Wang X, Zhong X, Li J et al (2021c) Inorganic nanomaterials with rapid clearance for biomedical applications. Chem Soc Rev 50:8669–8742. https://doi.org/10.1039/D0CS00461H

Wani SP (2023) New paradigm for transforming Indian agriculture to climate-resilient and sustainable agriculture is a must. J Agrometeorol 25:79–91. https://doi.org/10.54386/jam.v25i1.2011

Wu H, Tito N, Giraldo JP (2017) Anionic cerium oxide nanoparticles protect plant photosynthesis from abiotic stress by scavenging reactive oxygen species. ACS Nano 11:11283–11297. https://doi.org/10.1021/acsnano.7b05723

Yin H, Cao Y, Marelli B et al (2021) Soil sensors and plant wearables for smart and precision agriculture. Adv Mater 33:2007764. https://doi.org/10.1002/adma.202007764

Chapter 16
Application of X-Ray Spectroscopy to Monitor Absorption and Transport of Nanoparticles

Ankita Ghosh and Nidhi Chauhan

Abstract Nanotechnology is a transformative field due to its wide-ranging applications in biomedical, agricultural, and materials science. The unique physicochemical properties of nanoparticles, like high surface area and tunable size and shape, have emphasised their applications in targeted therapeutics, diagnostic devices, and improved nutrient delivery in crops. However, the increased application of nanoparticles has also raised concerns about toxicity, biodistribution, and environmental impact; thus, there is a need for analytical and regulatory approaches. Firstly, we have reviewed nanoparticles, their importance, and the toxicity. Following that, the chapter provides special emphasis on the role of X-ray-based techniques—X-ray diffraction (XRD) and X-ray fluorescence (XRF) techniques in characterising the structural parameters, particle size, crystallinity, phase composition, and adsorption behaviour in biological and plant materials. XRD and XRF are sophisticated tools that serve as an indispensable method for monitoring the interaction of nanoparticles at the molecular level, thereby advancing the effective application of nanoparticles.

Keywords Nanoparticles · X-ray diffraction (XRD) · X-ray fluorescence · Adsorption · Detection

16.1 Introduction

Nanotechnology is a new science to alter the size of existing structures and devices with organisational properties. The developed particles have an intermediate size up to 100 nm (Konwar and Ahmed 2016). The size of nanoparticles is much less than the bulk molecule; however, the size of nanoparticles is greater than the atomic range. The reduction of size produces novel properties compared to bulk materials. Nanotechnology is increasingly recognized as a key enabling technology of the twenty-first century, complementing established areas like information technology and biotechnology. This status arises from the cross-disciplinary combination of

A. Ghosh · N. Chauhan (✉)
School of Health Sciences & Technology (SoHST), UPES, Dehradun, Uttarakhand, India

R. N. Pudake et al. (eds.), *Phyto Nanotoxicology*, Nanotechnology in Plant Sciences 7, https://doi.org/10.1007/978-3-032-24518-2_16

physics, chemistry, biology, materials science, and engineering at the nanoscale, together with the key role played by nanoscale manipulation of matter across an extensive range of technologies. The synthesis of nanoparticles is a basic element of nanotechnology since the characteristic physicochemical properties inherent in nanomaterials develop primarily at the nanoparticle, nanocrystal, or monolayer scale. Thus, the synthesis and directed assembly of precursor nanoparticles and related structures rank among the most general and extensively used methods for the synthesis of nanostructured materials (Roco 1999). There are various types of nanoparticles, like inorganic, polymeric, solid lipid nanoparticles, liposomes (Bangham et al. 1965; Hartner et al. 2009; Torchilin 2005; Wang et al. 2010), and nanocrystals. Inorganic nanoparticles are generally made of inorganic compounds, metals, and metal salts. Metallic nanoparticles and semiconductors are generally used in biomedical fields for drug delivery, imaging and detection devices. In the present day, various chemical functional groups are used to modify the structure of nanoparticles. The modified nanoparticles can be conjugated with biological molecules, including antibodies, ligands, and drugs of interest, resulting it's the immense application of inorganic nanoparticles in biotechnology and the biomedical field. Polymeric nanoparticles (PNPs) have recently attracted widespread interest due to their unique physicochemical properties inherent in nanoscale sizes (Farokhzad and Langer 2009). Figure 16.1 shows the wide application and implementation of nanoparticles in various fields. Several studies have pointed to their potential in a broad range of applications, and especially in diagnostics and drug delivery systems

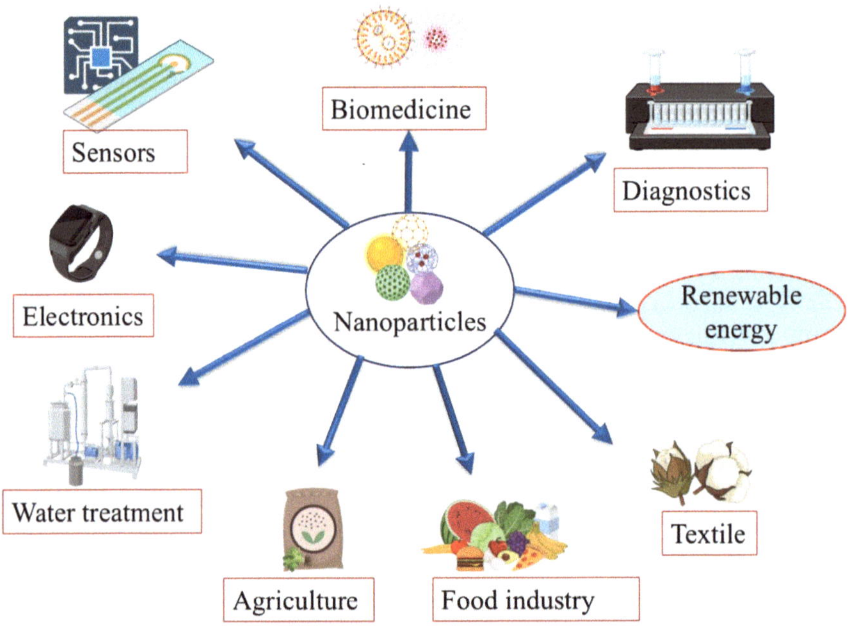

Fig. 16.1 Application of nanoparticles in various fields

(Crucho and Barros 2017; Patel et al. 2012). The advantages of using PNPs as drug delivery systems include drug release profiles, a combination of therapeutic and imaging modalities (theranostics), active pharmaceutical ingredients protection from degradation, and targeted delivery capacity, all of which lead to increasing the therapeutic index (Kamaly et al. 2012; Krasia-Christoforou and Georgiou 2013; Zhong et al. 2014).

16.2 Application of Nanoparticles

16.2.1 Food Industry

Nanotechnology has applications in all phases of production, processing, storage, and commercialization in the food industry. The most outstanding and widely researched applications in food science are directed towards flavor enhancement, nutrient encapsulation (Alex & Bodmeier, 1990) and delivery, introduction of antibacterial nanoparticles into food and packaging matrices, shelf-life extension, contamination detection, storage condition improvement, and effective tracking, tracing, and brand authentication (Duncan 2011). Nanoprocessing technologies enable the adjustment of several physicochemical properties of food products, giving rise to new characteristics. These encompass changes in color, taste, and sensory properties; nutritional functionality enhancement; and elimination of chemical residues or pathogen contaminants. Nanomaterials have facilitated the development of sophisticated packaging systems that can extend product shelf life through enhanced barrier properties, improve food safety, offer real-time contamination or spoilage warnings, repair small packaging flaws, and release active nanoparticles or preservative compounds to ensure product quality. Food commercialization and labeling have been enhanced by the development of nano barcoding technologies, which support accurate labeling, tracking, and distribution of foodstuffs. The innovative and highly adaptable field of nanoencapsulation, which supports the embedding of key nutrients, bioactive food molecules, and dietary supplements into nanoscale containers. Through this method, controlled and targeted release is possible, hence optimizing the bioavailability and stability of the encapsulated agents (Carrillo-Inungaray et al. 2018).

16.2.2 Agriculture

The growth of plants is dependent on light and water; however, plants need mineral nutrients for their healthy growth. These micronutrients get depleted after every harvest. In the present time, with increasing demand for food, the agricultural process is increasing without providing time for soil to get enriched with these micronutrients again. Improper application micronutrients will cause imbalance in microbial community in the plant roots (da Cruz et al. 2017). The application of nanotechnology overcomes the limitations (Giraldo et al. 2014). The basic principles like tunable surface charge, multiphase architecture (Janus, core-shell or onion-like arrangement), boosted surface reactivity and improved solubility provides innovative methods to develop new fertilizers and deliver the fertilizer in unique and target specific method. However, there are several studies regarding the negative impact of nanomaterials on the plant growth (Miralles et al. 2012; Mukherjee et al. 2016).

16.3 Transport of Nanoparticles in Plants

Nanoparticles enable site-specific delivery and controlled release of agrochemicals and macromolecules required for growth, thus enhancing resource efficiency and reducing environmental exposure for ecological safeguard (Rajput et al. 2020). Nanoparticle-assisted plant transformation provides a better way for genetic modification than traditional methodologies. Nanoparticles can penetrate plant cells and, at the same time, shield encapsulated molecules from breakdown due to their reduced size in nanoscale range (Gao et al. 2023). Besides, nanoparticles also offer new approaches to crop protection against various agro-associated problems (Singla et al. 2019). However, some nanoparticles have been shown to impose phytotoxicity, which could hinder the growth and development of plants (Oliveri Conti et al. 2020). Therefore, it is important to evaluate the chemical, biological, and environmental factors affecting the behavior and destiny of nanoparticles after their application to plants. The plants are capable of absorbing and translocating nanoparticles, and efficiency in uptake and transport is determined by several factors such as particle size, charge, concentration, exposure time, and plant species. Nanoparticles enter plant tissues through a variety of routes such as stomatal pores, root hair, and microcracks on leaf surfaces. Once they are inside, they can move within plant systems through diffusion, bulk flow, and phloem loading. Their mobility is also affected by the morphology, surface chemistry of the nanoparticles, the ambient pH, and the presence of other coexisting ions or compounds in the exposure medium. Experimentally, nanoparticles have been given to plants using different modes, including foliar spraying, root exposure, branch injection, and seed treatment, with several studies verifying nanoparticle uptake (Alimi et al. 2018; Khan et al. 2022). However, direct proof of their systemic transport in plants is still scarce.

The mechanisms of nanoparticle uptake and translocation need to be understood for developing optimized nanomaterials for agricultural uses. Additionally, understanding their mechanism of action and bioaccumulation characteristics will help to evaluate their biological safety and provide guidance for using agricultural nanomaterials responsibly (Khan et al. 2022). While other research has reported that nanoparticles have been found to accumulate in edible plant tissues and potentially pass on through the food chain to threaten human health and food security, the present review does not discuss transfer of nanoparticles to animals through plants. Rather, it addresses the uptake and transport behavior of agricultural nanoparticles in plants and their physiological effects, with emphasis on the effect of particle size, surface charge, and plant species on nanoparticle absorption and transport (Wang et al. 2023).

Nano-fertilizers, designed at the nanoscale, are a cutting-edge method of enhancing the efficiency of nutrient uptake by plants. Nano-fertilizers provide controlled release and prolonged release of nutrients, coordinating the supply of nutrients with the physiological needs of crops to ensure minimal nutrient loss and minimize negative environmental impacts (Nongbet et al. 2022). Studies have indicated that nitrogen- and zinc-based nano-fertilizers, combined with organic farming practices, were found to improve cereal and oilseed crop yields by leaps and bounds over traditional chemical fertilizers (Kumar et al. 2022). Additionally, hybrid nano-fertilizers (HNFs) made of copper, iron, and zinc incorporated with urea-modified hydroxyapatite have also been identified to promote fertilizer efficiency in *Abelmoschus esculentus* (Okra) by providing a slow and steady release of nutrients, improving water retention, and nutrient uptake. Such HNFs performed better than conventional fertilizers as far as growth and yield factors (Tarafder et al. 2020). Likewise, the use of micronutrient nanoparticles (MN-NPs) has been reported to greatly enhance common bean growth and productivity. The Egyptian field trials in 2019–2020 demonstrated that 40 mg L^{-1} MN-NPs, especially ZnO, MnO_2, and MoO_3, increased vegetative growth, flowering, and yield of the crop, indicating the tremendous potential of nanotechnology in enhancing agricultural productivity (Salama et al. 2022).

Nano-pesticides have been designed to enhance the accuracy and efficiency of pest management programs (Chaud et al. 2021). Unlike traditional pesticides, which tend to necessitate high quantities and carry risks of environmental contamination and human health risks, nano-pesticides leverage their great surface area-to-volume ratio and increased chemical reactivity to provide effective, targeted control of pests with much lower amounts. This technology lowers the overall chemical load on ecosystems while preserving or enhancing the effectiveness of pest control (Das et al. 2025).

16.4 Detection of Nanoparticles

The intensive interest in nanomaterials in recent decades has been primarily driven by the appearance of new properties displayed by nanoparticles with decreasing dimensions or sizes of their structural units. Therefore, innovations and accessibility of accurate analytical methods for measuring the sizes of these structural units are pivotal in carrying out research and characterization in the nanomaterials field. Recent views concerning difficulties and methods of nanoparticle size determination are discussed in detail in a new monograph Some of the commonly used methods for determining the sizes of nanoparticles and crystallites within polycrystalline materials on the nanoscale are transmission electron microscopy (TEM), scanning electron microscopy (SEM), atomic force microscopy (AFM) (Aziz et al. 2025;

Hodoroaba et al. 2016), Brunauer Emmett Teller (BET) measurement, X-ray diffraction (XRD) and X-ray fluorescence. TEM represents a direct visualization method and is especially appreciated for studying crystallites' fine details of structure (Asadi Asadabad and Jafari Eskandari 2015). Nonetheless, it has some technical disadvantages and limitations, such as time-consuming sample preparation and possible issues in achieving enough contrast to image very tiny particles, which might be electron transparent. Moreover, the fine nanostructures within relatively big polycrystalline specimens are difficult to observe. BET method, conversely, measures the specific surface area of particles by means of the physical adsorption and desorption of gases. Particle size approximations are subsequently calculated from the measured specific surface area, considering the density of the material as well as assumptions about particle shape (Dazon et al. 2020). X-ray diffraction (XRD) is extremely useful for nano-structural analysis since the width and shape of diffraction peaks give information on material substructures, such as dimensions of microcrystallites, microstrain in the lattice, and dislocation organization.

16.4.1 X-Ray Spectroscopy

X-ray spectroscopy is confined to a wide range of analytical techniques that utilise the interaction of X-rays with matter to determine the elemental structure and chemical state. The various types of X-ray spectroscopy are X-ray diffraction (XRD), X-ray absorption spectroscopy (XAS), and X-ray emission spectroscopy, which includes X-ray fluorescence (XRF) and X-ray photoelectron spectroscopy (XPS). The comparison table between various types of X-ray spectroscopy is given in Table 16.1.

16.4.1.1 X-Ray Diffraction (XRD)

X-ray diffraction (XRD) is an analytical method used for studying the structural characteristics of compounds and determining the crystalline nature of the materials. The XRD technique relies on constructive interference among monochromatic X-rays scattered from the periodic distribution of atoms in a crystal lattice. As shown in Fig. 16.2, when it collides with a crystalline specimen, the atoms serve as scattering centres that deflect the X-rays in particular directions. Constructive interference, which gives rise to observable diffraction peaks, is only possible when the path difference between rays scattered by consecutive lattice planes fulfils Bragg's law:

$$n\lambda = 2d \sin \theta$$

Table 16.1 Various types of X-ray spectroscopy

S. no.	Type of X-ray spectroscopy	Principle	Application	Merits	Demerits	References
1.	XRD	Measures the diffraction of X-rays	Phase identification, crystallinity	Structure information, non-destructive	Not suitable for amorphous samples	Biftu and Ravindhranath (2024); Dorofeev et al. (2012)
2.	XRF	Measures the secondary fluorescent X-rays emitted after excitation	Analysis of elemental composition	Non-destructive, sample preparation is simple	Lower sensitivity for the light sample	Shepp and Vardi (1982)
3.	XAS	Measures the adsorption of X-ray energies present near the adsorption edges	Atomic structure, oxidation state	Element detection, applicable to all states of matter	Need a synchrotron source for high resolution	Glasner and Frenkel (2007)
4.	XPS	Measures the photoelectron emitted after X-ray irradiation	Chemical composition of the surface	Surface sensitive	Requires high vacuum	X. Zhang and Cresswell (2016)
5.	EDX	Measures the emitted X-rays using a semiconductor	Analysis of elemental composition	Rapid detection	Lower resolution	Scimeca et al. (2018)

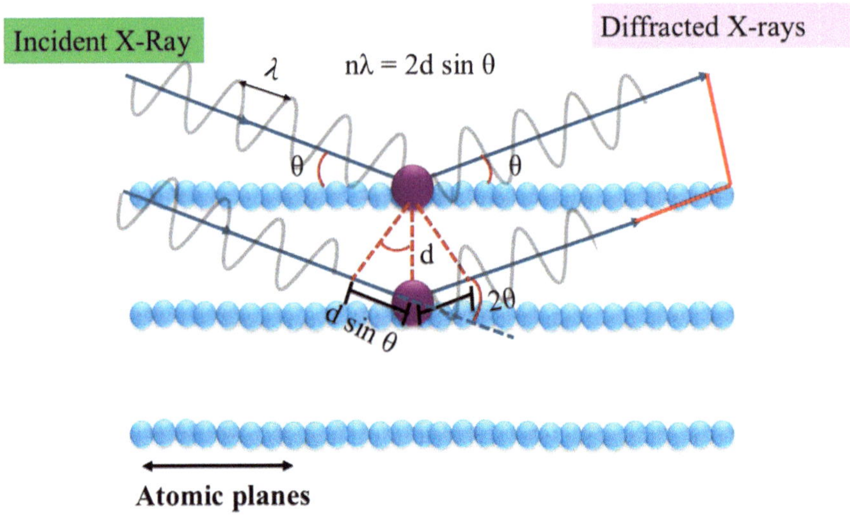

Fig. 16.2 Principle of X-ray diffraction

n is an integer that indicates the order of diffraction, λ is the wavelength of the incident X-rays, d is the interplanar spacing between crystallographic planes, and θ is the angle of incidence.

By varying the incident angle systematically and determining the intensity of diffracted X-rays, an XRD pattern is obtained, made up of peaks representing lattice spacings. Direct information on the crystal structure in terms of lattice parameters, symmetry, etc., is obtained from the positions of these peaks, while peak intensity depends on the arrangement of atoms within the unit cell as well as on the nature of the atoms. In addition, peak width and profile shape hold essential information regarding microstrain, crystallite size, and defects in material substructures. For example, wider peaks frequently relate to nanoscale crystallite dimensions or lattice distortions (Dong et al. 2019; Ghasempour-Mouziraji et al. 2024; Müller and Behler 2024). The Scherrer equation may be used to calculate crystallite size through peak broadening. XRD is widely applied across the field of materials science, chemistry, physics, geology, and nanotechnology for phase identification, quantitative analysis of mixtures, determination of crystallographic orientation, and study of residual stress and texture in materials. Its non-destructive capability, high precision, and capacity to deliver complete structural information render XRD a foundational technique in the characterization of crystalline materials (Reddy and Reddy 2025).

Several methods are used for X-ray diffraction line profile analysis, among them the most frequently applied being the Scherrer, Williamson–Hall, and Warren–Averbach methods (Naidu and Houska 1971; Rahman et al. 2013; Williamson and Hall 1953). The methods vary according to the kind and depth of structural data they offer, their robustness in returns, as well as computational intensity. An appropriate method for analyzing materials is still a great challenge for scientists to make a choice. The Scherrer technique is based on the estimation of crystallite size from only the width of a single diffraction peak but no other contributions to broadening such as lattice microstrain, stacking faults, or other structural defects. On the other hand, the Williamson–Hall technique allows the size and strain effects on-line broadening to be deconvolved and usually needs at least two diffraction peak data to carry out a linear extrapolation. However, for better accuracy, scientists tend to study an extensive range of diffraction peaks of the investigated phase. Both Scherrer and Williamson–Hall approaches only pay attention to peak widths and not to the fine features of the diffraction profiles. The Warren–Averbach procedure, by contrast, is based on full-profile analysis, measuring the whole diffraction peak shape, not just its width. In its traditional application, this procedure requires knowledge of diffraction profiles from two reflections of various orders corresponding to the same planes of crystals. Analyzing the whole diffraction profile provides much more detailed insight into the microstructural properties of the material than analyzing peak widths (Biftu and Ravindhranath 2024; Dorofeev et al. 2012).

16.4.1.2 X-Ray Fluorescence

X-ray fluorescence is a powerful tool used to determine the chemical composition of samples. Every atom has characteristic electron energy levels defined by its nuclear charge. Figure 16.3 shows electron transitions between these discrete energy levels

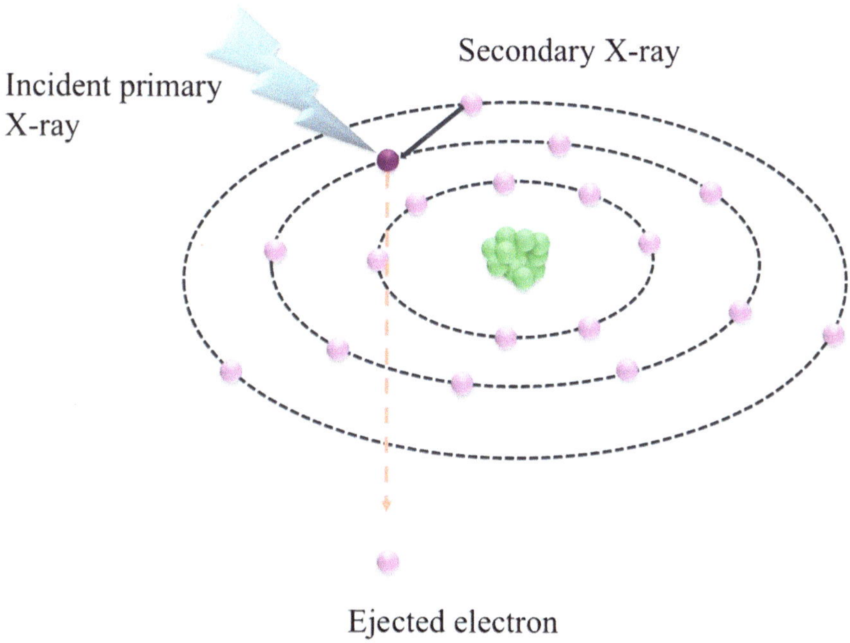

Fig. 16.3 Principle of X-ray fluorescence

that are accomplished by the absorption or emission of a photon whose energy is equal to the difference between the corresponding shells, generating element-specific fluorescence X-ray energies that facilitate the identification of precise elements (Shepp and Vardi 1982). Analytical methods like energy-dispersive X-ray spectroscopy (EDS), which utilise electron excitation, and X-ray fluorescence (XRF), utilise photon excitation, take advantage of this principle (Rawat et al. 2022). When this technique is combined with tomography imaging, such as in X-ray fluorescence computed tomography (XFCT), then these techniques have the capability of precisely localizing and measuring fluorescent materials like nanoparticle tracers, inside an object by illuminating the sample with a beam of photons, measuring the resulting fluorescence with energy-sensitive photon detectors, and reconstructing spatial distributions from tomographic analysis of fluorescence intensities at various angles (Castillo and Strivay 2012; Janssens 2013; Prasher and Kumar 2024).

16.4.1.3 X-Ray Absorption Spectroscopy (XAS)

The catalytic activity of nanomaterials is controlled by their size and shape, and these are defined by their surface topography. The underlying mechanisms of catalytic performance can be understood using a strong and accurate tool for the characterization of nanoparticles. Standard laboratory methods like chemisorption, XRD, and TEM have difficulty in detecting nanoparticles in the 1–5 nm diameter range (Corain et al. 2008). The use of TEM is increasing rapidly in the evaluation of size distributions; however, the electron beam damages the sensitivity of nanoclusters. Additionally, most electron microscopes have a spatial resolution (>0.1 nm), which is more than 20 times worse than what can be achieved with X-ray absorption spectroscopy (XAS) (Glasner and Frenkel 2007). Technological progress, in instrumentation, data processing, and modeling, has brought XAS and its derivative techniques, extended X-ray absorption fine structure (EXAFS) and X-ray absorption near edge structure (XANES), to a leading role in the investigation of catalysis (Calvin et al. 2003). XAS provides analysis capabilities comparable to, and, in some instances, superior to those of other methods for particle size, morphology, and internal structure determination at the nanometer scale. Its unique strengths are the capacity to: (i) measure interatomic distances, (ii) assess average particle size in the range below ~2 nm, (iii) clarify nanoparticle shape, and (iv) function under in situ or in operando conditions. In addition, the element specificity of EXAFS allows atomic mole fraction quantification inside nanoparticles (Frenkel 2007).

16.4.1.4 X-Ray Photoelectron Spectroscopy (XPS)

X-ray photoelectron spectroscopy (XPS) is also known as electron spectroscopy for chemical analysis (ESCA). It is a powerful and quantitative method for analyzing the surface chemistry of the materials (X. Zhang and Cresswell 2016). In XPS, the sample is excited by monochromatic X-rays, commonly from Al Kα or Mg Kα sources, while the kinetic energy and the quantity of emitted photoelectrons from surface atoms are measured in tandem. The kinetic energies are measured, and the binding energies are calculated to produce the XPS spectrum. Measurements are carried out under ultra-high vacuum (UHV) to achieve maximum electron detection efficiency, considering the analyzer is usually placed approximately 1 m away from the irradiated surface (Watts 2010). XPS allows both qualitative and quantitative analysis of elemental composition for atoms with atomic numbers ≥ 3 (lithium), as well as identification of chemical bonding environments and oxidation states on the surface of the material (Korin et al. 2017). XPS has been widely used across various research areas such as electronics, forensic science, and biological investigations as this method has flexible application.

16.4.2 Role of XRD in Nanoparticles Detection

X-ray scattering techniques are robust techniques for exploring crystal structure, particle size, and phase composition, all of which have a crucial bearing on the material's physical properties. These methods are largely non-destructive and permit analysis of statistical averages over large populations of particles, in contrast to direct imaging techniques like electron microscopy, which inspect only a small subset of the particles and will not always represent the material's properties, especially in the presence of moderate to severe polydispersity of size distribution. X-ray methods provide direct particle size and lattice parameter measurements, while indirect methods, such as UV–visible spectroscopy, derive particle size from changes in absorption peak positions (J. Park et al. 2008; Qu and Peng 2002; Tuinenga et al. 2008). X-ray diffraction (XRD) helps in determining the atomic structure in crystalline material and helps in monitoring the presence of novel nanostructures that do not exist in bulk materials, like icosahedral or decahedral structures, or metastable phases that are characterized by new physical properties, unlike the stable bulk phase (Ingham 2015).

There are several possible harmful effects of nanomaterials (NMs) on human health. Nanotoxicology research has grown to evaluate nanomaterials and their safety in a comprehensive way. Several reviews, viewpoints, and *in situ* toxicological studies have been carried out in this area (Augustine et al. 2020; do Nascimento et al. 2010; Eidi et al. 2010; Holgate 2010; E.-J. Park et al. 2010; Zhu et al. 2010). To assess the risks that nanoparticles pose in aquatic systems, various studies have investigated analytical methods and conducted ecotoxicological testing (Baun et al. 2008; Carl Englert 2007; Chen and Elimelech 2009; Grieger et al. 2010; Hassellöv et al. 2008). Various publications have also outlined the use of nanomaterials within industries like food, ranging from food additives to packaging materials and cosmetic products (Chaudhry et al. 2008; Stamm et al. 2012). Nanomaterials have high mobility, even when they are situated within matrices, resulting in direct and mostly deliberate exposure to the human body (Tiede et al. 2009). The increased industrial manufacturing and incorporation of nanoparticles in consumer products may lead to the accumulation of more enduring phases of these substances, which would become a component of concern regarding the eventual entry into the food chain (Summers 2010). To appropriately assess and regulate, institutions such as the OECD, as well as standardizing organizations like ISO and the CEN (European Committee for Standardization), have developed measurement standards, confirmed testing procedures, and made guidelines related to nanomaterials (Peña et al. 2016).

16.5 Detection of Adsorption of Nanoparticles

Techniques of X-ray diffraction (XRD) are extensively used to examine the phase composition and structural characteristics of materials by interpreting mainly the positions, shapes, and intensities of Bragg reflections. Average atomic arrangement and substructural parameters are often inferred from full-profile analysis (FPA), which examines several diffraction peaks, and from harmonic analysis focusing on selected peak profiles. There are, however, challenges in applying these methods to nanomaterials. Nanomaterials are generally characterized by high phase dispersion and structural disorder and therefore yield broad, sometimes indistinguishable diffraction peaks that merge into the background. Structural examination of nanocomposites is made even more challenging by issues of quantifying the proportions of phases, for example, separating catalytically active metal nanoparticles from supporting matrices in heterogeneous catalysts. This problem becomes significant in very dilute systems, in which the contribution of the dispersed active phase to the total diffraction pattern is weak and difficult to distinguish. Moreover, diffraction behaviour in nanocomposites can be non-additive because of interactions among various nanophases. In addition, nanocrystalline materials tend to have high defect densities, local lattice strain, and intergranular grain boundaries, which all introduce internal strain and contribute to additional distortion of diffraction patterns, making structural interpretation more difficult. Phase composition and local structural characteristics in nanoscale materials require advanced methods that consider the peculiar X-ray scattering properties of these materials. The increased relevance of nanomaterials has stimulated the development of structural diagnostic methods, leading to efforts from different research communities in developing novel methodologies and software specifically designed for nanocrystal examination based on basic principles of X-ray scattering. Among the available methods, the most detailed structural information comes from full-profile analysis of X-ray diffraction (XRD) patterns. Consequently, the description of highly dispersed materials must utilize methods capable of taking full advantage of this information. In polycrystalline materials containing coherent scattering domains less than about 1–3 nm in size, the diffraction pattern is that of amorphous materials, exhibiting broad and diffusing halos instead of sharp Bragg peaks. Since such similarity exists, structural studies of such nanomaterials are aided by analytical techniques generally used in the study of X-ray amorphous systems like glasses, liquids, and metallic alloys. When nanoparticles adsorb onto or intercalate inside host materials, they tend to cause the crystal lattice parameters to change, which is observed as displacements of the positions, intensities, and shapes of Bragg reflections of XRD patterns (Yang et al. 2021). The differences can indicate interlayer spacings, crystallite sizes, and strain, allowing for explanations of adsorption behaviors and nanoparticle distribution (Singh et al. 2019). XRD in porous or layered hosts can identify intercalation effects by changes in basal spacings, reflecting successful incorporation of nanoparticles (Liu et al. 2024). Further, new full-profile analysis methods improve the sensitivity of small structural changes related to adsorption, giving information on the nanoparticle-host matrix interaction. In such a way, XRD offers an effective means for investigating the structural implications of nanoparticle adsorption in various material systems.

16.6 X-Ray Reflectivity (XRR) for Analyzing Adsorption of Nanoparticles

The realistic detection of the adsorption process can be obtained using in-situ measurement. XRR is a powerful and accurate technique that uses high proton energies for detecting the electron densities, thickness and the roughness of the surface of the thin film using sub-angstrom resolution

$$q_z = (4\pi/\lambda)\,\sin\theta$$

θ = scattering angle, λ = wavelength.

The above-mentioned XRR equation is used for detecting a component perpendicular to the sample's surface (z-axis). This scattering method is sensitive to the laterally averaged electron density profile along the z direction (Tolan 1999).

$$R(qz) = R_F(qz)\frac{1}{\sigma}\left[\int \frac{d\rho(z)}{dz} e^{iq_z z} dz\right]^2$$

$R(qz)$ = measured reflectivity as a function of momentum transfer qz, $R_F(qz)$ = Fresnel reflectivity for an ideal flat interface, $\rho(z)$ = electron density profile along the surface (z-axis), σ = normalization constant, $\frac{d\rho(z)}{dz}$ = electron density gradient, $e^{iq_z z}$ = phase factor in the Fourier transform.

The above-mentioned equation is used for detecting reflectivity.

T. Brenner and team in 2012 used in situ XRR measurements for investigating the adsorption of differently charged nanoparticles at the negatively charged surface of silicon dioxide. The study showed the strong electrostatic interaction between the γ-Fe_2O_3 nanoparticles and the surface of silicon dioxide (Brenner et al. 2012).

16.7 Role of X-Ray Fluorescence Microanalysis in Plants

The *in-vivo* adsorption rate in plants can be monitored by micro-XRF. It helps in determining the nanoparticle size, presence or absence of surfactants and the concentration of the dispersion on uptake velocity (da Cruz et al. 2017). The size of the nanoparticles affects the adsorption rate. The size within 100 nm is better adsorbed, whereas the particle size more than 100 nm showed decreased adsorption. μ-XRF images provide advanced results as compared to SEM-EDS. It generates a lower X-Ray scattering background and has a lower detection limit, thus improving the signal-to-noise ratio.

16.8 Limitations of X-Ray Spectroscopy

X-ray spectroscopy comprises several powerful and accurate instruments for detecting the nanoparticle. However, there are limitations in using X-ray spectroscopy. Firstly, the instrumentation is generally expensive and sophisticated devices that need high maintenance, resulting in limited availability. Some X-ray spectroscopies need restricted experimental conditions like high vacuum, thus limiting the range of analyzable samples. The sensitivity of the light elements is comparatively lower due to the low X-ray adsorption cross-section. Therefore, the accurate quantification of low atomic elements is challenging. Prolonged exposure of the biological samples or very sensitive samples to an intense X-ray beam can cause radiation damage to the sample. The data interpretation needs complex fitting procedures and theoretical modelling; the interpretation needs expert personnel to avoid misinterpretation.

16.9 Conclusion

Nanotechnology has emerged as a transformative scientific discipline that provides an innovative solution in diverse fields, including biomedical, food technology, material science and agriculture. Due to the unique physicochemical properties of nanoparticles, it is widely involved in drug delivery and the development of smart fertilisers for agriculture. The increasing use also raises the concern of the potential toxicity, thus highlighting the need for evaluation and regulation. Advanced analytical technologies like XRD and μ-XRF help in characterizing the properties of nanoparticles. The structural modification of nanoparticles can be analyzed using XRD, whereas the adsorption and dispersion can be tracked using μ-XRF. These methods not only help in determining the particle size, crystallinity, and phase composition; however, it also help in determining the nanoparticle-host interaction at the molecular level.

References

Alex R, Bodmeier R (1990) Encapsulation of water-soluble drugs by a modified solvent evaporation method. I. Effect of process and formulation variables on drug entrapment. J Microencapsul 7(3):347–355. https://doi.org/10.3109/02652049009021845

Alimi OS, Farner Budarz J, Hernandez LM, Tufenkji N (2018) Microplastics and nanoplastics in aquatic environments: aggregation, deposition, and enhanced contaminant transport. Environ Sci Technol 52(4):1704–1724. https://doi.org/10.1021/acs.est.7b05559

Asadi Asadabad M, Jafari Eskandari M (2015) Transmission electron microscopy as best technique for characterization in nanotechnology. Synth React Inorg Metal Org Nano-Metal Chem 45 (3):323–326. https://doi.org/10.1080/15533174.2013.831901

Augustine R, Hasan A, Primavera R, Wilson RJ, Thakor AS, Kevadiya BD (2020) Cellular uptake and retention of nanoparticles: insights on particle properties and interaction with cellular components. Mater Today Commun 25:101692. https://doi.org/10.1016/j.mtcomm.2020.101692

Aziz A, Shaikh H, Abbas A, Zehra KE, Javed B (2025) Microscopic techniques for nanomaterials characterization: a concise review. Microsc Res Tech 88(5):1599–1614. https://doi.org/10.1002/jemt.24799

Bangham AD, Standish MM, Watkins JC (1965) Diffusion of univalent ions across the lamellae of swollen phospholipids. J Mol Biol 13(1):238-IN27. https://doi.org/10.1016/S0022-2836(65)80093-6

Baun A, Hartmann NB, Grieger K, Kusk KO (2008) Ecotoxicity of engineered nanoparticles to aquatic invertebrates: a brief review and recommendations for future toxicity testing. Ecotoxicology 17(5):387–395. https://doi.org/10.1007/s10646-008-0208-y

Biftu WK, Ravindhranath K (2024) Novel adsorptive methods for the effective arsenic(III) removal from polluted water. Biomass Convers Biorefinery 14(13):14477–14489. https://doi.org/10.1007/s13399-022-03540-8

Brenner T, Paulus M, Schroer MA, Tiemeyer S, Sternemann C, Möller J, Tolan M, Degen P, Rehage H (2012) Adsorption of nanoparticles at the solid–liquid interface. J Colloid Interface Sci 374(1):287–290. https://doi.org/10.1016/j.jcis.2012.02.010

Calvin S, Miller MM, Goswami R, Cheng S-F, Mulvaney SP, Whitman LJ, Harris VG (2003) Determination of crystallite size in a magnetic nanocomposite using extended x-ray absorption fine structure. J Appl Phys 94(1):778–783. https://doi.org/10.1063/1.1581344

Carl Englert B (2007) Nanomaterials and the environment: uses, methods and measurement. J Environ Monit 9(11):1154. https://doi.org/10.1039/b705988d

Carrillo-Inungaray ML, Trejo-Ramirez JA, Reyes-Munguia A, Carranza-Alvarez C (2018) Use of nanoparticles in the food industry: advances and perspectives. In: Impact of nanoscience in the food industry. Elsevier, pp 419–444. https://doi.org/10.1016/B978-0-12-811441-4.00015-7

Chaud M, Souto EB, Zielinska A, Severino P, Batain F, Oliveira-Junior J, Alves T (2021) Nanopesticides in agriculture: benefits and challenge in agricultural productivity, toxicological risks to human health and environment. Toxics 9(6):131. https://doi.org/10.3390/toxics9060131

Chaudhry Q, Scotter M, Blackburn J, Ross B, Boxall A, Castle L, Aitken R, Watkins R (2008) Applications and implications of nanotechnologies for the food sector. Food Addit Contamin Part A 25(3):241–258. https://doi.org/10.1080/02652030701744538

Chen KL, Elimelech M (2009) Relating colloidal stability of Fullerene (C_{60}) nanoparticles to nanoparticle charge and electrokinetic properties. Environ Sci Technol 43(19):7270–7276. https://doi.org/10.1021/es900185p

Corain B, Schmid G, Toshima N (2008) Metal nanoclusters in catalysis and materials science. Elsevier. https://doi.org/10.1016/B978-0-444-53057-8.X5001-6

Crucho CIC, Barros MT (2017) Polymeric nanoparticles: a study on the preparation variables and characterization methods. Mater Sci Eng C 80:771–784. https://doi.org/10.1016/j.msec.2017.06.004

da Cruz TNM, Savassa SM, Gomes MHF, Rodrigues ES, Duran NM, de Almeida E, Martinelli AP, de Carvalho HWP (2017) Shedding light on the mechanisms of absorption and transport of ZnO nanoparticles by plants *via in vivo* X-ray spectroscopy. Environ Sci Nano 4(12):2367–2376. https://doi.org/10.1039/C7EN00785J
Das D, Ingti B, Paul P, Jamatia JP, Ete T, Khan T, Kalita J (2025) Soil microbial dynamics in response to the impact of nanoparticles on agricultural implications. Discov Soil 2(1):58. https://doi.org/10.1007/s44378-025-00087-8
Dazon C, Fierro V, Celzard A, Witschger O (2020) Identification of nanomaterials by the volume specific surface area (VSSA) criterion: application to powder mixes. Nanoscale Adv 2 (10):4908–4917. https://doi.org/10.1039/D0NA00395F
del Castillo HC, Strivay D (2012) X-Ray methods. In: Analytical archaeometry. The Royal Society of Chemistry, pp 59–113. https://doi.org/10.1039/9781849732741-00059
do Nascimento GM, de Oliveira RC, Pradie NA, Lins PRG, Worfel PR, Martinez GR, Di Mascio P, Dresselhaus MS, Corio P (2010) Single-wall carbon nanotubes modified with organic dyes: synthesis, characterization and potential cytotoxic effects. J Photochem Photobiol A Chem 211 (2–3):99–107. https://doi.org/10.1016/j.jphotochem.2010.01.019
Dong G, Marleau-Finley J, Zhao YF (2019) Investigation of electrochemical post-processing procedure for Ti-6Al-4V lattice structure manufactured by direct metal laser sintering (DMLS). Int J Adv Manuf Technol 104(9–12):3401–3417. https://doi.org/10.1007/s00170-019-03996-5
Dorofeev GA, Streletskii AN, Povstugar IV, Protasov AV, Elsukov EP (2012) Determination of nanoparticle sizes by X-ray diffraction. Colloid J 74(6):675–685. https://doi.org/10.1134/S1061933X12060051
Duncan TV (2011) Applications of nanotechnology in food packaging and food safety: barrier materials, antimicrobials and sensors. J Colloid Interface Sci 363(1):1–24. https://doi.org/10.1016/j.jcis.2011.07.017
Eidi H, Joubert O, Attik G, Duval RE, Bottin MC, Hamouia A, Maincent P, Rihn BH (2010) Cytotoxicity assessment of heparin nanoparticles in NR8383 macrophages. Int J Pharm 396(1–2):156–165. https://doi.org/10.1016/j.ijpharm.2010.06.006
Farokhzad OC, Langer R (2009) Impact of nanotechnology on drug delivery. ACS Nano 3(1):16–20. https://doi.org/10.1021/nn900002m
Frenkel A (2007) Solving the 3D structure of metal nanoparticles. Zeitschrift Für Kristallographie Crystal Mater 222(11):605–611. https://doi.org/10.1524/zkri.2007.222.11.605
Gao M, Chang J, Wang Z, Zhang H, Wang T (2023) Advances in transport and toxicity of nanoparticles in plants. J Nanobiotechnol 21(1):75. https://doi.org/10.1186/s12951-023-01830-5
Ghasempour-Mouziraji M, Lagarinhos J, Afonso D, Alves de Sousa R (2024) A review study on metal powder materials and processing parameters in Laser Metal Deposition. Opt Laser Technol 170:110226. https://doi.org/10.1016/j.optlastec.2023.110226
Giraldo JP, Landry MP, Faltermeier SM, McNicholas TP, Iverson NM, Boghossian AA, Reuel NF, Hilmer AJ, Sen F, Brew JA, Strano MS (2014) Plant nanobionics approach to augment photosynthesis and biochemical sensing. Nat Mater 13(4):400–408. https://doi.org/10.1038/nmat3890
Glasner D, Frenkel AI (2007) Geometrical characteristics of regular Polyhedra: application to EXAFS studies of nanoclusters. AIP Conf Proc:746–748. https://doi.org/10.1063/1.2644651
Grieger KD, Baun A, Owen R (2010) Redefining risk research priorities for nanomaterials. J Nanopart Res 12(2):383–392. https://doi.org/10.1007/s11051-009-9829-1
Hartner WC, Verma DD, Levchenko TS, Bernstein EA, Torchilin VP (2009) ATP-loaded liposomes for treatment of myocardial ischemia. WIREs Nanomed Nanobiotechnol 1(5):530–539. https://doi.org/10.1002/wnan.46
Hassellöv M, Readman JW, Ranville JF, Tiede K (2008) Nanoparticle analysis and characterization methodologies in environmental risk assessment of engineered nanoparticles. Ecotoxicology 17 (5):344–361. https://doi.org/10.1007/s10646-008-0225-x

Hodoroaba V-D, Rades S, Salge T, Mielke J, Ortel E, Schmidt R (2016) Characterisation of nanoparticles by means of high-resolution SEM/EDS in transmission mode. IOP Conf Ser Mater Sci Eng 109:012006. https://doi.org/10.1088/1757-899X/109/1/012006
Holgate ST (2010) Exposure, uptake, distribution and toxicity of nanomaterials in humans. J Biomed Nanotechnol 6(1):1–19. https://doi.org/10.1166/jbn.2010.1098
Ingham B (2015) X-ray scattering characterisation of nanoparticles. Crystallogr Rev 21(4):229–303. https://doi.org/10.1080/0889311X.2015.1024114
Janssens K (2013) X–Ray based methods of analysis. In: Modern methods for analysing archaeological and historical glass. Wiley, pp 79–128. https://doi.org/10.1002/9781118314234.ch5
Kamaly N, Xiao Z, Valencia PM, Radovic-Moreno AF, Farokhzad OC (2012) Targeted polymeric therapeutic nanoparticles: design, development and clinical translation. Chem Soc Rev 41 (7):2971. https://doi.org/10.1039/c2cs15344k
Khan I, Awan SA, Rizwan M, Hassan ZU, Akram MA, Tariq R, Brestic M, Xie W (2022) Nanoparticle's uptake and translocation mechanisms in plants via seed priming, foliar treatment, and root exposure: a review. Environ Sci Pollut Res 29(60):89823–89833. https://doi.org/10.1007/s11356-022-23945-2
Konwar R, Ahmed AB (2016) Nanoparticle: an overview of preparation, characterization and application. Int Res J Pharm 4(4):47–57. https://doi.org/10.7897/2230-8407.04408
Korin E, Froumin N, Cohen S (2017) Surface analysis of nanocomplexes by X-ray photoelectron spectroscopy (XPS). ACS Biomater Sci Eng 3(6):882–889. https://doi.org/10.1021/acsbiomaterials.7b00040
Krasia-Christoforou T, Georgiou TK (2013) Polymeric theranostics: using polymer-based systems for simultaneous imaging and therapy. J Mater Chem B 1(24):3002. https://doi.org/10.1039/c3tb20191k
Kumar A, Singh K, Verma P, Singh O, Panwar A, Singh T, Kumar Y, Raliya R (2022) Effect of nitrogen and zinc nanofertilizer with the organic farming practices on cereal and oil seed crops. Sci Rep 12(1):6938. https://doi.org/10.1038/s41598-022-10843-3
Miralles P, Church TL, Harris AT (2012) Toxicity, uptake, and translocation of engineered nanomaterials in vascular plants. Environ Sci Technol 46(17):9224–9239. https://doi.org/10.1021/es202995d
Mosqueira VCF, Legrand P, Morgat J, Vert M, Mysiakine E, Gref R, Devissaguet J, Barratt G (2001) Biodistribution of long-circulating PEG-grafted nanocapsules in mice: effects of PEG chain length and density. Pharm Res 18(10):1411–1419. https://doi.org/10.1023/A:1012248721523
Mukherjee A, Majumdar S, Servin AD, Pagano L, Dhankher OP, White JC (2016) Carbon nanomaterials in agriculture: a critical review. Front Plant Sci 7. https://doi.org/10.3389/fpls.2016.00172
Müller E, Behler J (2024) Determination of residual stress by X-Ray diffraction with the Cos-Alpha method and its application. In: Residual stress measurements [working title]. IntechOpen. https://doi.org/10.5772/intechopen.1003795
Naidu SVN, Houska CR (1971) X-ray studies of cold-worked ternary alloys containing Pd-Ag-Au. Metallur Trans 2(1):309–313. https://doi.org/10.1007/BF02662677
Nongbet A, Mishra AK, Mohanta YK, Mahanta S, Ray MK, Khan M, Baek K-H, Chakrabartty I (2022) Nanofertilizers: a smart and sustainable attribute to modern agriculture. Plants 11 (19):2587. https://doi.org/10.3390/plants11192587
Oliveri Conti G, Ferrante M, Banni M, Favara C, Nicolosi I, Cristaldi A, Fiore M, Zuccarello P (2020) Micro- and nano-plastics in edible fruit and vegetables. The first diet risks assessment for the general population. Environ Res 187:109677. https://doi.org/10.1016/j.envres.2020.109677
Park J, Lee KH, Galloway JF, Searson PC (2008) Synthesis of cadmium selenide quantum dots from a non-coordinating solvent: growth kinetics and particle size distribution. J Phys Chem C 112(46):17849–17854. https://doi.org/10.1021/jp803746b
Park E-J, Bae E, Yi J, Kim Y, Choi K, Lee SH, Yoon J, Lee BC, Park K (2010) Repeated-dose toxicity and inflammatory responses in mice by oral administration of silver nanoparticles. Environ Toxicol Pharmacol 30(2):162–168. https://doi.org/10.1016/j.etap.2010.05.004

Patel T, Zhou J, Piepmeier JM, Saltzman WM (2012) Polymeric nanoparticles for drug delivery to the central nervous system. Adv Drug Deliv Rev 64(7):701–705. https://doi.org/10.1016/j.addr.2011.12.006

Peña MDPS, Gottipati A, Tahiliani S, Neu-Baker NM, Frame MD, Friedman AJ, Brenner SA (2016) Hyperspectral imaging of nanoparticles in biological samples: simultaneous visualization and elemental identification. Microsc Res Tech 79(5):349–358. https://doi.org/10.1002/jemt.22637

Prasher S, Kumar M (2024) Exploring X-Ray techniques for comprehensive material characterization and analysis, pp 272–296. https://doi.org/10.4018/979-8-3693-1306-0.ch013

Qu L, Peng X (2002) Control of photoluminescence properties of CdSe nanocrystals in growth. J Am Chem Soc 124(9):2049–2055. https://doi.org/10.1021/ja017002j

Rahman MA, Halfar J, Shinjo R (2013) X-Ray diffraction is a promising tool to characterize coral skeletons. Adv Mater Phys Chem 03(01):120–125. https://doi.org/10.4236/ampc.2013.31A015

Rajput V, Minkina T, Mazarji M, Shende S, Sushkova S, Mandzhieva S, Burachevskaya M, Chaplygin V, Singh A, Jatav H (2020) Accumulation of nanoparticles in the soil-plant systems and their effects on human health. Ann Agric Sci 65(2):137–143. https://doi.org/10.1016/j.aoas.2020.08.001

Rawat K, Sharma N, Singh VK (2022) X-Ray fluorescence and comparison with other analytical methods (AAS, ICP-AES, LA-ICP-MS, IC, LIBS, SEM-EDS, and XRD). In: X-Ray fluorescence in biological sciences. Wiley, pp 1–20. https://doi.org/10.1002/9781119645719.ch1

Reddy KS, Reddy S (2025) X-ray diffraction: a non-destructive technique for crystalline structure analysis. Int J Chem Res Dev 7(1):63–66. https://doi.org/10.33545/26646552.2025.v7.i1a.81

Roco MC (1999) Nanoparticles and nanotechnology research. J Nanopart Res 1(1):1–6. https://doi.org/10.1023/A:1010093308079

Salama DM, Abd El-Aziz ME, Shaaban EA, Osman SA, Abd El-Wahed MS (2022) The impact of nanofertilizer on agro-morphological criteria, yield, and genomic stability of common bean (Phaseolus vulgaris L.). Sci Rep 12(1):18552. https://doi.org/10.1038/s41598-022-21834-9

Scimeca M, Bischetti S, Lamsira HK, Bonfiglio R, Bonanno E (2018) Energy Dispersive X-ray (EDX) microanalysis: a powerful tool in biomedical research and diagnosis. Eur J Histochem. https://doi.org/10.4081/ejh.2018.2841

Shepp LA, Vardi Y (1982) Maximum likelihood reconstruction for emission tomography. IEEE Trans Med Imaging 1(2):113–122. https://doi.org/10.1109/TMI.1982.4307558

Singh S, Nalwa HS (2007) Nanotechnology and health safety—toxicity and risk assessments of nanostructured materials on human health. J Nanosci Nanotechnol 7(9):3048–3070. https://doi.org/10.1166/jnn.2007.922

Singh P, Kumar R, Singh RK (2019) Progress on transition metal-doped ZnO nanoparticles and its application. Ind Eng Chem Res 58(37):17130–17163. https://doi.org/10.1021/acs.iecr.9b01561

Singla R, Kumari A, Yadav SK (2019) Impact of nanomaterials on plant physiology and functions. In: Nanomaterials and plant potential. Springer International Publishing, pp 349–377. https://doi.org/10.1007/978-3-030-05569-1_14

Stamm H, Gibson N, Anklam E (2012) Detection of nanomaterials in food and consumer products: bridging the gap from legislation to enforcement. Food Addit Contamin Part A 29(8):1175–1182. https://doi.org/10.1080/19440049.2012.689778

Summers H (2010) Can cells reduce nanoparticle toxicity? Nano Today 5(2):83–84. https://doi.org/10.1016/j.nantod.2010.01.003

Tarafder C, Daizy M, Alam MM, Ali MR, Islam MJ, Islam R, Ahommed MS, Aly Saad Aly M, Khan MZH (2020) Formulation of a hybrid nanofertilizer for slow and sustainable release of micronutrients. ACS Omega 5(37):23960–23966. https://doi.org/10.1021/acsomega.0c03233

Tiede K, Hassellöv M, Breitbarth E, Chaudhry Q, Boxall ABA (2009) Considerations for environmental fate and ecotoxicity testing to support environmental risk assessments for engineered nanoparticles. J Chromatogr A 1216(3):503–509. https://doi.org/10.1016/j.chroma.2008.09.008

Tolan M (1999) X-Ray scattering from soft-matter thin films, pp 1–4. https://doi.org/10.1007/BFb0112835

Torchilin VP (2005) Recent advances with liposomes as pharmaceutical carriers. Nat Rev Drug Discov 4(2):145–160. https://doi.org/10.1038/nrd1632

Tuinenga C, Jasinski J, Iwamoto T, Chikan V (2008) *In situ* observation of heterogeneous growth of CdSe quantum dots: effect of indium doping on the growth kinetics. ACS Nano 2(7):1411–1421. https://doi.org/10.1021/nn700377q

Wang T, D'Souza GG, Bedi D, Fagbohun OA, Potturi LP, Papahadjopoulos-Sternberg B, Petrenko VA, Torchilin VP (2010) Enhanced binding and killing of target tumor cells by drug-loaded liposomes modified with tumor-specific phage fusion coat protein. Nanomedicine 5(4):563–574. https://doi.org/10.2217/nnm.10.30

Wang X, Xie H, Wang P, Yin H (2023) Nanoparticles in plants: uptake, transport and physiological activity in leaf and root. Materials 16(8):3097. https://doi.org/10.3390/ma16083097

Watts JF (2010) The potential for the application of X-ray photoelectron spectroscopy in forensic science. Surf Interface Anal 42(5):358–362. https://doi.org/10.1002/sia.3192

Williamson GK, Hall WH (1953) X-ray line broadening from filed aluminium and wolfram. Acta Metall 1(1):22–31. https://doi.org/10.1016/0001-6160(53)90006-6

Yang J, Wang H, Zhou J, Chen D-R, Kong D, Yu F, Gui H, Liu J, Chen M (2021) A build-in data inversion method to retrieve aerosol size distributions for a Portable Ultrafine Particle Sizer (PUPS). IEEE Access 9:2879–2889. https://doi.org/10.1109/ACCESS.2020.3047627

Zhang X, Cresswell M (2016) Materials characterization of inorganic controlled release. In: Inorganic controlled release technology. Elsevier, pp 57–91. https://doi.org/10.1016/B978-0-08-099991-3.00003-X

Zhong Y, Meng F, Deng C, Zhong Z (2014) Ligand-directed active tumor-targeting polymeric nanoparticles for cancer chemotherapy. Biomacromolecules 15(6):1955–1969. https://doi.org/10.1021/bm5003009

Zhu X, Chang Y, Chen Y (2010) Toxicity and bioaccumulation of TiO2 nanoparticle aggregates in Daphnia magna. Chemosphere 78(3):209–215. https://doi.org/10.1016/j.chemosphere.2009.11.013

Index

R. N. Pudake et al. (eds.), *Phyto Nanotoxicology*, Nanotechnology in Plant Sciences 7, https://doi.org/10.1007/978-3-032-24518-2

T

U

V

MIX
Papier aus verantwortungsvollen Quellen
Paper from responsible sources
FSC® C105338

If you have any concerns about our products,
you can contact us on
ProductSafety@springernature.com

In case Publisher is established outside the EU,
the EU authorized representative is:
Springer Nature Customer Service Center GmbH
Europaplatz 3, 69115 Heidelberg, Germany

Printed by Libri Plureos GmbH
in Hamburg, Germany